Fluorescence Microscopy

Fluorescence Microscopy

From Principles to Biological Applications

Edited by Ulrich Kubitscheck

Second Edition

Editor

Prof. Dr. Ulrich Kubitscheck
Rheinische Friedrich-Wilhelms-Universität Bonn
Institut für Physikalische & Theoretische Chemie
Wegelerstr. 12
53115 Bonn
Germany

Cover
Cover graphic was created by
Max Brauner, Hennef, Germany.

All books published by **Wiley-VCH** are carefully produced. Nevertheless, authors, editors, and publisher do not warrant the information contained in these books, including this book, to be free of errors. Readers are advised to keep in mind that statements, data, illustrations, procedural details or other items may inadvertently be inaccurate.

Library of Congress Card No.: applied for

British Library Cataloguing-in-Publication Data
A catalogue record for this book is available from the British Library.

Bibliographic information published by the Deutsche Nationalbibliothek
The Deutsche Nationalbibliothek lists this publication in the Deutsche Nationalbibliografie; detailed bibliographic data are available on the Internet at <http://dnb.d-nb.de>.

© 2017 Wiley-VCH Verlag GmbH & Co. KGaA, Boschstr. 12, 69469 Weinheim, Germany

All rights reserved (including those of translation into other languages). No part of this book may be reproduced in any form – by photoprinting, microfilm, or any other means – nor transmitted or translated into a machine language without written permission from the publishers. Registered names, trademarks, etc. used in this book, even when not specifically marked as such, are not to be considered unprotected by law.

Print ISBN: 978-3-527-33837-5
ePDF ISBN: 978-3-527-68772-5
ePub ISBN: 978-3-527-68774-9
Mobi ISBN: 978-3-527-68775-6
oBook ISBN: 978-3-527-68773-2

Cover Design Formgeber, Mannheim, Germany
Typesetting SPi Global, Chennai, India

Printed on acid-free paper

Contents

List of Contributors *xv*
Preface *xix*

1 Introduction to Optics *1*
Rainer Heintzmann and Ulrich Kubitscheck
1.1   A Short History of Theories about Light *1*
1.2   Properties of Light Waves *2*
1.2.1 An Experiment on Interference *2*
1.2.2 Physical Description of Light Waves *3*
1.3   Four Effects of Interference *7*
1.3.1 Diffraction *7*
1.3.2 The Refractive Index *9*
1.3.3 Refraction *9*
1.3.4 Reflection *10*
1.3.5 Light Waves and Light Rays *11*
1.4   Optical Elements *13*
1.4.1 Lenses *13*
1.4.2 Metallic Mirrors *15*
1.4.3 Dielectric Mirrors *16*
1.4.4 Filters *17*
1.4.5 Chromatic Reflectors *18*
1.5   Optical Aberrations *20*
      References *22*

2 Principles of Light Microscopy *23*
Ulrich Kubitscheck
2.1   Introduction *23*
2.2   Construction of Light Microscopes *23*
2.2.1 Components of Light Microscopes *23*
2.2.2 Imaging Path *24*
2.2.3 Magnification *26*
2.2.4 Angular and Numerical Aperture *27*
2.2.5 Field of View *28*
2.2.6 Illumination Beam Path *28*

2.2.6.1	Critical and Köhler Illumination	28
2.2.6.2	Bright-Field and Epi-Illumination	31
2.3	Wave Optics and Resolution	32
2.3.1	Wave Optical Description of the Imaging Process	33
2.3.2	The Airy Pattern	37
2.3.3	Point Spread Function and Optical Transfer Function	40
2.3.4	Lateral and Axial Resolution	41
2.3.4.1	Lateral Resolution Using Incoherent Light Sources	41
2.3.4.2	Lateral Resolution of Coherent Light Sources	43
2.3.4.3	Axial Resolution	45
2.3.5	Magnification and Resolution	48
2.3.6	Depth of Field and Depth of Focus	49
2.3.7	Over- and Undersampling	50
2.4	Apertures, Pupils, and Telecentricity	50
2.5	Microscope Objectives	53
2.5.1	Objective Lens Design	53
2.5.2	Light Collection Efficiency and Image Brightness	57
2.5.3	Objective Lens Classes	61
2.5.4	Immersion Media	62
2.5.5	Special Applications	65
2.6	Contrast	67
2.6.1	Dark Field	68
2.6.2	Phase Contrast	69
2.6.2.1	Frits Zernike's Experiments	70
2.6.2.2	Setup of a Phase-Contrast Microscope	73
2.6.2.3	Properties of Phase-Contrast Images	74
2.6.3	Interference Contrast	74
2.6.4	Advanced Topic: Differential Interference Contrast	77
2.6.4.1	Optical Setup of a DIC Microscope	77
2.6.4.2	Interpretation of DIC Images	81
2.6.4.3	Comparison between DIC and Phase Contrast	81
2.7	Summary	82
	Acknowledgments	82
	References	83
3	**Fluorescence Microscopy**	**85**
	Jurek W. Dobrucki and Ulrich Kubitscheck	
3.1	Contrast in Optical Microscopy	85
3.2	Physical Foundations of Fluorescence	86
3.2.1	What is Fluorescence?	86
3.2.2	Fluorescence Excitation and Emission Spectra	89
3.3	Features of Fluorescence Microscopy	90
3.3.1	Image Contrast	90
3.3.2	Specificity of Fluorescence Labeling	93
3.3.3	Sensitivity of Detection	94
3.4	A Fluorescence Microscope	95
3.4.1	Principle of Operation	95

3.4.2	Sources of Exciting Light 99
3.4.3	Optical Filters in a Fluorescence Microscope 101
3.4.4	Electronic Filters 103
3.4.5	Photodetectors for Fluorescence Microscopy 104
3.4.6	CCD or Charge-Coupled Device 104
3.4.7	Intensified CCD (ICCD) 107
3.4.8	Electron-Multiplying Charge-Coupled Device (EMCCD) 109
3.4.9	CMOS 111
3.4.10	Scientific CMOS (sCMOS) 112
3.4.11	Features of CCD and CMOS Cameras 112
3.4.12	Choosing a Digital Camera for Fluorescence Microscopy 113
3.4.13	Photomultiplier Tube (PMT) 113
3.4.14	Avalanche Photodiode (APD) 114
3.5	Types of Noise in a Digital Microscopy Image 114
3.6	Quantitative Fluorescence Microscopy 119
3.6.1	Measurements of Fluorescence Intensity and Concentration of the Labeled Target 119
3.6.2	Ratiometric Measurements (Ca^{++}, pH) 121
3.6.3	Measurements of Dimensions in 3D Fluorescence Microscopy 121
3.6.4	Measurements of Exciting Light Intensity 122
3.6.5	Technical Tips for Quantitative Fluorescence Microscopy 123
3.7	Limitations of Fluorescence Microscopy 124
3.7.1	Photobleaching 124
3.7.2	Reversible Photobleaching under Oxidizing or Reducing Conditions 125
3.7.3	Phototoxicity 125
3.7.4	Optical Resolution 126
3.7.5	Misrepresentation of Small Objects 127
3.8	Summary and Outlook 128
	References 130
	Recommended Internet Resources 131
	Fluorescent Spectra Database 132
4	**Fluorescence Labeling** *133*
	Gerd Ulrich Nienhaus and Karin Nienhaus
4.1	Introduction 133
4.2	Key Properties of Fluorescent Labels 133
4.3	Synthetic Fluorophores 138
4.3.1	Organic Dyes 138
4.3.2	Fluorescent Nanoparticles 140
4.3.3	Conjugation Strategies for Synthetic Fluorophores 142
4.3.4	Non-natural Amino Acids 146
4.3.5	Bringing the Fluorophore to Its Target 147
4.4	Genetically Encoded Labels 149
4.4.1	Phycobiliproteins 149
4.4.2	GFP-Like Proteins 150

4.5	Label Selection for Particular Applications	155
4.5.1	FRET to Monitor Intramolecular Conformational Dynamics	155
4.5.2	Protein Expression in Cells	159
4.5.3	Fluorophores as Sensors Inside the Cell	160
4.5.4	Live-Cell Dynamics	160
4.5.5	Super-Resolution Imaging	160
4.6	Summary	161
	References	162

5 Confocal Microscopy 165
Nikolaus Naredi-Rainer, Jens Prescher, Achim Hartschuh, and Don C. Lamb

5.1	Evolution and Limits of Conventional Widefield Microscopy	165
5.2	Theory of Confocal Microscopy	166
5.2.1	Principle of Confocal Microscopy	166
5.2.2	Radial and Axial Resolution and the Impact of the Pinhole Size	173
5.2.3	Scanning Confocal Imaging	179
5.2.3.1	Stage Scanning	179
5.2.3.2	Laser Scanning	180
5.2.3.3	Spinning Disk Confocal Microscope	181
5.2.4	Confocal Deconvolution	184
5.3	Applications of Confocal Microscopy	186
5.3.1	Nonscanning Applications	186
5.3.1.1	Fluorescence Correlation Spectroscopy	186
5.3.1.2	Fluorescence Cross-Correlation Spectroscopy	190
5.3.1.3	Pulsed Interleaved Excitation	191
5.3.1.4	Burst Analysis with Multiparameter Fluorescence Detection	193
5.3.2	Scanning Applications beyond Imaging	195
5.3.2.1	Number and Brightness Analysis	195
5.3.2.2	Raster Image Correlation Spectroscopy	198
	Acknowledgments	200
	References	200

6 Two-Photon Excitation Microscopy for Three-Dimensional Imaging of Living Intact Tissues 203
David W. Piston

6.1	Introduction	203
6.2	What is Two-Photon Excitation?	205
6.2.1	Nonlinear Optics and 2PM	206
6.2.2	History and Theory of 2PM	207
6.3	How Does Two-Photon Excitation Microscopy Work in Practice?	211
6.3.1	The Role of Light Absorption in 2PM	212
6.3.2	The Role of Light Scattering in 2PM	213
6.4	Instrumentation	216
6.4.1	Lasers for 2PM	216
6.4.2	Detection Strategies for 2PM	219

6.4.3	The Advantages of 2PM for Deep-Tissue Imaging	*220*
6.5	Limitations of Two-Photon Excitation Microscopy	*222*
6.5.1	Limits of Spatial Resolution in 2PM	*222*
6.5.2	Potential Sample Heating by the High Laser Powers in 2PM	*223*
6.5.3	Difficulties in Predicting and Measuring Two-Photon Excitation Spectra	*224*
6.5.4	Accelerated Photobleaching (and Associated Photodamage) in the Focal Plane	*227*
6.5.5	Expensive Lasers Create a Practical Limitation for Some Experiments	*228*
6.6	When is 2PM the Best Option?	*229*
6.6.1	Thick Specimen including *In Vivo* Imaging	*229*
6.6.2	Imaging Fluorophores with Excitation Peaks in the Ultraviolet (UV) Spectrum	*231*
6.6.3	Localized Photochemistry	*231*
6.7	Applications of Two-Photon Microscopy	*231*
6.7.1	Imaging UV-Excited Fluorophores, such as NADH for Metabolic Activity	*231*
6.7.2	Localized Photoactivation of "Caged" Compounds	*233*
6.7.3	Imaging Electrical Activity in Deep Tissue	*236*
6.7.4	Light Sheet Microscopy Using Two-Photon Excitation	*237*
6.7.5	Other Applications of 2PM	*238*
6.8	Other Nonlinear Microscopies	*239*
6.9	Future Outlook for 2PM	*240*
6.10	Summary	*240*
	Acknowledgment	*241*
	References	*241*

7	**Light Sheet Microscopy**	*243*
	Gopi Shah, Michael Weber, and Jan Huisken	
7.1	Principle of Light Sheet Microscopy	*244*
7.2	Light Sheet Microscopy: Key Advantages	*245*
7.3	Construction and Working of a Light Sheet Microscope	*246*
7.4	Theory of Light Sheet Microscopy	*247*
7.5	Light Sheet Interaction with Tissue	*251*
7.6	3D Imaging	*253*
7.7	Multiview Imaging	*255*
7.8	Different Lens Configurations	*257*
7.9	Sample Mounting	*258*
7.10	Recent Advances in Light Sheet Microscopy	*259*
7.11	Outlook	*260*
7.11.1	Big Data	*260*
7.11.2	Smart Microscope: Imaging Concept of the Future	*261*
7.11.3	High-Throughput Imaging	*261*
7.12	Summary	*262*
	References	*262*

8	**Localization-Based Super-Resolution Microscopy** *267*
	Markus Sauer and Mike Heilemann
8.1	Super-Resolution Microscopy: An Introduction *267*
8.2	The Principle of Single-Molecule Localization Microscopy *269*
8.3	Photoactivatable and Photoconvertible Probes *272*
8.4	Intrinsically Photoswitchable Probes *272*
8.5	Photoswitching of Organic Fluorophores by Chemical Reactions *273*
8.6	Experimental Setup for Localization Microscopy *273*
8.7	Optical Resolution and Imaging Artifacts *276*
8.8	Fluorescence Labeling for Super-Resolution Microscopy *278*
8.8.1	Label Size versus Structural Resolution *278*
8.8.2	Live-Cell Labeling *280*
8.8.3	Click Chemistry *280*
8.8.4	Three-Dimensional SMLM *281*
8.8.5	Astigmatic Imaging *281*
8.8.6	Biplane Imaging *282*
8.8.7	Double Helix PSF *282*
8.8.8	Interferometric Imaging *282*
8.9	Measures for Improving Imaging Contrast *283*
8.10	SMLM Software *283*
8.11	Reference Structures for SMLM *285*
8.12	Quantification of SMLM Data *286*
8.13	Summary *287*
	References *287*
9	**Super-Resolution Microscopy: Interference and Pattern Techniques** *291*
	Udo Birk, Gerrit Best, Roman Amberger, and Christoph Cremer
9.1	Introduction *291*
9.1.1	Review: The Resolution Limit *292*
9.2	Structured Illumination Microscopy (SIM) *293*
9.2.1	Image Generation in Structured Illumination Microscopy *295*
9.2.2	Extracting the High-Resolution Information *298*
9.2.3	Optical Sectioning by SIM *299*
9.2.4	How the Illumination Pattern is Generated? *301*
9.2.5	Mathematical Derivation of the Interference Pattern *302*
9.2.6	Examples for SIM Setups *304*
9.3	Spatially Modulated Illumination (SMI) Microscopy *307*
9.3.1	Overview *307*
9.3.2	SMI Setup *309*
9.3.3	Excitation Light Distribution *309*
9.3.4	Object Size Estimation with SMI Microscopy *311*
9.4	Application of Patterned Techniques *313*
9.5	Conclusion *317*
9.6	Summary *317*
	Acknowledgments *317*
	References *318*

10   STED Microscopy   *321*
Travis J. Gould, Lena K. Schroeder, Patrina A. Pellett, and Joerg Bewersdorf
10.1   Introduction   *321*
10.2   The Concepts behind STED Microscopy   *322*
10.2.1   Fundamental Concepts   *322*
10.2.1.1   Switching between Optical States   *322*
10.2.1.2   Stimulated Emission Depletion   *322*
10.2.1.3   Stimulated Emission Depletion Microscopy   *324*
10.2.2   Key Parameters in STED Microscopy   *326*
10.2.2.1   Pulsed Lasers and Fluorophore Kinetics   *326*
10.2.2.2   Wavelength Effects   *328*
10.2.2.3   PSF Shape and Quality   *328*
10.3   Experimental Setup   *330*
10.3.1   Light Sources and Synchronization   *330*
10.3.2   Scanning and Speed   *331*
10.3.3   Multicolor STED Imaging   *332*
10.3.4   Improving Axial Resolution in STED Microscopy   *333*
10.4   Applications   *334*
10.4.1   Choice of Fluorophore   *334*
10.4.2   Labeling Strategies   *335*
10.5   Summary   *336*
References   *337*

11   Fluorescence Photobleaching Techniques   *339*
Reiner Peters
11.1   Introduction   *339*
11.2   Basic Concepts and Procedures   *340*
11.2.1   One Principle, Several Modes   *340*
11.2.2   Setting up an Instrument   *343*
11.2.3   Approaching Complexity from Bottom up   *344*
11.3   Fluorescence Recovery after Photobleaching (FRAP)   *345*
11.3.1   Evaluation of Diffusion Measurements   *345*
11.3.2   Binding   *348*
11.3.3   Membrane Transport   *349*
11.4   Continuous Fluorescence Microphotolysis (CFM)   *352*
11.4.1   Theoretical Background and Data Evaluation   *352*
11.4.2   Combination of CFM with Other Techniques   *355*
11.4.3   CFM Variants   *355*
11.5   CLSM-Assisted Photobleaching Methods   *356*
11.5.1   Implementation   *356*
11.5.2   New Opportunities   *357*
11.5.2.1   Multiple ROPs   *357*
11.5.2.2   Arbitrarily Shaped ROPs   *359*
11.5.2.3   Spatially Resolved Bleaching and Recovery   *359*
11.5.2.4   Millisecond Time Resolution   *359*
11.5.2.5   Three-Dimensional Photobleaching   *359*
11.5.3   Two Common Artifacts and Their Correction   *360*

11.6	Summary and Outlook *360*
	References *361*

12 Single-Molecule Microscopy in the Life Sciences *365*
Markus Axmann, Josef Madl, and Gerhard J. Schütz

12.1	Encircling the Problem *365*
12.2	What is the Unique Information? *367*
12.2.1	Kinetics Can Be Directly Resolved *367*
12.2.2	Full Probability Distributions Can Be Measured *367*
12.2.3	Structures Can Be Related to Functional States *369*
12.2.4	Structures Can Be Imaged at Super-Resolution *370*
12.2.5	Bioanalysis Can Be Extended Down to the Single-Molecule Level *372*
12.3	Building a Single-Molecule Microscope *372*
12.3.1	Microscopes/Objectives *373*
12.3.1.1	Dual View *373*
12.3.1.2	Objective *374*
12.3.2	Light Source *377*
12.3.2.1	Uniformity *377*
12.3.2.2	Intensity *378*
12.3.2.3	Illumination Time *380*
12.3.2.4	Polarization *380*
12.3.2.5	Wavelength *381*
12.3.2.6	Collimation *382*
12.3.3	Detector *382*
12.3.3.1	Pixel Size *382*
12.3.3.2	CCD Cameras *384*
12.3.3.3	Electron-Multiplying CCD Cameras *385*
12.3.3.4	CMOS Detectors *386*
12.4	Analyzing Single-Molecule Signals: Position, Orientation, Color, and Brightness *387*
12.4.1	Localizing in Two Dimensions *388*
12.4.2	Localizing along the Optical Axis *389*
12.4.2.1	Analysis of the Shape of the Point Spread Function *389*
12.4.2.2	Intensity Patterns along the Optical Axis *391*
12.4.3	Brightness *392*
12.4.4	Orientation *392*
12.4.4.1	Polarization Microscopy *392*
12.4.4.2	Defocused Imaging *393*
12.4.5	Color *393*
12.5	Learning from Single-Molecule Signals *394*
12.5.1	Determination of Molecular Associations *394*
12.5.2	Determination of Molecular Conformations via FRET *395*
12.5.3	Single-Molecule Tracking *400*
12.5.4	Detecting Transitions *401*
	Acknowledgments *402*
	References *402*

13	**Förster Resonance Energy Transfer and Fluorescence Lifetime Imaging** *405*
	Fred S. Wouters
13.1	General Introduction *405*
13.2	Förster Resonance Energy Transfer *406*
13.2.1	Physical Basis of FRET *406*
13.2.2	Historical Development of FRET *406*
13.2.3	Spectral and Distance Dependence of FRET *416*
13.2.4	FRET is of Limited Use as a Molecular Ruler *420*
13.2.5	Special FRET Conditions *422*
13.2.5.1	Diffusion-Enhanced FRET *422*
13.2.5.2	Multiple Acceptors *422*
13.2.5.3	FRET in a Plane *423*
13.3	Measuring FRET *426*
13.3.1	Spectral Changes *426*
13.3.1.1	FRET from Donor Quenching *426*
13.3.1.2	FRET-Induced Acceptor Emission *427*
13.3.1.3	Contrast in Intensity-Based FRET Measurements *430*
13.3.1.4	Full Quantitation of Intensity-Based FRET Measurements *431*
13.3.1.5	Occupancy Errors in FRET *432*
13.3.2	Decay Kinetics *432*
13.3.2.1	Photobleaching Rate *432*
13.3.2.2	Fluorescence Lifetime Changes *435*
13.4	FLIM *439*
13.4.1	Frequency-Domain FLIM *441*
13.4.1.1	Operation Principle and Technical Aspects *441*
13.4.2	Time-Domain FLIM *442*
13.4.2.1	Time-Correlated Single-Photon Counting *442*
13.4.2.2	Time Gating *443*
13.5	Analysis and Pitfalls *444*
13.5.1	Average Lifetime, Multiple Lifetime Fitting *444*
13.5.2	From FRET/Lifetime to Species *444*
13.6	Summary *448*
	References *448*

A	**Appendix A: What Exactly is a Digital Image?** *453*
	Ulrich Kubitscheck
A.1	Introduction *453*
A.2	Digital Images as Matrices *453*
A.2.1	Gray Values as a Function of Space and Time *453*
A.2.1.1	Parallel Data Acquisition *454*
A.2.1.2	Sequential Data Acquisition *455*
A.2.2	Image Size, Bit Depth, and Storage Requirements *456*
A.3	Look-up Table *457*
A.4	Intensity Histograms *457*
A.5	Image Processing *458*
A.5.1	Operations on Single Pixels *458*

A.5.2	Operations on Pixel Groups *459*
A.5.3	Low-Pass Filters *460*
A.6	Pitfalls *460*
A.7	Summary *461*
	References *461*

B **Appendix B: Practical Guide to Optical Alignment** *463*
Rainer Heintzmann

B.1	How to Obtain a Widened Parallel Laser Beam? *463*
B.2	Mirror Alignment *465*
B.3	Lens Alignment *466*
B.4	Autocollimation Telescope *466*
B.5	Aligning a Single Lens Using a Laser Beam *466*
B.6	How to Find the Focal Plane of a Lens? *469*
B.7	How to Focus to the Back Focal Plane of an Objective Lens? *470*

Index *473*

List of Contributors

Roman Amberger
Heidelberg University
Applied Optics and Information
Processing Kirchhoff-Institute
for Physics
Im Neuenheimer Feld 227
69120 Heidelberg
Germany

Markus Axmann
Institute of Medical Chemistry and
Pathobiochemistry
Center for Pathobiochemistry and
Genetics Medical University of Vienna
Währinger Straße 10
1090 Vienna
Austria

Gerrit Best
Heidelberg University
Applied Optics and Information
Processing
Kirchhoff-Institute for Physics
Im Neuenheimer Feld 227
69120 Heidelberg
Germany

Heidelberg University Hospital
Department of Ophthalmology
Im Neuenheimer Feld 400
69120 Heidelberg
Germany

Joerg Bewersdorf
Yale University
Department of Cell Biology
333 Cedar Street
New Haven
CT 06520-8002
USA

Udo Birk
Institute of Molecular Biology
GmbH (IMB)
Ackermannweg 4
55128 Mainz
Germany

Mainz University
Department of Physics, Mathematics
and Computer Science
Institute of Physics
Staudingerweg 9
55128 Mainz
Germany

Heidelberg University
Applied Optics and Information
Processing
Kirchhoff-Institute for Physics
Im Neuenheimer Feld 227
69120 Heidelberg
Germany

Christoph Cremer
Institute of Molecular Biology
GmbH (IMB)
Ackermannweg 4
55128 Mainz
Germany

Mainz University
Department of Physics, Mathematics
and Computer Science
Institute of Physics, Staudingerweg 9
55128 Mainz
Germany

Heidelberg University
Applied Optics and Information
Processing
Kirchhoff-Institute for Physics
Im Neuenheimer Feld 227
69120 Heidelberg
Germany

Jurek W. Dobrucki
Jagiellonian University
Faculty of Biochemistry, Biophysics
and Biotechnology
Department of Cell Biophysics
ul Gronostajowa 7
30-387 Krakow
Poland

Travis J. Gould
Bates College
Department of Physics & Astronomy
44 Campus Ave
Lewiston
ME 04240
USA

Achim Hartschuh
Ludwig-Maximilians-Universität
München
Department of Chemistry
Butenandtstr. 5-13
81377 München
Germany

Mike Heilemann
Goethe-University Frankfurt
Single Molecule Biophysics
Institute of Physical and Theoretical
Chemistry
Max-von-Laue-Str. 7
60438 Frankfurt
Germany

Rainer Heintzmann
Friedrich Schiller-Universität
Institut für Physikalische Chemie und
Abbe Center of Photonics
Helmholtzweg 4
07743 Jena
Germany

Leibniz Institute of Photonic
Technology
Albert-Einstein Str. 9
07745 Jena
Germany

Jan Huisken
Max Planck Institute of Molecular
Cell Biology & Genetics
Pfotenhauerstr. 108
01307 Dresden
Germany

Department of Medical Engineering
Morgridge Institute for Research
330 N Orchard Street
Madison, WI 53715
USA

Ulrich Kubitscheck
Rheinische
Friedrich-Wilhelms-Universität Bonn
Institut für Physikalische &
Theoretische Chemie
Wegelerstr. 12
53115 Bonn
Germany

Don C. Lamb
Ludwig-Maximilians-Universität
München
Department of Chemistry
Butenandtstr. 5-13
81377 München
Germany

Josef Madl
Faculty of Biology and BIOSS
Albert-Ludwigs University Freiburg
Schänzlestraße 18
79104 Freiburg
Germany

Nikolaus Naredi-Rainer
Ludwig-Maximilians-Universität
München
Department of Chemistry
Butenandtstr. 5-13
81377 München
Germany

Gerd Ulrich Nienhaus
Karlsruher Institut für Technologie
Institut für Angewandte Physik
Wolfgang-Gaede-Str. 1
76131 Karlsruhe
Germany

Karin Nienhaus
Karlsruher Institut für Technologie
Institut für Angewandte Physik
Wolfgang-Gaede-Str. 1
76131 Karlsruhe
Germany

Patrina A. Pellett
Yale University
Department of Cell Biology
333 Cedar Street
New Haven
CT 06520-8002
USA

Reiner Peters
The Rockefeller University
Laboratory of Mass Spectrometry and
Gaseous Ion Chemistry
1230 York Avenue
New York, 10065 NY
USA

David W. Piston
Washington University in St. Louis
Department of Cell Biology and
Physiology
School of Medicine
660 S. Euclid Avenue
St. Louis
MO 63110-1093
USA

Jens Prescher
Ludwig-Maximilians-Universität
München
Department of Chemistry
Butenandtstr. 5-13
81377 München
Germany

Markus Sauer
University Würzburg
Department of Biotechnology and
Biophysics
Am Hubland
97074 Würzburg
Germany

Lena K. Schroeder
Yale University
Department of Cell Biology
333 Cedar Street
New Haven
CT 06520-8002
USA

Gerhard J. Schütz
Institute of Applied Physics
TU Wien, Wiedner Hauptstraße 8-10
1040 Wien
Austria

Gopi Shah
Max Planck Institute of Molecular
Cell Biology & Genetics
Pfotenhauerstr. 108
01307 Dresden
Germany

Cancer Research UK Cambridge
Institute University of Cambridge
Robinson Way CB20RE Cambridge
UK

Michael Weber
Max Planck Institute of Molecular
Cell Biology & Genetics
Pfotenhauerstr. 108
01307 Dresden
Germany

Harvard Medical School
Department of Cell Biology
200 Longwood Ave
LHRRB 113, Boston, MA 02115
USA

Fred S. Wouters
University Medical Center Göttingen
Laboratory for Molecular and Cellular
Systems
Department of Neuropathology
Waldweg 33
37073 Göttingen
Germany

Preface to the Second Edition

What is This Book?

This book is both a high-level textbook and a reference work for researchers applying high-performance microscopy. It provides a comprehensive yet compact account of the theoretical foundations of light microscopy, the large variety of specialized microscopy techniques, and the quantitative utilization of light microscopy data. It will enable the user of modern microscopy equipment to fully exploit the complex instrumental features with knowledge and skill. These diverse goals were approached by recruiting a collective of leading scientists as authors. We applied a stringent internal reviewing process to achieve homogeneity, readability, and optimal coverage of the field. Finally, we took care to reduce redundancy as far as possible.

Why This Book?

Meanwhile, there are numerous books on light microscopy on the market. At a closer look, however, many available books are written at an introductory level with regard to the physics behind the often sophisticated techniques. Or, they represent rather a collection of review articles on advanced topics. Books *introducing* a wide range of techniques such as light sheet microscopy, fluorescence resonance energy transfer, stimulated emission depletion, or structured illumination microscopy, together with the required basics and theory, are rare. Even the basic optical theory such as the Fourier theory of optical imaging or topics such as the sine condition are seldom introduced from scratch. With this book, we fill this gap.

Is This Book for You?

The book is aimed at advanced undergraduate and graduate students of the biosciences and researchers entering the field of quantitative microscopy. Since they are usually recruited mostly from natural sciences, that is, physics, biology, chemistry, and biomedicine, we addressed the book to this readership. Readers would definitely profit from a sound knowledge of physics and mathematics. This allows diving much deeper into the presented material than without such knowledge. However, all authors are experienced in teaching university and summer courses on light microscopy and for many years explored the best ways to present the required knowledge. Hopefully, you will find that they have come up with good solutions. In case you see room for improvement, please let me know.

How Should You Read the Book?

Students who require an in-depth knowledge should begin at their level of knowledge, either Chapter 1 (Introduction to Optics) or Chapter 2 (Principles of Light Microscopy). Beginners should initially omit advanced topics, for example, the section on Differential Interference Contrast (Section 2.6.4). Principally, the book is readable without the "text boxes," which present either historical summaries or theoretical derivations. They are meant to provide a good understanding of theory and scientific reasoning. Then, they should proceed through Chapters 3 (Fluorescence Microscopy), 4 (Labeling Techniques), and 5 (Confocal Microscopy). Chapters 6 (Two-photon Microscopy) and 7 (Light Sheet Microscopy) are on alternative approaches to achieving optical sectioning, and more. Chapters 8–10 cover the most important super-resolution techniques, and Chapters 11–13 cover advanced topics and special techniques. They should be studied according to the reader's interest and requirement.

What Updates Done to the First Edition?

Most importantly, we added three new chapters and therefore cover now all major currently used approaches to fluorescence microscopic imaging. These chapters focus on two-photon microscopy, localization-based super-resolution microscopy, and light sheet microscopy. These topics were really missing in the first edition. I am happy that very distinguished authors have now contributed these materials. Of course, numerous typos and mistakes were corrected. All chapters were carefully revised to improve the accessibility, and updated; many figures were redrawn or exchanged. Some chapters were almost completely rewritten. Also, we shortened the lists of references as much as possible to reduce the volume of the book. The chapters were rearranged to follow the logic described above: (i) foundations, (ii) general fluorescence microscopy, (iii) optical sectioning approaches, (iv) high-resolution imaging, and (v) advanced quantitative techniques. Finally, I added a second appendix, which gives a very short introduction into the computational treatment of images. It is meant to serve as a bridge to the extensive existing literature on this topic.

Website of the Book

There is a website containing material extending the contents of this book www.wiley-vch.de/home/fluorescence_microscopy2. There you will find all figures as JPGs for use in courses and additional illustrative material such as movies.

Personal Remarks on the History of This Book

I saw one of the very first commercial laser scanning microscopes in the lab of Tom Jovin at the Max-Planck Institute of Biophysical Chemistry, Göttingen, at the end of the 1980s, and was immediately fascinated by the images produced by that instrument. In the beginning of the 1990s, confocal laser scanning microscopes began to spread over biological and medical labs. At that time, they usually required really dedicated scientists for proper operation, filled a small laboratory room, and were governed by computers as big as today's refrigerators. The required image processing demanded substantial investments. At that time, Reiner Peters and I noticed that biologists and medical scientists needed an introduction to the physical background of optics, spectroscopy, and image analysis for coping with the new techniques. Hence, we offered a lecture series entitled "Microscopes, Lasers and Computers" at the Institute of Medical Physics and Biophysics, University of Münster, Germany, which was very well received. We began to write a book on microscopy containing the material we had presented, which we thought should not be as comprehensive as Jim Pawley's "Handbook" but should offer an accessible path to modern quantitative microscopy. We invested almost 1 year into this enterprise, but then gave up ... in view of the numerous challenging research topics that kept us busy, the insight into the dimension of this task, and the reality of career requirements. We realized we could not do it alone.

In 2009, Reiner Peters, now at The Rockefeller University, New York, organized a workshop on "Watching the Cellular Nanomachinery at Work," and gathered some of the current leaders in microscopy to report on their latest technical and methodological advances. On that occasion, he noted that the book that had been in our minds 15 years ago was still missing ... and contacted the speakers of his meeting. Like many years before, I was excited by the idea to create this book, and together we directly addressed the lecturers of the meeting and other experts in the field and asked for introductory book chapters in the areas of their methodological expertise. Most of them responded positively, and thus began the struggle for an introductory text. Unfortunately, Reiner could not keep his position as an editor of the book due to further obligations, so I had to finish our joint project. Here is the result, and I very much hope that the authors have succeeded in transmitting their ongoing fascination for advanced light microscopy. To me, microscopy appears as a century-old tree that began another phase of growth about 40 years ago, and since then has shown almost every year a new branch with a surprising and remarkable technique offering exciting and fresh scientific fruits.

Acknowledgments

Finally, I would like to thank some people who have contributed directly and indirectly to this book. First of all, I would like to name Prof. Dr Reiner Peters. As mentioned, he invited me to the first steps to teach microscopy and to the first attempt to write this book. Finally, he launched the initiative to create this book as an edited work. Furthermore, I would like to thank all authors who invested their expertise, time, and energy in writing, correcting, and finalizing their respective chapters. They are all very respected colleagues, and some of them became my friends during this project. Also, I thank some people who were earlier collaborators or colleagues and helped me to learn more and more about microscopy: Prof. Dr Reinhard Schweitzer-Stenner, Dr Donna Arndt-Jovin, Prof. Dr Tom Jovin, Dr Thorsten Kues, and Prof. Dr David Grünwald. Likewise, I acknowledge the project editors responsible at Wiley-VCH for this project, Dr Andreas Sendtko, Dr Gregor Cicchetti, and Anne du Guerny, who wholeheartedly supported this project and showed very professional patience when yet another delay occurred, but also pushed when required. Last but not least, I would like to thank my collaborator Dr Jan Peter Siebrasse and my wife Martina, who patiently listened to my concerns when yet another problem occurred.

Bonn, Germany
January 20th, 2017

Prof. Dr. Ulrich Kubitscheck

1

Introduction to Optics

Rainer Heintzmann[1,3] *and Ulrich Kubitscheck*[2]

[1] *Friedrich Schiller-Universität, Institut für Physikalische Chemie und Abbe Center of Photonics, Helmholtzweg 4, 07743 Jena, Germany*
[2] *Rheinische Friedrich-Wilhelms-Universität Bonn, Institut für Physikalische & Theoretische Chemie, Wegelerstr. 12, 53115 Bonn, Germany*
[3] *Leibniz Institute of Photonic Technology, Albert-Einstein Str. 9, 07745 Jena, Germany*

In this chapter, we introduce the wave nature of light by discussing interference, which is then used to explain the laws of refraction, reflection, and diffraction. We then discuss light propagation in the form of rays, which leads to the laws of lenses and ray diagrams of= optical systems. Finally, the working principles of the most common optical elements are outlined.

1.1 A Short History of Theories about Light

For a long time, scientists like Pierre Gassendi (1592–1655) and Sir Isaac Newton (1643–1727) believed that light consisted of particles named *corpuscles* traveling along straight lines, the so-called *light rays*. This concept explains brightness and darkness, and effects such as shadows or even the fuzzy boundary of shadows due to the extent of the sun in the sky. However, in the sixteenth century, it was discovered that light can sometimes "bend" around sharp edges by a phenomenon called *diffraction*. This phenomenon was not compatible with the predictions of the ray theory, but rather, light must be described as a *wave*. In the beginning, the wave theory of light – based on Christiaan Huygens' (1629–1695) work and expanded later by Augustin Jean Fresnel (1788–1827) – was not accepted. Siméon Denis Poisson, one of the judges for evaluating Fresnel's work in a science competition, tried to ridicule it by showing that Fresnel's theory would predict a bright spot in the center of the dark shadow of a round, illuminated obstacle. Poisson considered this to be nonsense. Another judge, Arago, however, then demonstrated that this spot does indeed exist and can be observed when measurements are done very carefully. This was a phenomenal success of the wave description of light, and the spot was named *Poisson's spot*.

Only at the beginning of the twentieth century quantum mechanics fused both theories and suggested that light has a dual character: a wave nature and a particle nature. Since then, the description of light has maintained this dual view.

Fluorescence Microscopy: From Principles to Biological Applications, Second Edition.
Edited by Ulrich Kubitscheck.
© 2017 Wiley-VCH Verlag GmbH & Co. KGaA. Published 2017 by Wiley-VCH Verlag GmbH & Co. KGaA.

When it is interacting with matter, one often has to consider the quantum or particle nature of light. However, the propagation of these particles is described by equations written down by James Clerk Maxwell (1831–1879). The famous Maxwell wave equations of electrodynamics identify oscillating electromagnetic fields as the waves responsible for what we call *light* and still serve today as an extremely precise description of most aspects of light.

The wave concept is required to understand the behavior of light in the context of microscopy. Light propagation in the form of rays can explain refraction, reflection, and even aberrations, but fails to explain diffraction and interference. Therefore, we start out by introducing interference, an effect that is observed only when experiments are designed very carefully.

1.2 Properties of Light Waves

1.2.1 An Experiment on Interference

Suppose that we perform the following experiment: We construct the instrument shown in Figure 1.1 consisting of only two ordinary mirrors and two 50/50 beam splitters. These beam splitters reflect 50% of the incoming light and transmit 50%. Now we use a laser of any color to illuminate beam splitter BS1. It is absolutely crucial that the distances along the two different light paths between the two beam splitters are exactly equal with a precision better than 1/10 000 of a millimeter. Then we observe that something surprising is happening: the light entering the device will leave only through exit 1. Exit 2 will be completely dark. This is very strange, as one would expect 50% of the light exiting on either side. Even more surprising is what happens if one blocks one of the two light paths inside the instrument. Now, 25% of the light will emerge from both exits.

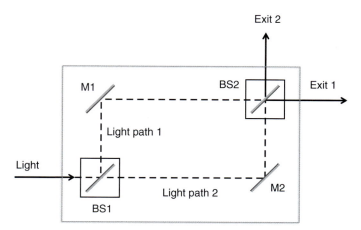

Figure 1.1 Interference experiment. In a simple arrangement of mirrors M1 and M2 and 50/50 beam splitters BS1 and BS2, an incoming light beam is split by BS1 and after reflection at M1 or M2 passes through BS2. Constructive interference occurs in direction of exit 1 if the optical path lengths of the two beams are exactly equal. But the light in exit 2 cancels by destructive interference.

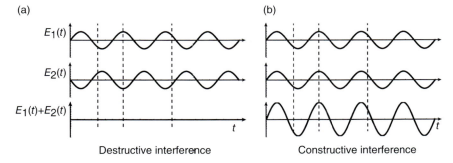

Figure 1.2 Destructive and constructive interference. (a) Two waves cancel each other if the electric fields of two interfering light waves $E_1(t)$ and $E_2(t)$ are always of opposite value, that is, they have a phase shift of π. (b) Constructive interference occurs if the two waves are completely in phase.

The explanation for this effect of interference lies in the wave nature of light. According to the wave description, light is an electromagnetic field oscillating with time. If two electromagnetic waves with identical amplitudes spatially superimpose but oscillate in exactly opposite directions, they will completely cancel each other (Figure 1.2a). Exactly this happens at exit 2 of our instrument. At exit 1, constructive interference of the two beams coming along light path 1 and 2 occurs. In constructive interference, the two waves oscillate in phase and therefore add up (Figure 1.2b). That is why all light is exiting at exit 1. Of course, the exact reasons for the asymmetric interference processes are not immediately obvious, since they are a consequence of subtle asymmetries in light path 1 and 2 for the two beams reaching the different exits. The effect will be explained later in detail after we covered some basic features of light (see Box 1.4).

The explanation for the second part of the experiment – the effect of blocking one of the light paths – is simple. Blocking one light path will prevent interference and yield 25% brightness at either exit, as expected when splitting 50% of the total input light again in two equal amounts.

The discussed instrument is called a *Mach–Zehnder interferometer*. Such interferometers are extremely sensitive instruments capable of detecting extremely small path length differences in the two arms of the interferometer.

1.2.2 Physical Description of Light Waves

In wave-optical terms, a light ray is an oscillating and propagating electromagnetic field. What does that mean?

Figure 1.3 provides a graphical representation of such a light wave in the simplest case – a plane wave in vacuum. The wave comprises an electric and a magnetic field component. Both oscillate perpendicular to each other and also to the propagation direction of the wave. Therefore, light is called a *transverse wave*. Since most of the effects of light on matter are caused by its electric field, we often neglect for simplicity the magnetic field altogether. In the figure, several important parameters describing such waves are indicated. The wavelength, λ, describes the spatial distance between two electric field maxima. The direction

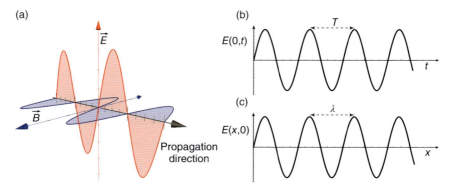

Figure 1.3 Sketch of a linearly polarized electromagnetic wave. (a) Wave with electric and magnetic field components, $\vec{E}$ and $\vec{B}$. (b) Temporal oscillation at a fixed place in space. (c) Still image of the wave.

and strength of the oscillating electrical field is given by the vector $\vec{E}$, and the propagation direction is characterized by the vector $\vec{k}$. The oscillation direction of the electric field in Figure 1.3 is constant in space. We call this the *polarization direction* of the light wave and such a wave *linearly polarized*. The polarization direction is not necessarily constant, for example, it may also rotate around the propagation direction. Such waves are called *elliptically* or *circular polarized* (see Box 1.1 for details).

The oscillation of the electric field as a function of time at a specific position in space is shown in Figure 1.3b. The time period until the electric field assumes again an identical profile is designated as oscillation duration T. The frequency $\nu = 1/T$ at which the electric field oscillates at a given position defines the *color* of the light wave. Electromagnetic waves exist over a vast range of frequencies, out of which only a very small range is perceived as "light" and detected by our eyes or cameras (Figure 1.4). Yet, the same wave theory of light governs

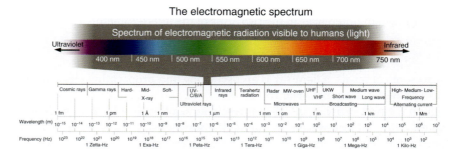

Figure 1.4 Electromagnetic spectrum. Different types of radiation are essentially electromagnetic waves with oscillation frequencies or vacuum wavelengths ranging over many orders of magnitude. English version of a graphic by Horst Frank (https://de.wikipedia.org/wiki/Elektromagnetisches_Spektrum, https://en.wikipedia.org/wiki/GNU_Free_Documentation_License).

all wavelength ranges of the electromagnetic spectrum, from cosmic waves to gamma rays. However, light microscopy uses only the visible and near-infrared range. Blue light has a higher frequency v and higher energy $E = hv$ per photon than green, yellow, red, and infrared light. Here, h denotes Planck's constant. Figure 1.3c shows a still image of the wave. The spatial distance between two positions in which the electric field assumes identical values is designated as wavelength λ. In vacuum, the oscillation frequency v and wavelength λ are related to each other as follows:

$$\lambda v = c \tag{1.1}$$

with c denoting the speed of light in vacuum. The wavelength λ is short for blue (about 450 nm), and longer for green (~520 nm), yellow (~580 nm), red (~630 nm), and infrared (~800 nm) light.

A short summary of the mathematical description of waves in space and time using trigonometric functions is given in Box 1.1. A mathematically more advanced description uses complex numbers and functions, a notation that is used in several later chapters of this book and considerably simplifies calculations.

Box 1.1 Mathematical Description of Waves

A harmonic, linearly polarized plane wave in space traveling in the x-direction is described as

$$\vec{E}(x,t) = \vec{E}_0 \sin\left(\frac{2\pi}{\lambda}x - \frac{2\pi}{T}t\right) \tag{1.2}$$

where $\vec{E}_0$ describes the amplitude and direction of the oscillating electric field, λ is the wavelength, T is the oscillation duration, and t the time. A still image of the wave, for example, at time $t = 0$, showing the spatial profile of the wave

$$\vec{E}(x,0) = \vec{E}_0 \sin\left(\frac{2\pi}{\lambda}x\right) \tag{1.3}$$

is shown in Figure 1.3c.

After the distance $x = \lambda$, the wave profile is repeated. The ratio $k = 2\pi/\lambda$ is designated as the wave number. At a specific position in space, for example, at $x = 0$, we see a periodic change of the electric field with the oscillation duration T (Figure 1.3b):

$$\vec{E}(0,t) = \vec{E}_0 \sin\left(-\frac{2\pi}{T}t\right) \tag{1.4}$$

The frequency of the oscillation is given by $v = 1/T$, whereas the ratio $\omega = 2\pi/T$ is called the *angular frequency*. We see that $\omega = 2\pi v$. Inserting the respective definitions into Eq. (1.2) yields

$$\vec{E}(x,t) = \vec{E}_0 \sin(kx - \omega t) = \vec{E}_0 \sin\left[k\left(x - \frac{\omega}{k}t\right)\right] \tag{1.5}$$

We see that the wave moves in the direction of the positive x-axis. When time t advances, x must correspondingly advance such that argument and function

(Continued)

Box 1.1 (Continued)

value remain constant. Thus the speed c of, for example, a wave crest is given by

$$c = \frac{\omega}{k} = \frac{2\pi\nu}{2\pi/\lambda} = \nu\lambda \qquad (1.6)$$

For electromagnetic waves in a homogeneous isotropic medium, the oscillation direction $\vec{E_0}$ is perpendicular to the propagation direction. In this case, the direction $\vec{E_0}$ is constant in space and time. Such a wave is designated as *linearly polarized*, and the polarization direction is identical to the direction of $\vec{E_0}$. However, there are other cases possible. In the most general case, the wave can be described by

$$\vec{E}(x,t) = \begin{pmatrix} E_y \sin(kx - \omega t + \phi_y) \\ E_z \sin(kx - \omega t + \phi_z) \end{pmatrix} \qquad (1.7)$$

where E_y, E_z, ϕ_y, and ϕ_z denote the amplitudes and phases with regard to the y- and z-directions. There are several cases possible. If the phase difference between both wave components is 0, that is, $\phi_y = \phi_z$, the light is linearly polarized for arbitrary choices of E_y and E_z. Then the polarization direction is given by the vector

$$\begin{pmatrix} 0 \\ E_y \\ E_z \end{pmatrix}.$$

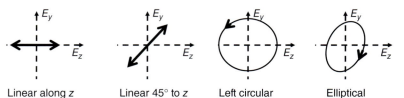

Linear along z Linear 45° to z Left circular Elliptical

If $E_y = E_z$ and the phase difference $\phi_y - \phi_z = \pi/4$, we have a *circular polarized* wave. This means that the polarization vector of constant length is rotating around the propagation direction clockwise or counterclockwise depending on the sign of the phase difference (see sketch and online material). The so-called elliptically polarized light is obtained if $E_y \neq E_z$, or in the case where $E_y = E_z$ but $\phi_y - \phi_z \neq \pi/4$.

Water waves are well-known everyday examples that illustrate many properties of waves quite plainly. They provide, however, only a two-dimensional analogy to electromagnetic waves. However, it is merely an analogy to compare light to waves in a medium, such as water, where the particles of the medium actually are displaced. In the case of light, there is really no displacement of matter necessary for its description, as the basic oscillating quantity is the electric field, which can even propagate in vacuum. The moving line of a wave crest is a good conception of a *phase front*. A phase front of a light wave is the surface in space formed by the local maxima of the electric field or, more generally, the surface formed of any equal phase of the wave. Such phase fronts travel with the propagation speed of the wave. The waves we observe in water close to the shore that form straight lines approaching the beach are an analogy to what is called a *plane wave* as described in Box 1.1, whereas the waves seen in a pond, when we throw a stone into the water, are a two-dimensional analogy to a spherical wave.

Finally it should be noted that when discussing the properties of light, we often neglect the direction of the oscillating electric field, that is, the direction of the vector $\vec{E}$, and rather just use the scalar amplitude of the wave. This is just a bit careless, but convenient way of describing light when polarization effects do not matter for the experiment under consideration.

1.3 Four Effects of Interference

The wave nature of light can explain four fundamental aspects of light: diffraction, the refractive index, refraction, and reflection. Reflection at a mirror has the famous property of the incident angle corresponding to the angle of the light leaving the mirror. Refraction refers to the effect where light *rays* seem to change their direction when they pass from one medium to another. This is, for example, seen when trying to look at a scene through a glass full of water. It is connected with the fact that different media often show different refractive indices. Finally, diffraction is a phenomenon occurring when light interacts with very fine structures or openings, which are often quasi-periodic.

Even though these effects may seem very different at a first glance, all of these effects are ultimately based on interference. Diffraction is most prominent when light illuminates structures of a feature size similar to the wavelength of light. In contrast, reflection and refraction, for example, the bending of light rays caused by a lens, occur when different media such as air and glass comprise elements – molecules – that are far smaller than the wavelength of light but cover isotropic domains that are much more extended than the wavelength.

1.3.1 Diffraction

To describe diffraction, it is useful to first look at the light emitted by a point-like source. Let us consider an idealized source, which is infinitely small and emits only a single light color. This source then emits a spherical wave.

Christiaan Huygens (1629–1695) had a clever idea: to find out how a wave will spread in space, he proposed choosing an arbitrary border surface – meaning a surface of equal phase – then placing virtual point emitters everywhere on this surface, and letting the light of these emitters interfere. The resulting interference pattern will reconstitute the original wave beyond that surface. This "Huygens' principle" can well explain that parallel waves remain parallel, because we find constructive interference only in the propagation direction of the wave. Strictly speaking, one would also find a backward-propagating wave. However, when Huygen's idea is formulated in a mathematically rigorous way, the backward wave is not obtained. Huygens' principle is very useful when predicting the spreading of a wave hitting a structure with feature sizes comparable to its wavelength, for example, a small slit aperture or a periodic diffraction grating. In Figure 1.5, we consider the example of diffraction at a grating of transmitting

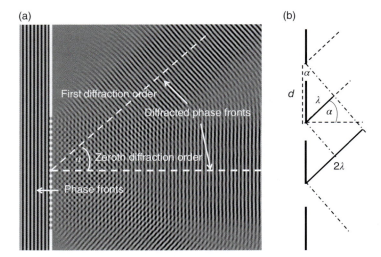

Figure 1.5 Diffraction at a grating. (a) A plane wave hits perpendicularly on a grating. The directions of constructive interference where maxima and minima of one wave interfere constructively with the maxima and minima of the second wave are shown for the zeroth- and first-order diffraction. (b) Magnified view of the grating geometry with the first-order diffraction direction. The condition for constructive interference is that the pathlength difference is equal to multiples of the wavelength λ.

slits. The distance between the slits is designated as the grating constant d. The figure shows how circular waves corresponding to Huygens' wavelets originate at each aperture and join to form new wave fronts, finally giving rise to plane waves traveling in various distinct directions. At most places, they interfere with a great diversity of phases, which means that they cancel in those locations, but in certain directions the wavelets add up constructively. This is seen best at large distances from the grating, where the individual waves join to form uniform plane phase fronts again (see the online supplemental material for an animated version of Figure 1.5). The directions of constructive interference can be characterized by their angle α with regard to the propagation direction of the incident wave. Then α fulfills the condition (Figure 1.5)

$$\sin \alpha = N \frac{\lambda}{d} \tag{1.8}$$

with N denoting an integer multiple of wavelengths λ to yield the same phase for creating constructive interference. N is called the *diffraction order*. Note that the angle α of the diffracted waves depends on the wavelength and thus on the color of light. In addition, note that the crests of the waves form connected lines, which are called *phase fronts* or *wave fronts*, whereas the dashed lines perpendicular to these phase fronts can be thought of as corresponding to the light rays of geometrical optics.

A compact disk is a good example of such a diffractive structure. White light is a mixture of light of many visible wavelengths. Thus, illuminating a compact disk with a white light source from a large distance will cause only certain colors to be diffracted from certain places on the disk into our eyes. This leads to the observation of the beautiful rainbow-like color effect when looking at it.

1.3.2 The Refractive Index

Understanding the behavior of light inside materials is not at all trivial. For a detailed treatment on an introductory level, we refer the reader to [1].

Somewhat simplified, Huygens' idea can also explain what happens when light traverses a homogeneous medium. Here, the Huygens emitters correspond to the real molecules inside the material. The incoming wave with its electric field induces an oscillation of the electrons with respect to their atomic nuclei. These oscillating electrons constitute accelerated charges that radiate a new electromagnetic wave. It turns out that the emission from each molecule is slightly phase-shifted with respect to the incoming wave. Notably, the magnitude of the phase shift as well as the amplitude of the emitted wave depends on the nature of the material (see Box 1.4). Even though each scattering molecule generates a spherical wave, the superposition of all the scattered waves from atoms at random positions will interfere constructively only in the propagation direction of the incoming wave. Thus, each very thin layer of molecules generates another parallel wave, which differs in phase from the original wave. The sum of the original sinusoidal wave and the interfering sinusoidal wave of scattered light results in a forward-propagating parallel wave with sinusoidal modulation, but typically lagging slightly in phase. In a dense medium, this phase delay is continuously occurring in every new layer of material throughout the medium, giving the impression that the wave has "slowed down" in this medium compared to vacuum. This reduction in the phase propagation speed is described by the refractive index n

$$c_{medium} = \frac{c_{vacuum}}{n} \tag{1.9}$$

According to Eq. (1.6), it can also be interpreted as an effectively reduced wavelength λ_{medium} inside the medium:

$$\lambda_{medium} = \frac{\lambda_{vacuum}}{n} \tag{1.10}$$

with λ_{vacuum} denoting the wavelength in vacuum. Note that the oscillation frequency of the electric field does not depend on the medium. Usually, n is dependent on the wavelength of the incoming light, $n = n(\lambda)$. This wavelength dependence of the refractive index is called *dispersion*.

1.3.3 Refraction

Now we analyze what happens when a plane light wave hits the interface between two materials with different refractive indices n_1 and n_2 at an angle α_1 to the surface normal, as shown in Figure 1.6. The wave directly at the interface is the incident wave, and the first layer of molecules of material 2 still "feels" this wave. This means that the phases of the light wave along the boundary must be identical in both materials. However, the wavelength of the light wave is $\lambda_1 = \lambda_{vacuum}/n_1$ in material 1 and $\lambda_2 = \lambda_{vacuum}/n_2$ in material 2. Both conditions can be fulfilled only if

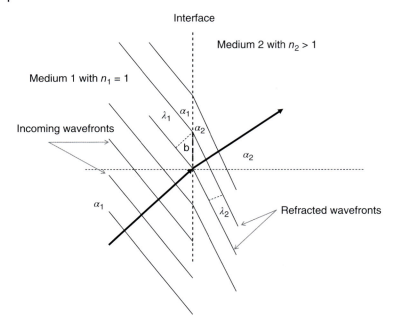

Figure 1.6 Snell's law of refraction. The phases of the electric field along the interface between the two materials must be identical. The wavelengths inside the materials are given by λ_{vacuum}/n_1 and λ_{vacuum}/n_2. We note that $\sin \alpha_1 = \lambda_1/b$, where b denotes the distance between two wave crests at the interface, and also that $\sin \alpha_2 = \lambda_2/b$. Eliminating b yields Eq. (1.12).

the wave changes its propagation direction in the second material. We denote the angles between the direction of propagation of the plane waves in material 1 and 2 and the perpendicular line onto the medium's surface by α_1 and α_2, respectively. We find (Figure 1.6) that

$$\frac{\lambda_1}{\lambda_2} = \frac{n_2}{n_1} = \frac{\sin \alpha_1}{\sin \alpha_2} \tag{1.11}$$

or

$$n_1 \sin \alpha_1 = n_2 \sin \alpha_2 \tag{1.12}$$

This is Snell's famous law of refraction, which forms the foundation of geometrical optics. One important use is to trace rays on their way through an optical system with multiple transitions between optical glasses and air.

1.3.4 Reflection

Light waves hitting surfaces are also reflected. There is an immediate consequence of considering the behavior of the Huygens wavelets at an interface. The law of reflection states that the angle of the incoming light is equal to the angle of the outgoing light. This can immediately be deduced by applying the reasoning used for the derivation of Snell's law by assuming that the wave is continuous along the interface and the reflected wave is traveling back into medium 1. It turns out that this simplified picture is a little bit too crude when it comes to fully understanding reflection at different materials such as metals and dielectrics, since a little bit of scattering volume in medium 2 is needed to provide enough power for the reflected wave. Even though the law of reflection is always valid, the phase of the reflected wave may change, which has to do with the penetration depth of the wave into the material. For a more detailed discussion on polarization and reflection, see Box 1.2.

1.3.5 Light Waves and Light Rays

How can we connect the concept of light rays as commonly sketched in geometrical optics with the concept of light as a wave phenomenon? A light ray represents a plane wave with a lateral extension small enough to be looked upon as a ray but extended enough not to show substantial broadening by diffraction. The light beam emitted by a laser pointer is a good example for such a ray. When light rays encounter interfaces between two media with different refractive indices, we can apply Snell's law to calculate what happens to the ray when it hits the interface at different angles. The ray bends toward the surface normal at the transition from an optically thin medium such as air with $n_1 = 1$ to an optically thicker medium such as glass with $n_2 = 1.52$ and $n_2 > n_1$. The opposite happens at a glass–air interface. Lenses have curved surfaces, and therefore the fate of the incoming beams depends on the position at which they hit the interface. Such calculations can rapidly be performed with high spatial resolution for complete lens surfaces as a function of the wavelength using computers. Such ray-tracing computations serve to exactly predict the effect of lenses of any shape and are a key tool to design modern optical systems.

Box 1.2 Polarization and Reflection

Light in a homogeneous isotropic medium is a transverse electromagnetic wave. This means that the vector of the electric field is perpendicular to the direction of propagation. Sometimes we can simplify this view by ignoring the direction of the electric field vector and treating the amplitude as a scalar. However, there are situations in which the direction of the electric field vector, that is, its polarization direction, is important. Ultimately polarization-dependent effects can be traced back to the fact that scattered light emitted from individual molecules ("dipole emission") has an angular dependence rather than being a spherical wave with uniform strength.

Transparent materials such as glass reflect a certain but small amount of light at their surface, even though they are 100% transparent once the light is inside the material. At perpendicular incidence, ~4% of the incident light is reflected. Under oblique incidence, the amount and the phase of the reflected light strongly depend on the polarization of the incident light. This is quantified by the so-called Fresnel reflection coefficients. These coefficients are plotted in Figure 1.7 for light hitting an air/glass and a glass/air interface. The plane that is formed by the propagation direction of the incident beam and the surface normal is called the *plane of incidence*. The Fresnel coefficients differ largely for the polarization component oscillating within the plane of incidence, the p-component (from the

(Continued)

Box 1.2 (Continued)

German word "parallel"), and the polarization component perpendicular to it, the so-called s-polarization (from the German word "senkrecht" for perpendicular). There is a specific angle at which the p-component has a zero reflectivity coefficient. At this so-called Brewster angle, this component is entirely transmitted into the glass without any reflection (Figure 1.7a). Let us assume that we illuminate the surface with unpolarized light precisely at the Brewster angle. Then the reflected light will be completely polarized in the direction perpendicular to the incidence plane, the s-direction. The p-component completely enters the glass. Interestingly, the Fresnel coefficients for a glass–air interface predict a range of total reflection when the inclination angle of the incoming light is beyond a certain angle, the so-called critical angle (Figure 1.7b). This effect is employed in total internal reflection microscopy, which is abbreviated as TIRF microscopy (see Chapter 9).

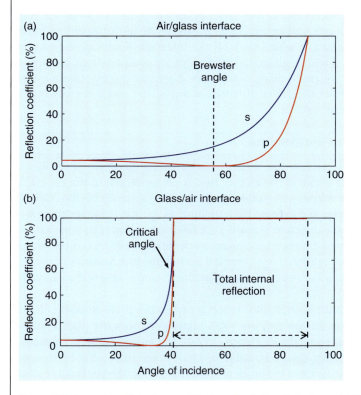

Figure 1.7 Fresnel coefficients quantify the reflectivity of interfaces. (a) Reflectivity of an air/glass interface for p (red) and s (blue) polarization directions of the incoming light. "p" means that the polarization vector is in the plane of incidence, whereas "s" means that the polarization vector is perpendicular to this plane. The angles are given with respect to the surface normal of the interface. The refractive indices for air and glass are $n_{air} = 1.0$ and $n_{glass} = 1.52$. (b) Reflectivity of a glass/air interface. Note the total reflection at supercritical angles.

> There are crystalline materials that are not isotropic: that is, their unit cells have an asymmetry that leads to different refractive indices for different polarization directions. This effect is designated as *birefringence*. A beam entering such a material will usually be split into two beams within the birefringent crystal that travel into different directions. By cutting crystal wedges along different directions and joining them together, one can create the so-called Wollaston or Normaski prisms. The p- and s-polarized light components leaving the crystal will be slightly tilted with respect to each other. Such prisms are used for differential interference contrast (DIC) microscopy (Chapter 2).

1.4 Optical Elements

With the knowledge that light should ultimately be described as a wave, we can now move to more practical aspects of what can be done with light. In the context of this book, we need to understand various optical elements that are used in microscopes: lenses, mirrors, pinholes, filters, and chromatic reflectors. Although the wave picture is essential in microscopy, it can nevertheless be sometimes useful to approximate light propagation using the ray picture. This is the realm of "geometrical optics," which will be used for some of the following considerations.

1.4.1 Lenses

Here we will analyze a few situations to understand the general behavior of lenses. In principle, we should use Snell's law to calculate the shape of an ideal lens with perfect focusing ability. However, this would be beyond the scope of this chapter. Rather, we assume that a spherical lens made from glass with refractive index n and radii of curvature R_1 and R_2 – both positive for a convex surface – focuses parallel incoming light rays into a point at the focal distance f behind the lens. Then f is given by [2]

$$\frac{1}{f} = (n-1)\left[\frac{1}{R_1} + \frac{1}{R_2} - \frac{(n-1)d}{nR_1R_2}\right] \tag{1.13}$$

where d denotes the thickness of the lens measured at its center on the optical axis. The above equation is called the *lensmaker's equation* for air. If the lens is thin and the radii of curvature are large, the term containing $d/(R_1R_2)$ can be neglected, yielding the equation for "thin" lenses:

$$f = \frac{1}{(n-1)\left(1/R_1 + 1/R_2\right)} \tag{1.14}$$

This approximation is usually made, and the rules of geometrical optics as stated below apply.

The beauty of geometrical optics is that one can construct *ray diagrams* with pencil and paper and can graphically work out what happens in an optical system. Figure 1.8a shows how all the rays parallel to the optical axis are focused

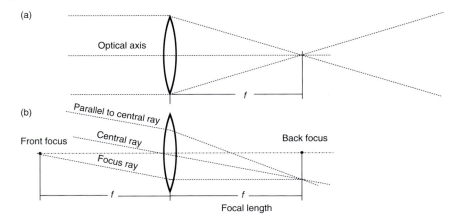

Figure 1.8 Focus of a lens under parallel illumination. (a) Illumination parallel to the optical axis. (b) Parallel illumination by beams inclined with respect to the optical axis, yielding a focus in the same plane but at a distance from the optical axis.

onto the focus of the lens. The optical axis is the symmetry axis of the lens and describes the general direction of the rays. Using the wave picture, we can say that a plane incoming wave is transformed by the lens to a spherically converging wave behind the lens, which converges at the lens focus. Figure 1.8b shows that this is also true for parallel rays entering the lens at an oblique angle. They are also focused onto the same plane. There are two more basic rays used for the geometrical construction of ray diagrams. The ray traversing the center of a thin lens is always unperturbed; it is called the *center ray*. This is easily understood, as the material is oriented at the same angle on its input side and exit side. For a thin lens, this "slab of glass" is considered to be infinitely thin. Thus, we can follow the ray right through the center of the lens.

The other key ray is traversing the front focal point of the lens at any angle and leaving the lens as a ray parallel to the optical axis. In geometrical optics of thin lenses, lenses are assumed to be symmetrical. Thus, the front focal distance of a thin lens is the same as the back focal distance.

The principle of *optical reciprocity* of geometrical optics states that we can always retrace the direction of light rays and yield identical paths of the rays. Obviously, this is not strictly true for any optical component. For example, absorption filters will not lead to amplification when the rays are propagated backward! However, it follows from this principle that any ray traversing a lens focus on the optical axis will leave the lens parallel to the optical axis.

Parallel incoming light is often referred to as coming from sources *at infinity*, as this is the limiting case when moving a source further and further away from the lens. Let us now consider what happens to the light emitted from an object located at "infinity." The starry night sky is a good example of light sources at practically infinite distance. A lens will map such an object to its focal plane because each object point source at an "infinite distance" generates a parallel wave with its unique direction. A lens will "image" such sources to unique positions in its focal plane. Therefore, telescopes produce images of the night sky in their focal plane.

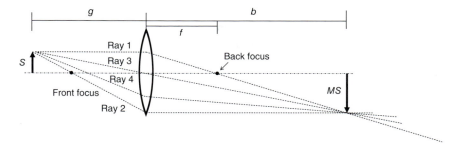

Figure 1.9 A single lens imaging an object as an example for drawing optical ray diagrams. For discussion of the ray paths, see text.

In Figure 1.9 we consider what happens to light emitted from an object of height S located at a distance g from a lens with focal length f. We assume that $g > f$. The trick now is to follow those key light rays for which we know their path through the lens:

- A ray parallel to the optical axis will pass through the back focus on the optical axis (Ray 1).
- A ray traversing the front focal point on the optical axis will end up being parallel to it behind the lens (Ray 2).
- A ray going through the center of the lens will pass through it unperturbed (Ray 3).

We know that lenses generate images, that is, if two special rays emerging from the same object point cross each other at a certain point on the image side, all other rays emerging from that object point will cross at this same point. Thus, if one such crossing is found, we are free to draw further rays from the same source object point to the same crossing on the image side, for example, Ray 4 in Figure 1.9.

We see that the image is flipped over and has generally a different size MS compared to the original size S, with M denoting the *magnification factor*. We find two conditions for similar triangles: $MS/b = S/g$ and $S/(g-f) = MS/f$.

This leads to the description of the imaging properties of a single lens:

$$\frac{1}{f} = \frac{1}{g} + \frac{1}{b} \quad \text{with the magnification} \quad M = \frac{b}{g} \tag{1.15}$$

where b is the distance between the lens and the image.

Using the aforementioned steps of optical construction, even complicated optical setups can be treated with ease.

1.4.2 Metallic Mirrors

In the discussion of refraction, we considered materials in which the electrons are bound to their nuclei. This leads to a phase shift of the scattered wave with respect to the incident wave. The situation is slightly different when an electromagnetic wave encounters a metal because the valence electrons of the metal are essentially free and very mobile. At the metal surface, the wave induces oscillations of the electrons, which oscillate such that they emit a wave that is phase-shifted by 180° (or π radians) with respect to the incoming one. This causes destructive interference along the propagation direction, and there is no transmitted wave. The reason for the 180° phase shift is that the conducting metal will always compensate for any electric field in its interior by moving charges such that the field vanishes. Thus, there is only a reflected wave from the surface, which means that very good mirrors can be made using planar metal surfaces. Application of Huygens' principle then leads to the law of reflection as discussed previously.

However, there is a small current induced by the electric field of the light wave wiggling at the electrons. Also, the material of the mirror has some electrical resistance. This will then lead to absorption losses. The reflectivity of a good-quality front-silvered mirror is typically 97.5% for wavelengths between 450 nm and 2 µm. Such unprotected, front-silvered mirrors are quite delicate and easy to damage, which often leads to much higher losses in praxis.

1.4.3 Dielectric Mirrors

Nonmetallic materials also show a certain amount of reflection. For example, glass with a refractive index of $n = 1.52$ shows about 4% reflection from its surface at perpendicular incidence of light owing to the interference effects of the waves generated in the material. Through careful deposition of multiple layers of materials with different refractive indices at well-defined thicknesses, it is possible to reach a reflectivity of almost 100% for very defined spectral ranges or, alternatively, over large wavelength bands and specific incidence angles.

The working principle of dielectric mirrors is sketched in Figure 1.10. A glass substrate is coated with a carefully designed series of layers of materials with different refractive indices. The layer thickness can be controlled such that only a

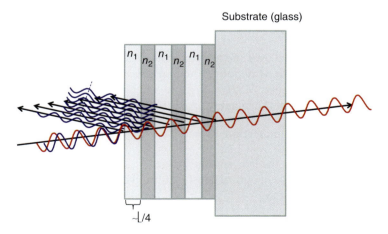

Figure 1.10 Principle of reflection by multilayer coatings. Dielectric mirrors, interference filters, and chromatic reflectors are all based on this principle. The exact order, thickness, and refractive index of the various layers of dielectric materials lead to a wavelength-dependent constructive or destructive interference for the reflected or transmitted light. In this manner, from very narrow-band notch filters to broadband dielectric mirrors with >99.9% reflectivity over a considerable spectral range can be constructed.

specific wavelength is reflected or transmitted. By varying the thickness of the layers, it is also possible to design the coating such that it reflects very well over a broad range of wavelengths. These layers yield multiple reflections and build up standing optical waves inside the material, which leads to the name "cavities" of these layers. In wavelength-selective mirrors, typically a thin layer of a material with a high index of refraction alternates with a thicker layer of a material with a low index of refraction such that the optical path for a reflected wave is an integer multiple of the wavelength, leading to constructive interference for the reflected light (Figure 1.10).

Such mirrors with selective or optimized reflectivity are employed in microscopes to steer the beam, to keep the microscope compact, to allow precise optical adjustment, and to switch between its various modes of operation. Dielectric mirrors are also used in lasers to form their cavity for amplifying the light. Such laser mirrors can have a reflectivity well above 99.999%.

The light losses of dielectric mirrors can be much smaller than for metallic mirrors, leading to overall higher performances. In addition, scratches on the mirror surface have less impact on the performance of these mirrors. However, one should be aware of the fact that waves with different polarizations penetrate the mirror material to different extents when they hit the mirror surface not perpendicularly. This leads to different phase shifts for p- and s-polarized waves. A linear polarization, when oriented at 45° to the plane containing the incident and reflected beams, for instance, is typically not conserved but is converted to elliptical polarization. This effect is particularly noticeable when the wavelength of the incident wave is close to the edge of the specified wavelength range of the coating.

Finally, such multilayer coatings can also be used to reduce the reflectivity. The usual 4% reflection at each air–glass and glass–air surface of optical elements can be significantly reduced by coating the lenses with layers of well-defined thicknesses and refractive indices. These are the so-called antireflection coatings. When viewed at oblique angles, antireflection-coated surfaces often display a blue, oily shimmer as can be noticed on photographic camera lenses.

1.4.4 Filters

There are two different types of optical filters and combinations of these. Absorption filters consist of a thick piece of glass in which a material with strong absorption in a certain wavelength band is embedded. Such filters have the advantage that a scratch will not significantly influence the filtering characteristics. A problem is that the spectral edge of the transition between absorption and transmission is usually not very steep and the transmission in the transmitted band of wavelengths is not very high. Note that when one uses the term *wavelengths* a bit carelessly, as in this case, one usually refers to the corresponding wavelength in vacuum and not to the wavelength inside the material.

The so-called interference filters are always coated on at least one side with a multilayer structure of dielectric materials as discussed previously. These coatings can be tailored precisely to reflect exactly a range of wavelength while effectively transmitting another well-defined range. Such coatings function by interference, and therefore there is an inherent angular and wavelength dependence.

This means that a filter placed at a 45° angle will transmit a different range of wavelengths than when placed at perpendicular incidence of the incoming light. This has consequences for the practical use of such filters in microscopy, as outlined in Box 1.3.

> **Box 1.3 Practical Aspects of Using Filters in Microscopy**
>
> We will see later that in a fluorescence microscope the exact object position in the front focal plane of the objective lens will define the specific angle at which the light from this point leaves the objective. Fluorescence filters are typically placed in the space between the objective and the tube lens. Therefore, this angle is also the angle of incidence on the filter. Theoretically, there is therefore a position-dependent color sensitivity. However, the fluorescence spectra are rather broad, and the angular spread for a typical field of view is in the range of only ±2°. Therefore, this effect can be completely neglected for fluorescence microscopy.
>
> In microscopic imaging, it is important to reduce the background light as much as possible. Background can stem from the generation of residual fluorescence or Raman excitation even in glass. For this reason, optical filters always need to have their coated side facing the incident light. Because the coatings usually never reach completely the edge of the filter, one can determine the coated side by careful inspection. When building setups in-house, one also has to ensure that no light can possibly pass through an uncoated portion of a filter, as this would have disastrous effects on the suppression of unwanted scattering.

1.4.5 Chromatic Reflectors

Chromatic reflectors are designed to reflect a specific wavelength range, which is usually shorter than a critical wavelength, and to transmit another range. Like normal interference filters, chromatic reflectors are manufactured by multilayer dielectric coating. The comments about the wavelength and angular dependence given in Section 1.4.4 also apply here. Actually, they are especially valid for chromatic filters, which are mostly used at an incidence angle of 45°. In this case, the angular dependence is much stronger and the spectral steepness of the edges is much softer than for angles closer to perpendicular incidence.

Chromatic reflectors are often designated as *dichroic mirrors*. This is a confusing and potentially misleading term because the physical effect of "dichroism", which refers to a polarization-dependent absorption, has nothing to do with the functioning of these special mirrors.

> **Box 1.4 Beam Paths and Interference in the Mach–Zehnder Interferometer**
>
> Finally, we can examine the two beam paths in the Mach–Zehnder interferometer shown in Figure 1.1. Of course, the key to the working of the instrument is the phases of the waves traveling through various paths. Having discussed some basic properties of light in materials and at surfaces, we first summarize some detailed facts about the used components.

BS1 and BS2 reflect half the incident light and refract the other half through them. The speed of light in air is approximately the speed of light (c) in vacuum. However, the speed of light inside glass is significantly smaller than that in vacuum, namely about $(2/3)c$. Thus, when a light ray traverses a medium such as a glass plate, its phase will be altered by an amount that depends on the index of refraction of the medium and the path length within the medium. When a light ray hits an interface, and the material on the other side of the interface has a *higher* index of refraction (i.e., a *lower* speed of light) than the medium the light is traveling in, then the reflected light ray is shifted in its phase by exactly π. This is the case when light is reflected by a metallic mirror. When a light ray hits an interface and the material on the other side of the interface has a *lower* index of refraction, the reflected light ray does not change its phase. When a light ray passes from one medium into another, its direction changes as a result of Snell's law of refraction but there is no phase change at the interface (Figure 1.6).

Now we can consider the two light paths leading to exit 1 (see Figure 1.1).

Light path 1:
1. Reflection at the front of BS1 yields a phase change of π.
2. Reflection by M1 yields a further phase change of π.
3. Transmission through BS2 yields some constant phase change $\Delta\varphi$.

Light path 2:
1. Transmission through BS1 yields some constant phase change $\Delta\varphi$.
2. Reflection at M2 yields a phase change of π.
3. Reflection at the front of BS2 yields a phase change of π.

Summing up the phase shifts for the two paths shows that they are identical, $2\pi + \Delta\varphi$. The light leaving the instrument at exit 1 via the two paths is in phase, which results in constructive interference.

Now we consider the two light paths leading to exit 2:

Light path 1:
1. Reflection at the front of BS1 yields a phase change of π.
2. Reflection by M1 yields a further phase change of π.
3. Transmission through BS2 yields some constant phase change $\Delta\varphi$. Reflection at the inner surface of BS2 yields no phase change. Transmission through BS2 a second time yields again some constant phase change $\Delta\varphi$.

Light path 2:
1. Transmission through BS1 yields some constant phase change $\Delta\varphi$.
2. Reflection at M2 yields a phase change of π.
3. Transmission through BS2 yields some constant phase change $\Delta\varphi$.

The sum of the phase shifts for light path 1 yields $2\pi + 2\Delta\varphi$. The sum of the phase shifts for light path 2, however, is $\pi + 2\Delta\varphi$. Thus, they differ exactly by π, which results in destructive interference in the direction of exit 2. Notably, this result does not depend on the wavelength! This explanation followed the presentation of David M. Harrison [3].

1.5 Optical Aberrations

High-resolution microscopes are difficult to manufacture and usually come with quite tight demands on experimental parameters (e.g., the specified temperature and the refractive index of the mounting medium) that should be used. Thus, in the real world, the microscope will almost always show imperfections because the demands are not perfectly fulfilled. Usually, this leads to aberrations in the optical path. For a microscopist, it is very useful to be able to recognize such optical aberrations and classify them. One can then try to remove them through modifications to the setup or, in the extreme case, by using *adaptive optics*, where such aberrations are corrected by using special elements such as deformable mirrors that can modify and adjust the wave front before it passes through the optics and the sample.

The most basic modes of aberration called *tip*, *tilt*, and *defocus* are usually not noticed because they simply displace the image slightly along the x-, y-, and/or z-direction. They are not a problem because the image quality remains unaffected.

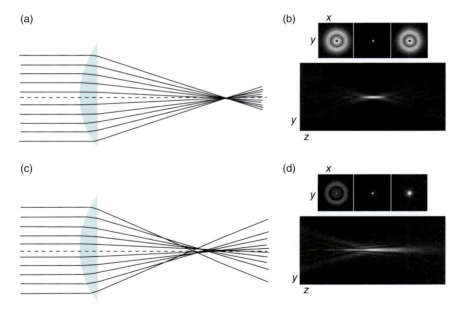

Figure 1.11 Spherical aberration. (a) Perfect lens without spherical aberration. (b) Lateral (top) and axial (bottom) beam profiles. The lateral profiles show the intensity before (left), at (center), and behind the focus (right). (c) An uncorrected lens showing positive spherical aberration: off-axis rays are refracted too much and miss the nominal focus. (d) This causes a focus asymmetry: overly sharp rings and a fine peak appear before the nominal focus, while at equal distance behind the focus a fluffy spot is seen. To improve visibility, the axial distributions show the square-root of the intensity and the lateral distributions show the intensity normalized to the brightest point. Negative spherical aberration occurs under certain conditions, and then the effect is reversed. (The graphics was inspired by http://en.wikipedia.org/wiki/Spherical_aberration.)

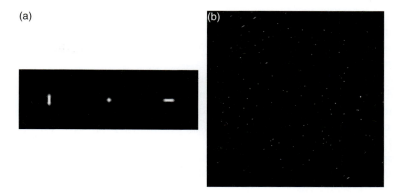

Figure 1.12 Astigmatism. (a) Image of a point object above, at, and below the focus displaying astigmatism. A vertical ellipse, a cross, and a horizontal ellipse can be observed. (b) Astigmatism as a field-dependent aberration in a microscopic image.

The most important aberration that indeed affects microscopic imaging is *spherical aberration*. In fact, a normal thick lens with spherical surfaces to which the lensmaker's equation would apply shows spherical aberration. Rays from the outer rim of the lens focus to a slightly different position than rays from the inner area, as shown in Figure 1.11. This blurs the focus and decreases the brightness when pointlike objects are imaged. A modern microscope objective can achieve an image free of spherical aberration when its optics is carefully designed. However, even for quality objectives, spherical aberration is commonly observed in praxis. This is because of a mismatch between the room temperature and the design temperature, or because the objective is not used according to its specifications. This happens, for example, when a sample is embedded in water with a refractive index of $n = 1.33$, and imaged by an oil-immersion objective designed for samples embedded in materials having a refractive index similar to that of oil, namely $n = 1.52$. Another common reason for spherical aberrations is the use of coverslips with a different thickness than that for which the objective was designed. A convenient way to notice spherical aberration is to pick a small, bright object and manually defocus up and down. If both sides of defocus essentially show the same fuzzy pattern of rings, the setup is free of spherical aberrations. However, if an especially nice pattern of rings is seen on one side and no rings or very fuzzy rings are seen on the other side, there are strong spherical aberrations present.

Another common aberration is *astigmatism*. When defocusing a small pointlike object up and down, one sees an ellipse oriented in a particular direction above the plane of focus and a perpendicular orientation of the ellipse below the focus (Figure 1.12a). Astigmatism is often seen in images at positions far away from the center of the ideal field of view (Figure 1.12b). A close observation of the in-focus point objects shows a star- or cross-shaped appearance. In astigmatism, there is essentially a disparity between the focus along the x-direction and the focus along the y-direction.

The last aberration to mention here is *coma*. Coma can arise from misaligned – tilted – lenses. At the position of best focus, a point object imaged with a coma aberration looks like an off-center dot, as opposed by an asymmetric quarter-moon-shaped comet when in focus. The farther one goes from the focus, the more prominent is the asymmetry. What should look like a fluffy ring seems to be missing a part.

Aberrations can also result in a position-dependent displacement of image points called *radial distortion*. Depending on the way that the positions are distorted, one discriminates between a barrel distortion, in which x- and y-axes are pushed to the outside, or a pin-cushion distortion, where the x- and y-axes are squeezed to the center.

A final monochromatic distortion is designated *curvature of field* or *Petzval field curvature*. The image of a flat, planar object cannot be sharply projected onto a plane in the image space. A perfectly sharp image can be produced only on a curved surface. This generates problems when objects are imaged with a flat two-dimensional image sensor such as a charge-coupled device (CCD) camera chip. Image points near the optical axis will be perfectly in focus, but off-axis rays come into focus before the image sensor. Effectively, the image cannot be sharp over the complete field of view.

We often use different colors for microscopic imaging, especially when in fluorescence microscopy (Chapter 3). Therefore, it is useful to examine how the light paths through lenses depend on the specific light color. The index of refraction of most materials depends on the wavelength. According to the lensmaker's equation, a single lens made out of one type of glass will therefore show a wavelength-dependent focal length f. As a consequence, the images of an object seen in different colors will have slightly different sizes and also different positions in the image space. The shift of the image along the optical axis is called *axial chromatic aberration*. The wavelength-dependent magnification is called *lateral chromatic aberration*. By combining different lenses of different materials into a single optical element, it is possible to at least partially compensate for this dispersion effect. A typical example is the readily available achromatic doublet lenses, where two lenses of different materials and curvatures are combined to yield a focus largely free of chromatic aberrations. Nevertheless, all microscopes show a fair amount of – predominantly axial – chromatic aberration. This can be demonstrated by imaging the same object in red and blue light. Usually, there is a significant shift along the axial direction between the respective positions of the two images.

References

1 Feynman, R.P. (2011) *The Feynman Lectures on Physics*, The New Millennium Edition: Mainly Mechanics, Radiation and Heat, vol. **1**, Basic Books, New York.
2 Hecht, E. (2002) *Optics*, 4th edn, Chapter 6.2, Addison-Wesley. ISBN: 0-321-18 878-0.
3 Harrison, D.M. (1999) *The Physics Virtual Bookshelf*, University of Toronto, http://www.upscale.utoronto.ca/GeneralInterest/Harrison/MachZehnder/MachZehnder.html (accessed 25 October 2016).

2

Principles of Light Microscopy

Ulrich Kubitscheck

Friedrich-Wilhelms-Universität, Institut für Physik & Theoretische Chemie, Wegelerstr. 12, 53115 Bonn, Germany

2.1 Introduction

In this chapter, we use the basic knowledge of geometrical optics and the properties of light waves to understand the construction and functioning of light microscopes. We focus more on the underlying physical principles of microscope operation and less on the technical details of the optical construction. The central goal is to achieve an understanding of the capabilities and limitations of microscopic imaging, which will serve as a basis for the subsequent chapters. The chapter also includes an introduction to the (currently) most important optical contrast techniques in the biosciences: phase contrast and differential interference contrast (DIC). These techniques allow the creation of contrast-rich images of almost all transparent specimens that are typically encountered in biological and medical research. This chapter presents the basic material that is needed to understand fluorescence microscopy and its wide range of applications, as well as new technical developments as detailed in the following text.

2.2 Construction of Light Microscopes

2.2.1 Components of Light Microscopes

An optical microscope consists of two functionally distinct sets of components the illumination elements and the imaging elements. Newcomers usually consider only the optical elements of the imaging beam path to form the "real" microscope and tend to overlook the importance of the illumination components. Indeed, the two parts are deeply interconnected with each other and of equal importance for the achievable optical resolution and the creation of image contrast. The two parts comprise their own light paths, which are in a certain sense complementary to each other. To begin with, we consider them according to the rules of geometrical optics, which means that we consider light rays and beam paths while neglecting the wave nature of light and any diffraction processes. These are treated in the later sections of this chapter.

Fluorescence Microscopy: From Principles to Biological Applications, Second Edition.
Edited by Ulrich Kubitscheck.
© 2017 Wiley-VCH Verlag GmbH & Co. KGaA. Published 2017 by Wiley-VCH Verlag GmbH & Co. KGaA.

2.2.2 Imaging Path

The basic element of the imaging part of a modern microscope is formed by two lenses. The specimen or object is located in the front focal plane of the first lens of the optical system, which is designated as the *objective lens* and produces an image of the object at infinity. This image is focused by a second lens – the *tube lens* – into the primary image plane, as shown in Figure 2.1.

The imaging process can be understood by the geometric imaging laws as follows: To this end, we follow the beams emitted from the two extreme ends of an object. The light emitted by the object point at the lens focus o1, is transformed by the objective lens into a set of light beams that travel parallel to the optical axis (Figure 2.1, full lines). This parallel light bundle enters the tube lens and is refracted into its focus. The fate of the light beams emitted from the off-axis point o2 can be determined by the following two special light beams: the ray rc, hitting the lens center, passes the objective unrefracted; the ray rp, emitted parallel to the optical axis, is focused into the back focal point of the objective lens. We note that both beams are parallel to each other in the space between the objective lens and the tube lens. This space is designated as *infinity space*. All beams emitted from o2 will also be parallel to each other in this space, because o2 lies in the front focal plane of the objective (Figure 2.1, dashed lines). The position of the tube lens in relation to the objective can be chosen quite arbitrarily. However, usually it is positioned such that its front focal point coincides with the back focal point of the objective lens, at least approximately. In order to find out to which point the tube lens focuses these beams, we can follow the known special beams of the tube lens. We follow the beam passing through the front focus of the tube lens. This is translated into a beam parallel to the optical axis behind the tube lens, rp'. The central beam rc' passes the tube lens again unrefracted. Hence, the point o2 is projected to the intersection of these two beams, b2, in the back focal plane

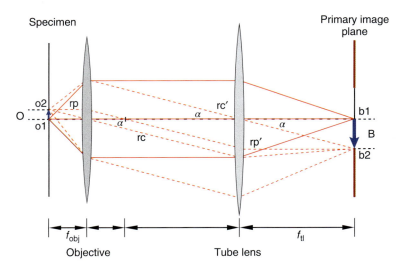

Figure 2.1 Infinity-corrected imaging process using a light-sensitive array image detector. f_{obj} and f_{tl} denote the focal length of the objective and tube lens, respectively.

of the tube lens. This plane contains a real image of the object and is designated as the *primary image plane*. The image size corresponds to the vertical distance between this point and the optical axis where the image point b1 of o1 was found. It can be seen that

$$\tan \alpha = \frac{O}{f_{obj}} \tag{2.1}$$

where α is the angle between the optical axis and beam rc in the object space, which lies to the left of the objective lens. O is the distance between o1 and o2 corresponding to the object size, and f_{obj} is the focal length of the objective lens. Examining the beams that form the image with the size B on the right-hand side of the tube lens, we observe also that

$$\tan \alpha = \frac{B}{f_{tl}} \tag{2.2}$$

where f_{tl} is the focal length of the tube lens. It follows that the magnification factor M of this two-lens system is

$$M = \frac{B}{O} = \frac{f_{tl}}{f_{obj}} \tag{2.3}$$

The ratio of the focal lengths of the tube lens and the objective lens determines the magnification factor. The tube lens is fixed within the microscope, and therefore the choice of the objective lens determines the respective magnification. The minimum focal length of the objective lens is limited for practical and technical reasons, but the focal length of the tube lens is relatively large. Magnification factors usually range between 1 and 100.

For visual observation of microscopic samples, two magnification steps are arranged in sequence (Figure 2.2). The objective lens and tube lens form the first magnification step and create together the *primary* or *intermediate image*. Its absolute size has a diameter of 2–3 cm depending on the type and quality of the objective lens. The second magnification step is also a two-lens infinity optical system working in a way equivalent to the first magnification step. It is formed by the eyepiece and the eye lens of the observer, which project the final image onto the retina. The retina in the eye of the observer is the *secondary image plane*.

In principle, there are two different ways in which a lens system with a focal length f may produce a real image: when the object is placed at a distance s_o from the lens with $2f > s_o > f$, a real, inverted, and magnified image is produced directly at a fixed distance from the lens in the primary image plane. This type of configuration is called the *finite optics setup*. The other possibility is to place the object at the focal plane of the lens and use a second lens in an infinity setup as discussed earlier. In microscopy, both magnification steps have been realized using finite and infinite setups.

Modern research microscopes feature infinite optics as shown in Figures 2.1 and 2.2. They are designated as infinity-corrected microscopes. The object is located exactly in the front focal plane of the objective lens, which therefore creates its image at an infinite distance. This requires an additional lens to form a real primary image, the tube lens, which may also be used to compensate

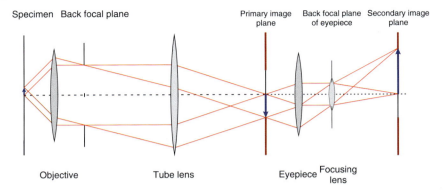

Figure 2.2 Microscope with two infinity-corrected magnification steps. The magnification factor of the sequential arrangement of two magnification steps is the product of the two single factors. Multiple creation of infinity spaces should be appreciated.

for residual image aberrations. The great advantage of this setup is that the light path behind the objective contains only sets of parallel rays that originate from the various object locations. There is ample space to insert additional flat optical elements without distorting the beam path, since parallel beams are only displaced by plane-parallel optical elements. Optical components such as the fluorescence filter cube, polarization beam splitter, analyzer, or additional filters are required for various contrast techniques. They are mostly inserted between the objective lens and the tube lens. Another advantage of infinity-corrected objective lenses is that one can focus by moving the objective rather than the specimen stage.

2.2.3 Magnification

In the finite optics setup, the objective lens produces directly a real and inverted primary image. For common transmitted-light microscopes of small to medium magnifications, there is a commercial advantage in employing finite optics. Its major handicap is that the insertion of optical components, even plane-parallel optical devices, always distorts the converging imaging beam path. To prevent this, an additional lens system must be inserted in the space between the objective lens and the primary image plane.

In many modern microscopic systems, a light detection device such as a charge-coupled device (CCD) camera is positioned already at the primary image plane, which means that the image is registered after a single magnification step. For visual observation, however, the primary image is magnified once more by the eyepiece system. Then, the primary image is located (almost) in the front focal plane of the eyepiece, which maps its image to infinity, and the human eye focuses the final image onto the retina.

Thereby, the eye becomes an intrinsic part of the imaging system (Figure 2.2). As the primary image is located in the front focal plane of the eyepiece, the emerging rays leave the eyepiece parallel to each other, such that the eye is looking at an object in infinity, which is physiologically optimal for relaxed viewing. The eyepiece is used as a magnifying glass by which the eye observes the primary image. As shown in Figure 2.2, the eyepiece and eye together form a second infinity setup, which projects the final image onto the retina. In this figure and further drawings of this chapter, the eyepiece is sketched as a single lens. This is a simplification, as the eyepiece – similar to the objective – is a system comprising several internal lenses that are needed to correct imaging artifacts. Further details on the optical construction of various types of eyepieces can be found in the references to this chapter.

What is the magnification of this second stage alone? To answer this, we need the magnification factor of a magnifying glass. The effect of a magnifying glass is usually quantified by relating the image size of an object with size O when seen with (B') and without magnifying glass (B), $M_{mg} = B'/B$. To calculate that ratio, we can use the magnification of the magnifying glass–eye lens system when looking at it as infinity system, $M_{is} = f_{eye}/f_{mg}$. Here, f_{eye} and f_{mg} denote the focal lengths of the eye lens and the magnifying glass, respectively. The image size B' is then $B' = O \times f_{eye}/f_{mg}$. When we place the object in the normal viewing distance of 250 mm in front of the eye, and observe it without the magnifying glass, the image on the retina has a size of $B = O \times d_{eye}/250$ mm, where d_{eye} is the distance between the eye lens and the retina, that is, the eye diameter. This value is close to the focal length of the eye, $f_{eye} \approx d_{eye}$. Relating the two image sizes yields therefore a magnification of $M_{mg} = B'/B \approx 250$ mm$/f_{mg}$.

Consequently, we can write for the magnification factor of a microscope eyepiece, M_{ep}

$$M_{ep} = \frac{250 \text{ mm}}{f_{ep}} \quad (2.4)$$

here f_{ep} designates the focal length of the eyepiece. Finally, the product of the magnification factors of the two magnification stages yields the total magnification of the compound microscope, M_{total}:

$$M_{total} = M_{obj} \times M_{ep} \quad (2.5)$$

The total magnification can theoretically be infinite; however, the meaningful magnification that should be chosen can be well defined. The determination of the optimal magnification factor is discussed in Section 2.3.6.

2.2.4 Angular and Numerical Aperture

An important characteristic of an objective lens is the maximum angle at which the light emitted from the object can be collected by the objective lens. Only the light contained in the respective cone contributes to the final image. The total angular opening – also designated as angular aperture – of the objective lens is the maximum angle at which the beams emitted from the focus F finally contribute to the image (Figure 2.3). The angle α in the figure designates half of the angular aperture. The sine of this angle α plays a central role in the theory of microscopy, because the *numerical aperture* of an objective lens, the NA, is defined as

$$\text{NA} = n \sin \alpha \quad (2.6)$$

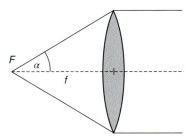

Figure 2.3 Angular aperture of an objective.

where n is the refractive index of the medium between the lens and the object. Biological samples are often observed in an aqueous buffer solution whose refractive index is $n_w = 1.33$. Often, the so-called oil immersion is used, for which $n_{oil} = 1.518$. Generally objectives that are designed and optimized for the respective media must be used. The crucial importance of the NA will be made clear in the course of this chapter.

2.2.5 Field of View

The observable object field is limited in the primary image plane by a field stop, which is often located inside the eyepiece. The diameter of this aperture in millimeters is designated as *field number* or *field-of-view number*, s. With this number and the magnification of the objective lens, M_{obj}, the diameter of the object field, D_{obj}, that can maximally be imaged is given by

$$D_{obj} = \frac{s}{M_{obj}} \tag{2.7}$$

Clearly, the quality of the image depends on the field performance of the objective lens, which refers to the image quality in the peripheral regions of the image field. In older microscopes, the usable field number was up to 18, whereas nowadays for plan achromats 20 is standard, and modern plan apochromats feature a corrected maximum field number of up to 28. Objectives with a poor field performance will produce a blur and chromatic aberrations in the peripheral regions of the image.

2.2.6 Illumination Beam Path

The illumination system of a light microscope does more than simply illuminate nonluminescent specimen. Rather, it defines the contrast mode, the resolution of the instrument, and the overall brightness in an equally important way as the imaging optics. For the illumination system, two principally different optical setups are in use. The simpler one is the *source focus* or *critical illumination* and the other, which is by far more prevalent, is called *Köhler illumination*.

2.2.6.1 Critical and Köhler Illumination

The illumination beam path for critical illumination is shown in Figure 2.4a. The light source is imaged directly by a lens, the so-called condenser lens, into the object plane. A variable aperture is located in the front focal plane of the condenser lens, whose variation allows the control of the aperture of the lens

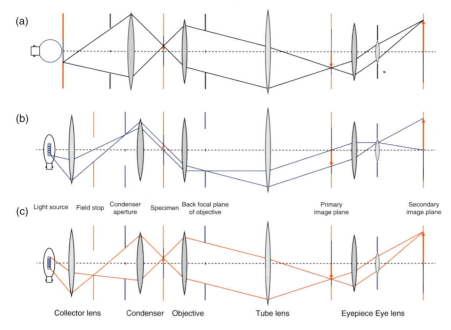

Figure 2.4 Illumination and imaging beam path. (a) Critical illumination. (b) Köhler illumination and (c) the corresponding imaging beam path.

and thereby the amount of light that illuminates the specimen. At the same time, however, the size of the illuminated field is affected. The illumination field size is fully controlled by a variable aperture located immediately in front of the light source. Images of the light source are located in the object plane, the primary image plane, and the retina. Therefore, the light source must be as large and featureless as possible; otherwise, the object brightness will be modulated by the uneven illumination. Usually, a light source with an opal glass bulb is used, or a pane of opal glass illuminated from the back. Critical illumination is used in routine instruments that are not equipped with special contrast techniques. It is simple, easy to set up, and requires few optical components.

Köhler illumination is the standard illumination mode for scientific microscopes and is named after its developer August Köhler who introduced it in 1893 [1]. The illumination components are

- the light source
- two diaphragms, namely the field stop and the condenser aperture stop
- two lenses, namely the collector and the condenser.

For transmitted-light illumination, usually a white light source is used, for example, a halogen lamp or a light-emitting diode (LED). The Köhler illumination beam path is shown in Figure 2.4b. The filament of the light source is imaged by the collector lens on the front focal plane of the condenser lens where the condenser aperture is located. This aperture should completely be filled with the image of the filament. Since the image of the filament is in the front focal plane of the condenser, all illumination light rays leaving the condenser toward the object plane are parallel bundles and traverse the specimen in a parallel manner. The undiffracted, parallel beams are focused by the objective onto its back focal plane and form there an image of the light source. Planes that contain images of each other are designated as *conjugate planes.* Thus, the *back focal plane of the objective lens* and the *front focal plane of the condenser lens* are conjugate to each other. This is of utmost importance because it is used in the key contrast techniques as, for example, phase contrast and DIC. Behind the back focal plane of the objective lens, the illumination light diverges, but it is again collected by the tube lens. The distance between tube lens and back focal plane of the objective matches the focal length of the tube lens, at least approximately. Thereby, a second infinity space for the illumination beam path is generated that contains the primary image plane. The illumination light traverses this space in parallel rays. Finally, the eyepiece focuses the incoming parallel light beams onto its back focal plane, thus generating another image of the lamp filament, which is the so-called *exit pupil* or *Ramsden disk* of the microscope. Only light passing this pupil can contribute to image formation. Here, the pupil of the observing eye must be placed, as it represents the ultimate limiting pupil. Hence, the illumination rays cannot be focused onto the retina. Consequently, the illuminated object field appears homogeneously illuminated to the observer.

In summary, optically conjugated planes of the *illumination beam path* are located in the plane of the lamp filament, the condenser aperture, the back focal plane of the objective lens, and finally the exit pupil of the eyepiece. The condenser produces an image of the object back at a position close to the collector lens of the lamp. In this plane, a variable field stop is located that defines the extension of the illuminated object field. Therefore, we have a second set of conjugate planes comprising the field stop, the object plane, the primary image plane, and the retina, called the *secondary image plane*. This set of conjugate planes is called the *imaging optical path*. The two sets of conjugate planes are complementary to each other (Figure 2.4b,c).

Köhler illumination has a number of important features:

- The specimen is evenly illuminated, although a structured light source, for example, a lamp filament, is used. This is due to the fact that all illuminating beams traverse the specimen in a parallel manner.
- The size of the illuminated field is determined by the field stop alone, because it is located in a conjugate plane of the specimen.
- Changing the size of the condenser diaphragm changes the opening angle and the NA of the light incident on the specimen. Thereby, the specimen contrast and the optical resolution (see later) can be adjusted without affecting the size of the illumination field.
- Illumination field size and condenser aperture are completely independent of each other.
- The front focal plane of the condenser and the back focal plane of the objective are conjugate to each other. This principle is used in important light microscopic contrast techniques.

2.2.6.2 Bright-Field and Epi-Illumination

In the illumination beam paths discussed earlier, the specimen is placed between the light source and the objective lens. This configuration is called *dia-illumination, transmitted-light illumination*, or simply *bright-field illumination*. The illuminating light that has traversed the specimen and the light that is diffracted by the specimen are collected by the objective lens. Various light microscopic techniques use this type of illumination.

In many cases, however, it is advantageous to illuminate the specimen from the side of observation, for instance, when looking at the reflection of opaque or fluorescent samples. In that case, some optical components of illumination and imaging are identical, for example, the objective lens. The specimen is illuminated through the objective lens by means of a semitransparent mirror, as shown in Figure 2.5.

The semitransparent mirror directs the light through the objective lens, which is simultaneously taking over the role of the condenser. Such setups are

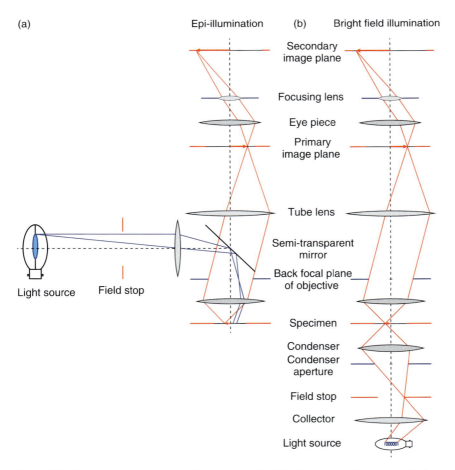

Figure 2.5 Illumination modes in a microscope. (a) Epi-illumination. (b) Transmitted-light mode.

designated as *incident-light* or *epi-illumination* configurations. Köhler illumination can also be realized in epi-illumination. To this end, it is only necessary to project an image of the lamp filament onto the back focal plane of the objective lens as shown in the figure.

In epi-illumination, the illumination field size is adjusted by a field stop that is located between the light source and the semitransparent mirror. The front focal plane of the condenser lens corresponds in epi-illumination to the back focal plane of the objective lens. However, the aperture diameter is not adjusted in this plane – as it is usually inaccessible – but in an additional conjugate plane, which is not shown in Figure 2.5 for the sake of simplicity.

The traditional microscope configuration is upright. The object is placed on an object glass that is illuminated from below and viewed from above. Upright microscopes have a direct optical path with great optical efficiency. A disadvantage is, however, that it is difficult to manipulate the object during observation.

On *inverted* microscopes, specimens are positioned above the objective lens and therefore observed from below. This is a convenient configuration when samples are examined in open chambers containing liquid media, for example, cell culture flasks or dishes. Specimen manipulation during microscopic examination such as microinjection can easily be performed. In addition to the advantage of better specimen accessibility, inverted microscopes are also generally more stable. Microscopes that feature a camera port at the bottom of the instrument have the highest possible optical efficiency for this base port.

2.3 Wave Optics and Resolution

The path of light rays in the microscope is determined by the law of refraction and the position and curvature of lens surfaces. Using simple geometrical optics, we are able to deduce many fundamental imaging characteristics of optical systems, for example, magnification factors and lens aberrations.

The origin of the limited microscopic resolution and image contrast, however, can be understood only by considering the wave nature of light. Before going into details, it is important to carefully discriminate between magnification and resolution. It is possible to increase the magnification by adding more and more magnification stages to a microscope. However, it is not possible to improve the resolution of an imaging system beyond a fundamental limit.

Resolution refers to the level of detail that can be recognized in an image, such as small and intricate structures or the distance between closely placed small objects. Indeed, such a distance is used to define and quantify the optical resolution. Using light microscopy, minute objects can be discriminated from each other when they are positioned at a minimum distance of $\sim 0.25\,\mu m$ from each other and green light and a high-quality oil-immersion objective lens are used. This obviously means that proteins and supramolecular complexes occurring in living cells cannot be recognized in detail. In later chapters, we discuss how the principal optical resolution limit can be overcome or circumvented by advanced optical techniques.

2.3.1 Wave Optical Description of the Imaging Process

There is a rigorous approach to the optical resolution limit that follows the experiments and argumentation of Ernst Abbe [2]. We indicate here the principal line of argument and examine what happens when light illuminates an object and it is imaged by a lens system [3]. We begin with a simple object, and then generalize the basic ideas.

Let us assume that the object is a simple rectangular grating formed by multiple lines within a distance d from each other, which is designated as grating constant and of the order of micrometers – somewhere near the wavelength of visible light. We illuminate the grating in transmission with coherent light of wavelength λ. For simplicity, we consider only illumination parallel to the optical axis. The grating diffracts the incoming light and forms a diffraction pattern due to the Huygens wavelets that emanate from the various grating lines. Such grating diffraction patterns are discussed in any textbook on optics [4]. It can be shown that the Huygens wavelets originating from the grating lines interfere constructively only in very distinct directions α_n.

Only in directions at which the angle α_n fulfills the so-called Bragg condition

$$d \sin \alpha_n = n\lambda \tag{2.8}$$

with n being an integer, interference is constructive. In all other directions, the waves cancel each other by destructive interference. Hence, behind the grating, we will observe diffracted wave fronts – or light beams – in specific directions α_n. If we place a screen at a great distance behind the grating, we will observe distinct spots of diffracted light corresponding to the various diffraction orders n. However, when we place a lens in the space behind the object, it will focus the diffraction pattern into its back focal plane. All diffraction orders are plane-parallel wave fronts that will be focused onto points at positions p_n in the focal plane of the lens. Positive and negative p_n denote positions above and below the optical axis, respectively (Figure 2.6). In order to create the conditions of infinity optics introduced earlier, we place the lens – thought to represent the objective lens – exactly a focal length f behind the object. If the lens satisfies the *von Bieren condition* (see Box 2.5 for details on the *sine* and *von Bieren conditions*), we get

$$\sin \alpha_n = \frac{p_n}{f} \tag{2.9}$$

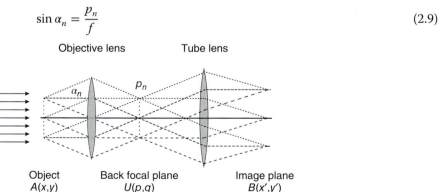

Figure 2.6 Imaging a grating structure. (Source: Modified from [3], with kind permission by Carl Zeiss Microscopy GmbH.)

Inserting the Bragg condition (Eq. (2.8)) in Eq. (2.9), we can calculate the positions p_n:

$$p_n = \frac{n\lambda f}{d} \tag{2.10}$$

Starting at the focus at $p=0$, we find the diffraction orders above and below the optical axis corresponding to increasing orders n of the light diffracted by the grating. Interestingly, the diffraction image of finer gratings with smaller constants d displays points at positions p_n that have greater distances from the optical axis, because d is in the denominator in Eq. (2.10). In addition, p_n are linearly dependent on the wavelength λ. Using white light for illumination will thus produce points that have red edges away from the axis and blue edges toward the axis. It is a very general conclusion that a lens produces an image of the diffraction pattern of the object in its back focal plane, and it is not required that the object is in the front focal plane, as we chose to place it. For microscopy, the extremely important role of the back focal plane was noted by Ernst Abbe, and it was extensively discussed in his famous, classic article that was published in 1873 [2].

Starting from here, we can build a microscope by simply using a second lens – the tube lens – as discussed earlier. The diffraction pattern presents again a repetitive structure. We will find a second diffraction pattern in the back focal plane of the tube lens, which is the diffraction pattern of the first one. This is the plane that we introduced earlier as the primary image plane. The second diffraction pattern will be magnified according to the ratio of the focal distances of objective and tube lens as given by Eq. (2.3). Actually, Abbe quite logically designated the structure seen in the back focal plane of the objective lens as the "primary" image, and the final, projected image as the "secondary" image. However, we will stick to our earlier definition.

What happens if we image finer and finer gratings? Equations (2.8) and (2.10) reveal that α_n and the distances p_n from the optical axis, where we find the foci of the diffraction orders, increase more and more for decreasing d. However, the lens represents an aperture of finite size that limits the angle to a value α_{max}, for which the diffraction pattern can be collected. Soon, some diffraction orders will be missed even by large lenses because, for practical reasons, we cannot construct a lens with an aperture angle approaching 90° ($\pi/2$). This leads to an incomplete first diffraction pattern that is used to form the final image, which will not be any longer a true image of the initial structure. And when not even the first diffraction order is transferred by the lens, that is

$$d < \frac{\lambda}{\sin \alpha_{max}} \tag{2.11}$$

all information of the grating structure is lost. This is the underlying physical reason for the *resolution limit of microscopy*. Finally, when we consider that there may be an immersion medium with a refractive index n in the object space that reduces the vacuum wavelength of the light λ_0, we finally get the resolution limit for periodic objects such as gratings:

$$d < \frac{\lambda_0}{n \sin \alpha_{max}} \tag{2.12}$$

This expression is valid if the illumination direction is parallel to the optical axis. In case we illuminate the object under an oblique angle, the interference conditions are slightly different, and we obtain a further factor of 2 in the denominator of Eq. (2.12) for the maximum possible oblique illumination [2]. Thus, a high degree of oblique illumination improves the resolution of the microscope. A similar expression is given later for the resolution limit of a microscope when imaging a point object and not a repetitive grating structure. The essential point of the presented considerations is that even an ideal, aberration-free objective lens that does not modify phases or amplitudes of the first diffraction pattern in its back focal plane will nevertheless have severe consequences on the imaging process. Obviously, the outer points of the diffraction pattern represent the finest object details, because a reduction of the grating constant d will increase α_n and cause the points p_n to move away from the optical axis. Hence, the image that can be produced by the microscope will often miss the finest details present in the original structure (Box 2.1).

Box 2.1 General Theory of Image Formation

In a more general approach, we image a two-dimensional object of size $2a$ in the x-direction and $2b$ in the y-direction. It may be described by an amplitude transmission function $A(x, y)$, for which then $A(x, y) > 0$ only for $|x| \leq a$ and $|y| \leq b$. In general, the diffraction pattern U is given as a function of the angle α_x and α_y away from the optical axis by the Huygens–Fresnel integral equation:

$$U(\alpha_x, \alpha_y) = c \int\int_{-\infty}^{+\infty} A(x, y)\, e^{i\frac{2\pi}{\lambda}(x \sin \alpha_x + y \sin \alpha_y)}\, dx\, dy \tag{2.13}$$

This expression is nothing but the mathematical representation of the Huygens wavelets that originate at the object structure, and describes the diffraction pattern far away from the diffracting object. Next, we place the object at the focal plane of the objective lens with a focal length f. If it satisfies the von Bieren condition (see Box 2.5 for details), we can write

$$\sin \alpha_x = \frac{-p}{f}$$
$$\sin \alpha_y = \frac{-q}{f} \tag{2.14}$$

The negative sign was chosen for convenience; we are free to define the axis. We can substitute that into Eq. (2.13) for the diffraction pattern, which we can now write as a function of p and q:

$$U(p, q) = c \int\int_{-\infty}^{+\infty} A(x, y)\, e^{-i\frac{2\pi}{\lambda f}(px + qy)}\, dx\, dy \tag{2.15}$$

This is the field of the diffraction pattern in the back focal plane of the objective lens. Indeed, this expression mathematically represents a Fourier integral with the

(Continued)

Box 2.1 (Continued)

Fourier frequencies ω_x and ω_y:

$$\omega_x = 2\pi R_x = \frac{2\pi p}{\lambda f} = -\frac{2\pi}{\lambda} \sin \alpha_x$$

$$\omega_y = 2\pi R_y = \frac{2\pi q}{\lambda f} = -\frac{2\pi}{\lambda} \sin \alpha_y \quad (2.16)$$

R_x and R_y are the spatial frequencies. If we identify the constant c with $1/2\pi$, we see that U represents the Fourier transform of $A(x)$ [5]:

$$U(\omega_x, \omega_y) = \frac{1}{2\pi} \int\!\!\!\int_{-\infty}^{+\infty} A(x,y)\, e^{-i(\omega_x x + \omega_y y)} dx\, Ey = \tilde{F}[A] \quad (2.17)$$

We note that the Fourier frequencies are linearly related to the positions p and q in the back focal plane. This shows that the diffraction image in the back focal plane represents a principally undistorted Fourier spectrum of the object function. The inverse Fourier transformation leads back to the object function:

$$\tilde{F}^{-1}\{U(\omega_x, \omega_y)\} = \frac{1}{2\pi} \int\!\!\!\int_{-\infty}^{+\infty} U(\omega_x, \omega_y)\, e^{i(\omega_x x + \omega_y y)} d\omega_x\, d\omega_y = A(x,y) \quad (2.18)$$

We could imagine this back projection to occur by simply placing a mirror at the objective's back focal plane. Or, as indicated Figure 2.6, we could simply add a second lens – the tube lens – to regenerate an image out of the diffraction pattern.

The crucial point is that even an ideal objective lens, which does not modify any phases or amplitudes of the Fourier frequency components in the back focal plane, still *limits* the Fourier frequencies due to its NA. An alteration of the Fourier spectrum, for example, by modifying the Fourier frequencies by multiplication with a function $G(\omega_x, \omega_y)$, to $U'(\omega_x, \omega_y)$ would result in a modified object function $A'(x, y)$

$$\tilde{F}^{-1}\{U'(\omega_x, \omega_y)\} = \tilde{F}^{-1}\{U(\omega_x, \omega_y) G(\omega_x, \omega_y)\} = A'(x,y) \quad (2.19)$$

As discussed earlier, there is always a limiting collecting angle of the objective, that is, α_{max}, beyond which the diffracted light cannot be collected by the lens. The lens represents a circular aperture of finite size that naturally limits the angle for which the diffraction pattern can be collected to α_{max}. This leads to a cut-off frequency, an upper limit of Fourier frequencies that can be transmitted by the lens, ω_{max}. It is given by

$$\omega_{max} = 2\pi R_{max} = -\frac{2\pi}{\lambda} \sin \alpha_{max} \quad (2.20)$$

This basically means that even an ideal lens represents a low-pass filter for the frequency information of an object. The image that can be produced by a lens will always miss the highest frequencies present in the original structure. The spatial frequency is directly related to the Fourier frequency by multiplication with 2π. The limiting spatial frequency can be inverted to yield the smallest distance that can be resolved.

> This argument allows some very important insights into the imaging process. First, the back focal plane contains the diffraction image of our object structure. The spatial coordinates in that plane can be directly related to the spatial frequencies or Fourier frequencies of our object. Second, even the ideal lens will distort the spatial frequencies contained in our object function, namely, by restricting it to certain values. The exact limiting value depends on the collection efficiency of the lens, which is quantified by the NA.

2.3.2 The Airy Pattern

The earlier discussion of imaging grating structures illustrates the physical reason for the resolution limit of microscopy. Let us now place a point object instead of a grating in the focus in front of the objective lens, as depicted in Figure 2.7. Also, we will assume that the angles of the light beams with regard to the optical axis are small, and we will neglect the vectorial properties of light. The point object diffracts the incoming light into a spherical wave w_o. A segment of the spherical wave is transformed by the lens, which is assumed to be ideal and aberration free, into a parallel wave front w_i. As discussed earlier, this wave front of constant amplitude corresponds to the diffraction image of the point object. However, it is modified by the limited angular aperture of the objective lens. According to the Huygens principle, we can imagine each point of the parallel wave front w_i in the back focal plane of the lens to be the origin of a Huygens wavelet. All wavelets have identical phases and are therefore completely coherent. If we place a tube lens behind the objective lens to project the diffraction pattern, the wavelets interfere with each other constructively exactly at the focus in the image plane because only for this point the optical path length is identical for all wavelets. This point also corresponds to the position of the image of the point object, as it can be determined by geometrical optics or by ray tracing. However, because this image spot is the result of an interference process of a finite range of wavelets, it has a finite diameter. Around that spot in the image plane at a certain distance from the optical axis is a circular zone where the wavelets interfere destructively owing to their *optical path length differences* (OPDs). This is indicated as the dark

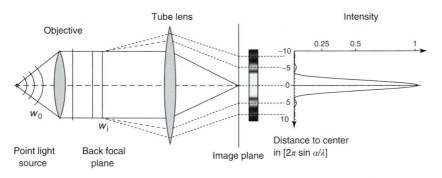

Figure 2.7 Interference of wavelets originating from the exit pupil of the objective lens.

stripe in Figure 2.7. At an even greater distance from the focus, we find a second circle where a majority of wavelets interferes constructively. Here appears a bright ring around the dark zone of destructive interference (indicated by the dotted line in Figure 2.7). This ring is followed again by a ring of destructive interference, then again of constructive interference, and so forth. Altogether, the interference process of the wavelets results in the intensity distribution sketched on the right-hand side of Figure 2.7. The sketch shows a central cut through the two-dimensional, rotationally symmetric intensity distribution. This dark–bright intensity profile is also designated as an Airy pattern, and the central bright spot is the Airy disk. As discussed earlier for the grating, the final image is formed by the interference of the Huygens wavelets originating in the diffraction structure in the back focal plane of the objective lens.

Hence, it is a consequence of the interference that the image of a point object is not a point but a smeared spot with a characteristic intensity profile. This profile is called the *point spread function* (PSF) of the imaging lens. Its analytical derivation is not difficult but lengthy and cannot be discussed here in detail. Thus, we only indicate the derivation.

Let us consider again the circular plane wave segment in the back focal plane of the objective lens, which is the lens's aperture-limited diffraction image of the point object. It corresponds to the field behind a circular aperture of radius a that is illuminated by a plane wave. This situation is discussed in optics textbooks under the topic "diffraction at a circular aperture" [4]. For the imaging process, only the field at a great distance from the aperture is of relevance. Then we can neglect the curvature of the outgoing waves and consider only the planar waves (Figure 2.8a).

In order to compute the far field, we have to integrate over all wavelet contributions originating at the aperture and calculate the total field. The situation is radially symmetric and the field strength is only a function of the angle θ to the optical axis. For $R \gg a$, the integration of the wavelets over the aperture surface results in the following expression for the field $E(\theta)$ [4]:

$$E(\theta) \propto E_0 \frac{J_1(2\pi a \sin\theta / \lambda)}{2\pi a \sin\theta / \lambda} \tag{2.21}$$

where $J_1(x)$ represents the Bessel function of the first kind of order 1. A tube lens would project this Fraunhofer diffraction pattern from the far field into its focal plane (Figure 2.8b). Parallel light falling on a lens with an angle θ is focused at the focal plane at a distance r from the optical axis, as shown in Box 2.2. There, it is also shown that for small values of angle θ

$$a \sin\theta \approx r \sin\beta \tag{2.22}$$

where β is half of the opening angle of the tube lens. The approximation in Eq. (2.22) is certainly justified for the tube lens when we consider that its focal length is, in practice, between 160 and 250 mm, the lens aperture up to 20 mm, and the maximum image size 28 mm.

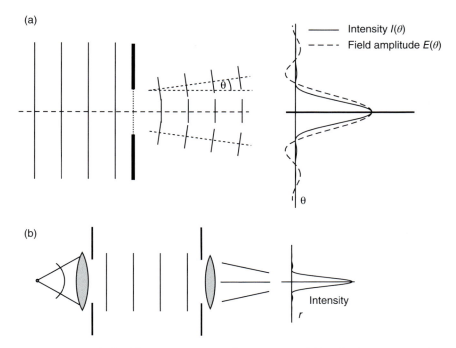

Figure 2.8 Fraunhofer diffraction (a) at a circular aperture and (b) with a lens positioned before and behind the aperture.

Box 2.2 Proof of Eq. (2.22)

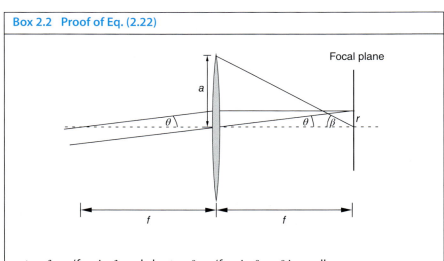

$\tan \beta = a/f \approx \sin \beta$, and also $\tan \theta = r/f \approx \sin \theta$, as θ is small.
Solving both expressions for f and equating the results yields $r/\sin \theta \approx a/\sin \beta$, and after rearranging

$$r \sin \beta \approx a \sin \theta$$

In this context, see also Box 2.5.

When we use this expression in Eq. (2.21), the amplitude of the field in the focal plane of the tube lens is obtained as a function of the distance to the optical axis, r:

$$E(r) \propto E_0 \frac{J_1(2\pi r \sin\beta/\lambda)}{2\pi r \sin\beta/\lambda} \quad (2.23)$$

The light intensity $I(r)$ in the focal plane of the lens as a function of the distance to the optical axis is found by calculating $I(r) = [E(r)]^2$. This results in the following expression:

$$I(r) = I_0 \left[\frac{J_1(2\pi r \sin\beta/\lambda)}{2\pi r \sin\beta/\lambda}\right]^2 \quad (2.24)$$

where I_0 is the light intensity at the focus of the lens. The two functions $E(r)$ and $I(r)$ are plotted in Figure 2.9.

2.3.3 Point Spread Function and Optical Transfer Function

The earlier discussion leads directly to central concepts of imaging. Box 2.3 explains that an optical system such as a microscope is nothing but a low-pass

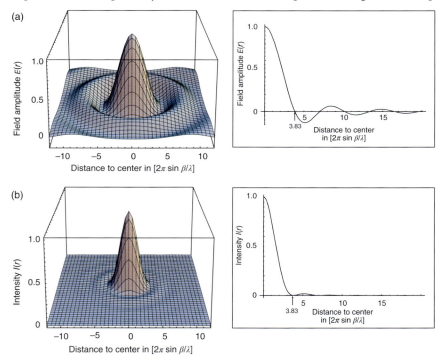

Figure 2.9 Intensity and field distribution at the focal plane of the tube lens when imaging a self-luminous point object. The surface plots illustrate the distribution of the (a) field amplitude $E(r)$ and (b) the light intensity $I(r)$ in the image plane as a function of the distance to the optical axis in optical units when a single point object is imaged. The latter function, $I(r)$, is called the *point spread function* (PSF) of the lens system. Notably, the intensity distribution in (b) is identical to the light distribution in the focal plane of a lens when illuminated with a plane wave.

filter for the spatial frequencies of the object. Fine object details are smeared because point objects are imaged as Airy patterns. Microscopic imaging means to smooth reality.

Box 2.3 PSF and OTF

Let us recapitulate the imaging of a point object. Illumination produces a Fraunhofer diffraction pattern. Mathematically, this corresponds to the Fourier transform of the point object. In Fourier theory, it is trivial to show that the Fourier transform of a point – or a δ-function – is a constant, because it contains all frequencies. The objective lens collects only a limited fraction of the diffraction pattern and projects it into its back focal plane. Therefore, the spatially extended, complete Fourier transform in the back focal plane is restricted to the set of frequencies below the limiting frequency. It contains only those frequencies that can be transmitted by the lens. The inverse Fourier transformation of this is done by the tube lens and corresponds to the image of a point. This image is the so-called PSF. Its Fourier transform in the back focal plane of the objective lens is the instrument function of the microscope, which is also called the *optical transfer function* (OTF). As indicated in Box 2.1, the imaging process of an arbitrary object $A(x, y)$ can be formulated as follows: The objective lens projects the object onto its back focal plane which corresponds to the "frequency space." Here, the complete transform of the object function, $\tilde{F}[A]$, is limited to the transmittable frequencies. Mathematically, this can be formulated as a multiplication with the OTF, which contains the information about the transmittable frequencies.

The multiplication yields a truncated and modified version of $\tilde{F}[A]$, which is designated as $\tilde{F}'[A]$:

$$\tilde{F}'[A] = \tilde{F}[A] \cdot \text{OTF} \qquad (2.25)$$

According to Fourier theory, a multiplication in frequency space corresponds to a convolution in object space. Thus, convolution of object function $A(x, y)$ and PSF yields the image $A'(x, y)$:

$$A'[x, y] = A[x, y] \otimes \text{PSF}(x, y) \qquad (2.26)$$

The multiplication of $\tilde{F}[A]$ and OTF leads to a deletion of the high frequencies in the Fourier spectrum of the object. Thus, the imaging process corresponds to a low-pass filtering of the object. An objective lens is a low-pass filter that blurs fine structures of the object. This is further illustrated in Appendix A.

2.3.4 Lateral and Axial Resolution

2.3.4.1 Lateral Resolution Using Incoherent Light Sources

Diffraction at the aperture of a lens system is the reason for the resolution limit of optical imaging. The image of a luminous point object corresponds to the intensity distribution $I(r)$ (Eq. (2.24) and Figure 2.9b), which is known as the *Airy pattern*. In that pattern, the circular area with a radius defined by the distance between the central maximum and the first zero is designated as the Airy disk.

The zero of the Bessel function $J_1(x)$ is at $x = 3.83$, therefore the Airy disk's radius in the image plane, $r_{im,0}$, is given by

$$r_{im,0} \approx \frac{0.61\lambda}{\sin\beta} \tag{2.27}$$

In the image plane, 84% of the total light of the complete distribution is found within the distance $0 < r < r_{im,0}$. To find out to what distance $r_{obj,0}$ corresponds in the object plane, Eq. (2.27) must be divided by the magnification factor M. The value of $r_{im,0}$ corresponds in object space to $r_{obj,0} = r_{im,0}/M$. When taking the sine condition of the objective/tube lens system into account (see box 2.5)

$$M \sin\beta = \sin\alpha \tag{2.28}$$

and considering that there often is an immersion medium with a refractive index n in object space, which requires the use of λ/n as wavelength, we obtain for $r_{obj,0}$

$$r_{obj,0} \approx \frac{0.61\lambda}{n \sin\alpha} \tag{2.29}$$

where α denotes the opening angle of the objective lens divided by 2. Note that $n \sin\alpha$ is the NA of the objective lens.

Now, we look at the image that is produced by two neighboring point objects at a distance d. First, we consider incoherent light sources, for example, two single fluorescent molecules. Their light cannot interfere, as the emitters are independent of each other. The two point objects are not imaged as dots but rather as "smeared" intensity distributions, as shown in Figure 2.10. The total resulting light distribution is obtained by adding up the two intensity distributions given

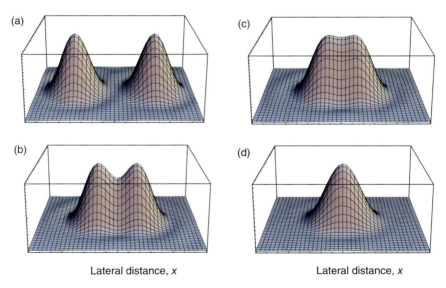

Figure 2.10 Surface plot of the intensity distribution in the image plane of two incoherent point light sources. Images of two point objects separated from each other by (a) a distance much greater than the Rayleigh distance d_R (Eq. 2.30), (b) a distance equal to d_R, (c) a distance corresponding to the full-width at half-maximum (FWHM) of the PSF, which is designated as the "Sparrow criterion," and (d) a distance of $d_R/10$, at which the images cannot be discriminated from each other.

by Eq. (2.24). Two well-separated maxima can still be observed when $d > r_{obj,0}$ (Figure 2.10a). If the point objects are situated close to each other with a separation d, and $d = r_{obj,0}$ in object space, their Airy disks clearly overlap in the image plane (Figure 2.10b). In this case, the signal intensity in the center between the two maxima is about 75% of the maximum intensity. We say that two objects are resolvable when they can still be discriminated from each other in the image. The exact distance at which the resolution is lost depends on the ability of the detector or observer to discriminate intensity levels and on the signal-to-noise ratio (Figure 2.10c). Therefore, the exact definition of a resolution distance is somewhat arbitrary. Lord Rayleigh suggested the distance $d = r_{obj,0}$ as a criterion for optical resolution, because in that case the human eye can still perceive two separate image points. This condition is called the *Rayleigh criterion*. If the distance between the objects is much smaller than that value, they can no longer be perceived as separate objects (Figure 2.10d). The optical resolution d_R according to the Rayleigh criterion is therefore

$$d_R = \frac{0.61\lambda}{n \sin \alpha} \tag{2.30}$$

The smaller the distance d_R, the higher the resolving power of an optical system. This formula explains why using light of short wavelengths – blue or even ultraviolet (UV) – gives a superior resolution compared to long wavelengths – red or infrared. Utilization of high NA objectives and immersion media with higher refractive indices such as water, glycerol, or oil in combination with suitable objectives enhances the resolution further.

2.3.4.2 Lateral Resolution of Coherent Light Sources

The situation is different for two coherent point light sources. Such light sources could be formed by two small structures in the same plane of a microscopic sample that are illuminated by a plane wave. Then, the diffracted light waves from the two sources have identical phases and may interfere with each other in the image plane, and the resulting field distribution is determined by the sum of the two field distributions according to Eq. (2.23), as shown in Figure 2.9a. The observed light intensity distribution is now obtained by squaring the sum of the field contributions, and not by summing the squares of the individual field contributions as in the case of self-luminous point objects as discussed earlier. This produces a very different image than for incoherent point objects, as shown in Figure 2.11.

In Figure 2.11, the image of two coherent point light sources is shown, which are separated in the object plane by a distance d from each other. The image is quite different from the incoherent case (Figure 2.10). In particular, for $d = r_{obj,0}$, as defined in Eq. (2.30), the two image points are clearly not resolvable. Rather, a larger distance d with

$$d = \frac{1.22\lambda}{NA_{obj}} \tag{2.31}$$

is required to observe a minimum of $\sim 0.75 I_0$ between the two maxima. This value might be used as a "Rayleigh-type" resolution limit for two coherent point light sources. Obviously, the optical resolution is reduced for coherent objects by a

(a)

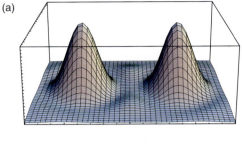

(b)

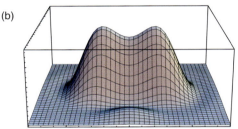

(c)

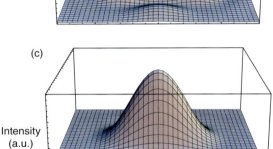

Intensity (a.u.)

Lateral distance, x

Figure 2.11 Surface plot presentation of the intensity distribution in the image plane of two coherent point light sources. Airy disks and lateral resolution: (a) large distance more than twice the Rayleigh distance ($2d_R$), (b) a well-resolved distance, $d_R = 1.3$, and (c) a distance that corresponds to the Rayleigh distance d_R (Eq. (2.30)). Coherent point objects at that distance cannot be resolved anymore.

factor of 2. This is an important fact because coherent objects are frequent in microscopic imaging. Closing the condenser aperture in bright-field illumination produces a pointlike light source. The light passing through the specimen is coherent, and so are the diffracted waves. The lateral resolution in this case would be given by Eq. (2.31).

So far, we assumed that the illuminating wave propagates along the optical axis and hits the structures in the object plane simultaneously. In that case, the diffracted waves have no phase shift with respect to each other. However, when the incoming wave propagates at a certain angle in relation to the optical axis, the situation is quite different. Then, the diffracted Huygens wavelets originating in the object plane have a specific phase shift with respect to each other because they are produced at different time points depending on their positions in the sample. This phase shift in the diffracted light is still present in the image plane where the light is focused and the image is formed by the interference of the coherent and phase-shifted waves. Now, the two field amplitudes must be summed as earlier, but their relative phase shift must be taken into account.

When a sample is illuminated by transmitted light with an extended light source, for example, by a completely opened condenser aperture, then we have numerous distinct illumination point light sources. The illumination by these sources is of varying angles when approaching the limit of the aperture. Each source on its own illuminates the sample and creates coherent, diffracted light from the sample. With respect to each other, however, these sources are incoherent, because they correspond to different loci on the illumination lamp filament, and therefore the light distributions of all these produced in the image plane must be summed together. The sum must be performed in an appropriate way over the intensity distributions produced by the various oblique illumination angles. It must be taken into account that the contribution of the more oblique illumination directions is greater owing to their larger number in the outer part of the diaphragm.

The final image depends on the detailed mode of illumination or, in other words, on the degree of coherence in the illumination light, which can be regulated by the size of the condenser aperture. In exact terms, it is the NA of the condenser illumination, NA_{cond}, that determines the degree of coherence of the illumination light. The final resolution, according to the Rayleigh criterion, can be formulated as follows:

$$d = \frac{1.22\lambda}{NA_{obj} + NA_{cond}} \tag{2.32}$$

For bright-field illumination with a condenser, the achievable resolution depends on the properties of the illuminating lens and its NA. Closing the condenser aperture – which reduces light scattering in the sample – diminishes the attainable optical resolution because the overall coherence of the illumination light increases. One has to choose between reduced stray light and higher depth of focus on one hand, and a reduced lateral optical resolution on the other.

2.3.4.3 Axial Resolution

So far, we discussed the light intensity distribution exactly within the image plane. However, image formation is a 3D process. The light waves travel toward the image plane through the space in front of the objective lens – the object space – through the objective lens, the space between the objective and tube lens, and the tube lens. Then, the waves are focused within the image space into the image.

They not only interfere with each other exactly in the image plane but also fill all the space behind the tube lens producing a complex 3D intensity distribution of light. The Airy pattern describes the lateral intensity distribution exactly in the primary image plane. Its mathematical formulation (Eq. (2.24)) can be derived analytically in the so-called paraxial approximation. This means that it is valid only in a small region close to the optical axis of the object and image space and that the considered angles with regard to the optical axis must be small. Using paraxial theory, the complete 3D light intensity distribution near the image center of a self-luminous point can be analytically calculated [6]. A graphical representation is shown in Figure 2.12.

The distribution is given as a function of the distances to the image center in normalized optical units. These correspond to the expressions occurring in the functions describing the lateral and axial light distribution (for details,

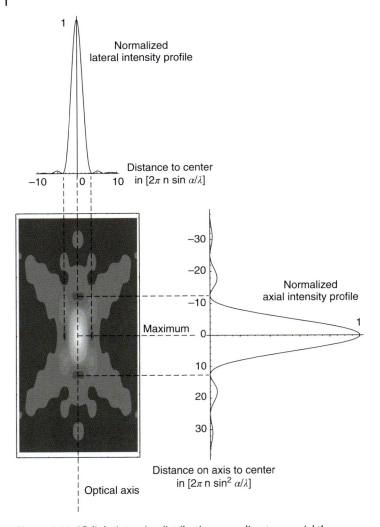

Figure 2.12 3D light intensity distribution according to paraxial theory.

see Box 2.4). The intensity distribution along the optical axis can be described analytically, and the distance between the intensity maximum and the first minimum on the optical axis, $z_{0,\text{im}}$, can be used as an axial resolution criterion in analogy to the Rayleigh criterion. In image space, for low NA_{obj} $z_{0,\text{im}}$ is given by

$$z_{0,\text{im}} = \frac{2M^2 \lambda}{NA_{obj}^2} \qquad (2.33)$$

Because $z_{\text{im}} = z_{\text{obj}} M^2/n$, this corresponds to an axial distance $z_{0,\text{obj}}$ in the object space of

$$z_{0,\text{obj}} = \frac{2\lambda n}{NA_{obj}^2} \qquad (2.34)$$

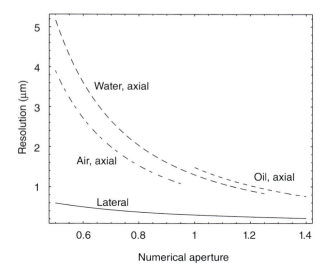

Figure 2.13 Axial and lateral resolution as a function of numerical aperture for air-, water-, and oil-immersion objectives. The curves were calculated according to Eqs (2.30) and (2.34) for indices of refraction of 1 for air, 1.33 for water, and 1.515 for oil, and for $\lambda = 488$ nm.

Practically, this is the vertical distance by which the object stage with a self-luminous point object must be moved so that the image changes from the central intensity maximum to its first minimum. While the lateral resolution depends on the reciprocal of the NA, the axial resolution is related to the reciprocal of NA^2. The dependence of the lateral and axial resolutions on the NA for different immersion media is shown in Figure 2.13. Indeed, water- and air-immersion objective lenses exhibit a higher axial resolution than oil-immersion lenses at identical NA, whereas their radial resolutions do not differ.

The given formula for the axial resolution is valid for lenses with a small NA because it was derived on the basis of paraxial optics. A more general description can be obtained by using the Fresnel–Huygens theory, which works with fewer approximations and gives a more realistic 3D distribution of light intensity in the focal region for high NA lenses.

Box 2.4 Lateral and Axial Optical Units

The definition of optical units results from the derivation of the intensity distribution at the focus of a lens as derived by paraxial theory in Born and Wolf [6]. The intensity distribution in the lateral plane was given by Eq. (2.24) and is plotted in Figure 2.9. The argument of the function is the dimensionless variable v

$$v = r\frac{2\pi}{\lambda_0} n \sin \alpha \qquad (2.35)$$

(Continued)

> **Box 2.4 (Continued)**
>
> The intensity distribution along the optical axis can be derived as
>
> $$I(u) = I_0 \frac{\sin^2 u}{u^2} \qquad (2.36)$$
>
> The function $\sin u/u$ is also designated as the *sinc function*. The argument u of this function in the object space is [6]
>
> $$u = z \frac{2\pi}{\lambda} n \sin^2 \alpha \qquad (2.37)$$
>
> It was shown, however, that the latter expression leads to erroneous results for deviations from paraxial theory, for example, in the case of a large NA. The large discrepancy can be corrected by a simple redefinition of u as follows [7]:
>
> $$u = z \frac{8\pi}{\lambda} n \sin^2 \frac{\alpha}{2} \qquad (2.38)$$
>
> The coordinates u and v are designated as "optical units" and allow the description of the intensity distribution in the PSF independent of the NA of the lens and of the refractive index n of the medium.

2.3.5 Magnification and Resolution

Earlier we have seen that the lateral image of a point object corresponds to an Airy pattern. Often the image is recorded by a digital device with a certain detector element size d_d, for example, by a CCD camera (see Chapter 3). It can be shown that the overall magnification should be chosen such that the radius of the Airy disk, d_{Airy}, in the image plane exceeds the size of a single detector element d_d at least by a factor of 2 if the resolution of the microscope is to be fully exploited. As demonstrated in Figure 2.14, all available information in the image can be recorded sufficiently well by the image detector when $d_d \approx d_{Airy}/2$.

The optimal magnification factor, therefore, depends on the resolving power of the optical system and the type of the detection system, more exactly on the size of a single detector element, for example, the edge length d_d of the photodiodes of a CCD.

Thus the optimal magnification $M_{opt,d}$ should be approximately such that

$$M_{opt,d} \approx \frac{2d_d}{d_R} \qquad (2.39)$$

where d_R is the optical resolution attainable by the lens system as defined in Eq. (2.30). Hence, using a microscope with a lateral optical resolution $d_R = 0.25\,\mu m$ and a detector comprising elements with a size of $6.25\,\mu m$, the optimal magnification would be $M_{opt,d} \geq 50$.

The situation is different for visual inspection of the microscopic image. The minimum visual opening angle for discriminating two object points by the eye should be ≥ 2 arc min or 10^{-3} rad. Placing an object with the size $d_R = 0.25\,\mu m$ at the comfortable viewing distance of 25 cm would produce only a viewing

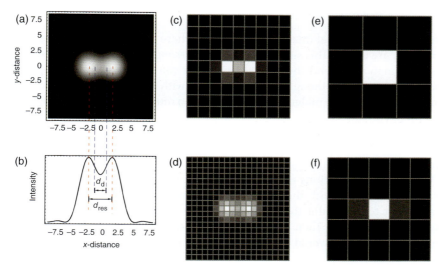

Figure 2.14 Optical resolution and detector element size. (a) Image of two point objects separated by the Rayleigh distance d_{res}. (b) Line profile through the horizontal center of the image. The points are optically well resolved. (c) Image created with a detector element size $d_d = M \times d_{res}/2$. M is the overall magnification factor. In the digitized image, the two objects are separated. (d) Detector size $d_d = M \times 0.24 \times d_{res}$. Details of the intensity distribution of the two object can be recognized. This situation is designated as "super resolution." If $d_d > M \times d_{res}/2$, the digitized image does not preserve the optical resolution as shown in (e) $d_d = 1.5 \times M \times d_{res}$ and (f) $d_d = 0.9 \times M \times d_{res}$.

angle of 10^{-6} rad. Therefore, the total magnification for visual observation M_{eye} should be

$$M_{eye} \approx \frac{10^{-3}}{10^{-6}} \text{rad} = 1000 \qquad (2.40)$$

This magnification can be generated by a combination of two magnification stages, where the first stage comprising objective and tube lens magnifies 100-fold and the second comprising the eyepiece and the eye magnifies 10-fold. Magnifications above the values defined by Eqs (2.39) and (2.40) are designated as *empty magnifications* because no new information is obtained.

2.3.6 Depth of Field and Depth of Focus

The axial extension of the PSF represents the achievable resolution along the optical axis. It is also called the *depth of field*, D_{ob}, and is twice the value that is returned by Eq. (2.34). It corresponds to the axial distance by which a thin object can be moved axially without significantly losing its sharpness. The depth of field depends on the NA of the objective lens and the refractive index of the medium. The *depth of focus*, D_{of}, is the corresponding distance in the image space. It is obtained from D_{ob} by multiplication with the square of the magnification factor M and, therefore, is usually relatively large, namely, in the range of several millimeters for a magnification of 100.

2.3.7 Over- and Undersampling

There exist some special imaging applications where magnification factors well above M_{opt} are useful. This situation is called *oversampling*. Obviously, no structural information beyond the optical resolution can be obtained from optical imaging. It is possible, though, to locate objects with sizes far below the resolution limit such as single molecules with a precision much smaller than the optical resolution. For this purpose, it is necessary to have well-resolved images of the complete PSF. Usually, 3–7 pixels are used to sample the diffraction-limited signals in each direction with sufficient detail in these super-resolution approaches (Chapters 8 and 12).

Other imaging situations exist where it is not essential to acquire data visualizing all structures with the utmost possible resolution. For example, it may be sufficient to just count the number of cells within a given field of view for a specific experiment. Then, it is not required to image at high resolution. A relatively crude image would be sufficient, which is designated as *undersampled*.

2.4 Apertures, Pupils, and Telecentricity

In the last sections, we discussed the capability of optical microscopes to produce magnified images of specimen. This is only one part of the use of microscopes. In order to generate high-quality images, we have to ensure that the light waves forming the image finally reach the image plane. This is not a trivial task in such a complex optical system such as a microscope, which comprises several magnification stages and often additional relay stages.

The amount of light that is collected by a lens is obviously limited by the physical diameter of the lens. Such apertures define the amount of light that is available for image formation. Apertures are principally independent of the focal length of lenses, which defines the imaging magnification as we saw earlier.

In simple cases, it is the edge of a lens or its mounting that defines the "entrance" aperture which *a priori* limits the amount of light that can pass to the image plane. In complex imaging systems such as an objective lens constructed of multiple lenses, it is usually the diameter of the smallest lens or an extra, sometimes adjustable aperture stop that limits the lateral extension of the beam path. The image of that aperture as seen from the object defines the maximum amount of light that contributes to the image. The image of the aperture as seen from the on-axis point of the object is called the *entrance pupil*. Similarly, the limiting aperture as seen from the on-axis point of the image is called the *exit pupil*. In case the aperture is located in front of the lens in the object space, the virtual image of the aperture represents the exit pupil (Figure 2.15a). All light passing through the entrance pupil also passes through the exit pupil, because they are conjugate to each other and to the aperture stop in cases where it is not identical with one of the pupils.

Objective lenses are constructed in a special way: their beam-limiting aperture stops are located in their back focal plane, which therefore are simultaneously their exit pupils. Accordingly, the entrance pupil is located in infinity (Figure 2.15b). This construction has a number of very significant consequences.

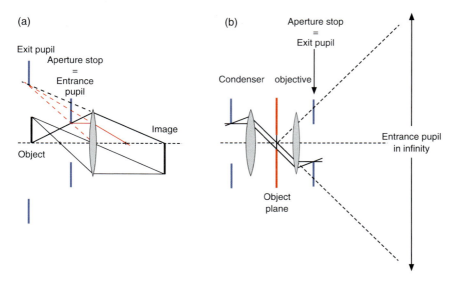

Figure 2.15 Aperture stop and pupils of the objective lens. (a) The aperture stop is in front of the lens, and hence identical to the entrance pupil. The exit pupil is the virtual image of the aperture as seen from the on-axis point of the image (note the black full dashed line from that point to the aperture image). (b) The aperture stop and exit pupil of an objective are located at its back focal plane. This causes the entrance pupil to be at infinity in the image space. In addition, the entrance pupil of the condenser lens and the exit pupil of the objective lens are conjugate to each other in Köhler illumination.

For all microscope systems, the position of the image plane is well defined and located in the front focal plane of the eyepiece. In order to understand the importance of the exit pupil's position, we first consider the case where the aperture stop is *not* in the back focal plane of lens but, for example, in front of the lens (Figure 2.16a). In that case, defocusing the specimen has two simultaneous effects: the image becomes blurred in the shape of the aperture stop, and the blurred image is reduced or magnified in size. This is demonstrated in Figure 2.16b. Clearly, such a magnification change due to defocusing would be inconvenient and very confusing for the observer.

Now we modify the optical setup and place the exit pupil at the back focal plane of the objective lens and thus the entrance pupil at infinity. Now the chief ray runs parallel to the optical axis in the object space. Such a lens configuration is called telecentric in the object space (Figure 2.16c). We note the important consequence that defocusing of the object results in a blurred image in the (fixed) image plane, but no magnification change occurs (Figure 2.16d). This effect is of great importance for the practical work with microscopes.

In transmission microscopy, usually a Köhler illumination mode is employed. For Köhler illumination, the entrance pupil of the illumination system is realized by the condenser aperture, which is located in the front focal plane of the condenser lens. We already noted that this aperture is imaged onto the back focal plane of the objective lens. This plane also contains the exit pupil of the imaging system. Obviously, it is optimal to match the sizes of these two pupils, because

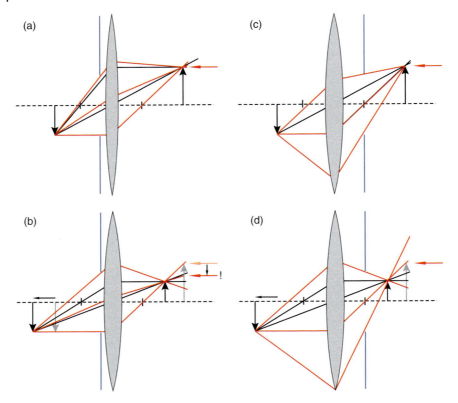

Figure 2.16 Telecentricity. (a) Imaging process of a lens with an aperture stop – the entrance pupil – in the object space. The chief ray is the one that passes from the outmost object position (tip of black arrow) through the center of the aperture. The chief ray represents the cone of light creating the image of that point, and its course in the image space marks the center of the defocused image. In the sketches, the chief ray is the red central ray. (b) When the object is defocused, the chief ray moves downward on the fixed image plane compared to its position in (a), as indicated by the arrows. (c) Imaging by a lens that is telecentric in the object space. According to the definition for such lenses, the aperture stop is located at the back focal plane. The chief ray now passes through the back focus and is parallel to the optical axis in the object space. (d) For telecentric lenses, defocusing the object does not alter the position of the chief ray, as it remains parallel to the optical axis. It hits the image plane at the same position as in (c) (see position of red arrow in (c) and (d)).

only then all incoming illumination light will pass the exit pupil and be available for image formation. Matching the size of these pupils, however, implies the existence of a defined position of the exit pupil of the objective lens and of its entrance pupil, which is in infinity. The position of these pupils should be identical for all objective lenses in order to avoid an axial repositioning of the condenser lens when switching objectives. Therefore, the location of the entrance pupil at infinity, or correspondingly of the exit pupil at the back focal plane of the respective objective lens, is very convenient because it provides the required fixed position. Switching from one objective lens to another with a different magnification does not require a repositioning of the condenser. Only the size of the condenser aperture has to be readjusted to the new exit pupil diameter.

A further very important advantage of the exit pupil being at the back focal plane – and its image being in the front focal plane of the condenser – is its use in several important transmitted light contrasting techniques, as discussed in Section 2.2.6. In epi-illumination setups such as in fluorescence microscopes, the entrance pupil of the illumination light path and the exit pupil of the imaging path both fall together at the same position and are at the back focal plane of the objective lens. Therefore, their sizes are intrinsically matched.

2.5 Microscope Objectives

The *objective* is the central component of a microscope. It comprises up to 20 different single convex and concave lenses with different radii of curvature. For optimal performance, different glass substrates and cements are used within one objective. To avoid any aberrations on the resulting image, all these single lenses have to be mounted with very high precision. While this is already demanding in high-end objectives, it becomes decisive when designing and manufacturing objectives with movable lens groups as one can find in objectives with a correction collar. Therefore, objective design is closely associated with precision mechanics.

The objective is the most sophisticated optical component of the microscope. Its function is to collect light from the object and to form the primary real image in combination with the tube lens. The image should be as free from aberration and the resolution as high as possible. The objective achieves the major part of the magnification of the microscope and often creates a specific image contrast. Today, there exists a large variety of different objectives with widely differing properties for different purposes. This makes the choice of the appropriate objective for the specimen of interest essential for satisfactory imaging results.

2.5.1 Objective Lens Design

The objective lens determines the overall imaging quality of the complete microscope. The formation of an ideal image, however, is in practice impaired by monochromatic and chromatic aberrations. Monochromatic aberrations occur because most lenses have spherical surfaces. In addition, chromatic aberrations basically cannot be avoided because the refractive index of transparent media is a function of the wavelength, and therefore aberrations arise when light with different wavelengths is used. The type and extent of aberrations in imaging is dependent on the object field size, the focal length, and therefore the magnification, along with the width and inclination of the incident beams.

The naturally occurring aberrations can be reduced by combining different types of glass materials with different dispersion properties, radii of curvature, thicknesses, and spacing between the different lens elements. These are the parameters that define the imaging properties of the composite system. For each single aberration type, there exists a strategy to reduce its effect by combining

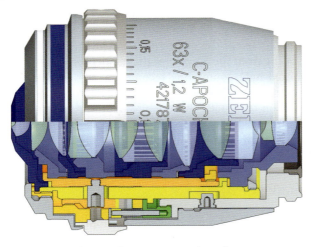

Figure 2.17 Objective lens system. Modern objectives are master pieces of optical and precision mechanical engineering. The graphics shows a 63 × /1.2 NA C-Apochromat objective. This objective is apochromatically corrected from the UV through the visible to the near-infrared spectral range. It features a correction collar for compensating differences in cover glass thickness and/or the refractive index within the specimen. The path of specific rays and further details are shown in Figure 2.18. (Reprinted by permission from the Carl Zeiss Microscopy GmbH.)

lenses with different parameters. For example, axial chromatic aberration can be corrected by combining two lenses made of different glasses with different color dispersions. This strategy brings two selected colors to exactly the same focus, but other colors will show slight deviations from this focal point. These residual aberrations are called the *secondary spectrum*.

The combined elimination or reduction of multiple aberrations is partly a science and partly a genuine art, which requires a lot of optical knowledge, intuition, and experience. For high-power objective lenses, this results in an arrangement of many different lenses into a complex lens system (Figure 2.17).

The first step of designing a lens is to define the desired properties such as magnification, field number, NA, wavelength region, immersion medium, working distance (WD), and the degree of aberration correction. Then the principal setup of the lens is selected. This is done by referring to existing lens designs with similar desired properties or by the creation of a completely new design. This step represents the most creative and important phase, but also very difficult one. The arrangement of each lens element, the corresponding allotted refractive powers, and the choices of the materials define not only the image quality but also the time required for the design, which might extend over weeks and months, and the costs of manufacturing.

The detailed imaging properties of the lens system are studied by computerized ray tracing. Objectives built from several thick lenses cannot be treated any more by the simple Gaussian imaging laws which are valid only for paraxial rays. In the design and analysis of objectives, these simplifying assumptions are not valid and must be replaced by considering the thickness of lenses and their aberrations. For ray tracing, representative points of the object plane with a number of light ray bundles of appropriate colors emerging from them are chosen. Then, the rays are traced by calculating how they are refracted at each single lens surface according to the law of refraction. Here, the exact optical properties of the respective lens materials and the wavelength-dependent index of refraction are considered. The rays in the bundles are chosen such that the complete exit pupil of the system is scanned, and any ray can exactly be traced through the optical system. Therefore, ray tracing with dedicated computer programs is today the chief tool of optical designers.

By ray tracing, the imaging process can be simulated and evaluated. In the early stage, deviations from the desired imaging properties may be substantial, which must then be reduced by adjusting the lens parameters or by the addition of new elements. Sometimes, the start design may have to be dropped. If a design comes close to the desired one, several iterations of parameter optimization to achieve further aberration reductions are performed. For microscope objective lenses, the correction of most aberrations has to be done rigorously up to the resolution limit. Especially, the correction of spherical and chromatic aberrations is important. At this stage, the lens design requires detailed observation of the changes in the various aberrations on the computer until the design is finally optimized, which can be achieved by automatic lens design computer programs. The accumulation of many small aberration corrections completes the final design.

Example designs for five different 63× objectives are shown in Figure 2.18. The number "63×" means that the lens systems yield a 63-fold magnification when used with the corresponding tube lens. The magnification of all objectives shown is identical, but the level of aberration correction increases from left to right for the N-Achroplan, Plan-Neofluar, and the Plan-Apochromat. The last two objectives have special qualities. C-Apochromat has a high NA and is a high-performance lens designed for use with water immersion. The Alpha Plan-Apochromat has a lower correction degree, but it is a complex lens system featuring a very high NA and is designed for use in total internal reflection fluorescence microscopy (see Box 12.2).

To evaluate and compare the correction level of different objectives, quantitative measures are required. There are numerous different parameters for this purpose, for example, the mentioned secondary spectrum or the curvature of field that reports the deviations in the axial position of the images of point objects that are located in the object plane on axis and at the border of the field of view.

A comfortable general measure are the OPDs in the exit pupil. For a perfect objective, all monochromatic rays that originate at the focus have identical phases at the exit pupil. Their optical path lengths are identical, and they form an ideal parallel bundle of light rays when leaving the objective. As discussed earlier, such a performance cannot be achieved by real lenses. Rather, there occur OPDs between rays traversing the lens system in different regions of the lenses, for example, centrally or marginally. To quantify these differences, the OPDs for all traced rays in relation to a reference ray are squared and then averaged. The root of the average represents the root-mean-square-optical path length difference (rms-OPD) for a given wavelength. Obviously, residual chromatic aberrations will have an impact on the rms-OPD. Hence, rms-OPDs are normalized to the

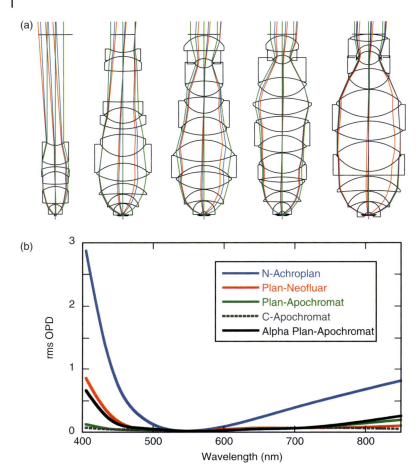

Figure 2.18 Correction of aberrations in objectives. (a) Optical design of five different 63× objective lens systems. From left to right: N-Achroplan with NA 0.85, air immersion; Plan-Neofluar with NA 1.25, oil immersion; Plan-Apochromat with NA 1.4, oil immersion; C-Apochromat with NA 1.2, water immersion; and Alpha Plan-Apochromat with NA 1.46, oil immersion. The different colors label ray bundles emerging from different positions in the focal plane. (b) Comparison of the degree of correction for the objectives shown in (a) in terms of the root-mean-square deviations of the optical path length difference (rms-OPD) in the exit pupil of the objective plotted as a function of the wavelength. N-Achroplan has the simplest design and is achromatic; Plan-Neofluar is optimized for fluorescence applications and shows good correction of chromatic aberrations in the visible wavelength region, whereas the Plan-Apochromat shows excellent chromatic aberration and a very low curvature of field. C-Apochromat has excellent chromatic correction from near-UV to the near-infrared spectral range, and shows an especially low curvature of field. Alpha Plan-Apochromat features an especially large NA and very good chromatic and planar correction. (Reprinted by permission from the Carl Zeiss Microscopy GmbH.)

wavelength and plotted for the wavelength range for which the objective was designed. Therefore, the rms-OPD curves represent a quantitative measure of the existing residual aberrations of an objective, such as chromatic aberrations, spherical aberration, and curvature of field, and are an excellent tool to compare the quality of different objectives.

Figure 2.18b shows the significant improvement of the corrections for the various lens systems sketched in Figure 2.18a. The increasingly complex design of the first three objectives (blue, red, and green curves) results in a significant reduction in the rms-OPD over the spectrum from near-ultraviolet to near-infrared light. C-Apochromat shows the smallest aberrations and appears to be the lens of the highest imaging quality over a wide range of wavelengths. Finally, it is obvious that the optical designers of the Alpha Plan-Apochromat had to sacrifice some imaging quality to create an objective with an NA of 1.46.

No real lens system is perfect, but an important question is whether the residual aberrations are below or above the intrinsic diffraction-limited effects. For pure visual observation, an aberration is not relevant when its effect is below the diffraction limit. However, such aberrations are significant for applications in the field of super-resolution microscopy. Usually, only a careful test and comparison of objectives reveals their suitability for the intended imaging tasks.

2.5.2 Light Collection Efficiency and Image Brightness

We saw earlier that any objective can capture only a fraction of the elementary waves emitted or diffracted by the object owing to its finite opening angle. This results in the limited resolution of optical systems. Furthermore, it limits the light collection efficiency of the objective, which is of course a central feature of every microscope. Ultimately, it determines the brightness of the final image.

Surprisingly, it is well known that image brightness is related to the NA of the objective – meaning to the sine of the opening angle α – and not to the opening solid angle of the objective. This would be the first intuitive guess. Hence, is this true, and why? A careful examination of this question reveals the reason and also leads to an important concept in microscopy, the so-called sine condition.

First, we consider a luminous point object that emits light into the surrounding space. The light flux that is collected by a lens is directly proportional to the surface of a spherical segment filling the opening of the lens or, in mathematical terms, to the opening solid angle as shown in Figure 2.19a. Considering the rotational symmetry around the optical axis, which may be labeled by an angle φ, we can calculate the total solid angle of the surface segment by integrating over all surface elements $dS = d\varphi \sin\theta \, d\theta$ according to

$$\Phi_{\text{point}} \propto \int_0^\alpha \left[\int_0^{2\pi} d\varphi \right] \sin\theta \, d\theta = 2\pi(1 - \cos\alpha) \qquad (2.41)$$

The integration can be performed directly because the light of the point object falls perpendicular on all surface elements regardless of their position in space, which is characterized by the angles φ and θ. Hence, the light collection efficiency

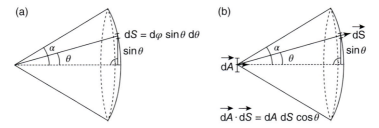

Figure 2.19 Light efficiency of objectives. (a) Imaging of a luminous point object. (b) Imaging of a luminous 2D object, a Lambert radiator.

of an objective when examining point objects is *not* related to the sine of the opening angle respectively the NA of the lens, but given by Eq. (2.41).

The situation is different, however, when a 2D extended luminous object is imaged. Now we must determine the amount of light emitted by a surface element dA and collected by the lens. Such a luminous surface is called a *Lambert radiator*. However, the effective surface of dA that is visible under different angles θ varies (Figure 2.19b). For increasing angles θ, the effectively perceivable surface becomes smaller until it is 0 for a hypothetical $\theta = 90°$. Mathematically speaking, we must now consider the scalar product $\vec{dA} \cdot \vec{dS}$.

This adds a factor of $\cos\theta$. If we take this into account and perform the integration over the total surface of the spherical segment, we find now

$$\Phi_{dA} \propto dA \int_0^\alpha \left[\int_0^{2\pi} d\varphi \right] \sin\theta \cos\theta \, d\theta = dA \cdot \pi \sin^2\alpha \qquad (2.42)$$

We note that the collected light flux is related to the size of the emitting surface element dA and the square of the sine of the opening angle of the objective. This expression is valid for air-immersion objective lenses. When a material with a higher index of refraction is located in front of the lens, such as water or oil, an appropriate immersion lens must be used. This situation is sketched in Figure 2.20.

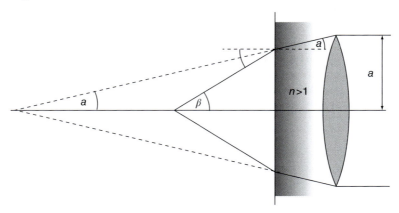

Figure 2.20 Light flux with and without an immersion medium. (Modified from [8].)

We see that the light flux entering the lens with a maximum angle α in the material with an index of refraction n would be identical to the light flux enclosed by the angle β when working in air. According to the law of refraction, $\sin \beta = n \sin \alpha$. The light flux, however, does not change. Thus, when using optically denser materials, we must multiply the sine of the opening angle by the index of refraction to get the correct expression for the collected light flux:

$$\Phi_{n,dA} \propto dA \cdot \pi \cdot n^2 \sin^2 \alpha = dA \cdot \pi \cdot NA^2 \tag{2.43}$$

Hence, we see that the amount of light collected by an objective is proportional to the square of its NA if a flat 2D specimen is imaged. Of course, this is most often the case in microscopy.

Box 2.5 The Sine Condition

Considering the light flux through a lens allows further insights. To achieve a faultless image, all light collected from the surface element dA must obviously be projected onto the corresponding, magnified surface element dA' in the image plane. This is equivalent to stating that the surface element dA must be mapped exactly onto dA'. This requirement sounds trivial, but indeed it cannot be achieved by a simple spherical lens but rather requires elaborate lens systems. However, we can turn the tables: we can use this as a condition that must be fulfilled by an optimally working microscope. Considering that the image space is usually in air with a refractive index of 1, the condition can be formulated as follows:

$$dA \cdot \pi \cdot n^2 \sin^2 \alpha = dA' \cdot \pi \cdot \sin^2 \alpha' \tag{2.44}$$

For the definition of α', see the following sketch. We can simplify this expression by considering that the size of the surface dA is related to its image by the square of the magnification factor M:

$$dA' = M^2 dA \tag{2.45}$$

Inserting this above, we obtain

$$n \sin \alpha = M \sin \alpha' \tag{2.46}$$

This expression is known as *sine condition* and was formulated by Ernst Abbe as a condition that a good imaging system must fulfill. Now we can also reverse the argument: the light flux for a Lambert radiator in an immersion medium with refractive index n must be given by Eq. (2.43).

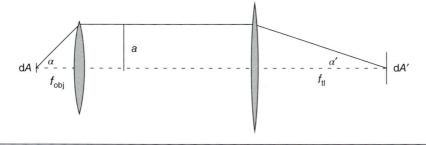

(Continued)

Box 2.5 (Continued)

A further condition (used in Eqs (2.9) and (2.14)) is closely related to the sine condition. It can be illustrated as follows: For an air imaging system, as shown in Figure 2.1, M is given by the ratio of the focal lengths of tube lens and the objective (Eq. (2.3)). If we insert this into the sine condition Eq. (2.46), we have

$$\sin \alpha = \frac{f_{tl} \sin \alpha'}{f_{obj}} \quad (2.47)$$

Now let us increase f_{tl} more and more. The image size increases, and $\sin \alpha'$ decreases accordingly. As α is not altered, the product $f_{tl} \sin \alpha'$ must be a constant, and we can call it a. In a system fulfilling the sine condition, this is valid for all angles α, and if we consider small α, we can identify a. For small α, we have $\sin \alpha \approx \tan \alpha = a/f_{obj}$ (see sketch) and thus Eq. (2.47) yields in general

$$\sin \alpha = \frac{a}{f_{obj}} \quad (2.48)$$

Hence, an incoming beam through the front focus with an inclination angle α proceeds behind the lens in the infinity space parallel to the optical axis at a distance a, where a is given by $f_{obj} \sin \alpha$ (Eq. (2.47)). This insight is central for Fourier optics (Box 2.1), where it is the starting point for realizing that the back focal plane of a lens contains the Fourier transform of the object function [9]. It is called the *von Bieren condition*.

In summary, the light collection efficiency is normally proportional to $\sin^2 \alpha$ and NA^2, and translates directly into image brightness. In the special case where single isotropic point emitters are examined, as may occur in single-molecule imaging (Chapters 8 and 12), the light collection efficiency is higher and related to $(1 - \cos \alpha)$. In Figure 2.21, the differing light collection efficiencies are illustrated

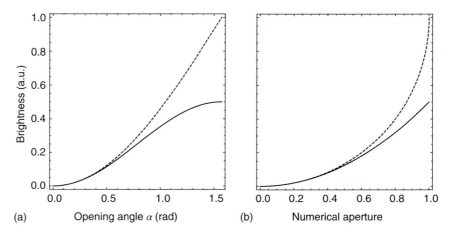

Figure 2.21 Light collection efficiency in air for a luminous point emitter (dashed line) and a flat luminous object (full line) as a function of (a) the opening angle and (b) the NA.

as a function of the opening angle α and of the NA. The importance of using high NA lenses is obvious in both cases, but especially when imaging point objects.

The collected light passes the objective's exit pupil and is projected onto the image plane. How is image brightness related to the magnification? This becomes obvious when considering the image of a 2D self-luminous object: increasing the magnification enlarges the image but reduces its luminous density proportional to the square of the magnification, because the total integrated luminosity of the image remains constant.

Combining the earlier considerations, we find that the image brightness B is proportional to the square of the NA and inversely proportional to the square of the magnification:

$$B \propto \frac{NA^2}{M^2} \tag{2.49}$$

Obviously, these considerations neglect absorption and reflection effects of the lenses forming the objective, or special lens features such as, for example, the absorbing ring in phase contrast objectives. Thus, objectives with the same magnification and NA may produce images with quite different brightness values.

Finally, it should be noted that the impact of NA in fluorescence microscopy is even more significant. In this case, the illumination is achieved via the objective, which introduces a further factor of NA^2 in the irradiance in the sample plane. Consequently, the brightness of the final image is related to the used NA as

$$B \propto NA^4 \tag{2.50}$$

We see that using a high NA is extremely important for achieving bright images. However, in fluorescence imaging, the dependence of the brightness on the magnification vanishes, as its effects on the illumination and imaging cancel each other.

2.5.3 Objective Lens Classes

We have seen that a very important feature of an objective is its degree of aberration correction. Two of the monochromatic aberrations, namely, spherical aberration and coma, have a diffraction-limited correction in modern objectives, which means that these aberrations are below the visibility threshold for all objective types. Curvature of field and chromatic aberrations are still the most significant aberrations, and objective lenses are classified according to the degree of correction of these aberrations. The first part of the objective class name refers to the degree of correction of the curvature of field, and the second part to the degree of correction of the chromatic aberrations. A "plan" designation is added to the lens class with a particular low curvature of field and distortion. Such objectives have a large field over which the image is well corrected. Lateral and longitudinal chromatic aberrations are of particular significance to achieve good bright-field contrast, because objects are imaged with colored fringes when these aberrations are present. With regard to chromatic correction, microscope objectives are classified into achromats, semi-apochromats or fluorites, and apochromats. These specifications are inscribed on the objective lens barrel. Only achromats are usually not designated as such.

Unfortunately, each manufacturer denotes their objective lenses in a different way, which makes distinctions and comparisons difficult. Nevertheless, the name indicates the degree of correction. Aside from the correction level, further important parameters are inscribed color-coded or abbreviated on the objectives (Table 2.1). These indicate the magnification, immersion medium, and suitability for special imaging applications. Furthermore, one or sometimes two colored rings indicate the achievable magnification and the required immersion medium (Tables 2.2 and 2.3).

2.5.4 Immersion Media

Objectives may be used in air or in combination with a special immersion medium such as oil or water. Air objectives are also designated as dry lenses. They are often used with or without a cover glass, but an NA of 0.25 is the upper threshold value for their universal use. Departure from the standard cover glass thickness of 0.17 mm is not critical for objectives with an NA below 0.4. The maximum NA of an air objective is 0.95, which corresponds to an aperture angle of ~72°. Such objectives are very difficult to use because the large aperture angle results in a very short free WD, which can be as short as 100 µm. Air objective lenses that are designed for use with cover glasses are even more critical in their handling.

Many objective lenses are designed for use with a medium of higher refractive index than air between the cover glass and the front lens of the objective. Mostly, media such as immersion oil, glycerol, or water are used. This increases the NA and thus the achievable brightness and resolution. A homogeneous immersion is beneficial, that is, a situation where the refractive indices of mounting medium, cover glass, and immersion fluid, and – ideally – the front optical element of the objective lens are matched as well as possible. This is favorable for avoiding spherical and chromatic aberrations at these interfaces. The construction of such objectives is more straightforward, because all media may be looked upon as an extension of the front lens element. At the same time, reflections at the interfaces are reduced, which improves image contrast. This implies the use of an immersion oil with a refractive index close to that of glass, the use of a glass cover slip, and the mounting of the specimen in a resin with its refractive index matched to glass and oil.

Imaging using high NA oil-immersion objectives is especially critical and requires experience and precautions. First, the refractive indices of the cover glass and oil are not perfectly matched. The objectives are optimized for the standard cover glass thickness. However, there are cover glasses with thicknesses between 120 and 200 µm. The most common is 170 µm. Even in a single batch, the thickness may vary considerably. The actual thickness should not deviate by more than ±5 µm from the indicated value on the objective for an NA > 1.3; otherwise, aberrations occur. In addition, immersion objectives are designed to be used at 23 °C. Deviation from this temperature causes minute thermal movements of the lens arrangement in the objective, which may affect the image quality. Some objectives have a mechanical correction collar that allows the adjustment of the internal position of a few lenses of the system to correct for

Table 2.1 Inscriptions on objective barrels.

Position/type	Abbreviation	Meaning
1. Correction of image curvature	—	Lower class of image curvature correction
	NPL, N plan	Normal field of view plan (20 mm)
	PL, plan, plano	Plan field of view (25 mm)
2. Longitudinal chromatic correction	—	Achromat (lower class of chromatic correction)
	FL, fluotar, neofluar, fluor, neofluor	Fluorite or semi-apochromat
	Apo	Apochromat
3. Magnification	2.5–160	Magnification in the primary image plane when used with appropriate tube lens
4. Numerical aperture	0.1–1.46	In case a range is given (e.g., 1.25–0.65), the objective has an adjustable aperture built in
5. Immersion	—	Air
	Water, W, W.I.	Water immersion
	GLY, Gly	Glycerol immersion
	Oil, HI, H	Oil immersion
	IMM	Can be used with air, water, or oil immersion (old)
	LCI	Can be used with several media, for example, glycerol, water, or oil immersion
6. Mechanical tube length	∞	Infinite mechanical tube length
	160–250	Tube length in millimeters
7. Cover glass thickness	—	May be used with or without cover glass
	0	To be used without cover glass
	0.17	Stipulated specimen cover glass thickness
8. Working distance	WD + number	Working distance in millimeters
9. Other (examples)	Phase, Ph, phaco, PC PH1, PH2, and so on	Phase-contrast objective using respective phase condenser annulus
	D	Dark-field objective
	I, IRIS, W/IRIS	Iris diaphragm built in for adjustment of NA
	UV, UV-A	UV-transmitting, down to 340 nm

(Continued)

Table 2.1 (Continued)

Position/type	Abbreviation	Meaning
	Ultrafluar	Fluorite objective for imaging down to 250 nm in the UV as well as in the visible region
	DIC	Differential (Nomarski) interference contrast
	L	Long working distance
	LL, LD, LWD	Very long working distance

Inscriptions on objectives usually indicate the degree of correction, magnification, numerical aperture, immersion medium, cover slip thickness, and often the working distance. Magnification and numerical aperture as well as tube length and cover glass thickness are usually separated by a slash, respectively. The table gives a selection of common abbreviations. These are not exhaustive, as every manufacturer uses their own designation code.

Table 2.2 Color coding for magnification.

Color code	Magnification
White	100–160×
Dark blue	60–63×
Light blue	40–50×
Green	16–32×
Yellow	10×
Orange	6.3×
Red	4–5×
Brown	2.5×
Gray	1.6×
Black	1–1.25×

If the objective shows a single colored ring, it refers to the magnification factor. Many objectives have two colored rings. In this case, the ring located near the mounting thread indicates the magnification factor, and the one near the front lens indicates the immersion medium. The magnification factor is also printed in numbers. Special features are sometimes indicated by a special color of the text inscriptions.

Table 2.3 Color coding for immersion media.

Color code	Immersion
White	Water
Orange	Glycerol
Red	Multi-immersion
Black	Oil

If two rings exist, the ring near the front lens indicates the immersion medium.

different temperatures or cover glass thicknesses. To keep the warming of the objective – and the specimen – minimal, the intensity of the illumination light should be kept as low as possible using neutral density filters or by reducing the lamp current. In case no correction collar is present, such deviations may be compensated by using specific immersion oils that counter the temperature effects inside the objective.

Several options are available for microscopic analysis of specimen in aqueous solution. The so-called water-dipping objectives are designed for direct use in aqueous media for observation of live biological specimen from above. However, an important and very widespread application of microscopy is the imaging of living cell samples using inverted microscopes. Obviously, this requires the use of cover glasses that are mounted on the microscope stage and separate specimen and immersion fluids. In this case, the inhomogeneous optical system between the front lens and specimen is unavoidable. The use of dedicated water-immersion lenses is highly recommended, which provide very good images from planes deep in the specimen. For these objectives, the significant and inevitable spherical aberration at the cover glass–water interface is corrected by the intrinsic objective design. For water-immersion objectives, distilled or at least demineralized water should be used, as it is troublesome to remove sediments from water that has dried on the objective front lens. When dedicated high-quality water-immersion objectives are not readily available, oil-immersion lenses are often used instead. This works surprisingly well for thin objects that are very close to the cover glass/buffer interface. However, image quality degrades very rapidly when focusing deeper into a water-mounted specimen with an oil-immersion objective lens. Figure 2.22 illustrates how the PSF is degraded in this case.

2.5.5 Special Applications

The WD of an objective is the distance between the objective front mount and the closest surface of the cover glass when the specimen on its other side is in focus. In case no cover glass is used, the WD is the distance to the specimen itself. As long as the immersion conditions are maintained, this also corresponds to the depth inside the specimen that can sharply be imaged.

For objectives with high resolution, the WD is quite short, because the NA and the opening angle must be as high as possible. For high-power objective lenses of 100× magnification and 1.4 NA, WD is shorter than 200 µm, whereas it goes up to 6 mm for a 10× objective with NA 0.2. For high-magnification objectives, the front part is usually spring-mounted such that it may slide into the objective barrel when the specimen is hit, in order to avoid damage to the front lens and the specimen. For some applications, very long WDs are required, for example, when the specimen is manipulated during microscopic observation, for instance, by microinjection of living cells. These objectives are called *long-distance* or LD objectives. For such purposes, dedicated objectives were developed despite the difficulty involved in achieving large NAs and a good degree of correction. As such objectives are most often built with a correction collar, they are usually relatively expensive.

Besides accomplishing pure magnification, objectives must have certain characteristic features for specific contrast techniques. Objectives for use in fluorescence microscopy must have low intrinsic auto-fluorescence, low intrinsic reflectivity, and high light transmittance down to the near-UV. These properties depend on the employed glass types and careful coating of the lens surfaces. However, today's objectives have transmission coefficients in the range of 90% to at least 360 nm. Objectives featuring high transmission below this wavelength, which are suited for work with UV light, are usually specially labeled.

Objectives that employ specific contrast techniques are specifically designed for this task. Usually this requires highly modified optical properties. These are discussed in the following section.

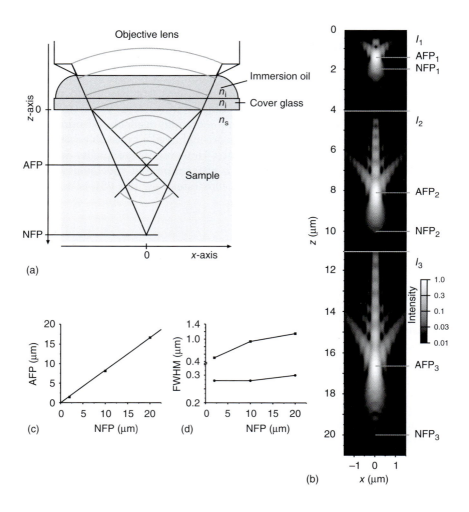

Figure 2.22 Spherical aberration at the cover glass when using an oil-immersion objective for imaging in water. The degradation of the PSF as a function of the distance to the cover glass was computed by Dr Christian Verbeek according to [10]. (a) The immersion oil and the cover glass have a matched refractive index of $n_1 = 1.518$. Incoming parallel light waves are focused by the objective lens into water with a refractive index of $n_s = 1.33$. The resulting light intensity distribution represents the PSF. The distance of the intensity maximum from the interface as it would be for a medium with a matched refractive index is designated the nominal focus position (NFP). The sketch illustrates the problem that occurs in a mismatch situation: the incoming light is refracted at the cover glass/water interface if the refractive indices of both materials are not identical. Therefore, the light waves do not interfere constructively in the NFP, but rather above it in the so-called actual focus position (AFP). (b) The resulting intensity distribution in water was numerically computed for a 100× oil-immersion objective lens with an NA of 1.3 for different exemplary NFPs, namely $NFP_1 = 2\,\mu m$, $NFP_2 = 10\,\mu m$, and $NFP_3 = 20\,\mu m$, respectively. The computation showed that the refraction shifted the PSF maxima to $AFP_1 = 1.45\,\mu m$, $AFP_2 = 8.1\,\mu m$, and $AFP_3 = 16.6\,\mu m$, respectively. The resulting actual PSFs were calculated in regions around the AFP_i, namely, I_1 (0–4 µm), I_2 (4–11 µm), and I_3 (11–21 µm), respectively. The results are shown as contour plots with a logarithmic intensity scale (see inset). (c) The relationship between NFPs and AFPs is almost linear. Since $AFP_i < NFP_i$, the axial extension of thick samples appears elongated when it is determined by measuring the stage movement when focusing to the top and bottom of the sample. Therefore, spheres appear axially stretched when imaged by confocal microscopes. (d) Furthermore, the PSF is more and more smeared out for increasing NFPs, which is shown by the calculation of the FWHM values along and perpendicular to the optical axis. This leads to a depth-dependent reduction of lateral and axial resolution. A further effect is the reduction of the maximum of the distribution, which is not quite visible in (b) because the shown distributions were normalized to unity for a better representation of low intensity values. (Hell *et al*. 1993 [10]. Reproduced with permission of John Wiley & Sons.)

2.6 Contrast

Light is characterized by different properties – phase, polarization, wavelength (or color), and intensity (or brightness). Objects in the beam path usually alter one or more of these properties. Both the human eye and electronic light detectors are sensitive only to wavelength and intensity. Therefore, we can directly detect only objects that change one or both of these properties. Such objects are called *amplitude objects*, because they alter the amplitude of the light waves traversing them. If this occurs in a wavelength-dependent manner, the respective objects will appear in color. Objects that alter only the phase of light waves – possibly in a polarization-dependent manner – are called *phase objects*. Typical examples are transparent objects made of glass or crystals. Such objects effect a phase delay on the traversing light wave owing to a different index of refraction as compared to the surrounding medium. In addition, biological objects such as thin layers of cells or tissue influence the phase, but much less the amplitude of traversing light waves by absorption.

A phase object that is well resolved and magnified is not necessarily visible for the detection device. For an object to be visible, either its color or its light intensity must differ sufficiently from those of the surroundings. In other words, the contrast between the object and the background must be detectable. Contrast, C, means here the degree to which the object differs from its background in terms of intensity or wavelength. For light intensity I, one of several possible definitions is

$$C_I = \frac{I_{obj} - I_{sur}}{I_{obj} + I_{sur}} \qquad (2.51)$$

where I_{obj} and I_{sur} denote the intensity of the object and of the surroundings, respectively. For electronic light detection devices, the detectable contrast depends on the overall signal-to-noise ratio and the number of discretization steps used when measuring the signal, which corresponds to the bit-depth of the detection device. These questions are dealt with in detail in the next chapter.

For human perception in bright light, the contrast in intensity may be as little as 0.02, but in poor light the contrast required for perception of the object is 0.05. For very small objects, it needs to be up to 0.2. Contrast difference between the object and its background is essential for seeing fine details and should be as large as possible for a clear perception of the object.

In the history of microscopy, numerous different techniques were developed to render phase objects visible, which are referred to as *optical contrast methods*. Obviously, such techniques play a major role in the microscopy of biological objects. In principle, there are two different types of contrast techniques: chemical and physical approaches. Chemical techniques introduce a color contrast by selective labeling of object structures with dyes. In modern microscopy, mostly fluorescent dyes are employed, which are the focus of Chapters 3 and 4. The physical techniques turn the initially unperceivable phase contrast into an amplitude contrast. Today, the most important optical contrast techniques for biomedical applications are phase contrast and DIC. Dark-field contrast represents an easy access to the physical contrast techniques.

2.6.1 Dark Field

The simplest of the physical contrast techniques is *dark-field contrast*, also called *dark-field illumination*. The principle is to retain only the object-diffracted light and eliminate all direct illumination light by means of a special illumination system. This can easily be achieved by illuminating the specimen from the side [11]. Alternatively, when illuminating the specimen from below in a transmitted-light dark field, the inclination of the incident radiation can be chosen so high that no direct light is captured by the objective lens. To achieve this, the angle of the incident light rays with respect to the optical axis must exceed the maximum objective aperture angle. Then only the light scattered by the specimen enters the objective, as shown in Figure 2.23, and creates an image with a high contrast. The object background is completely dark, and only light-scattering object structures appear bright.

When using low-power objectives with NA ≤ 0.4, dark-field illumination can easily be achieved by inserting a ring-shaped mask of an appropriate size into the front focal plane of the condenser. For this purpose, even phase-contrast condensers can be employed: as discussed later, a phase-contrast condenser contains

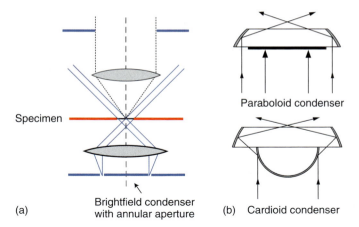

Figure 2.23 Dark-field contrast. (a) Principle of dark-field illumination and (b) dark-field condensers.

an annular aperture in its front focal plane. The central portion of the illumination light is blocked by a circular, axially centered mask (Figure 2.23a). The advantage is that the mask is positioned in the correct front focal plane of the condenser and can be axially centered by adjusting screws. A disadvantage, however, is that reflections at the lens surfaces of the condenser will diminish the dark-field effect. When the use of higher NA objectives is intended, dedicated dark-field condensers of a reflective type must be employed (Figure 2.23b). Such condensers have additional advantages. In comparison to ring condensers, they yield higher irradiance and result in identical inclinations for different colors, resulting in a sharper dark-field illumination cone. It is important to exactly position these dark-field condensers on the optical axis of the objective. When a cardioid condenser is used, an immersion fluid is added between the object slide and the condenser surface to avoid refraction into the object slide, which would lower the effective inclination angle. The use of high NAs for the condenser (e.g., NA 1.4) is especially advantageous because the illumination light is totally reflected at the upper surface of the cover glass when an air objective is used for observation. In that case, the elimination of the illumination light is particularly complete. The use of an objective lens whose NA can be adjusted by means of an internal iris diaphragm is practical because its effective NA may be adjusted until a completely dark background is created.

In principle, a dark-field effect is also obtained by blocking the illumination light in the back focal plane of the objective lens. This, however, would mean the loss of an appreciable amount of diffracted light and require the use of a specific objective for dark-field observation.

2.6.2 Phase Contrast

The principle of phase contrast and some of the interference contrast methods described later is the separation of light that is diffracted by the object from the light that is not altered by the object, that is, the background or undiffracted light. The amplitudes and phases of these two light components are modified in such a way that finally an amplitude contrast due to constructive or destructive interference between the two light components occurs. Frits Zernike introduced in 1932 a versatile and useful microscopic technique to visualize phase objects that are almost invisible in bright-field microscopy [12]. He performed a series of experiments that elegantly illustrated the principle of his contrast technique – and simultaneously the wave theory of image formation in the microscope. His experiments are very instructive, and therefore we will follow his argument in detail.

2.6.2.1 Frits Zernike's Experiments

We begin the series of experiments by examining a tiny amplitude object using bright-field illumination. With a wide open condenser aperture, a dark spot on a bright ground is visible in the image plane – the standard bright-field image of a small amplitude object. Next, we close the condenser aperture to a small spot and thus illuminate the sample now by a point light source. Simultaneously, we place a small light-blocking mask in the back focal plane of the objective lens (Figure 2.24c). What happens? The condenser lens transforms the light emanating from the point source in its front focal plane into plane waves. These waves encounter the specimen, which they partly pass undiffracted and are partly diffracted. The undiffracted light is focused again by the objective lens into its back focal plane, which is conjugate to the front focal plane of the condenser. Here it is blocked by the above-mentioned mask. Its shape is complementary to the mask in the condenser front focal plane, and selectively prevents the undiffracted background light from passing the back focal plane of the objective lens. Blocking the undiffracted light passing by the object leads to a dark background in the image plane. The diffracted light, however, is not focused at the back focus of the objective because it is not a plane wave but rather a diverging spherical wave in the object space as discussed in Section 2.3 of this chapter. Therefore, it is not blocked in the back focal plane by the mask. It passes the back focal plane and is focused by the tube lens into the primary image plane, creating a bright spot on a dark background. Actually, this construction with the two complementary masks in conjugate planes represents a special type of dark-planefield illumination, which was discussed in Section 2.6.1.

Figure 2.24 shows the two different sets of conjugate planes: the condenser front focal and the objective back focal plane, as well as the object and image planes. In the first experiment, both undiffracted and diffracted light reach the image plane and produce a dark object image. When we block the background light in the back focal plane of the objective lens, the image is formed by the diffracted light alone, and we see a bright spot. The physical explanation for these two quite different image patterns is that the diffracted light is phase-shifted by 180° compared to the undiffracted light. In the first experiment, the image is the result of the destructive interference of the object-diffracted and the undiffracted light waves. In the second experiment, the image is the result of the object-diffracted light waves *alone*. No destructive interference can take place (Figure 2.25).

In the third experiment we image a small phase object. A small transparent corn is placed on the object slide, the condenser aperture is opened, and the mask is removed from the back focal plane of the objective, thus returning to bright-field microscopy. Now we see a bright background field and practically no object, as it

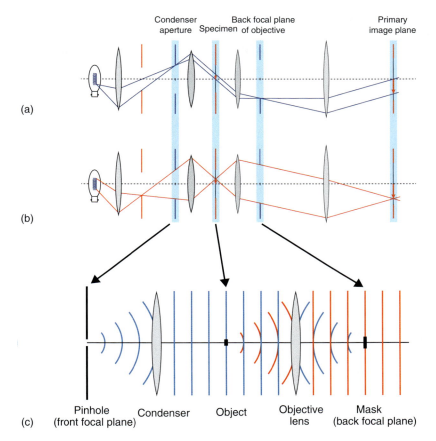

Figure 2.24 Zernike's second experiment. (a) Illumination beam path and (b) imaging beam path of a microscope. The arrows indicate the respective planes in the sketch underneath. (c) Magnification showing the space between the front focal plane of the condenser and the back focal plane of the objective lens. The undiffracted and diffracted waves are shown in blue and red, respectively. The four planes marked in light blue are further discussed in Figure 2.25.

is transparent. Next, we again close the condenser aperture and place the complementary mask back into the back focal plane of the objective lens. Again, we see a dark background because the undiffracted illumination light is blocked – and a bright spot from the object. Obviously, the transparent object does indeed diffract light similar to an amplitude object, but its phase is obviously not shifted by 180°. Otherwise, a dark object image on a bright background would be visible in the bright field.

Zernike assumed that the diffracted light of a phase object is shifted only by about 90°. A light wave with small amplitude that is phase-shifted by 90° when interfering with the background wave produces a wave of almost unaltered amplitude and phase compared to the undiffracted wave. This is demonstrated in the last column of Figure 2.25. It shows the amplitudes and phase relationships of the different wave components in all experiments. Amplitude and phase of the undiffracted wave are represented by the blue arrow pointing along the

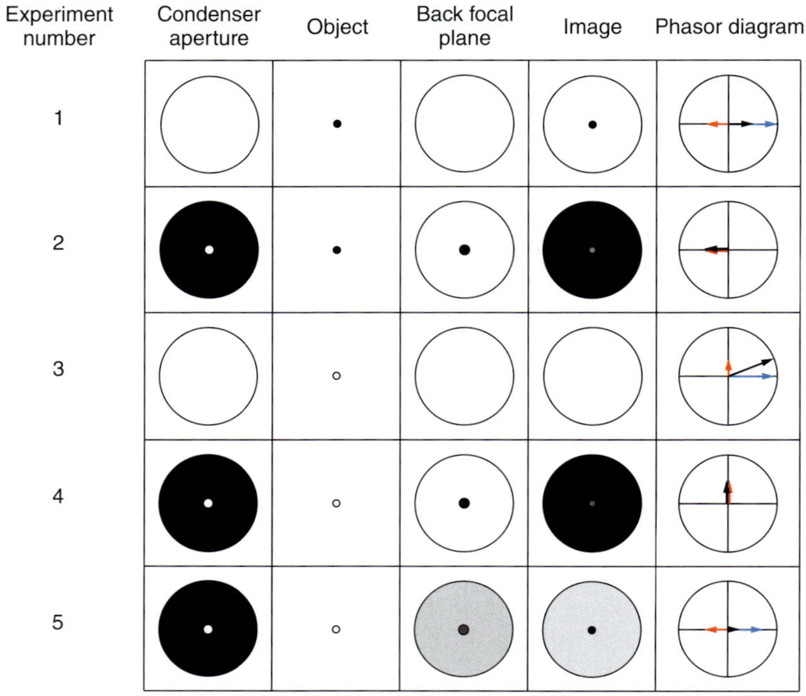

Figure 2.25 Masks, objects, images, phases, and amplitudes in Zernike's experiments. The first and third columns describe the masks in the front focal plane of the condenser and the back focal plane of the objective, respectively. In the upper two experiments, absorbing objects and in the lower three experiments transparent or phase objects were examined as indicated in the second column. The fourth column sketches the resulting image, and the fifth column indicates the relative phases and amplitudes of the involved waves in terms of so-called phasor diagrams (blue, undiffracted background light; red, diffracted object light; black, resulting wave at position of the object obtained by vector addition of the two other components).

x-direction. The amplitude and phase of the diffracted wave is indicated by the red arrow. For example, in the first experiment it is pointing into the $-x$-direction due to the phase shift of 180° with respect to the undiffracted wave. This arrow is also shorter, because the amplitude of the diffracted wave is smaller than that of the reference. The black arrow indicates amplitude and phase of the sum wave of the diffracted and undiffracted wave at the position of the spot image. It is obtained by vector addition of the first two arrows. Such diagrams are called *phasor diagrams*.

For each experiment, the produced contrast can be assessed by comparing the lengths of the blue and the black arrows. In the first experiment, they differ considerably, and in the second experiment there is no undiffracted wave in the image plane because it is blocked at the back focal plane of the objective. The small change in amplitude that occurs in the third experiment produces only very little contrast. Therefore, the transparent object is not perceptible on the background. In order to verify Zernike's assumption, we perform a final experiment that additionally demonstrates the principle of phase-contrast imaging. We insert a special mask into the back focal plane of the objective lens. This mask comprises a light-absorbing center which reduces the amplitude of the undiffracted background light to almost that of the diffracted light. This center is surrounded by a phase plate that adds an additional phase delay of 90° to the diffracted wave. This modification to the previous experiment yields a dark spot on a gray background in the image plane. This result verifies Zernike's assumption. The gray background is due to the reduction of the background light amplitude. The diffracted wave from the object traversing the outer region of the back focal plane of the objective lens is phase-delayed by additional 90°, resulting in a total phase delay of 180° compared to the background. The diffracted light now interferes destructively in the image plane with the dimmed undiffracted light, and creates a dark image of the phase object on a gray background. The trick to reduce the amplitude of the undiffracted light leads to a high image contrast.

In summary, the observable image contrast is dependent on the physical characteristics of the object, the illumination mode, and, in addition, selective manipulations of background and diffracted light. These two light components are spatially separated from each other and therefore independently accessible in the back focal plane of the objective lens.

2.6.2.2 Setup of a Phase-Contrast Microscope

The pinhole that was used for object illumination in the above experiments is now replaced by an annular mask. By this modification, more illumination light reaches the object, resulting in a brighter phase-contrast image. At the same time, the effective NA of the condenser is enlarged, which improves optical resolution. Of course, the mask at the back focal plane of the objective lens must be replaced by an appropriate complementary annular mask. This ring carries an absorbing layer and has a different thickness than the rest of the mask. It is called the *phase ring* . The undiffracted illumination light from the front focal plane of the condenser passes through this ring. The diffracted light from a specimen will pass through the central and outer regions of the mask, which is thicker – or thinner for the so-called negative phase contrast – than the absorbing ring such that the required total phase difference of about 180° is achieved.

The exact phase delay produced by the mask depends on the wavelength of the illumination light. Hence, the mask creates a 90° phase delay only for one specific wavelength. This is usually designed to be 550 nm or green – the color for which the human eye is most sensitive. A color filter in the illumination beam path may be inserted to select only this wavelength of the illumination light for imaging.

The image of the illumination ring at the objective's back focal plane must perfectly match the absorbing ring. Therefore, for different objective lenses with varying magnifications and NA, illumination rings of different diameters and widths must be used. These are located on a rotating turntable, which is positioned in the condenser's front focal plane. One position on the turntable is open for normal bright-field illumination. The coincidence of the illumination ring image and phase ring in the back focal plane of the objective can be checked and adjusted using a special auxiliary microscope, which is inserted instead of the eyepiece. It comprises two lenses by which the back focal plane can directly be observed. An alternative to the auxiliary microscope is the "Bertrand" lens, which can be switched into the beam path projecting an image of the back focal plane onto the principle image plane, and hence allows viewing the back focal plane using the existing eyepiece. In case the two ring structures are not perfectly matched, the lateral position of the illumination ring in the condenser turntable may be adjusted by two screws.

The back focal plane of the objective is located inside the objective housing. Therefore, objectives for phase-contrast imaging are specific products. Clearly, the absorbing phase ring also diminishes the image brightness when used in normal bright-field mode. Phase-contrast objectives are especially detrimental in epi-illumination techniques such as fluorescence microscopy.

2.6.2.3 Properties of Phase-Contrast Images

A phase-contrast image does not simply show a magnified view of the object. The generated contrast is due to the special manipulation of the phase of the undiffracted and the diffracted light. Therefore, its interpretation is not straightforward.

The phasor diagram in Figure 2.25 (experiment 5) shows that phase objects are seen dark on a bright background. However, if the phase delay due to the object becomes too large, for example, it exceeds 45° – in addition to the inherent 90° – the contrast is reduced again, and upon approaching 90° it almost vanishes. Then the phase relationships for bright-field observation of phase objects are recovered. Hence, the use of phase contrast is advisable only for thin objects producing a relatively small phase delay up to ~30°. This is the case for thin biological specimens such as cell monolayers or tissue sections. If the phase object produces only a small contrast because the phase shift is too small – or already too large – it may be improved by changing the refractive index of the mounting medium. The phase shift is proportional to the path length in the object and to the difference of the refractive indices of the object and the mounting medium. Therefore, the phase shift may be modified by choosing a suitable mounting medium.

The production of the phase contrast is based on the fact that the direct illumination light and the diffracted light are spatially separated from each other in the back focal plane of the objective lens. However, this separation is not complete owing to the finite width of the phase ring. Light diffracted from coarse specimen features may not be diffracted strongly enough to pass outside the phase ring, but rather traverse it. Likewise, some of the light diffracted by fine specimen features passes through the ring as well. The result of these effects is the bright halo that surrounds objects that appear dark, making the identification of object borders difficult.

2.6.3 Interference Contrast

The solution of the problem of phase objects is to translate the phase difference between the object and its surroundings into an amplitude difference. For this purpose, the phase of the diffracted wave is modified such that destructive interference between the diffracted and undiffracted waves occurs in the image plane. Zernike accomplished this by manipulating both wave components spatially separated in the back focal plane of the objective lens.

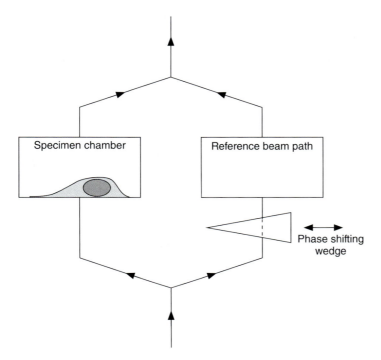

Figure 2.26 Principle of interference contrast.

There are further techniques that effectively reach the same goal. Actually, some of the so-called interference contrast techniques are even more versatile than phase contrast. The principal idea of interference contrast is shown in Figure 2.26: before encountering the object, the illuminating light is separated by a beam splitter into two coherent wave trains. One traverses the specimen chamber containing the object. The second wave – the *reference wave* – travels along an identical optical path, but without the object. Furthermore, a defined phase delay can be introduced in the reference beam path by inserting a glass wedge. After passing the specimen and the reference chamber, both wave trains are recombined and form the image by interference in the image plane.

This setup enables the microscopist to do the same as – or more than – what is achieved by Zernike's phase-contrast setup [13]. It is possible to manipulate the amplitudes and phases of the light diffracted by the object and the primary undiffracted light. The final image is created by the interference of three different waves: those diffracted by the object, the undiffracted light passing the object chamber, and the undiffracted reference waves.

The phase of the reference wave can be delayed in a defined manner by the glass wedge. As a result, the phase of the undiffracted waves can be shifted relative to the object-diffracted waves such that their final total phase difference is 180°, no matter how large the actual phase shift by the object is. To see how this is achieved, we consider first what happens without an object in the specimen chamber. To begin with, we move the glass wedge completely out of the beam path (Figure 2.27a). As both beams take identical optical paths, the two

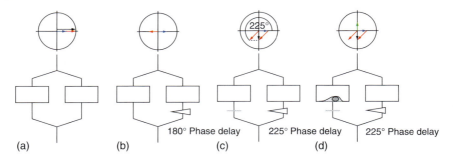

Figure 2.27 Interference contrast. The top sketches illustrate the phase relationships between the involved waves. The phasor of the background wave of the specimen chamber is shown in blue and the reference wave phasor in red. The phasor of the object-diffracted wave is represented by a green arrow, and the phasor of the sum background wave as a black one. (a) Constructive interference of the waves and (b) destructive interference due to 180° phase difference. (c) A phase shift of ~225° in the reference wave leads to an overall phase shift of ~270° in the sum of the two waves compared to the background wave in the specimen chamber. Note that the brightness was also slightly decreased in the specimen chamber. (d) A phase object is introduced in the specimen chamber. It diffracts a 90° phase-shifted object wave. The 270° phase-shifted background will interfere destructively with the object wave.

wave fronts will reach each point in the image plane simultaneously and interfere constructively.

This results in a homogeneous, bright background. Next, we move the glass wedge into the reference beam path and create a phase delay of 180° compared to the wave passing the specimen chamber (Figure 2.27b). Now, the two undiffracted waves interfere destructively everywhere in the image plane, which results in a completely dark background. The reference phase delay may also be adjusted such that the sum of the two waves in the image plane suffers effectively a 270° phase delay compared to the undiffracted wave from the specimen chamber (Figure 2.27c). When we finally add an object to the specimen chamber, the phases of the undiffracted reference and specimen chamber waves remain unchanged. The phase of the object-diffracted wave, however, shows the usual 90° phase delay. The phasor diagram shows that the phase difference between the sum of the undiffracted background waves and the object-diffracted wave is now 180° (Figure 2.27d). The undiffracted background light and the diffracted light will interfere destructively, resulting in a distinct phase contrast. We will see a dark object on a bright background. Even more, as we can choose the phase of the background sum wave with reference to the object wave arbitrarily, we can change from negative to positive contrast, or adjust the phase difference in such a way that we get a 180° phase difference also for objects whose diffraction wave phase is shifted by more or less than 90°. The physical separation of the two undiffracted waves allows free definition of their sum wave phase in relation to the object wave phase. Furthermore, by introducing a known phase shift into the background wave, the phase delay produced by the object may even be measured, and thus information gained on its refractive index or – if that is known – its physical thickness.

However, there is a price for this versatility. Interference microscopy requires a very special optical setup. There must be special elements to separate the specimen chamber and the reference beams, to recombine them again, and to adjust a defined phase delay between them. This requires very precise matching of the optical path length of the specimen and reference beam paths, which is not trivial to achieve. However, we will not focus here on the optical setups of interference microscopy, because it is actually another but related technique that is most important today for the observation of medical and biological samples.

2.6.4 Advanced Topic: Differential Interference Contrast

DIC microscopy is a very powerful method for contrast enhancement of phase objects. DIC can be used with transmitted or reflected light. For biomedical applications, epi-illumination DIC is of minor importance. How transmission DIC works and how it is realized can best be seen by considering the optical setup of a DIC microscope [14]. Its understanding requires a detailed knowledge of the wave properties of light such as interference, coherence, and polarization, as well as the function of polarizers and analyzers.

2.6.4.1 Optical Setup of a DIC Microscope

The optical setup of the DIC microscope is a further example of exploiting the optical conjugation of the front focal plane of the condenser and the back focal plane of the objective lens. The beam path is relatively complex (Figure 2.28). A polarizer is placed between the illumination field stop and the condenser. Let us assume that it produces light with a polarization direction of 45° (Figure 2.28b). This light encounters the *Wollaston prism* in the front focal plane of the condenser. A Wollaston prism is an optical device that splits the vertical and parallel polarization components of polarized light into two distinct waves that diverge by a small angle. We call these two coherent waves polarized in vertical and parallel direction (Figure 2.28a,c). In the object space, the two differently polarized wave fronts move parallel to each other, as the prism is located in the front focal plane of the condenser, but the two wave trains have a small lateral offset from each other (Figure 2.28b). The offset or *shear* is defined by the construction of the Wollaston prism, and chosen to be smaller than the optical resolution of the employed objective. The two spatially shifted or *sheared* beams are recombined again by a second Wollaston prism in the back focal plane of the objective. This Wollaston prism may be adjusted in its lateral position such that an additional phase delay Δ between the two beams with the perpendicular polarization to each other can be introduced. A lateral movement of the prism changes the optical path lengths inside the Wollaston prism for each of the two beams differently.

An analyzer is placed behind the second Wollaston prism. It transmits only a selected polarization component of the recombined beam. Altogether, the image is created by the interaction of four different wave components, namely, the diffracted and undiffracted waves of the two displaced wave fronts with their different polarization directions. The contrast generation depends on the OPDs between the two split wave fronts. It is relatively complex owing to the multiple parameters that modify these, which include the illumination light polarization,

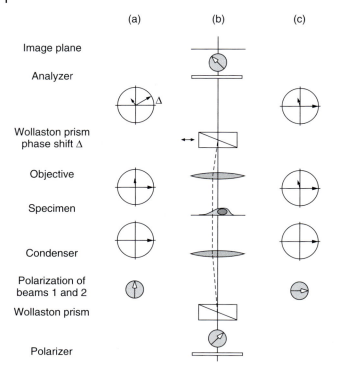

Figure 2.28 Optical setup of a DIC microscope. (a) Phases of undiffracted (long arrow) and diffracted waves (short arrow) of the offset beam (dashed lines in (b)) with vertical polarization, (b) sketch of beam path, and (c) phases of undiffracted (long arrow) and diffracted waves (short arrow) of the beam with parallel polarization. The circles with gray background indicate the *polarization* of the respective beam path – not the phase!

the lateral position of the second Wollaston prism (which determines the phase delay Δ between the two polarization directions), the setting of the analyzer, and finally the sample properties themselves.

An understanding of the contrast generation process is best achieved by discussing an example as given in Figure 2.29. The figure also directly relates to Figure 2.28. The upper panel of Figure 2.29 shows the specimen chamber picturing a cell with nucleus as an example of a phase object. This chamber is traversed by the two laterally shifted, perpendicularly polarized wave trains indicated by the two sheared vertical lines (dashed and full). To understand the contrast generation, we evaluate the wave components at four different positions in the sample (Figure 2.29a–d). The polarization state of the sum wave in the image at the indicated four positions is shown for two different values of the phase shift Δ (0° and 30°) in the middle and lower panels of Figure 2.29.

We proceed from Figure 2.29a–d, and assume that the analyzer above the second Wollaston prism is crossed to the polarizer in front of the first Wollaston prism (as indicated in Figure 2.28). At background positions, there are no diffracted waves from the object (Figure 2.29a). Hence, in the image, the background around the object is dark if the second Wollaston prism is adjusted

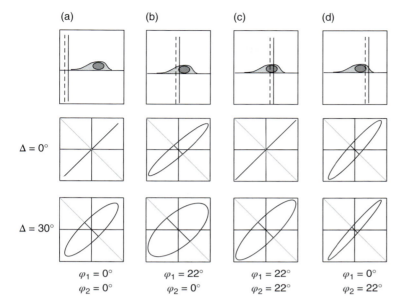

Figure 2.29 (a–d) Contrast generation, polarization, and phase relationships in DIC microscopy. For details, see the text.

such that the optical path lengths of the undiffracted background waves for both polarization directions are identical. For $\Delta = 0°$, this is the case. The polarization for this value of Δ is shown in the middle panel of Figure 2.29. Polarization and phase of the two shifted background waves are not altered at this sample point. Therefore, the background light is completely blocked by the analyzer, because it is crossed with the polarizer. The diagonal gray lines in the middle and lower panels of Figure 2.29 indicate the direction of the analyzer. A phase shift between the two polarization directions, which can be introduced by a lateral move of the upper Wollaston prism, results in an elliptical polarization of the combined background waves. The lower panel in the figure shows exemplarily the elliptical polarization, when $\Delta = 30°$. The analyzer lets pass only the polarization component along its orientation. The size of this component is indicated in the middle and lower panels of Figure 2.29 by the thick, black line in diagonal direction. Thus, we can modify the polarization state of the background light by moving the Wollaston prism in the back focal plane of the objective.

Let us move it back to $\Delta = 0°$. What happens at an object position where the parallel-polarized beam encounters the edge of the phase object, that is, the cell nucleus, but not the vertical-polarized beam (Figure 2.29b)? A wave with parallel polarization is diffracted. The light is diffracted by a phase object, which means that it gets an intrinsic phase shift of 90° in relation to the undiffracted wave plus a small additional phase delay φ_1, which is due to the difference in optical path length caused by the higher optical density of the nucleus. As an example, we assume $\varphi_1 = 22°$, but it could be any value. The crucial point now is that the diffracted wave component with its phase shift of (90° + 22°) changes the polarization of the sum wave that meets the analyzer. It is no longer completely linearly polarized as in specimen regions without the object, but elliptically polarized. This has the important consequence that the amplitude of the sum wave when passing the analyzer is not zero anymore, which results in some light at this position in the image. A phase difference is turned into a visible brightness difference.

In regions of the object where the sheared beam with the vertical polarization encounters the same object feature – the nucleus – a second vertically polarized object-diffracted wave with a phase retardation $90° + \varphi_2 \approx 90° + \varphi_1$ is created, which reduces the ellipticity of the sum wave back to a linear polarization because the parallel- and vertical-polarized waves have again identical phases and amplitudes (Figure 2.29c). The parallel- and the vertical-polarized waves with zero phase shift together form a linearly polarized wave that is completely blocked by the analyzer, leading to zero intensity at this image position. Thus, a contrast was just created at the edge of the object.

At the other side of the nucleus (Figure 2.29d), only the wave with vertical polarization interacts with the object whereas the parallel-polarized beam passes already undiffracted. Therefore, again a certain ellipticity results and the analyzer lets pass some light to the corresponding image point.

In summary, the final contrast in a DIC image is determined by the OPD at laterally slightly displaced positions along a certain direction in object space. Therefore, it is named "differential." The direction along which the path length difference occurs is defined by the orientation of the first Wollaston prism. A difference in the optical path lengths of adjacent object positions results in a pronounced contrast. Even more importantly, the contrast can be adjusted by the choice of Δ in such a way that a positive contrast occurs at one edge of the object and a negative contrast at another edge. This is demonstrated in the lower panel of Figure 2.29, which shows the elliptical polarizations that occur if the vertical-polarized waves acquire an additional phase shift of 30°. Then, the left boundary of the nucleus is brighter than the background, the nuclear interior, and, notably, the opposing boundary of the nucleus. In this manner, a visual impression of a 3D object, which is illuminated from one side and casts a shadow on the other side, is created. This asymmetrical contrast is typical for DIC.

DIC is a complex technique and not easy to realize. The second Wollaston prism must be located in the back focal plane of the objective lens. As it has a considerable thickness, this calls for special objectives. As prisms are incorporated in the objectives, they are unsuitable for other tasks. To overcome this problem, specifically constructed prisms were introduced that place the plane of beam recombination outside the prism itself [15]. These so-called Nomarski prisms may therefore be located above the focal plane, which can be easily accomplished. In addition, they are removable. This modification of classical DIC is known as *Nomarski DIC*, and has become one of the most widespread interference contrast techniques in the biosciences.

2.6.4.2 Interpretation of DIC Images

The strength of DIC is defined by the lateral amount of the shear between the two beams, the respective orientations of the prisms, polarizer, and analyzer, and the sample features. DIC has the advantage that the way of contrast generation and image appearance can widely be adjusted. However, this also leads to ambiguities in image interpretation.

The main feature of DIC is its excellent edge contrast. This results in a strong depth dependence of the effect. DIC varies very sensitively upon the axial movements of the specimen and therefore allows acquiring very thin optical sections. It is well suited for thick specimens. The pronounced relief structure shown by phase objects is rather a side effect of the technique. One edge can be bright, and the opposing edge can be dark. The object appears to be illuminated from one side and cast shadows on the opposite side. Thereby the images have a 3D appearance. However, all contrast effects are solely due to differences in optical path length. They may be related to the object topology, but may also be caused by refractive index differences, while the true object thickness is constant. The true specimen topology is not visualized. Furthermore, the image of a circular object is not rotationally symmetrical because we see OPDs only along the direction of the beam shear. It is helpful to use a rotating stage in order to understand the full structure of the object. Notably, it is also possible to turn the bright–dark contrast into a color contrast, which makes it even more difficult to interpret DIC images.

2.6.4.3 Comparison between DIC and Phase Contrast

The applications of phase-contrast and transmission DIC overlap. Both techniques convert phase differences between the object-diffracted light or between light traversing different regions of the object into perceivable intensity changes. However, there are several important differences between both techniques.

In phase contrast, the object is imaged in a rotationally symmetrical way. DIC shows optical path length *gradients* that occur along one spatial orientation only. DIC images show a typical 3D surface-shading appearance, which is an artifact of contrast generation. DIC images display a higher resolution compared to phase-contrast images, not least because they do not show the halo encircling object features with large phase differences to the surroundings. Contrast in DIC images is pronounced for small objects, and it works with thin and thick objects. Importantly, DIC shows excellent axial resolution. Object features in planes other than the focal plane do not appear in the image, which allows one to make optical sections. This is in clear contrast to phase contrast, which is not very effective for thick objects. Phase contrast works best with very thin objects. The depth of field is very large, and object features in planes outside the focal plane are quite disturbing. Problems occur when the path difference becomes too large, particularly if it exceeds the limit of 30°, which is the upper limit for a good phase contrast. In summary, DIC is technically more demanding but has a wider range of applications than phase contrast.

2.7 Summary

The key points of this chapter are the following:

- The basic imaging elements of a microscope are the objective and tube lenses. Typically they form an infinity-corrected setup.
- Magnification steps may be arranged sequentially.
- A microscope contains two complementary beam paths, the illumination path and the imaging beam path. Each comprises specific components and a different set of conjugate planes.
- Köhler illumination provides even illumination, separates the illumination field from the illumination aperture, and creates conjugate optical planes in the front focal plane of the condenser and objective.
- The back focal plane of the objective contains the diffraction pattern of the object. It corresponds to the truncated Fourier transform of the object function.
- The image is the convolution of the object and the PSF.
- A lens is a low-pass filter for object structures.
- The lateral optical resolution is given by the Rayleigh criterion $0.61 \lambda / n \sin \alpha$.
- Coherent and incoherent objects have different resolution limits.
- The optimal magnification is twice the ratio of the detector size and the optical resolution.
- Objectives should be telecentric.
- Objectives are of varying types and correction levels. They always exhibit residual aberrations.
- Image brightness is typically proportional to the square of the NA and inversely proportional to the square of the magnification.
- Image brightness in fluorescence microscopy increases with the fourth power of NA.
- The objective must be carefully selected for a given application. Only a proper objective will give satisfactory imaging results.
- The most important contrast techniques for biomedical applications are phase contrast and DIC.
- Optical contrast techniques are based on the selective manipulation of background and object-diffracted light.

Acknowledgments

The author thanks Dr Rolf Käthner, Dr Thorsten Kues, Manfred Matthä, and Dr Michael Zölffel from Carl Zeiss Microscopy GmbH, Göttingen, for helpful discussions and for critical reading of the text and numerous constructive suggestions. In addition, he thanks Manfred Matthä for providing Figures 2.17 and 2.18a and for the data illustrated in Figure 2.18b. Finally, he thanks Lisa Büttner, Jana Bürgers, Alexander Harder, Sahand Memarhosseini, Nicolai Pechstein, and Prof. Dr Rainer Heintzmann for critical comments on this text.

References

1 Köhler, A. (1893) Ein neues Beleuchtungsverfahren für mikrophotographische Zwecke. *Z. Wiss. Mikrosk. Mikrosk. Tech.*, **10**, 433–440.
2 Abbe, E. (1873) Beiträge zur Theorie des Mikroskops und der mikroskopischen Wahrnehmung. *Arch. Mikrosk. Anat.*, **9**, 413–468.
3 Köhler, H. (1981/1982) A modern presentation of Abbe's theory of image formation in the microscope, part I. *Zeiss Inf.*, **26** (93), 36–39.
4 Hecht, E. and Zajac, A. (2003) *Optics*, Addison-Wesley, San Francisco, CA.
5 Butz, T. (2005) *Fourier Transformation for Pedestrians*, 1st edn, Springer, Berlin, Heidelberg.
6 Born, M. and Wolf, E. (1999) *Principles of Optics: Electromagnetic Theory of Propagation, Interference and Diffraction of Light*, 7th edn, Cambridge University Press, Cambridge.
7 Sheppard, G.J.R. and Matthews, H.J. (1987) Imaging in high-aperture optical systems. *J. Opt. Soc. Am. A*, **4** (8), 1354–1360.
8 Michel, K. (1950) *Die Grundlagen der Theorie des Mikroskops*, Wissenschaftliche Verlagsgesellschaft M.B.H., Stuttgart.
9 von Bieren, K. (1971) Lens design for optical Fourier transform systems. *Appl. Opt.*, **10**, 2739–2742.
10 Hell, S., Reiner, G., Cremer, C., and Stelzer, E.H.K. (1993) Aberrations in confocal fluorescence microscopy induced by mismatches in refractive index. *J. Microsc.*, **169**, 391–405.
11 Siedentopf, H. and Zsigmondy, R. (1902) Über Sichtbarmachung und Größenbestimmung ultramikoskopischer Teilchen, mit besonderer Anwendung auf Goldrubingläser. *Ann. Phys.*, **315**, 1–39.
12 Beyer, H. (1965) *Theorie und Praxis des Phasenkontrast-verfahrens*, Akademische Verlagsgesellschaft Geest und Portig KG, Frankfurt am Main, Leipzig.
13 Beyer, H. (1974) *Theorie und Praxis der Interferenzmikroskopie*, Akademische Verlagsgesellschaft Geest und Portig KG, Frankfurt am Main, Leipzig.
14 Lang, W. (1968–1970) Differential-Interferenzkontrast-Mikroskopie nach Nomarski. I. Grundlagen und experimentelle Ausführung. *Zeiss Inf.*, **16**, 114–120 (1968); II. Entstehung des Interferenzbildes. *Zeiss Inf.* (1969) **17**, 12–16; III. Vergleich Phasenkontrastverfahren und Anwendungen. *Zeiss Inf.* (1970) **18**, 88–93.
15 Allen, R., David, G., and Nomarski, G. (1969) The Zeiss-Nomarski differential interference equipment for transmitted-light microscopy. *Z. Wiss. Mikrosk. Mikrosk. Tech.*, **69** (4), 193–221.

3

Fluorescence Microscopy

Jurek W. Dobrucki[1] and Ulrich Kubitscheck[2]

[1] *Jagiellonian University, Faculty of Biochemistry, Biophysics and Biotechnology, Department of Cell Biophysics of Cell Biophysics, ul Gronostajowa 7, Krakow 30-387, Poland*
[2] *Rheinische Friedrich-Wilhelms-Universität Bonn, Institut für Physikalische & Theoretische Chemie, Wegelerstr. 12, Bonn 53115, Germany*

3.1 Contrast in Optical Microscopy

The human eye requires contrast to perceive details of objects. Several ingenious methods of improving the contrast of microscopy images have been designed, and each of them opened new applications of optical microscopy in biology. The simplest and very effective contrasting method is "dark field." It exploits the scattering of light on small particles that differ from their environment in refractive index – the phenomenon known in physics as the *Tyndall effect*. Color staining of a preparation can provide even more information. A fixed and largely featureless preparation of tissue may reveal a lot of its structure if stained with a proper dye – a substance that recognizes specifically some tissue or cellular structure and absorbs light of a specific wavelength. This absorption results in a perception of color. The pioneers of histochemical staining of biological samples were Camillo Golgi, an Italian physician, and Santiago Ramon y Cajal, a Spanish pathologist. They received the 1906 Nobel Prize for Physiology or Medicine. A revolutionary step in the development of optical microscopy was the introduction of phase contrast proposed by the Dutch scientist Frits Zernike. This was such an important discovery that Zernike was awarded the 1953 Nobel Prize in Physics. The ability to observe fine subcellular details in an unstained specimen opened new avenues of research in biology. Light that passes through a specimen may change polarization characteristics. This is a phenomenon exploited in polarization microscopy, a technique that has also found numerous applications in materials science. The next important step in creating contrast in optical microscopy came with an invention of differential interference contrast by Jerzy (Georges) Nomarski, a Polish scientist who had to leave Poland after World War II and worked in France. In his technique, the light incident on the sample is split into two closely spaced beams of polarized light, where the planes of polarization are perpendicular to each other. Interference of the two beams after they have passed through the specimen results in excellent contrast, which creates an impression of three-dimensionality of the object.

Fluorescence Microscopy: From Principles to Biological Applications, Second Edition.
Edited by Ulrich Kubitscheck.
© 2017 Wiley-VCH Verlag GmbH & Co. KGaA. Published 2017 by Wiley-VCH Verlag GmbH & Co. KGaA.

However, by far the most popular contrasting technique now is fluorescence. It requires the use of so-called fluorochromes or fluorophores, which absorb light in a specific wavelength range and re-emit photons of lower energy, that is, shifted to a longer wavelength. Today, a very large number of different dyes with absorption from the ultraviolet (UV) to the near-infrared (NIR) region are available, and more fluorophores with new properties are still being developed (see Chapter 4). The principal advantages of this approach are very high contrast, sensitivity, specificity, and selectivity. The first dyes used in fluorescence microscopy were not made specifically for research, but were taken from a collection of stains used for coloring fabrics. The use of fluorescently stained antibodies and the introduction of a variety of fluorescent heterocyclic probes synthesized for specific biological applications brought about an unprecedented growth in biological applications of fluorescence microscopy. Introduction of fluorescent proteins sparked a new revolution in microscopy, contributed to the development of a plethora of new microscopy techniques, and enabled the recent enormous growth of optical microscopy and new developments in cell biology. A Japanese scientist, Osamu Shimomura, and two American scientists, Martin Chalfie and Robert Y. Tsien, were awarded the Nobel Prize for Chemistry in 2008 for "the discovery and development of the green fluorescent protein, 'GFP', in biological research." The next milestone in the unstoppable progress of modern fluorescence microscopy was reached when the super-resolution (or "sub-diffraction" resolution) imaging techniques became available. Their potential and importance in biological studies were soon recognized, and a climax of these developments was a 2014 Nobel Prize in Chemistry awarded to Eric Betzig, Stefan W. Hell, and William E. Moerner for "the development of super-resolved fluorescence microscopy."

The use of fluorophores requires several critical modifications in the illumination and imaging beam paths of an optical microscope. Fluorescence excitation requires specific light sources, and their emission is often recorded with advanced electronic light detection devices. However, fluorescence exhibits important limitations: fluorescent dyes have a limited stability, that is, they photobleach, and may induce phototoxic effects. This requires special precautions to be taken. We discuss, first, the physical basis of fluorescence, and then present the major technical and methodological questions involved in this chapter.

3.2 Physical Foundations of Fluorescence

3.2.1 What is Fluorescence?

Fluorophores are molecules or nanocrystals with a special property. They absorb photons of a certain energy and emit photons of a lower energy. Considering the electromagnetic spectrum, this means that the color of the emitted light is red-shifted with respect to the absorbed light. As explained previously, this effect can elegantly be used to generate a superb contrast in microscopy. So, how does this effect work in detail? We focus here on fluorescent molecules because these are by far the most often used fluorophores in microscopy.

The absorbing fluorescent molecule comprises a set of atoms that share a number of delocalized electrons in molecular orbitals with specific spatial charge distributions. We may say that the negatively charged electrons represent the *glue* that keeps the positively charged atomic nuclei close together to stabilize the molecule. The electronic ground state is typically a singlet state, designated as S_0. The spins of both electrons in this electronic orbital are antiparallel to each other. Upon absorption of light, that is, a photon, one of the electrons is lifted into another orbital, typically into the first excited singlet state, S_1. By the excitation from the ground state the spin of the excited electron is not flipped, therefore the spin of the excited electron is still antiparallel to that of the single electron left behind in its original molecular orbital. Energy conservation requires that the energy of the absorbed photon equals the energy difference between the ground state and the excited state. In the excited state, the overall charge distribution of the electrons around the atomic nuclei is altered compared to the ground state. A key point is that the electrons change their spatial distribution almost instantaneously – within femtoseconds – whereas the much heavier atomic nuclei move more slowly and take significantly longer – picoseconds – to relax to the new charge distribution of the electrons. The excited electron is relocated into a *higher* orbital with a greater distance from the nuclei than before, which results in a relaxation of the molecular structure that usually initiates a vibration of the complete molecule about the new equilibrium positions of the atoms. However, the vibrational energy rapidly dissipates upon collisions with the surrounding solvent molecules as heat, and the molecule decays into the vibrational ground state of the electronically excited state S_1 (Figure 3.1). However, since

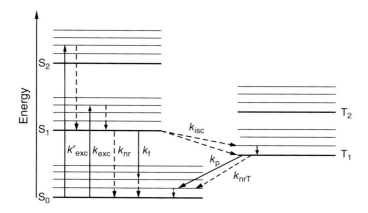

Figure 3.1 Simplified Jablonski diagram or molecular term scheme. Radiative and nonradiative transitions are indicated by full and dashed lines, respectively. "Energy" refers to the total potential and kinetic energy of electrons and atomic nuclei. For a discussion of the scheme, see text. Fluorescence studies on single molecules have shown that sometimes there are further nonfluorescent molecular configurations with lifetimes in the range of milliseconds to seconds. After leaving these long-lived states, the molecules may return to the normal fluorescence process. This effect is called "reversible" photobleaching, and, if occurring repeatedly, may result in molecular "blinking" (see Chapter 8). A further pathway to leave S_1 or T_1 is by photobleaching. Photobleaching is the irreversible destruction of the molecule, which is often accompanied by the production of reactive and toxic molecular species.

the energy of the molecule is still higher than in the electronic ground state, the S_1 state is not stable for a long time, but the molecule will return to the ground state S_0. The time spent in S_1 is designated as excited state lifetime τ. Highly fluorescent molecules return to the ground state upon emitting the excess energy as light – emission of a *fluorescence photon*. This again changes the electronic configuration around the nuclei and therefore this decay to the electronic ground state ends in a vibrational excited state. From here, the molecule again rapidly returns to the vibrational ground state of S_0 by dissipating the vibrational energy to the surroundings. Now the fluorescence cycle is completed.

It should be noted that the decay from S_1 to S_0 may also happen as a result of further collisions with solvent molecules, by which the electronic excess energy is lost. In this case, no fluorescence photon is emitted, and therefore this process is designated as *nonradiative decay*. Finally, a third decay pathway is relaxation via *intersystem crossing*. It exists because the excited electron has a certain, but usually very small, probability to flip its spin and thus to reach another state with a slightly lower energy, the so-called triplet state, designated as T_1. In order to return from T_1 back into the electronic ground state S_0, it needs to flip its spin a second time, because in S_0 it needs the antiparallel spin to the second electron to share the same molecular orbital. Again, this is an improbable and therefore slow process occurring on time scales of microseconds up to even hours. This may happen upon emission of a photon, a process that is designated as *phosphorescence*. However, often the radiative decay cannot compete with nonradiative pathways. A summary of the excitation and various decay pathways is usually given in form of a so-called Jablonski diagram (Figure 3.1).

It should be noted that even the longest absorption wavelength, corresponding to excitation from the lowest vibrational level of S_0 to the lowest vibrational level of S_1, and the shortest wavelength emission, corresponding to de-excitation between the same levels, occur at different wavelengths because reorientation of neighboring solvent molecules within the lifetime of the excited state lowers the energy of the excited state and raises that of the ground state, thereby decreasing the energy of the emitted photon. The process of the reorientation of the solvent molecules is designated as *solvent relaxation*. Altogether, the difference between the peak wavelengths of absorption and emission is designated as *Stokes shift*. It is due to solvent relaxation and to the loss of the vibrational energies in S_1 and S_0.

> **Box 3.1 Kinetics and Equilibrium Properties of Fluorescence Emission**
>
> A very powerful approach to the kinetics of fluorescence emission is to neglect the details of the vibrational states and just to analyze the transitions between the various electronic molecular states. These transitions can be formulated in the language of chemical kinetics. For an ensemble of molecules, we can describe the populations of molecules in S_0, S_1, and T_1 as $[S_0]$, $[S_1]$, and $[T_1]$. For simplicity, we consider only transitions from S_0 to S_1 and neglect excitation to higher states, for example, S_2. If we use the rates for the transitions between the various states as denoted in Figure 3.1, we can describe the time-dependent populations of the

states as follows:

$$\frac{d[S_0]}{dt} = -k_{exc}[S_0] + (k_f + k_{nr})[S_1] + k_p[T_1] + k_{nrT}[T_1] \quad (3.1a)$$

$$\frac{d[S_1]}{dt} = k_{exc}[S_0] - (k_f + k_{nr} + k_{isc})[S_1] \quad (3.1b)$$

where $k_{exc} = I_{exc}\sigma_a$, I_{exc} denotes the exciting photon flux, and σ_a is the absorption cross section of the fluorophore at the respective wavelength. This set of equations can be used to calculate the steady-state distribution of the ensemble by setting all derivatives to zero. Also, the same equations can be used to predict the time dependent decay of $[S_1]$ after a short excitation pulse, causing a certain population of S_1, $[S_1]_0$. In that case, integration of Eq. (3b) over time yields

$$[S_1(t)] = [S_1]_0 e^{-(k_f+k_{nr}+k_{isc})\,t} \quad (3.2)$$

and we see that the return to the ground state is described by a mono-exponential function $e^{-t/\tau}$, and the excited state lifetime τ is given as

$$\tau = \frac{1}{k_f + k_{nr} + k_{isc}} \quad (3.3)$$

The relative probability of a specific transition is given by that rate divided by the sum of all rates describing the decay from that state. For example, the probability of emitting a fluorescence photon after excitation into S_1, also called the *quantum efficiency* Φ, is given by

$$\Phi = \frac{k_f}{k_f + k_{nr} + k_{isc}} \quad (3.4)$$

For one of the most efficient fluorophores, phycoerythrin, this value is nearly unity (0.98). Finally, the amount of emitted fluorescence photons per second is given by the fluorescence rate k_f times the population $[S_1]$ of molecules in state S_1.

3.2.2 Fluorescence Excitation and Emission Spectra

The energy of the exciting photon must be high enough so that the molecule can cross the energy gap between ground state S_0 and excited state S_1. This energy gap is quite broad, however, because the molecule usually ends in a vibrational excited state of S_1. The energy gap is the reason why only absorption of photons with a certain minimum energy will lead to a molecular excitation. In a molecular ensemble, numerous effects further slightly alter the spectral transition energies, that is, additional translational and rotational energy levels and different molecular surroundings. Therefore, at room temperature molecules absorb and emit light within a continuous spectrum of energies rather than at sharp spectral lines. The energy or wavelength dependence of the fluorescence excitation is usually plotted in the form of a fluorescence excitation spectrum. This is a plot of the intensity of the emitted fluorescence measured at a specific emission wavelength as a function of the excitation wavelength. The fluorescence emission spectrum, on the other hand, shows the fluorescence intensity as a function of the emission wavelength when exciting a sample at a selected wavelength in the absorption

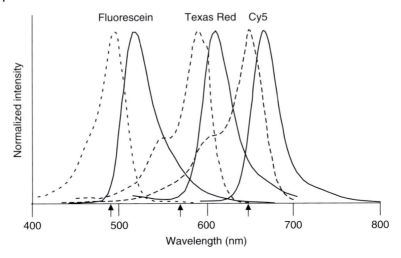

Figure 3.2 Fluorescence excitation (dashed lines) and emission (full lines) spectra of heterocyclic compounds fluorescein, Texas Red, and Cy5. The arrows indicate the wavelengths of laser lines that are often used in microscopy to excite these dyes.

band. For single molecules, both excitation and emission spectra represent probability densities of excitation and emission.

The fluorescence excitation and emission spectra for three typical dyes, fluorescein, Texas Red, and Cy5, are shown in Figure 3.2. There is one characteristic maximum in each excitation spectrum corresponding to the energy gap between the vibrational ground state of S_0 and the vibronic level of S_1 to which the transition is most probable. All three excitation spectra show a shoulder on the blue side, which is due to molecular transitions to a higher vibrational state of S_1. The fluorescence emission spectra are usually approximate mirror images of the excitation spectra because the decay occurs from the vibrational ground state of S_1 to levels of S_0 having overall a similar structure as in S_1. The magnitude of the respective Stokes shifts can be deduced directly from the spectra.

3.3 Features of Fluorescence Microscopy

3.3.1 Image Contrast

In full daylight, it is almost impossible to spot a firefly against the background of grasses, bushes, and trees at the edge of a forest. At night, however, the same firefly glows and thus becomes visible to an observer, while the plants and numerous other insects that might be active in the same area are completely undetectable. This example parallels, to some degree, the observation of a selected molecule under a fluorescence microscope. To be useful for an investigator, optical microscopy requires high *contrast*. This condition is fulfilled well by fluorescence microscopy, where selected molecules are incited to emit light and stand out against a black background (Box 3.2), which embraces a countless number of other molecules in the investigated cell (Figure 3.3).

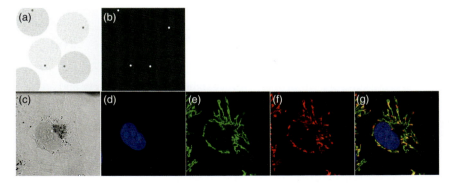

Figure 3.3 Importance of image contrast in microscopy. (a) Four dark-gray dots are almost undetectable against a light-gray background. (b) Four white dots, in the same positions as in (a), are easily discerned against a black background. (c) A transmitted light image of a live cell in culture – almost no internal features can be distinguished when no contrasting technique is used. (d) Nuclear DNA stained with DAPI; DAPI is a heterocyclic molecule with affinity for DNA. The major mode of binding to DNA is thought to be dependent on positioning a DAPI molecule in a minor groove of a double helix. (e,f) Low (green, e) and high (red, f) potential mitochondria fluorescently labeled with JC-1. JC-1 is a carbocyanine dye, which readily crosses plasma and mitochondrial membranes and is accumulated inside active mitochondria. JC-1 monomers emit green fluorescence. At high concentrations (above 0.1 μM in solution), JC-1 forms the so-called J-aggregates emitting red luminescence. A gradient of electric potential is responsible for the passage of JC-1 molecules across the membranes into active mitochondria. Mitochondria characterized by a high membrane potential accumulate JC-1 and the dye reaches the concentration that is sufficiently high to form J-aggregates. (g) Fluorescence signals from mitochondria and nucleus overlaid in one image. (Images (c)–(g) courtesy of Dr A. Waligórska, Jagiellonian University.)

Box 3.2 Discovery of Fluorescence

Fluorescence was first observed by an English mathematician and astronomer, Sir John Frederick William Herschel, probably around 1825. He observed blue light emitted from the surface of a solution of quinine. Sir John was a man of many talents. He made important contributions to mathematics, astronomy, and chemistry, authored inventions pertinent to early photography, published on the photographic process and was the first one to use the fundamental terms of analog photography – "a negative" and "a positive." He is thought to have influenced Charles Darwin. In 1845, Herschel described this observation in a letter to The Royal Society in London:

> A certain variety of fluor spar, of a green colour, from Alston Moor, is well known to mineralogists by its curious property of exhibiting a superficial colour, differing much from its transmitted tint, being a fine blue of a peculiar and delicate aspect like the bloom on a plumh…

(Continued)

> **Box 3.2 (Continued)**
>
> He found a similar property in a solution of quinine sulfate:
>
> > …Though perfectly transparent and colourless when held between the eye and the light, or a white object, it yet exhibits in certain aspects, and under certain incidences of the light, an extremely vivid and beautiful celestial blue color, which, from the circumstances of its occurrence, would seem to originate in those strata which the light first penetrates in entering the liquid… [1].
>
> In the next report, he referred to this phenomenon as *epipolic dispersion* of light [2]. Herschel envisaged the phenomenon he saw in fluorspar and quinine solution as a type of dispersion of light of a selected color.
>
> In 1846, Sir David Brewster, a Scottish physicist, mathematician, and astronomer, well known for his contributions to optics, and numerous inventions, used the term *internal dispersion* in relation to this phenomenon.
>
> In 1852, Sir John Gabriel Stokes, born in Ireland, a Cambridge University graduate and professor, published a 100 page long treatise "On the Change of Refrangibility of Light" [3] about his findings related to the phenomenon described by Sir John Herschel. He refers to it as "dispersive reflection." The text includes a short footnote, in which Stokes said:
>
> > I confess I do not like this term. I am almost inclined to coin a word, and call the appearance *fluorescence*, from fluor-spar, as the analogous term opalescence is derived from the name of a mineral.
>
> It was known at that time that when opal was held against light, it appeared yellowish red; however, when viewed from the side, it appeared bluish. Thus, the phenomenon was similar to the one observed by Herschel in fluorspar. We now know that opalescence is a phenomenon related to light scattering, while the phenomenon seen in a solution of quinine was not exactly a dispersion of light of a selected wavelength, but rather the emission of a light of a different color. Sir J. G. Stokes is remembered for his important contributions to physics, chemistry, and engineering. In fluorescence, the distance, in wavelength, between the maximum of excitation and emission is known as *Stokes shift*.
>
> The physical and molecular basis for the theory of fluorescence emission was formulated by a Polish physicist Aleksander Jabłoński [4]. Hence, the energy diagram that describes the process of excitation and emission of fluorescence is called the *Jabłoński diagram* (Figure 3.1). Aleksander Jabłoński was a gifted musician; during his PhD years, he played the first violin at the Warsaw Opera. In postwar years, he organized a Physics Department at the University of Toruń in Poland and worked there on various problems of fluorescence.

The range of applications of fluorescence microscopy was originally underestimated. Fluorescence microscopy was seen as just a method of obtaining nice, colorful images of selected structures in tissues and cells. The most attractive feature of contemporary fluorescence microscopy and many modern imaging and analytical techniques, which grew out of the original idea, is the ability to image and study quantitatively not only the structure but also the function, that is, physiology, of intact cells *in vitro* and *in situ*.

Figure 3.3 illustrates the fact that modern fluorescence microscopy should not be perceived merely as a technique of showing enlarged images of cells and subcellular structures but as a way of studying cellular functions. Figure 3.3e–g shows some selected structures within a cell (mitochondria and the cell nucleus) and demonstrate the ability to selectively convert a physiological (functional) parameter – in this case, mitochondrial potential – into a specific fluorescence signal.

3.3.2 Specificity of Fluorescence Labeling

Let us expand the analogy of the firefly. Although one does not see any features of the firefly – its size, shape, or color remain unknown – the informed observer knows that the tiny light speckle that reveals the position of a male firefly. It is most likely a European common glowworm, *Lampyris noctiluca*. No other insects are expected to emit light while flying in this region at this time of the year. Although there may be hundreds of insects hovering in this place, they blend into the black background and therefore are not visible. Thus, a tiny light label is characteristic of a firefly: it is specific. *Specificity* of fluorescence labeling of selected molecules of interest as well as specificity of translating some selected physiological and functional parameters such as membrane potential or enzyme activity into specific signals in a cell is another important advantage of fluorescence microscopy (Figures 3.3 and 3.4). Hundreds of fluorescent molecules, both small heterocyclic molecules and proteins, are available to be used as specific labels and tags in fixed and live cells. Also, a host of methods have been optimized for attaching a fluorescent tag to a molecule of interest; they are discussed in Chapter 4.

Specificity of labeling is an important advantage of fluorescence microscopy. However, this specificity is never ideal and should not be taken for granted. Let us consider the example given in Figure 3.4. DRAQ5 binds DNA fairly specifically; staining of RNA is negligible owing to either a low binding constant or a low fluorescence intensity of the DRAQ5 complex with RNA, as long as the concentration of DRAQ5 is low. At high DRAQ5 concentrations, when all binding sites for the dye on DNA are saturated, some binding to RNA can be detected. A similar phenomenon is observed with most, if not all, fluorescent labels that exhibit affinity for DNA. Thus, experimental conditions, especially the ratio between the available dye and the available binding sites, have to be optimized in order to fully exploit the advantages of the DNA-labeling techniques. As for Col-F, this small molecule binds to two major components of the extracellular matrix; in this respect, it is less specific than antibodies directed against selected epitopes on collagen or elastin in a selected species. The advantage of Col-F is the simplicity of labeling, deep tissue penetration, and low level of nonspecific staining. The advantage of immunofluorescence in the detection of collagen or elastin is its very high specificity, but it comes with limitations: shallow

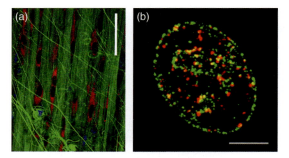

Figure 3.4 Specificity of fluorescence labeling. (a) Three fluorescent probes were used to stain different classes of molecules and structures in a fragment of live connective tissue – DNA in cell nuclei (DRAQ5, a deep-red emitting dye, shown here as blue), fibers of the extracellular matrix (Col-F, green), and active mitochondria (tetramethylrhodamine abbreviated as TMRE, red). Scale bar, 50 µm. DRAQ5 is an anthracycline derivative that has high affinity for DNA. The dye readily crosses plasma and nuclear membranes and binds to DNA in live cells. It is optimally excited by red light, emitting in deep red. Here, it is shown as blue to be distinguished from TMRE. Col-F is a dye that binds to collagen and elastin fibers (excitation – blue, emission – green). TMRE enters cells and is accumulated by active mitochondria (excitation – green, emission – red). (b) A DNA precursor analog, ethylenedeoxyuridine (EdU), was incorporated into nascent DNA during a short period of S-phase in the division cycle, in cells exposed to an inhibitor of topoisomerase type I, camptothecin. Cells were fixed and the EdU was labeled fluorescently using "click chemistry." Newly synthesized DNA is thus marked in green. In the same cell, γH2AX, a phosphorylated form of histone H2AX, which is considered a marker of DNA double-strand breaks (DSBs), was labeled using a specific antibody. Thus, DNA regions with DSBs are immunolabeled in red. When images of numerous foci representing DNA replication and DNA damage are overlaid, it becomes apparent that most damage occurred in replicating DNA (replication and γH2AX foci show large yellow areas of overlapping green and red signals). Scale bar, 5 µm. (Biela et al. [23] and Berniak et al. [24].)

penetration into tissue and some nonspecific binding of the fluorescently labeled secondary antibody used in the labeling method. An experimenter has to make a choice between the advantages and limitations of the low molecular weight label and the immunofluorescence approach. Similarly, Figure 3.4b shows two labeling methods that differ to some degree in their specificity. Incorporating precursors ethylenedeoxyuridine (EdU) into nascent DNA and subsequently labeling the incorporated molecules (the click reaction) leads to a very specific labeling of newly synthesized DNA, with very low or even no background staining. Labeling the phosphorylated moieties of histone H2AX with specific antibodies is very specific, but some nonspecific binding by a secondary antibody cannot be avoided. Consequently, a low-level fluorescence background is usually present.

A very high level of specificity is provided by genetic manipulation of cells, leading to expression of fluorescently tagged proteins (Chapter 4).

3.3.3 Sensitivity of Detection

Contemporary fluorescence microscopy offers highly sensitive detection of fluorescent species. The advent of highly sensitive light detectors and cameras made

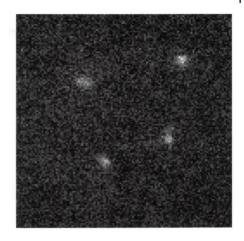

Figure 3.5 Single molecules of a fluorescent lipid tracer, TopFluor-PC, in lipid bilayers, imaged with an electron-multiplying CCD (EMCCD) camera. Each molecule produces a diffraction-limited signal as discussed in Chapter 2 (field of view, $10 \times 10\,\mu m^2$). (Image courtesy of Katharina Scherer, Bonn University.)

it possible to detect even single molecules in the specimen (Figure 3.5). Thus, the observation of single-molecule behavior, such as blinking, or the detection of interactions between individual molecules by Förster resonance energy transfer (FRET) has become possible.

The analogy between watching a fluorescent molecule under a microscope and a firefly at night still holds when one thinks of the size and shape of a firefly. A small spot of fluorescent light seen through a microscope may represent one molecule. Although the observer can identify the position of the molecule in space and can watch its movement, the shape and size of this molecule remain unknown (Figure 3.5). Despite this limitation, the ability to detect single molecules opened the way to a number of interesting new techniques, including speckle microscopy and, most importantly, super-resolution methods (see sections below and Chapters 8 and 12).

3.4 A Fluorescence Microscope

3.4.1 Principle of Operation

The analogy between a firefly and a fluorescent object, which was useful in introducing basic concepts, ends when one considers the instrumentation required to detect fluorescence in a microscope. While a firefly emits light on its own by a biochemical process called *bioluminescence*, which uses energy from adenosine triphosphate (ATP), but does not require light to be initiated, the fluorescence in a microscopic object has to be excited by incident light of a shorter wavelength. Thus, a fluorescence microscope has to be constructed in a way that allows excitation of fluorescence, subsequent separation of the relatively weak emission from the strong exciting light, and, finally, detection of fluorescence. An efficient separation of the exciting light from the fluorescence light, which eventually reaches the observer's eye or the electronic detector, is mandatory for obtaining high image contrast. A sketch of a standard widefield fluorescence microscope is shown in Figure 3.6.

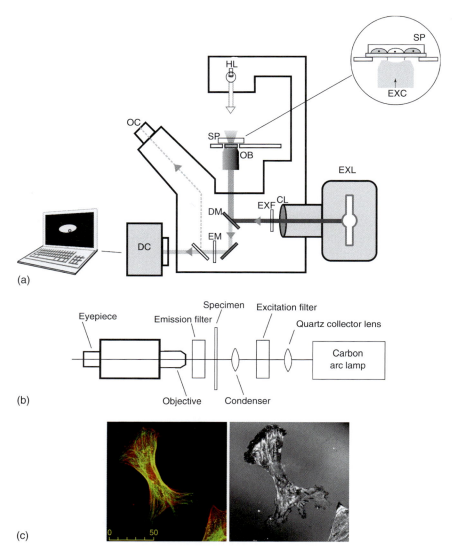

Figure 3.6 Fluorescence microscope – principle of operation. (a) A schematic diagram of an inverted fluorescence microscope (epifluorescence). This type of microscope enables studies of live cells maintained in a standard growth medium, in tissue culture vessels, such as Petri dishes. HL, halogen lamp; SP, specimen; OB, objective lens; OC, ocular lens (eyepiece); DC, digital camera; EXL, exciting light source; CL, collector lens; EXF, excitation filter; DM, dichroic mirror; EM, emission filter; EXC, exciting light incident on the specimen. (b) A schematic diagram of an early fluorescence microscope, with dia-illumination. (c) Fluorescence images of microtubules (green) and actin fibers (red) and an image of the same cell in reflected light (Box 3.3), demonstrating focal contacts (black). (Images by J. Dobrucki, Jagiellonian University, Kraków.)

Box 3.3 Reflected Light Imaging

The epifluorescence design makes it possible to detect reflected light. In this mode of observation, a dichroic mirror is replaced with a silver-sputtered mirror that reflects and transmits the incident light in desired proportions; also the emission filter has to be selected to allow the reflected exciting light to enter the detector (additional optical components are needed to minimize the effects of light interference). Imaging reflected light adds another dimension to fluorescence microscopy. Figure 3.6c shows an example of reflected light imaging – visualization of focal contacts in a fibroblast attached to a glass surface. The images were collected in a confocal fluorescence microscope that was set up for two-channel fluorescence and reflected light imaging.

Fluorescence is excited by light that is emitted by a mercury lamp. The exciting light is reflected toward the sample by a dichroic (two-color) mirror (Figures 3.6 and 3.7). This special mirror is positioned at 45° angle toward the incoming light and reflects photons of a selected wavelength but allows those of longer wavelengths to pass through. The dichroic mirror is selected for a given application; that is, it is made to reflect the chosen exciting wavelength and allow the passage of the expected fluorescence (Figure 3.8). It is important to realize that the efficiency of converting exciting light into fluorescence is usually low, that is, only one out of many exciting photons is converted into a photon

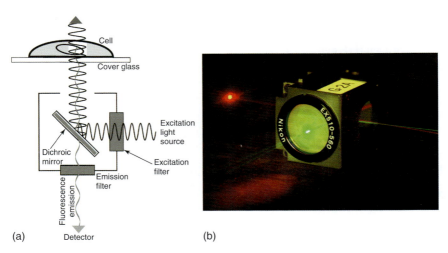

Figure 3.7 A fluorescence microscope filter block. (a) A schematic diagram of a typical filter block containing an excitation filter, a dichroic mirror, and an emission filter. The exciting light emitted by a light source is reflected by the dichroic mirror toward the specimen; the fluorescence emission has a longer wavelength than the exciting light and is transmitted by the dichroic mirror toward the ocular lens or a light detector. (b) A photograph of a microscope filter block, which reflects green but transmits red light. The excitation filter is facing the observer and the dichroic mirror is mounted inside the block; the emission filter (on the left side of the block) cannot be seen in this shot. Light beams emitted by laser pointers were used. In a fluorescence microscope, the intensity of the red emission would be significantly lower than the intensity of the exciting green light. (Photograph by J. Dobrucki.)

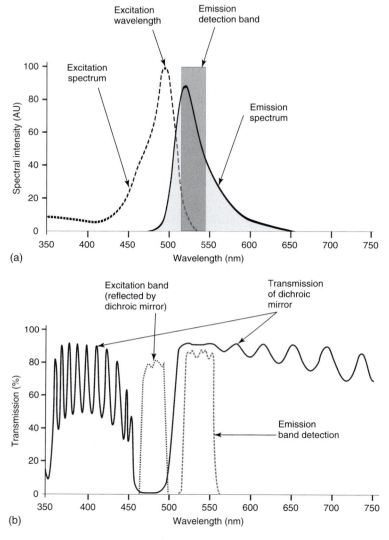

Figure 3.8 Spectral properties of a typical fluorescent label, fluorescein (a), and the characteristics of a filter set suitable for selecting the excitation band and detecting fluorescence emission (b).

of a longer wavelength and subsequently detected as fluorescence. Moreover, fluorescence is emitted by the sample in all directions, but only a selected light cone, that is, a fraction of this fluorescence, is collected by the objective lens. Consequently, fluorescence is weak in comparison with the exciting light and has to be efficiently separated and detected. High-quality optical filters are used to select exclusively the desired exciting wavelength (excitation filter) and the fluorescence emission bands (emission filter) (Figure 3.7). It is worth noting that the arrangement described here, called *epifluorescence* (which resembles the reflected light microscope that is used in studies of metal surfaces), makes it relatively easy to separate the fluorescence from the exciting light. It is also safer for the operator than an original dia-illumination system. In the early days of fluorescence microscopy, a direct (dia-)illumination of the sample was used (Figure 3.6b). The exciting light was prevented from reaching the observer only by virtue of placing an efficient blocking filter in the light path.

In the original fluorescence microscope design, removing the excitation filter from the light path allowed the intense exciting light into the eyepiece and the eyes of the observer. Even in the case where the objective lens does not allow UV light to pass through, this would be very dangerous. In the epifluorescence design, there is no danger of sending the exciting light directly into the eyes of the observer. Even when the dichroic and excitation filters are removed, the exciting light will not be incident directly onto the ocular lens. This does not mean, however, that it is safe to look through the microscope with the exciting light source turned on and all the filter blocks removed out of the light path. When filters are removed, the exciting light is still sufficiently reflected and scattered to pose a hazard to the eyes of the observer. To protect one's lab-mates, if filter blocks are removed, it is advisable to leave an appropriate note to warn others who might come to "only have a brief look at their sample."

3.4.2 Sources of Exciting Light

A typical fluorescence microscope contains two sources of light. One, usually a halogen lamp, is used for initial viewing of a specimen in the transmitted light mode. Another, often a mercury arc lamp, is used for exciting fluorescence. Halogen lamps belong to a class of incandescent (or "candescent") light sources, where the emission of photons occurs through the heating of a tungsten filament. High-pressure mercury vapor arc-discharge lamps (HBO) are a popular source of exciting light in standard fluorescence microscopes. The intensity of light emitted by a mercury arc lamp is up to 100 times greater than that of a halogen lamp. The spectrum of emission, shown in Figure 3.9a, extends from the UV to the infrared. The spectrum depends to some degree on the pressure of the mercury vapor and on the type of lamp. The desired excitation spectral band is selected by using an appropriate excitation filter. As the spectrum consists of numerous sharp maxima, the intensity of the exciting light strongly depends on the selected wavelength. Mercury arc lamps are very convenient sources of excitation for a number of typical fluorophores. Another popular source of exciting light is a xenon arc lamp. It emits an almost continuous spectrum of emission in the whole range of visible wavelengths (Figure 3.9b). Metal halide lamps are a modification of mercury vapor lamps, characterized by higher levels of emission between the major mercury arc spectral lines (Figure 3.9c) and a much longer lifetime. Here, the spectrum is dependent on the metal used for doping.

The stream of photons emitted by the filament in the bulb is not ideally uniform in space or constant in time. It has been demonstrated that a hot light source emits groups (bunches) of photons, rather than individual, independent photons. This intriguing phenomenon (photon bunching) was observed in a classical physics experiment performed by Hanbury Brown and Twiss (for a discussion

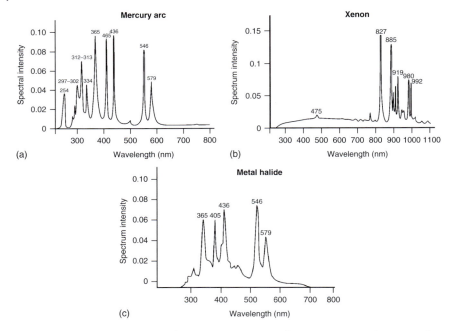

Figure 3.9 Schematic diagrams of the emission spectra of (a) a mercury arc, (b) a xenon lamp, and (c) a metal halide lamp. (Figure by M. Davidson, Microscopy Primer, University of Florida, with permission, redrawn.)

of photon bunching and the relation of this phenomenon to the corpuscular and wave nature of light, see [5]). There are also more trivial sources of spatial and temporal inhomogeneity of emitted light, including fluctuations in the temperature of the filament, instability of the electric current, and the influence of external electromagnetic fields. Consequently, the popular mercury arc lamps pose some problems for quantitative microscopy. The illumination of the field of view is not uniform, and the intensity of light fluctuates on a short time scale that is comparable with the times needed to record pixels and images and it diminishes over days of the lamp use because the electrodes are subject to erosion. Thus, quantitative fluorescence microscopy studies are hampered by the lack of short- and long-term stability of the exciting light. A higher stability of emission is offered by light-emitting diodes (LEDs) and laser light sources.

A typical HBO burner needs to be placed exactly at the focal point of the collector lens (when exchanging HBO burners, the quartz bulb must not be touched by fingers). Centering is important to ensure optimal illumination, that is, the highest attainable and symmetric illumination of the field of view. After being turned off, HBO burners should be allowed to cool before switching them on again. They should be exchanged upon reaching the lifetime specified by the manufacturer (usually 200 h). Using them beyond this time not only leads to a significantly less intensive illumination but also increases the risk of bulb explosion. Although relatively rare, such an event may be costly, as the quartz collector lens located in front of the burner may be shattered. HBO burners need to be properly disposed of to avoid contamination of the environment with mercury. Today's metal halide lamps are usually pre-centered and fixed in the optimal position in the lamp housing.

Lasers emit light of discrete wavelengths, characterized by high stability both spatially (the direction in which the beam is propagated is fixed; so-called beam-pointing stability), and temporally – both on a short and long timescale.

In the early days of fluorescence microscopy, lasers did not exist; subsequently, their use was limited to laser scanning microscopes, which were more advanced and more expensive than standard widefield fluorescence microscopes (Chapter 5). The stability of the light beams and the ability to focus them to a diffraction-limited spot were important advantages of lasers over mercury arc lamps. The disadvantages, however, were their high cost and the limited number of usable emission lines for exciting popular fluorophores. For instance, the popular 25 mW argon ion laser provided the 488 and 514 nm lines, which were not optimal for working with the most popular (at that time) pair of fluorescent dyes – fluorescein and rhodamine. Other gas lasers that are often used in confocal fluorescence microscopy include krypton–argon (488, 568, and 647 nm) and helium–cadmium (442 nm). Solid-state lasers, including a frequency-doubled neodymium-doped yttrium aluminum garnet (Nd:YAG, 532 nm), are also used.

Currently, the price of standard gas lasers has decreased, and a wide selection of various single-line lasers as well as lasers emitting a broad wavelength spectrum are available, including a "white light laser." Durable diode lasers have become available as well. Moreover, new low-cost sources of light, LEDs, have become available. LEDs are semiconductors that exploit the phenomenon of electroluminescence. Originally used only as red laser pointers, they now include a range of devices emitting in UV, visible, or NIR spectral region. The critical part of an LED is the junction between two different semiconducting materials. One of them is dominated by negative charge (n-type) and the other by positive charge (p-type). When voltage is applied across the junction, a flow of negative and positive charges is induced. The charges combine in the junction region. This process leads to a release of photons. The energy (wavelength) of these photons depends on the types of semiconducting materials used in the LED. A set of properly selected LEDs can now be used as a source of stable exciting light in a fluorescence microscope. LEDs are extremely durable, stable, and are not damaged by being switched on and off quickly. These advantages are making them very popular not only in scanning confocal microscopes but in standard fluorescence microscopes as well.

3.4.3 Optical Filters in a Fluorescence Microscope

In a fluorescence microscope, light beams of various wavelengths need to be selected and separated from each other by means of optical glass filters. The filter that selects the desired wavelength from the spectrum of the source of exciting light is called an *excitation filter*. There is also a need to control the intensity of the exciting light. This can be achieved by placing a neutral density filter in the light path, if it is not possible to regulate the emission intensity directly. Fluorescence is emitted in all directions; most of the exciting light passes straight through the specimen, but a large part is scattered and reflected by cells and subcellular structures. A majority of the reflected exciting light is directed by the dichroic mirror back to the light source, while a selected wavelength range of the emitted fluorescence passes through toward the ocular lens or the light detector. In a standard widefield fluorescence microscope, the emission filter is mounted on a filter block, which is placed in the light path to select the desired emission bandwidth. If only one set of spectral excitation and emission wavelengths were available in a microscope, the applicability of the instrument would be seriously limited, as only one group of spectrally similar fluorescent probes could be used. Therefore, a set of several filter blocks prepared for typical fluorescent dyes is usually mounted on a slider or a filter wheel. This allows for a rapid change of excitation and emission ranges. In fluorescence confocal microscopes, the dichroic filter is often designed so as to reflect two or three exciting wavelengths and transmit the corresponding emission bands. Additional dichroic mirrors split the emitted light into separate beams directed toward independent light detectors. This makes it possible to simultaneously excite and observe several fluorophores. Some widefield fluorescence microscopes incorporate electronically controlled excitation and emission filter wheels, which contain sets of several optical filters. The desired combination of filters can be quickly selected using a software package that drives the filter wheels and the shutters.

Although microscope manufacturers offer standard filter blocks for most popular fluorescent probes, it is useful to discuss and order custom-made filter sets optimized for the source of exciting light and for the user's specific applications. It is also prudent to buy an empty filter block (holder) when purchasing a new microscope, in order to make it possible to build one's own filter blocks when new applications are desired. "In-house" assembling one's own filter block may require some skill and patience, because the position of the dichroic mirror in relation to the beam of exciting light is critical. Microscope manufacturers provide filter blocks with dichroic filters fixed in the optimal position. Aligning the dichroic mirror by the user should be possible, but may be cumbersome in some microscopes. When buying individual filters, especially dichroics, it is important to choose high-quality filters, that is, low-wedge filters (flat, with both surfaces ideally parallel), with the smallest number of coating imperfections. Defects in coating will allow the exciting light to reach the fluorescence detector. The exciting light reaching the detector will be detected as a background, thereby reducing the contrast and degrading image quality. Also, when buying dichroics in order to build one's own filter combinations, it is important to ensure that not only the shape and diameter but also the thickness of the filter is right for a given microscope design. Aligning a dichroic filter of a different thickness may turn out to be impossible. The orientation of a dichroic filter in relation to the incident light is important – the surface of the interference filter that should face the light source is specified by the manufacturer (note that the arrow can mean the direction of light propagation or the direction toward the light source, depending on the manufacturer). Optical filters have to be handled with care to avoid leaving fingerprints, grease, and dust on the surface. Newer hard-coated filters are quite robust and easier to clean. If cleaning is required, it can be done by blowing air or using a soft brush (to remove the dust) and subsequently using a piece of optical tissue (the tissue is not to be reused) with a small amount of pure methanol or isopropyl alcohol. The manufacturer of the filter will usually recommend the solvent that is appropriate for their products. It is important to ensure that the antireflection coating is not scratched or damaged.

The state of current technology in the manufacturing glass optical filters is so advanced that filters of various optical characteristics can be custom-made by the optical companies.

3.4.4 Electronic Filters

A new class of versatile optoelectronic elements that serve as filters has become available in recent years. These include the acousto-optic tunable filters (AOTFs). They provide flexibility and speed in choosing light wavelength and intensity that cannot be achieved with glass optical filters. This flexibility and speed is indispensable in modern advanced fluorescence confocal microscopes.

AOTFs work in a manner similar (though not identical) to a diffraction grating. A specialized birefringent crystal is subjected to high-frequency (a few hundred megahertz) acoustic waves that induce a periodic pattern of compression (Figure 3.10). This, in turn, changes the local diffractive index of the crystal and imposes a periodic pattern of different refractive indices. Most AOTFs use tellurium dioxide (TeO_2). This material is transparent in the range of 450–4000 nm. Only a selected band from a whole range of light wavelengths incident on the crystal is deflected. In this respect, the crystal resembles a band-pass filter rather than a diffraction grating where a whole range of wavelengths are diffracted (at different angles). The direction in which the light beam is deflected is fixed and does not depend on the wavelength. The wavelength of the diffracted band depends on the frequency of the acoustic

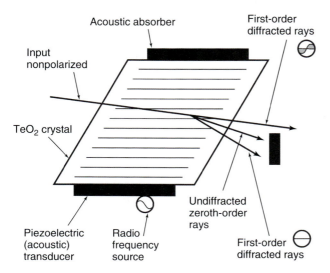

Figure 3.10 Architecture and principle of operation of an acousto-optic tunable filter (AOTF). (Lichtmikroskopie online, Vienna University, modified.)

wave. The intensity of the deflected light can be controlled by the amplitude of the acousto-mechanical wave incident on the crystal. The wave is impressed onto the crystal by a piezoelectric transducer, which is a device that expands and shrinks according to the applied voltage. AOTFs can be used to rapidly change the wavelength as well as the intensity of the deflected light by changing the frequency and the amplitude of the incident acoustic waves that drive the crystal. The spectral width of the deflected light can be controlled by delivering multiple frequencies to the AOTF. Moreover, using several widely spaced frequencies will allow a set of wavelength bands to be deflected at the same time. These characteristics make AOTF a flexible component of a fluorescence microscope, which can serve simultaneously as a set of fast shutters, neutral density filters, and emission filters, allowing simultaneous detection of several emission colors. Changing the wavelengths and their intensity can be achieved at a high speed, namely, tenths of microseconds, while filter wheels need time on the order of a second. An AOTF can also be used as a dichroic filter, which separates the exciting light from fluorescence emission. These advantages make AOTFs particularly valuable for multicolor confocal microscopy and techniques such as fluorescence recovery after photobleaching (FRAP, [6] chapter 11).

3.4.5 Photodetectors for Fluorescence Microscopy

In most microscopy applications, there is a need to record fluorescence images – often a large number of images within a short time – for subsequent processing, analysis, and archival storage. Originally, analog recording of fluorescence images on a light-sensitive film was extensively used, but in the past two decades methods of electronic detection, analysis, and storage have become efficient and widely available. Currently, fluorescence microscopes are equipped with suitable systems for light detection, image digitization, and recording. The most common light detector in a standard widefield fluorescence microscope is a charge-coupled device (CCD) camera, while photomultipliers are used in laser scanning confocal microscopes. Other types of light detectors, including intensified charge-coupled device (ICCD) and electron multiplied charge-coupled device (EMCCD) cameras, as well as avalanche photodiodes (APDs), are gaining increasing importance in modern fluorescence microscopy.

3.4.6 CCD or Charge-Coupled Device

In a camera based on a CCD, an image is projected onto an array of semiconducting, light-sensitive elements that generate an electric charge proportional to the intensity of the incident light (Figure 3.11). Two types of sensor architectures are currently produced: front-illuminated charge-coupled device (FI CCD) cameras, where light is incident on an electrode before reaching the photosensitive silicon layer, and back-illuminated charge-coupled device (BI CCD) cameras, where the incoming light falls directly on the silicon layer (Figure 3.12). Note that the structure of an FI CCD is similar to the anatomy of the human eye. In the retina, it is the nerve "wiring" that faces the incoming light. Only after light has passed the layer of nerve cells that are not sensitive to light can the photons interact with rhodopsin in the photoreceptors.

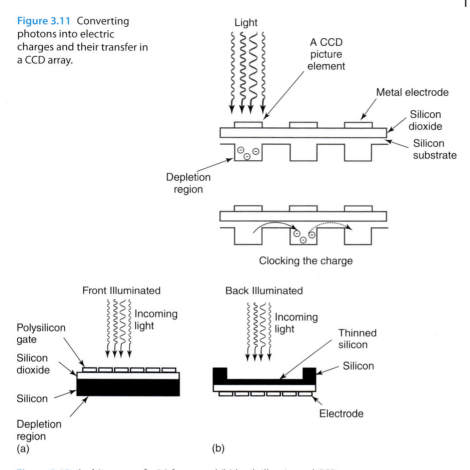

Figure 3.11 Converting photons into electric charges and their transfer in a CCD array.

Figure 3.12 Architecture of a (a) front- and (b) back-illuminated CCD.

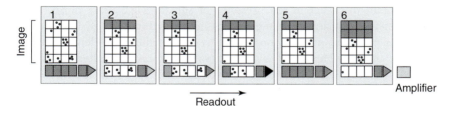

Figure 3.13 Charge transfer through a CCD array and into an amplifier.

When the array of light-sensitive elements (capacitors, photodiodes; in digital microscopy called *picture elements* or *pixels* for short) of a camera is exposed to light, each of them accumulates an electric charge (Figure 3.11). These charges are later read one by one. An electronic control system shifts each charge to a neighboring element, and the process is repeated until all charges are shifted to an amplifier (Figure 3.13). When the individual charges are dumped onto a charge amplifier, they are converted into corresponding voltage values. This process is

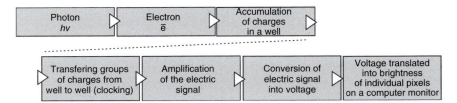

Figure 3.14 A schematic representation of the steps occurring from the absorption of a photon in a CCD array to the display of brightness on a computer screen. In some digital cameras, amplification of the signal is achieved already on the chip (see sections below).

repeated until all the light detected by a microscope in all pixels of the field of view is eventually translated, point by point, into voltage values. The values of these voltages are subsequently converted into discrete values (digitized) and translated into the levels of brightness (intensity of light) on the display screen (Figure 3.14). Sensitivity of a CCD camera and its ability to record strong signals depend on a number of factors, including the level of electronic noise and the ability to accumulate a large number of charges. Weak signals may be undetectable if they are comparable to the noise level. Strong signals may also be difficult to record faithfully. Exposing a CCD array to a large dose of light may result in filling the well with charges and reaching the maximum well capacity (saturation charge). This may cause charges spilling into the neighboring wells and result in deterioration of image quality.

In an ideal situation, only the photons that strike the silicone surface of a light-sensitive element of the camera should actually cause accumulation of an electric charge, which is later shifted and read out (Figure 3.14). However, some electrons may occur and be recorded even in the absence of any incident light. This contributes to the noise of the displayed image (see below). As fluorescence intensities encountered in microscopy are usually low, the fluorescence signal may be difficult to detect in the presence of substantial camera noise (for a comprehensive discussion of camera noise, see Section 12.3.3). In order to alleviate this problem, photons registered by the neighboring elements of an array can be combined and read as one entity. This procedure, called *binning*, is illustrated in Figure 3.15. A fluorescent structure in a specimen is symbolized by two black lines. It is overlaid with a fragment of a CCD array. The weak signals recorded in individual small pixels of a 6×6 array do not stand out well against the background noise generated by the electronics of the camera. When a 3×3 array of larger pixels is used, signals from the areas corresponding to four neighboring small pixels are summed up. Such signals become sufficiently strong to be detected well above the noise floor. The improvement in signal-to-noise ratio is achieved at the cost of spatial resolution. Note that Figure 3.15 is intended to explain the principle and benefit of binning, but it is a necessary oversimplification because it does not consider the dark noise generated in each pixel by thermal noise in the device; this noise will also be summed in the binning procedure.

Figure 3.15 A schematic diagram describing the principle of binning in a CCD sensor.

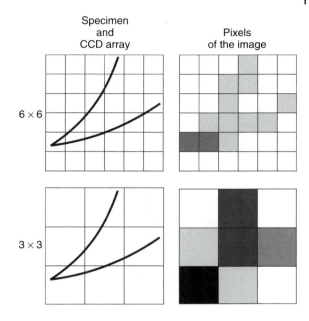

Longer recording times also help to isolate weak signals from random noise. Weak fluorescence signals that are below the detection level of a standard CCD camera may be detected by ICCD and EMCCD cameras (see below).

The light-sensitive element of a CCD camera cannot detect all incident photons because not all photons that arrive at a given pixel generate an electron. The ratio between the incident photons that generate an electric charge to all photons incident on an area of a detector is referred to as the *quantum efficiency* of a CCD camera. This efficiency is wavelength-dependent. FI CCD cameras reach 60–70% quantum efficiency in the visible range and are essentially unable to detect UV radiation (Figure 3.16). BI CCDs have a quantum efficiency (QE) of up to 95% and a better spectral response, including the ability to detect in the UV range. The difference arises from the fact that the BI CCD exposes light-sensitive elements, rather than an electrode structure, as in FI CCD, directly to the incoming light.

3.4.7 Intensified CCD (ICCD)

Fluorescence microscopy invariably struggles with weak signals. Image acquisition is particularly difficult when a high speed of data recording is required. A standard CCD camera is often insufficient for the detection of faint fluorescence due to a high readout noise. Moreover, high sensitivity of a standard CCD sensor is achieved only at relatively low image acquisition rates. Much better sensitivity is offered by an ICCD camera. While a standard CCD-based camera produces only one electron in response to one incoming photon, an ICCD generates thousands of electrons.

ICCDs use an amplifying device called a *microchannel plate*, which is placed in front of a standard CCD light sensor (Figure 3.17). A microchannel plate consists

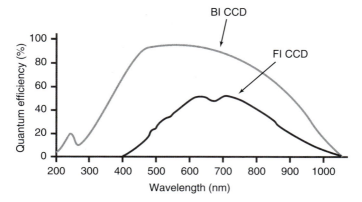

Figure 3.16 Quantum efficiency and spectral response of front- and back-illuminated CCDs. (Documentation from Andor Company, simplified.)

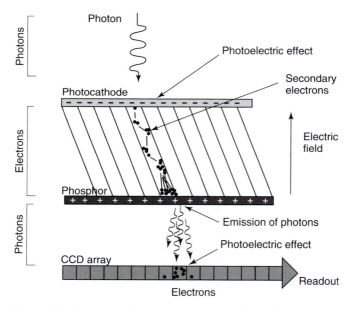

Figure 3.17 Signal amplification in an intensified CCD sensor.

of a photocathode, a set of channels, and the layer of a phosphor. A photon incident on a photocathode generates one electron (by means of the photoelectric effect) that travels inside a channel, toward the phosphor layer, and is accelerated by a strong electric field. Inside the channel, the electron hits the channel walls and generates more electrons that eventually reach the phosphor. There, each electron causes the emission of a photon which is subsequently registered by the sensor of a standard CCD camera. In this way, the microchannel plate amplifies the original weak light signal, and this amplified signal is eventually detected by a standard CCD sensor. In ICCDs, the signal-to-noise ratio can be improved by a factor of several thousand in comparison with a standard CCD camera.

ICCD cameras feature the high sensitivity that is required for the imaging of weak fluorescence signals (although the sensitivity of the primary photocathode is relatively low, not exceeding 50% quantum efficiency). However, the spatial resolution is low and background noise is usually high. Moreover, an ICCD chip can be damaged by excess light. When a high data registration speed is needed and the signal is weak, an EMCCD camera is a viable option.

3.4.8 Electron-Multiplying Charge-Coupled Device (EMCCD)

Another option available for detection of weak fluorescence signals is an EMCCD camera. This device is characterized by a very low read noise and offers sufficient sensitivity to detect single photons. It features high speed, high quantum efficiency, and high digital resolution. In this type of camera, amplification of the charge signal occurs before the charge amplification is performed by the electron-multiplying structure built into the chip (hence called *on-chip multiplication*).

An EMCCD camera is based on the so-called frame-transfer CCD (Figure 3.18; this technology is also used in some conventional CCDs) and includes a serial (readout) register and a multiplication register. The frame-transfer architecture is based on two sensor areas – the area that captures the image, and the storage area where the image is stored before it is read out. The storage area is covered with an opaque mask. The image captured by the sensor area is shifted to the storage area after a predefined image integration time, and subsequently read out. At the same time, the next image is acquired. The charge is shifted out through the readout register and through the multiplication register where amplification occurs prior to readout by the charge amplifier. The readout register is a type of a standard CCD serial register. Charges are subsequently shifted to the multiplication register, that is, an area where electrons are shifted from one element to the next one by applying a voltage that is higher than typically used in a standard CCD serial register. At an electric field of such a high value, the so-called secondary electrons are generated. The physical process that is activated by this voltage is called *impact ionization*. More and more electrons enter subsequent elements of the multiplication register, and a "snowball effect" ensues: the higher the voltage (for an operator – the *gain* value) and the larger the number of pixels in the multiplication register, the higher the overall multiplication of the original faint fluorescence signal (Figure 3.18). The process of multiplication is very fast, so an EMCCD camera is not only very sensitive but also very fast while preserving the high spatial resolution of a standard CCD chip.

The probability of generating a secondary electron depends on the voltage of the multiplication clock (Figure 3.19a) and the temperature of the sensor (Figure 3.19b). Although the efficiency of generating a secondary electron is quite low (~0.01 per shift), this process can occur in all elements of a long multiplication register, bringing the final multiplication value to several hundred or more. But this amplification is not for free. As the generation of the electrons is a stochastic process in each pixel, the extra gain increases the noise in the

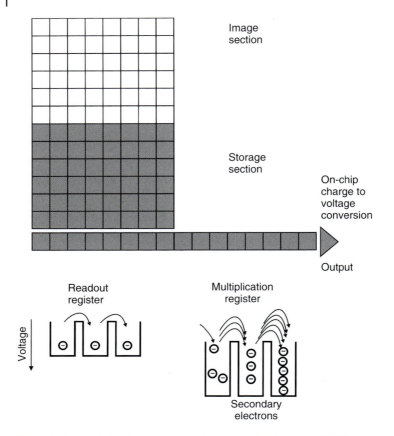

Figure 3.18 Principle of operation and signal amplification in an EMCCD sensor.

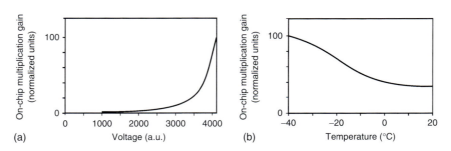

Figure 3.19 On-chip multiplication gain versus (a) voltage and (b) temperature in an EMCCD camera. (Source: Roper Scientific Technical Note #14.)

signal. This is accounted for by the so-called noise excess factor, F (see also Section 12.3.3).

The probability of generating a secondary electron decreases with temperature, as shown in Figure 3.19. Thus, cooling a chip gives an additional advantage in terms of a higher on-chip multiplication gain. Cooling a chip from room temperature to $-20\,°C$ increases the chance of generating a secondary electron by a factor of 2. Some EMCCD cameras are cooled to −100 °C. The lower temperature also results in a lower dark current (see below). Thus, even a very weak signal becomes detectable because it stands out above the noise floor. Reducing the noise level at low temperatures is particularly important because all signals that occur in a well or a serial register, and do not represent the fluorescence signal, will be multiplied with the signal of interest.

3.4.9 CMOS

An alternative to a CCD is a complementary metal–oxide–semiconductor (CMOS) image sensor. Note that the term CMOS actually refers to the technology of manufacturing transistors on a silicone wafer, not the method of image capture. Like the CCD, CMOS exploits the photoelectric effect. Photons interacting with a silicon semiconductor move electrons from the valence band into the conduction band. The electrons are collected in a potential well and are converted into a voltage that is different from that in a CCD sensor, where the charge is first moved into a register and subsequently converted into voltage. The measured voltage is then passed through an analog-to-digital converter and translated into a brightness value for an individual pixel on the computer monitor. CMOS sensors are manufactured in a process where the digital logic circuits, clock drivers, counters, and analog-to-digital converters are placed on the same silicon foundation and at the same time as the photodiode array. In this respect, the architecture of a CMOS sensor is distinctly different from that of a CCD device, where the charge of each photodiode is transferred first to the chip, and then read out in sequence outside of the chip.

The construction of a typical CMOS photodiode is presented in Figure 3.20. The actual light-sensitive element and the readout amplifier are combined into

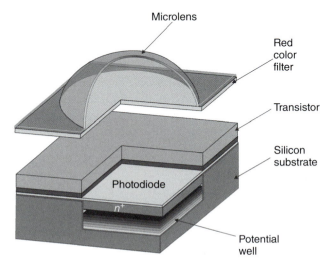

Figure 3.20 Architecture of a single CMOS photodiode. (A figure by M. Davidson, Microscopy Primer, University of Florida, modified.)

one entity. The charge accumulated by the photodiode is converted into an amplified voltage inside the pixel and subsequently transferred individually into the analog signal-processing portion of the chip. A large part of the array consists of electronic components, which do not collect light and are thus not involved in detecting the light incident on the sensor. The architecture of the array leaves only a part of the sensor available for light collection and imposes a limit on the light sensitivity of the device. This shortcoming is minimized by placing an array of microlenses over the sensor, which focus the incident light onto each photodiode.

The unique architecture of a CMOS image sensor makes it possible to read individual pixel data throughout the entire photodiode array. Thus, only a selected area of the sensor can be used to build an image (window-of-interest readout). This capability makes CMOS sensors attractive for many microscopy applications. CMOS technology has been refined in recent years so that today's sensors compete successfully even with high-end EMCCD cameras in many low-light microscopy applications. At the time of writing, 2048×2048 pixel CMOS sensors using $6.5 \times 6.5\,\mu m$ pixels with a quantum efficiency exceeding 70% and a 100 frames per second readout rate are available.

3.4.10 Scientific CMOS (sCMOS)

CMOS-based digital cameras were originally inferior to high-end CCD cameras. Until recently, EMCCD cameras were the best choice in terms of sensitivity, speed of data acquisition, and digital resolution. Yet the CMOS technology has made significant advances, and the newest design, called scientific CMOS (sCMOS), is challenging even the EMCCD sensors in many demanding microscopy applications. The advantages of the sCMOS sensor include a large-size array, small pixel size, low read noise, high frame rate, high dynamic range, and no multiplicative noise.

3.4.11 Features of CCD and CMOS Cameras

CMOS and CCD cameras are inherently monochromatic devices, responding only to the total number of electrons accumulated in the photodiodes and not to the color of light that gives rise to their release from the silicon substrate. In fluorescence microscopy, detection of two or more colors is often required. Sets of emission optical filters are then used to collect images in selected spectral bands sequentially. When simultaneous measurements are necessary, two digital cameras are mounted on a microscope stand.

Although the same type of CCD chip is used by many manufacturers of digital cameras, the ultimate noise level, speed of data acquisition, and dynamic range of a given camera may be quite different. These differences arise from the differences in electronics and software driving the camera. The user should explore the features and software options of the camera to identify the sources of noise, to calibrate the camera, and to optimize image collection for the specific type of experiment.

3.4.12 Choosing a Digital Camera for Fluorescence Microscopy

It might seem obvious that a researcher who is planning to purchase a new fluorescence microscopy system should buy, funds permitting, the best digital camera on the market. However, there is no "best digital camera" for fluorescence microscopy as such. The camera should be selected for a given application. Manufacturers provide important information about the chip and the software, including the pixel size, quantum efficiency, full well capacity, the size of the various noise contributions, speed of image acquisition and the corresponding achievable resolution, tools for calibration, and noise removal, and so on. Careful analysis of technical parameters of various cameras available on the market is essential. However, nothing can substitute testing the various cameras with a typical specimen, which is the researcher's prime object of investigation.

3.4.13 Photomultiplier Tube (PMT)

While the CCD and CMOS cameras briefly introduced above are used for recording the whole image of a field of view essentially at the same time in a parallel process, a photomultiplier is used as a point detector. In other words, it records the intensity of light only in one selected point of the image at a time. Thus, photomultiplier tubes (PMTs) are not used in standard widefield fluorescence microscopes, but serve as light detectors in laser scanning confocal microscopes.

A PMT is a signal-amplifying device that exploits the photoelectric effect and a secondary emission phenomenon, that is, the ability of electrons to cause the emission of other (secondary) electrons from an electrode in a vacuum tube. Light enters a PMT through a quartz (or glass) window and strikes a photosensitive surface (a photocathode) made of alkali metals (Figure 3.21). The photocathode releases electrons that subsequently strike the electrode (dynode), which releases a still larger number of electrons. These electrons hit the next dynode. A high voltage (1–2 kV) is applied between subsequent dynodes.

The electrons are accelerated and the process is repeated, leading to an amplification of the first electric current generated on the photocathode. The current

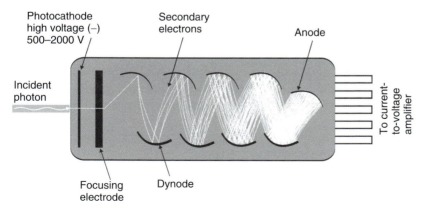

Figure 3.21 Schematic of a photomultiplier.

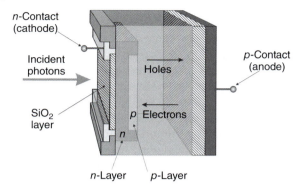

Figure 3.22 Schematic of an avalanche photodiode. (A figure by M. Davidson, Microscopy Primer, University of Florida, modified.)

measured at the last dynode is proportional to the intensity of light that was incident on the photocathode. The gain obtained by amplifying the electric current through subsequent dynodes can be as high as 10^7–10^8. However, the voltage that is applied to the dynodes causes a low-level electron flow through the PMT even in the absence of light. This is translated into a non-zero-level reading on a fluorescence image. The quantum efficiency of a PMT does not exceed 30%. PMTs are very fast detectors of UV and visible light. They can be used to detect single photons and follow extremely fast processes, as the response time can be as low as several nanoseconds. Typically, in laser scanning confocal microscopes, sets of 2–5 PMTs are used as fluorescence detectors in selected wavelength bands.

3.4.14 Avalanche Photodiode (APD)

An APD is also a signal-amplifying device that exploits the inner photoelectric effect. A photodiode is essentially a semiconductor p–n (or p–i–n) junction (Figure 3.22). When a photon of sufficient energy strikes the diode, it excites an electron, thereby creating a free electron (and a positively charged electron hole). This, in turn, creates a flow of electrons ("an avalanche") between the anode and the cathode, and electron holes between the cathode and the anode, because a high voltage is applied between the anode and the cathode. Electrons accelerated in the electric field collide with atoms in the crystalline silicon and induce more electron–hole pairs. This phenomenon amounts to a multiplication effect. In this respect, an APD bears similarity to a PMT. The quantum yield of an APD can reach 90%, that is, it is substantially higher than that of a PMT, and the response time is several times shorter. However, the gain is lower, in the range of 500–1000. Single-photon avalanche photodiodes (SPADs) are currently used as detectors in fluorescence lifetime imaging microscopy (FLIM).

3.5 Types of Noise in a Digital Microscopy Image

If a light detector were ideal, an image collected in the absence of any specimen should be completely black. However, even in the absence of any fluorescence in the sample, CCD sensors still generate certain readout values that are greater than zero. In laboratory vocabulary, these weak unwanted signals that do not represent fluorescence are generally called *background* or *noise*.

The adverse influence of noise on the quality of the recorded image is understandable. Let us assume that the noise signals have a value in the range between 1 and 10 on a scale of 1–100. If a signal representing fluorescence has an intensity of 80, it will be readily detected, but if a weak signal has an intensity of 10, it will not be distinguishable from the noise (Figure 3.23). Averaging a large number of image frames should make the signal detectable over the noise level, but an experimenter rarely has the luxury of collecting many images of the same field of view because photobleaching will inevitably diminish the fluorescence signal, while the noise level will remain the same. The range of intensities of fluorescence signals that can eventually be recorded above the level of noise is called the *dynamic range of the detector*. More precisely, dynamic range is the ratio between the maximum and the minimum level of signal that can be detected. Thus, the dynamic range of a CCD camera is equal to the saturation charge (full well capacity) divided by the readout noise (i.e., the noise generated in the absence of light), when both are expressed as the number of electrons. A higher dynamic range of a camera means a broader range of fluorescence intensities that can be faithfully recorded (Figure 3.24). It should be noted, however, that the dynamic range of a light detector defined that way does not provide any information about its absolute sensitivity.

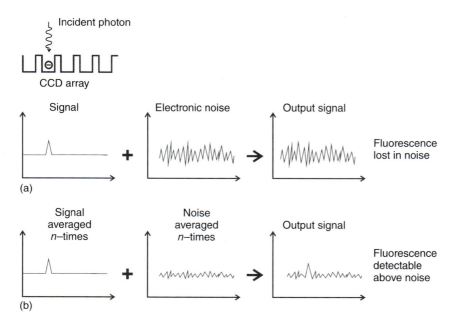

Figure 3.23 Signal averaging and detection of weak fluorescence signals. (a) A weak signal cannot be detected if it is comparable with the level of noise generated by the electronics of a camera. (b) When images of stable fluorescence signals are averaged, the noise is averaged out and becomes relatively low in comparison with the signal. This simplified scheme does not take dark noise into account.

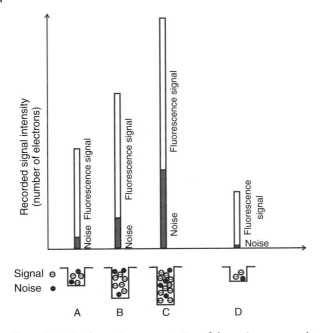

Figure 3.24 A schematic representation of dynamic ranges and sensitivities of four hypothetical digital cameras. The input signals received by cameras A, B, C, and D are such as to fill the charge wells; thus they are different in each case. The recorded signals consist of a noise contribution and fluorescence photons, as shown schematically by the bars. The electrons resulting from noise and fluorescence that fill the wells to a maximum capacity are shown symbolically below the bars. Cameras A, B, C, and D generate different levels of noise; therefore, their ability to detect weak signals differs. Sensors A, B, and C have similar dynamic ranges, that is, the ratio between the maximum recordable fluorescence signal and the level of noise is similar, but the ability to detect the strong signals differs – it is the best for camera C. Camera A is more sensitive than camera B or C. Camera D has a very low dynamic range but it has a very low noise level; thus, it is the most sensitive of the four. Camera D is not suitable for detecting strong signals. Different levels of the maximum recordable signal of these cameras are a consequence of different well depths of their sensors.

A weak fluorescence signal can be "fished out" of the noise by increasing the integration time (Figure 3.25). This simple procedure will be useful only if the rate of photobleaching does not offset the benefit of integration. Integration takes considerable time. Therefore, although weak signals can eventually be recorded by a CCD camera, the process may be relatively slow.

Another way to detect a weak signal is to use an ICCD or an EMCCD. These devices use two different ways of amplification of the signal before it is actually read out, that is, before an unavoidable addition of the read noise takes place. In order to fully appreciate different strategies that were used by digital camera developers aiming at enabling detection of weak signals, a brief discussion of various types of noise is required.

Some sources of noise were mentioned when speaking about the principles behind and construction of various camera types. Generally, one can identify

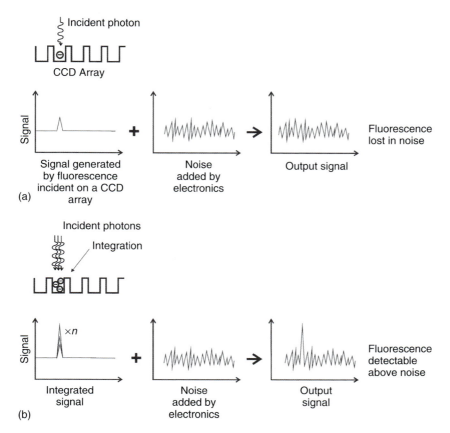

Figure 3.25 Integration elevates weak signals above the noise level. Integration, which is the summing up of incident fluorescence photons, improves the signal-to-noise ratio because the level of electronic noise added after signal integration remains constant. This simplified diagram does not take dark noise into account. (a) Signal collected as one frame, that is, without integration. (b) Integration of signals.

three major sources of noise in a digital microscopy image registered by a camera: (i) dark current noise, (ii) photon noise, and (iii) read noise.

The *dark current* (dark noise) arises from electrons that are generated in a well of a semiconductor sensor in the absence of any external light as a result of electron emission due to thermal motion. When the integration time on the CCD chip is increased, the accumulating thermal charge also increases. This leads to a detectable background in the image. As the electron emission is dependent on temperature, cooling the camera chip efficiently reduces the dark current. The dark current can be decreased from the final image by subtraction. Dark current noise should not be confused with background signal arising from low-level autofluorescence or fluorescence arising from nonspecific binding of an antibody in an immunofluorescence preparation.

The *photon noise* or *shot noise* results from the quantum nature of light. It is a term that refers to the temporal distribution of photons arriving at the surface of a sensor. Even if the fluorescence-emitting object is flat and uniformly stained, the frequency of photons arriving at the light-sensitive element of the sensor is governed by chance. The frequency of photon arrival follows the so-called Poisson statistics. This implies that the number of photons originating in a given small volume of a continuously illuminated specimen and subsequently reaching the detector varies in different time intervals (we ignore photobleaching for simplicity). This also implies that when the same (nonbleaching) voxel in the specimen is imaged repeatedly, and in a given measurement the number of detected photons is n, the subsequent determinations of the number of photons vary within a range of $\sqrt{n}$. Shot noise can be quite misleading. Inexperienced microscopists often take the grainy structure of an area in the image for a real variation of the fluorescence signal, or interpret differences between local signal intensities as evidence for a difference in the concentration of a fluorescent label. However, such features of the image may merely be a consequence of a very low number of photons that are typically collected in fluorescence microscopy studies.

The *read noise* arises in the process of converting a charge generated in the sensor well into voltage and digitization. Camera manufacturers provide information about the noise generated "on-chip" by specifying a root-mean-square (RMS) number of electrons per pixel. For instance, $10\,e^-$ RMS means that a read noise level of 10 electrons per pixel is expected. Thus, the signal obtained after readout of the charges would show a standard deviation of 10 electrons, even if all pixels contained identical numbers of electrons. At low signal levels, photon noise is the most significant noise contribution. As vendors may interface the same type of a light-sensitive chip with different electronics, the levels of electronic noise in cameras from different sources may also be different.

In addition to the major sources of noise mentioned above, other factors may result in unpredictable signal variability or nonzero levels. These factors include the nonuniformity of photoresponse, that is, noise values dependent on the location in the image sensor, and the nonuniformity of dark current. Nonuniformity of photoresponse is a consequence of the fact that individual pixels in a CCD chip do not convert photons to electrons with identical efficiency. Pixel-to-pixel variation is usually low in scientific-grade cameras and generally does not exceed 2%. The correction for differences between light sensitivity of individual pixels can be achieved by recording an image of an ideally uniform fluorescent object, for instance, a dye solution, and creating a correction mask using standard image processing tools. Dark current nonuniformity results from the fact that each pixel generates a dark current at a slightly different rate. The dark current may also drift over a longer period, for instance, owing to a change in the sensor's temperature.

In summary, different types of noise contribute differently to the final image. On-chip multiplication represented by ICCD, CMOS, and EMCCD digital cameras has opened new avenues of research by making it possible to detect very weak signals quickly and with a high spatial resolution.

3.6 Quantitative Fluorescence Microscopy

3.6.1 Measurements of Fluorescence Intensity and Concentration of the Labeled Target

A fluorescence microscopy image is usually treated as a source of information about structure, that is, about the spatial distribution of a molecule of interest. Let us use an immunofluorescently stained image of microtubules in a fibroblast as an example. On the basis of such an image, the observer can establish the presence or absence of tubulin in a given location of a cell. The local intensity of the fluorescence signal is not used here as a source of information about the local concentration of tubulin. Such information is not available because the strength of the fluorescence signal depends on the location of the microtubule in relation to the plane of focus. Using the fluorescence signal as a source of information about the local concentration of tubulin is also impossible because immunofluorescence staining may be nonuniform, as the antibody may be sterically hindered by other proteins from accessing the microtubule. Thus, in this example, the only information required and expected is the architecture of a network of microtubules. The interpretation of the image is based on the assumption that all microtubules stain with an antibody to a degree that makes them detectable.

Often, not just the presence or absence, but also the local *concentration* of the labeled protein or other molecular target is of interest. Extracting this information requires using a fluorescence microscope as an analytical device. In all honesty, however, one has to admit that a standard widefield fluorescence microscope is not made to be an analytical device capable of straightforward measurements of the quantities of fluorescently labeled molecules within a specimen. This task can be performed more adequately by confocal microscopes. A rough estimate of local concentrations can be made in a widefield fluorescence microscope; however, a microscopist should keep the following considerations in mind.

Any attempt to estimate *relative* (within the same field of view) or absolute concentrations (see below) of fluorescent molecules is based on the tacit assumption that the amount of the bound fluorescent probe is proportional to the amount of molecules of interest. This assumption is rarely true. Let us take DNA-binding fluorescent probes as an example. Among some 50 fluorescent probes that bind DNA, only a few bind it in an (almost) stoichiometric manner (4′,6-diamidino-2-phenylindole (DAPI), Hoechst, propidium, DRAQ5). Most DNA dyes also bind RNA, thus RNA has to be hydrolyzed before measuring the local concentration of DNA. Hoechst and DAPI have low affinity for RNA, but propidium is an example of a probe with high affinity for RNA. Nevertheless, propidium is a popular DNA probe used for measuring DNA content in cells by flow and laser scanning cytometry. Measurement of the absolute DNA amounts is also complicated by the fact that propidium as well as other DNA-affine dyes compete with proteins for binding to DNA. It has been demonstrated that the amount of propidium bound to DNA is higher in fixed cells, following removal of some of the DNA-associated proteins. This means that assessment of DNA

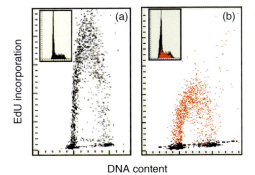

Figure 3.26 Laser scanning cytometry (LSC) determination of DNA content per cell, and the amount of newly synthesized DNA per cell (after delivering a pulse of DNA precursor, EdU; see also Figure 3.4) in a large population of cells. DNA was stained by DAPI; EdU was fluorescently labeled with AlexaFluor 488. (a) Untreated control culture and (b) cells exposed to hydrogen peroxide (200 μM) for 60 min, showing a reduced amount of nascent DNA. EdU-incorporating cells are marked with red and are shown in a DNA histogram in the inset. Each dot plot or histogram represents blue (DAPI) and green (EdU) fluorescence signals measured in over 3000 cells. The insets in (a) and (b) show DNA frequency histograms, based on DAPI signals from the respective cultures. (Zhao et al. [22].)

content requires careful calibration within the investigated system, including using the same type of cells, the same procedure of RNA removal, and so on. When these measures are taken, the assessment of the local concentration of DNA can be quite precise. Simple and reliable determinations of relative amounts of DNA in individual cells can be achieved by staining DNA with DAPI or Hoechst and recording images with a laser scanning cytometer (LSC, Figure 3.26).

Measurements of the *absolute* concentrations of a fluorescently labeled target are more difficult than measurements of relative concentrations within the same field of view. When attempting such measurements, a microscopist needs to consider the limitations described in the previous sections and the following factors that influence the estimate:

- linearity of the detector response;
- dynamic range of the detector;
- vignetting, that is, fluorescence intensity being lower at the edges than in the image center;
- light collecting efficiency and transmission of different objective lenses;
- the influence of the position of the focal plane in relation to the object;
- nonuniform bleaching of the fluorescent probe in different cellular compartments;
- dynamic exchange of the fluorescent probe that occurs in the case of equilibrium staining;
- the influence of collisional quenching of fluorescence (probe to probe or probe to oxygen);
- possible increase of fluorescence intensity resulting from bleaching of self-quenching probes;

- changes in the spectral properties of fluorescent probes upon exposure to exciting light;
- the inner filter effect.

Even this long list of factors may turn out not to be exhaustive in the case of some fluorescent probes and types of samples. It certainly illustrates the fact that measuring the concentrations of fluorescently labeled molecules in microscopy is a cumbersome task, to say the least.

There is at least one important example of relatively accurate measurements of absolute intracellular concentrations in fluorescence microscopy. These are the so-called ratio measurements of calcium and other ions in live cells. They are described briefly in the following section.

3.6.2 Ratiometric Measurements (Ca^{++}, pH)

Some fluorescent dyes respond to changes in calcium or hydrogen ion concentration not just by changing the intensity of fluorescence but by changing the spectral properties of two emission bands (Figure 3.27). In this case, a ratio, but not the absolute fluorescence intensity, can be linked to the calcium concentration or pH value. This approach has a very important advantage. In principle, the ratio between these two emissions is independent of the local concentration of the fluorescent probe. Thus, typically encountered changes of the dye concentration arising from influx, pumping out by multidrug resistance (MDR) mechanisms, photobleaching, and so on, should not influence the measurement. A detailed discussion of the advantages and limitations of ratiometric studies of intracellular calcium and pH goes far beyond the scope of this book. The reader is advised to also consider the limitations of these techniques, arising from the fact that the calcium dyes chelate these ions and thus may disturb the intracellular calcium balance and the fact that exposing intracellular calcium indicators to light is likely to generate singlet oxygen and cause various phototoxic effects [7]. A careful calibration and optimization of ratiometric studies is required.

3.6.3 Measurements of Dimensions in 3D Fluorescence Microscopy

Calibration and measurement of dimensions in the plane of the specimen is straightforward, as described in basic microscopy books. Modern widefield fluorescence microscopy, including microscopy with image deconvolution, deals with 3D rather than 2D objects, so accurate measurements of dimensions along the optical axis are important as well. This measurement is somewhat more complicated because a movement of the objective in relation to the specimen by a given distance does not necessarily shift the image plane by the same distance within the studied specimen. A possible difference may arise from the mismatch between the refractive indices of the immersion medium and the sample, and is caused by spherical aberration. This is most pronounced in the case of an oil-immersion objective lens used to image at a distance from the surface of a coverslip deep into a water-containing sample (Figure 2.22). The effect is less pronounced in the case of a water-immersion lens.

It is worth noting that *size* measurements of subcellular objects in fluorescence microscopy are limited by diffraction and dependent on wavelength. In the

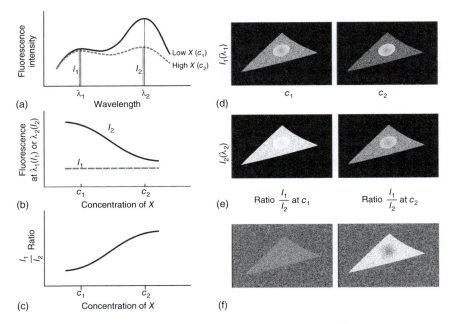

Figure 3.27 The basic principle of fluorescence ratio measurements. The emission spectrum of a fluorophore, shown schematically in (a), is a function of the concentration of a molecule X – for instance, calcium ions. At the wavelength λ_1, the emission intensity I_1 is independent of X, whereas at λ_2 the emission intensity I_2 decreases with concentration of X, as shown in (b). Different intensities of fluorescence of X detected at λ_1 within one cell reflect different local concentrations of the probe. At λ_2, intracellular fluorescence intensities are a function of local concentrations of the probe as well as the local concentrations of molecules of X. The ratio of the intensity of emission I_1 (at λ_1) and I_2 (at λ_2) can be used as a measure of the concentration of X, as shown in (c). A schematic representation of images collected at λ_1 and λ_2 at two different concentrations of $X(c_1, c_2)$ is shown in (d, e), and "ratio images" are shown in (f). In these images, each pixel depicts a ratio between values of I_1 over I_2. This value is independent of the local concentration of the probe and reveals a concentration of X in the cell. (Adapted from Dobrucki (2004), with kind permission of Elsevier)

plane of the specimen, dimensions of the object smaller than half a wavelength cannot be determined. Object length measurements along the optical axis are approximately 3 times less accurate. However, the *distances* between subresolution objects can be measured much more precisely. For instance, the distances between the barycenters of various small subcellular objects have been measured successfully in various microscopy experiments. Fluorescence microscopy also has another sophisticated tool to detect molecules that are less than a few nanometers apart. This approach is based on fluorescence resonance energy transfer (Chapter 13).

3.6.4 Measurements of Exciting Light Intensity

It is often necessary to know and compare the intensity of exciting light emerging out of the objective lens. Such measurements may be needed to calibrate the system in quantitative microscopy or to check the alignment of optical components.

Power meters that are calibrated for measurements of light intensities of various wavelengths are available. A note of caution is needed here because such meters typically have a flat entry window that is to be placed against the incoming light. The high-NA (numerical aperture) oil- and water-immersion lenses produce a cone of light that cannot be measured accurately by such power meters. The outermost light rays hit the meter's window at an angle, which results in reflection. Consequently, a simple flat-window meter placed in front of the high-NA objective lens cannot measure the intensity of the emerging light. Such measurements can be done, for example, for a 10× lens with a low NA.

3.6.5 Technical Tips for Quantitative Fluorescence Microscopy

The ability to perform quantitative fluorescence microscopy hinges upon one's capacity to recognize a number of factors that influence the performance of the microscope and the ability to perform measurements. Besides the parameters that were already mentioned (Section 3.4), these include factors relating to the microscope body, spectral properties, and photophysics of the fluorescent probe and the properties of a particular sample.

Critical issues concerning the microscope include the stability of the exciting light source; chromatic and spherical aberration; channel register, that is, a shift between the images of two color channels in the plane of focus arising from nonideal alignment of the microscope's optical components; the bleed-through between detection channels, that is, detection of photons emitted by one fluorophore in more than one detection channel; dependence of resolution on the wavelength of light emitted by the fluorescent label (i.e., lower resolution for red-emitting labels than blue-emitting ones); chromatic shift of some dyes that occurs upon binding to cellular components; mechanical stability of the microscope; and the choice of optimal excitation versus emission bands.

With regard to the fluorescent probe, one must consider its photobleaching kinetics [8–11]; possible collisional quenching between dyes or between dyes and oxygen, and self-quenching; the possible occurrence of FRET when using more than one color, which would result in an unexpected loss of a donor signal; and, finally, the phototoxic effects. Phototoxicity may result in a supposedly extracellular dye getting into damaged cells, lysosomes bursting during illumination, or calcium oscillations occurring as a result of the action induced by the fluorescent probe itself [7].

Finally, each particular sample has its caveats. To these belong existence of autofluorescence in selected spectral regions [11, 12], the consideration of the inner filter effect, scattering of exciting light [13], or sample aging. Also, one must consider the possible existence of different binding sites for specific labels, as has been discussed in the context of the DNA-staining dyes. This problem often occurs in immunolabeling strategies with antibodies, where cross-reactivities often exist.

One should always keep in mind that the microscope produces a 2D image of samples that are actually three dimensional. As the excitation light illuminates a relatively large area of the specimen in the plane of focus, as well as the regions above and below this plane, fluorescence is always excited in a large volume of the sample. Both the exciting light and the emitted fluorescence are scattered in this region. This causes substantial out-of-focus light, which is detected as background. Discrete fluorescing sample components appear to be surrounded by a "halo" of light, and even the areas between the fluorescing structures of the specimen seem to fluoresce. This blur and background is minimized in a scanning laser confocal microscope (Chapter 5).

3.7 Limitations of Fluorescence Microscopy

The limits of the range of applications of fluorescence microscopy, and its usefulness in studies of live cells, are delineated by three principal issues, namely, photobleaching of fluorescent probes, their toxicity and phototoxicity, and the limited spatial resolution of images.

3.7.1 Photobleaching

Photobleaching is a process of a gradual loss of fluorescence intensity of the specimen arising from interaction between the exciting light and the fluorescent compound. In this process, the fluorescent dye molecules are photochemically destroyed. As this loss of functional dyes occurs during the observation, photobleaching interferes with the collection of high-quality image data. Optimization of image recording parameters, especially the intensity of exciting light, can dramatically reduce the adverse effects of photobleaching (Figure 3.28).

The loss of a dye's fluorescence in a specimen is usually due to photoxidation, which is the oxidation of dye molecules in the presence of light. Photobleaching of fluorescent probes is generally an irreversible process. Other reactions that are

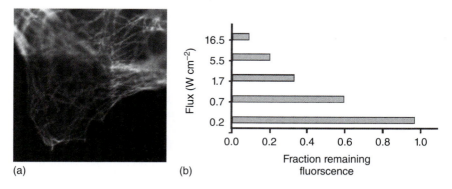

Figure 3.28 Photobleaching. (a) A square area of the fluorescent specimen was exposed to excitation light. The image shows a larger field of view, which embraces the originally illuminated field, thus demonstrating a region that was photobleached. (b) Intensity of exciting light strongly influences rates of photobleaching. In this example, a sample with eGFP was exposed to the same dose of light that was delivered by beams of different intensity. At high light fluxes, fluorescence was bleached almost entirely (only ∼10% of the initial signal remained), while using light of almost two orders of magnitude lower intensity necessitated a much longer data collection time but resulted in no detectable loss of signal. Figure from [8]. Reproduced with permission of John Wiley & Sons.

initiated by light and do not involve oxidation but lead to a change of the structure of a fluorescent molecule can also occur.

It is often possible to slow the loss of signal by adding reducing agents and/or scavengers of reactive oxygen species to the sample. These reagents include N-propyl gallate and mercaptoethanol. Although homemade or commercial preparations designed to slow down photobleaching can be very effective in preventing signal loss, their addition to the sample often causes a loss of the initial signal's intensity – a phenomenon rarely observed by the microscopist who adds the antioxidant solution to the sample before the first recording of any image [9].

Current applications of photobleaching are a nice example of turning a defeat into victory. Photobleaching, which is such an obstacle to performing successful fluorescence imaging, is exploited today in a number of cutting-edge microscopy techniques such as FRAP and related techniques, detection of FRET by acceptor photobleaching, as well as high-resolution techniques including photoactivation localization microscopy (PALM) [14] and stochastic optical reconstruction microscopy (STORM) [15, 16]. FRAP is based on permanently photobleaching a subpopulation of fluorescent molecules in a selected field of view and recording the rate of return (or lack thereof) of fluorescence due to dynamic exchange of these molecules in the photobleached area (for an extensive discussion, see Chapter 11). PALM, STORM, and related techniques use the ability of some fluorescent molecules to be made transiently nonfluorescent (Chapter 8).

3.7.2 Reversible Photobleaching under Oxidizing or Reducing Conditions

Usually, photobleaching is considered to be an irreversible loss of fluorescence by a fluorophore resulting from exposure to exciting light. As such, photobleaching is one of the main limitations of fluorescence microscopy. It has been discovered, however, that some fluorescent proteins can be reversibly bleached and made fluorescent again by exposure to a different wavelength of light [17]. This property is extremely useful, and is now exploited in modern super-resolution microscopy. Recently, it has been demonstrated that under selected conditions some low molecular weight dyes can also be bleached transiently, meaning that a loss of fluorescence is not permanent and emission can be regained over time or by exposing the sample to intense light of a specific wavelength [18, 19]. Apparently, loss of fluorescence can be made reversible under specific reducing/oxidizing conditions [20].

3.7.3 Phototoxicity

Live cells that are fluorescently labeled and exposed to exciting light are subject to the photodynamic effect. This effect is defined as damage inflicted on cells by light in the presence of a photosensitizer and molecular oxygen. A fluorescent dye that interacts with cell components during a live-cell imaging experiment acts as a photosensitizer and causes various types of damage. The adverse effects caused by low molecular weight dyes usually involve oxygen. Chromophores of fluorescent proteins, such as enhanced green fluorescent protein (eGFP), are shielded by the protein moiety, which prevents direct contact with molecular oxygen. This may explain why fluorescent proteins generally appear to be less phototoxic.

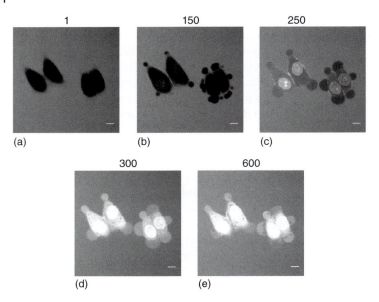

Figure 3.29 An example of phototoxic effects inflicted by a fluorescent probe on live cells. HeLa cells are incubated in culture medium supplemented with ruthenium phenanthroline complex Ru(phen)$_3^{2+}$, which acts as an extracellular photosensitizer (a). The complex is known to generate singlet oxygen when exposed to 458 nm light. Cells cultured on a confocal microscope stage are exposed to 458 nm light (0–600 frames, as indicated above). The first signs of damage are blebbing of the plasma membrane and a slow entry of Ru(phen)$_3^{2+}$ into cell interiors – this is revealed when the complex binds to DNA in nuclei (b, 150 frames). Subsequently, the integrity of plasma membranes is compromised and the complex rapidly enters cells, binds to intracellular structures, and accumulates in cells at a high concentration, which results in bright fluorescence of nuclei and the cytoplasm ((c–e), frames 250–600). (Zarębski et al. [25]).

Phototoxic effects may manifest themselves by readily detectable alterations in cell function, such as a loss of plasma membrane integrity (Figure 3.29), detachment of a cell from the substratum, blebbing, and a loss of mitochondrial potential. However, sometimes even lethal damage can be initially quite inconspicuous. Good examples are the photoxidation of DNA bases, DNA breaks, and so on. Such damage may lead to cell death, but it is not readily detectable during or immediately after a cell imaging experiment. Thus, it is important to recognize that light-induced damage may seriously alter the physiology of the interrogated cell but still remain undetectable for a considerable time. In particular, care is required to establish if the unavoidable phototoxic effects inflicted during a live-cell experiment have sufficient impact on cell physiology so as to influence the interpretation of the collected image data.

3.7.4 Optical Resolution

The Abbe formula (Chapter 2) describes the parameters that influence the optical resolution in the plane of the specimen. However, the formula only applies to images with negligible noise. In fluorescence microscopy, image noise is usually relatively high, while the signals are low and continue to decrease owing to photobleaching. In fact, under such conditions it is the noise level that becomes the decisive factor defining spatial resolution (Figure 3.30).

The optical resolution of a microscope in the horizontal plane of the specimen is ~250 nm, and depends on the wavelength of the light that builds the image (Chapter 2) and the NA of the lens (Figure 3.31). Resolution along the optical axis is proportional to NA^{-2}. With NA being greater than 1 for good lenses, it is worse than ~700 nm. The size of most subcellular structures is well below the resolution limit. For example, the thickness of biological membranes is 5–10 nm, the diameter of an actin filament is about 7 nm, the diameter of a microtubule is about 25 nm, and early endosomes can have a diameter of 100 nm. This means that a typical optical microscope cannot resolve folds of the inner mitochondrial membrane, closely spaced actin fibers, microtubules, or endosomes. Such structures are readily resolved by an electron microscope. However, live-cell studies cannot be performed with this technique owing to too large a thickness of intact cells and the damage induced by the beam of electrons. Thus, for the biologist there is a need for a substantially better spatial resolution to be achieved by fluorescence microscopy. This goal has now been attained by several new microscopy techniques. They are introduced in Chapters 8–10.

3.7.5 Misrepresentation of Small Objects

Even the smallest light-emitting object, such as a molecule of eGFP, will be represented on a fluorescence microscopy image by a diffraction-limited signal, which is a bright circle of a diameter not smaller than ~250 nm in the plane of the

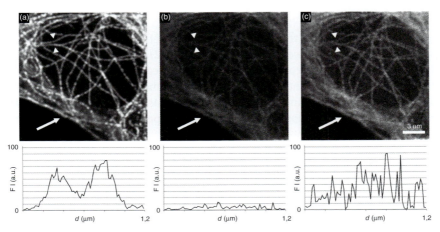

Figure 3.30 The influence of signal-to-noise ratio (S/N) on image quality. (a) An image of microtubules stained by immunofluorescence was recorded at a high fluorescence signal-to-noise ratio. Arrowheads point at the line along which the fluorescence profile (below), running across two microtubules, was drawn. (b) The same area imaged when the S/N ratio was made low due to photobleaching. (c) The image shown in (b) after increasing the brightness. The fluorescence profiles in panels (a), (b), and (c) demonstrate that image resolution becomes significantly lower when the S/N decreases. The two microtubules become visually unresolvable when the intensity of fluorescence signal is low in comparison to the noise. A loss of image resolution can also be seen clearly in the area marked with an arrow. (Images courtesy of Agnieszka Pierzyńska-Mach, Jagiellonian University, Kraków.)

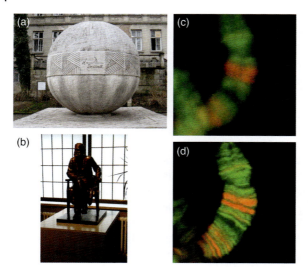

Figure 3.31 The Abbe formula and the role of numerical aperture in image resolution. (a) Abbe's formula engraved on a monument in Jena and (b) a statue of Ernst Abbe in the Optisches Museum of Jena. (c,d) Images showing a fragment of a polytene chromosome from *Chironomus tentans*, stained for condensed and relaxed chromatin. The images were recorded using a (c) 0.7 and (d) 1.4 NA lens, respectively, and demonstrate the substantially higher resolution at higher NA. The original image (d) is much brighter than image (c) owing to the higher NA; therefore, the brightness of (c) was increased for clarity. (Images by W. Krzeszowiec and J. Dobrucki, Jagiellonian University, Kraków.)

specimen. This means that the size of objects smaller than ∼250 nm will be misrepresented by a fluorescence microscope (Figure 3.32). Paradoxically, this means that an observer has to interpret fluorescence images carefully. The size of the cell and the nucleus are represented correctly, but the diameter of actin fibers, microtubules, early endosomes, and the thickness of the plasma membrane, all seen in the same image, will be seriously exaggerated and appear much larger and out of scale in comparison with the cell size.

3.8 Summary and Outlook

Fluorescence microscopy has witnessed an unprecedented growth in the past 20 years. Successful efforts to develop this area of technology have followed several avenues of research. These include studies of dynamic events using FRAP, FRET, speckle microscopy, fluorescence correlation spectroscopy (FCS), and FLIM techniques (Chapters 5, 11, and 13). New trends also include high-resolution stimulated emission depletion (STED) microscopy, structured illumination (SI) microscopy, PALM, and STORM, as well as high-sensitivity and large-area-of-view light sheet fluorescence microscopy [21] (Chapters 5, 6, 8–10) and a microscope equipped with a mesolens. Modern optical microscopy instrumentation has also been developed to study large numbers of cells (laser

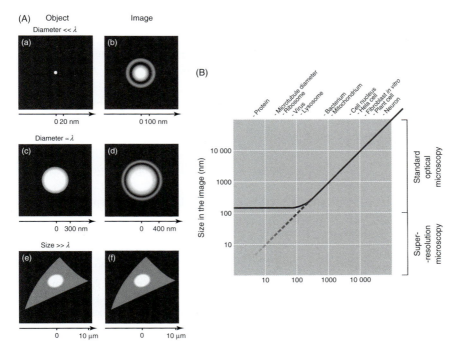

Figure 3.32 Misrepresentation of small objects in fluorescence microscopy. (A) A schematic representation of small objects (left) and their images (right) generated by a standard fluorescence microscope. An image of a bright 3 nm diameter sphere (shown in (a)) is a large circle ~250 nm in diameter, surrounded by interference fringes (b). When a sphere with a diameter comparable to the wavelength of light (a few hundred nanometers) (c) is imaged, the diameter of the image reflects the real size reasonably well (d), that is, in a correct proportion to the field of view; interference effects appear at the edges. When animal cells *in vitro* with a size of 10–30 μm (e) are imaged, their size and shape are reflected correctly and interference effects are less noticeable (f). Note that the object in (a) and the corresponding image (b) are not drawn to scale. If proportions between a 3 nm sphere and the corresponding image were to be maintained, the image (b) would have to be larger than the page on which it is printed. The intensity of interference fringes is exaggerated in diagrams (b) and (d). (B) Relationship between the true size of the object and the size of this object in a fluorescence microscopy image. The smaller the object, the greater (relatively) the distortion. The relative contribution of interference fringes is also greater in images of small objects. Cell components such as protein molecules (if detectable at all) or the thickness of the plasma membrane will appear too large in relation to bigger objects such as the nucleus, and to the whole cell in the same image. New super-resolution microscopy methods are aiming at increasing the spatial resolution and reducing the misrepresentation of small objects. (Adapted from Dobrucki [10], with kind permission of Elsevier.)

scanning cytometry (LSC), high-throughput screening, "image-in-stream" – an extension of flow cytometry). The expansion of these techniques has been facilitated by the development of new low molecular weight and fluorescent protein probes and labels (Chapter 4) and new image analysis tools. Thus, fluorescence microscopy has developed into an analytical tool capable of performing biochemical studies in intact living cells.

References

1 Herschel, J.F.W. (1845) No. I. On a case of superficial colour presented by a homogeneous liquid internally colourless. *Philos. Trans. R. Soc. London*, **135**, 143–145.
2 Herschel, J.F.W. (1845) No. II. On the epipolic dispersion of light, being a supplement to a paper entitled, "On a case of superficial colour presented by a homogeneous liquid internally colourless". *Philos. Trans. R. Soc. London*, **135**, 147–153.
3 Stokes, G.G. (1852) On the change of refrangibility of light. *Philos. Trans. R. Soc. London*, **142**, 463–562.
4 Jabłoński, A. (1945) General theory of pressure broadening of spectral lines. *Phys. Rev.*, **68** (3-4), 78–93.
5 Brown, R.H. and Twiss, R.Q. (1957) Interferometry of the intensity fluctuations in light. I. Basic theory: the correlation between photons in coherent beams of radiation. *Proc. R. Soc. London, Ser. A*, **242** (1230), 300–324.
6 Axelrod, D., Ravdin, P., Koppel, D.E., Schlessinger, J., Webb, W.W., Elson, E.L., and Podleski, T.R. (1976) Lateral motion of fluorescently labeled acetylcholine receptors in membranes of developing muscle fibers. *Proc. Natl. Acad. Sci. U.S.A.*, **73** (12), 4594–4598.
7 Knight, M.M., Roberts, S.R., Lee, D.A., and Bader, D.L. (2003) Live cell imaging using confocal microscopy induces intracellular calcium transients and cell death. *Am. J. Physiol. Cell Physiol.*, **284** (4), C1083–C1089.
8 Bernas, T., Zarębski, M., Cook, P.R., and Dobrucki, J.W. (2004) Minimizing photobleaching during confocal microscopy of fluorescent probes bound to chromatin: role of anoxia and photon flux. *J. Microsc.*, **215** (3), 281–296 (Erratum in: *J. Microsc.* (2004) **216** (Pt 2), 197).
9 Diaspro, A., Chirico, G., Usai, C., Ramoino, P., and Dobrucki, J. (2006) in *Handbook of Biological Confocal Microscopy*, 3rd edn (ed. J. Pawley), Springer, New York, pp. 690–702.
10 Dobrucki, J.W. (2004) Confocal microscopy: quantitative analytical capabilities. *Methods Cell Biol.*, **75**, 41–72.
11 Tsien, R.Y., Ernst, L., and Waggoner, A. (2006) in *Handbook of Biological Confocal Microscopy*, 3rd edn (ed. J. Pawley), Springer, New York, pp. 338–352.
12 Rajwa, B., Bernas, T., Acker, H., Dobrucki, J., and Robinson, J.P. (2007) Single- and two-photon spectral imaging of intrinsic fluorescence of transformed human hepatocytes. *Microsc. Res. Tech.*, **70** (10), 869–879.
13 Dobrucki, J.W., Feret, D., and Noatynska, A. (2007) Scattering of exciting light by live cells in fluorescence confocal imaging: phototoxic effects and relevance for FRAP studies. *Biophys. J.*, **93** (5), 1778–1786.
14 Betzig, E., Patterson, G.H., Sougrat, R., Lindwasser, O.W., Olenych, S., Bonifacino, J.S., Davidson, M.W., Lippincott-Schwartz, J., and Hess, H.F. (2006) Imaging intracellular fluorescent proteins at nanometer resolution. *Science*, **313** (5793), 1642–1645.

15 van de Linde, S., Löschberger, A., Klein, T., Heidbreder, M., Wolter, S., Heilemann, M., and Sauer, M. (2011) Direct stochastic optical reconstruction microscopy with standard fluorescent probes. *Nat. Protoc.*, **6** (7), 991–1009.
16 Rust, M.J., Bates, M., and Zhuang, X. (2006) Sub-diffraction-limit imaging by stochastic optical reconstruction microscopy (STORM). *Nat. Methods*, **3** (10), 793–795.
17 Sinnecker, D., Voigt, P., Hellwig, N., and Schaefer, M. (2005) Reversible photobleaching of enhanced green fluorescent proteins. *Biochemistry*, **44** (18), 7085–7094.
18 Baddeley, D., Jayasinghe, I.D., Cremer, C., Cannell, M.B., and Soeller, C. (2009) Light-induced dark states of organic fluorochromes enable 30 nm resolution imaging in standard media. *Biophys. J.*, **96** (2), L22–L24.
19 Żurek-Biesiada, D., Kędracka-Krok, S., and Dobrucki, J.W. (2013) UV-activated conversion of Hoechst 33258, DAPI, and Vybrant DyeCycle fluorescent dyes into blue-excited, green-emitting protonated forms. *Cytometry Part A*, **83** (5), 441–451.
20 Klein, T., van de Linde, S., and Sauer, M. (2012) Live-cell super-resolution imaging goes multicolor. *ChemBioChem*, **13** (13), 1861–1863.
21 Verveer, P.J., Swoger, J., Pampaloni, F., Greger, K., Marcello, M., and Stelzer, E.H. (2007) High-resolution three-dimensional imaging of large specimens with light sheet-based microscopy. *Nat. Methods*, **4** (4), 311–313.
22 Zhao, H., Dobrucki, J., Rybak, P., Traganos, F., Halicka, D., and Darzynkiewicz, Z. (2011) Induction of DNA damage signaling by oxidative stress in relation to DNA replication as detected using "click chemistry". *Cytometry A.*, **79** (11), 897–902.
23 Biela, E., Galas, J., Lee, B., Johnson, G., Darzynkiewicz, Z., and Dobrucki, J.W. (2013) Col-F, a fluorescent probe for ex vivo confocal imaging of collagen and elastin in animal tissues. *Cytometry A.*, **83** (6), 533–539.
24 Berniak, K., Rybak, P., Bernaś, T., Zarębski, M., Biela, E., Zhao, H., Darzynkiewicz, Z., and Dobrucki, J.W. (2013) Relationship between DNA Damage Response, initiated by camptothecin or oxidative stress, and DNA replication, analyzed by quantitative image analysis. *Cytometry A.*, **83** (10), 913–924.
25 Zarębski, M., Kordon, M., and Dobrucki, J.W. (2014) Photosensitized damage inflicted on plasma membranes of live cells by an extracellular generator of singlet oxygen–a linear dependence of a lethal dose on light intensity. *Photochem Photobiol.*, **90** (3), 709–15.

Recommended Internet Resources

http://micro.magnet.fsu.edu/primer/ by Michael W. Davidson and The Florida State University.
http://www.nobelprize.org/nobel_prizes/medicine/laureates/1906/golgi-lecture.html.
http://www.nobelprize.org/nobel_prizes/medicine/laureates/1906/press.html.
http://www.nobelprize.org/nobel_prizes/physics/laureates/1953/press.html.

http://www.nobelprize.org/nobel_prizes/physics/laureates/1953/zernike-lecture
.html.
http://www.nobelprize.org/nobel_prizes/chemistry/laureates/2008/shimomura-
lecture.html.
http://www.nobelprize.org/search/?query=chalfie.
http://www.nobelprize.org/search/?query=tsien.
http://www.univie.ac.at/mikroskopie/
http://www.nobelprize.org/nobel_prizes/chemistry/laureates/2014/advanced.html.
http://www.andor.com/learning-academy/ccd-spectral-response-%28qe%29-
defining-the-qe-of-a-ccd.

Fluorescent Spectra Database

http://www.spectra.arizona.edu/.

4

Fluorescence Labeling

Gerd Ulrich Nienhaus and Karin Nienhaus

Karlsruher Institut für Technologie, Institut für Angewandte Physik, Wolfgang-Gaede-Str. 1, 76131 Karlsruhe, Germany

4.1 Introduction

Optical imaging is one of the most important techniques in current life sciences research. Fluorescence microscopy, in particular, allows biological processes to be studied as they occur in space and time, at the cellular and molecular levels. The intrinsic fluorescence of biological molecules is weak and not specific and thus of limited use. Fortunately, over the years, a wide array of fluorescent labels and activatable probes have been developed for the investigation of biochemical processes at the molecular level in live and fixed cells. With these extrinsic probes, cells, tissues, and whole organisms can be labeled in very selective ways so as to visualize specific structures and (bio)molecules with high contrast. Selection of the appropriate fluorescent probe and labeling strategy can be a challenging task, however. This chapter gives an overview on fluorescence imaging agents and discusses practical considerations involved in probe selection.

4.2 Key Properties of Fluorescent Labels

The potential of fluorescence imaging is to a great extent determined by the physicochemical properties of the fluorophores employed, including their chemical nature and size, optical properties (see Chapter 3), biocompatibility, and interplay between the dye and a biological entity. In general, there is a large variety of fluorophores to choose from [1]. These include small organic dyes, nanoparticles (NPs) such as semiconductor quantum dots (QDs) and metal nanoclusters, and genetically encoded fluorescent proteins (FPs) (Figure 4.1).

Table 4.1 provides an overall comparison of the various classes of fluorescent labels, but it is not sufficient to enable selection of a particular fluorophore. To that end, key properties have to be considered that may depend on the specific application. These properties may determine the detection limit and the dynamic range of the method, the reliability of the readout for a particular target or event, and the possibility of parallel detection of different targets (multiplexing).

Fluorescence Microscopy: From Principles to Biological Applications, Second Edition.
Edited by Ulrich Kubitscheck.
© 2017 Wiley-VCH Verlag GmbH & Co. KGaA. Published 2017 by Wiley-VCH Verlag GmbH & Co. KGaA.

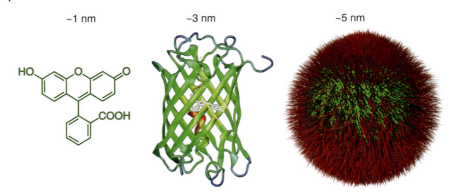

Figure 4.1 Fluorescent labels and their relative sizes. (from left to right) Organic dye, GFP-like fluorescent protein, and nanocrystal.

In general, a suitable label should be sufficiently (photo)stable, conveniently excitable by the available light sources without producing excessive fluorescence background from the biological matrix (intrinsic fluorescence), and have a high brightness to be detectable with conventional instrumentation. Evidently, for biological applications, solubility in buffer solvents, cell culture media, or body fluids is also required. Depending on the specific application, further important considerations might include steric and size-related effects of the label, the pH dependence of the emission properties, the possibility to deliver the label into cells, potential toxicity of the label, and the suitability of the label for multiplexing.

Table 4.2 summarizes the key characteristics of fluorescent labels and their significance for a potential experiment. Optical characteristics include excitation and emission spectra, extinction coefficient, and quantum yield (QY). From the "microscopy hardware" point of view, the optical properties should match the available excitation and detection equipment. As the excitation spectra of organic dyes and FPs consist of individual bands, the available light sources may restrict the selection. In contrast, excitation spectra of QDs do not consist of sharply confined bands but extend from the UV to just below the wavelength of the emission maximum. From the viewpoint of the sample, the excitation wavelength should coincide with the wavelength region that yields the best optical transmission through the sample and minimum autofluorescence. Because phototoxicity and light scattering are reduced at long wavelengths, fluorophores with excitation/emission maxima in the near-infrared (NIR) are preferred for live-cell imaging. Note that the most abundant endogenous fluorophores, for example, aromatic amino acids, and lipo-pigments, and pyridinic (NADPH), and flavin coenzymes, mainly emit in the blue-green region of the spectrum.

The extinction coefficient at the excitation wavelength quantifies the ability of a fluorophore to absorb photons and should be as high as possible. While organic dyes and FPs have extinction coefficients of 10^4–10^5 M^{-1} cm^{-1}, QDs may exhibit up to 100-fold higher values. Excitation and emission wavelengths are tightly coupled for organic dyes and FPs. The main emission band is red-shifted from the main excitation band (*Stokes shift*) by typically 20–40 nm, so once such a label has been selected for its particular excitation wavelength, the

Table 4.1 Fluorescent markers – comparison of key properties.

Class	Examples	Brightness[a]	Photostability[a]	Biocompatibility[a]	Environmental sensitivity[a]	Two-photon excitation[a]
AlexaFluor	AlexaFluor 488	xxxx	xxxx	xxxx	xx	xxx
Atto	Atto 488	xxxx	xxxxx	xxxx	xx	xxx
BODIPY	BODIPY TMR	xxx	xx	xxxx	xxx	xx
Coumarin	Coumarin 6	x	x	xxx	xxxx	x
Cyanines	Cy5, Cy7	xxx	xxxx	xxx	xx	x
Fluorescein	FITC	xxx	x	xxx	xxxx	xx
Rhodamines	Rhodamine 6G	xxx	xxxx	xx	xxx	xxx
Quantum dots	—	xxxxxx	xxxxxx	x	x	xxxxxx
GFP-like proteins	EGFP, EosFP	xx	xx	xxxxx	xx	xx
Phycobiliproteins	R-phycoerythrin	xxxx	xx	xx	xx	xxxx

a) The number of 'x' symbols encodes a coarse scale for the shown quantity.

Table 4.2 Key properties of fluorescent labels.

Property	Definition/impact
Optical properties	
Excitation spectrum	Wavelength dependence of the ability of light to induce fluorescence emission of a chromophore
Extinction coefficient ε_λ (M^{-1} cm^{-1})	Quantity characterizing a fluorophore's ability to absorb photons at a particular wavelength λ
Emission spectrum	Wavelength dependence of the emitted fluorescence
Stokes shift	Wavelength shift between excitation and emission bands
Fluorescence quantum yield (QY)	Ratio between the number of photons emitted and the number of photons absorbed
Molecular brightness	Product QY × ε quantifies the relative rate of photon emission from different fluorophores under identical excitation conditions
Photostability	Resistance to photobleaching, quantified by the quantum yield of photobleaching, that is, ratio of the number of photobleached molecules to the total number of photons absorbed in a sample during the same time interval
Physicochemical properties	
Size	Most often, the smaller the better
Material	May affect chemical stability, toxicity, options to functionalize the probe
Solubility	Most biological applications require aqueous solvents
Cytotoxicity	Effect on sample viability
Phototoxicity	Effect on sample viability
Conjugation chemistry	Determines labeling target
Localization of target	Affects the choice of conjugation
Metabolism of target	Determines temporal stability of labeled construct in live cells

emission range is already determined. Of note, a large Stokes shift is beneficial for efficient filtering of scattered excitation light from the emitted fluorescence. The emission wavelengths of QDs depend on their size. Their emission bands are symmetric and can be narrower than those of organic dyes and FPs (FWHM$_{QD}$ ~30–90 nm, FWHM$_{dye,FP}$ ~70–100 nm). In combination with their broadband absorption, the sharp emission bands make them particularly suitable for multicolor applications.

The efficiency of fluorescence emission is quantified by the *fluorescence quantum yield*, defined as the ratio of the number of photons emitted and the number of photons absorbed by the fluorophore. However, for a fluorophore to be bright, it must also have a high capacity to absorb photons. Consequently, a quantity called *molecular brightness* has been introduced; it is defined as the product of QY and extinction coefficient at the excitation wavelength, ε_{exc}.

In addition to these photophysical properties, basic physical and chemical properties also govern the choice of a particular fluorescence label. Normally, we assume that these labels are chemically stable, at least over the duration of the experiment. Organic dyes and QDs are intrinsically hydrophobic and have to be derivatized with functional groups such as carboxylates or sulfonic acids to render them soluble in aqueous solutions. FPs are proteins, and so they are intrinsically water-soluble. Some may show an aggregation tendency at higher concentration levels, but their solubility can be improved by suitable molecular engineering. The solubility of aromatic dyes also decreases with increasing size, so that additional charged groups have to be attached. The physical size by itself is a key property that varies considerably among fluorescent labels (Figure 4.1). In traditional diffraction-limited fluorescence microscopy, the size of the fluorescence marker and its way of attachment to the structure of interest introduce negligible positional errors. This is, however, no longer true for super-resolution imaging applications. In general, smaller labels may have a better ability to penetrate tissue and a weaker effect on the physiological activity of their biological targets. This property has to be thoroughly investigated for each and every application, though.

Chemical toxicity is another important parameter, especially for long-term experiments on live cells and organisms. Biomolecule-derived labels, including FPs, appear as the best choice in this regard. Incidences of pathological abnormalities related to the expression of FPs have been reported, but only sporadically. Importantly, their expression in the living cell should be kept at a moderate level. The systemic toxicity of fluorescent dyes is also low, although they were observed to accumulate at particular sites within a cell, where they may become locally toxic. CdSe semiconductor QDs are toxic to live specimens because of their chemical composition. However, if the toxic nanocrystalline core is tightly encapsulated in a chemically stable shell (e.g., silica, polymer), toxicity is hardly an issue. Solid assessments are still lacking, however, especially concerning long-term effects.

Beyond pure chemical toxicity, one also has to consider phototoxicity, which derives from the electronic properties of the photoexcited fluorophore. Most importantly, on excitation the fluorophore might react with molecular oxygen to yield a variety of reactive oxygen species (ROS). Phototoxicity results from the damage caused by those ROS on proteins, nucleic acids, and other biomolecules in the sample. Photobleaching is often caused by the same process: the generation of ROS due to the reaction between molecular oxygen and the excited dye in the triplet state. The resulting singlet oxygen is highly reactive and may attack a dye and irreversibly sever the conjugated π-electron system responsible for the fluorescence. It is not surprising that QDs show the highest photostability, as they are rigid nanocrystals containing hundreds and thousands of strongly bonded atoms. However, it is well known that unpassivated QDs promote ROS generation and eventually cause cell death.

Phototoxicity and photobleaching can be minimized by reducing the excitation power and also by applying oxygen scavenging systems, that is, redox-active substances that remove oxygen, and efficient triplet quenchers. Evidently, these agents can only be employed for *in vitro* assays. The most popular system to scavenge oxygen consists of glucose oxidase and catalase in a buffer containing millimolar concentrations of β-D-glucose. Another system uses protocatechuic acid (PCA) and protocatechuate-3,4-dioxygenase (PCD). Alternatively, reducing agents such as ascorbic acid (AA), N-propyl gallate, β-mercaptoethanol, and Trolox have been employed. Most recently, a combination of a reducing (e.g., ascorbic acid) and an oxidizing (e.g., methylviologen) agent was introduced (ROXS, reducing and oxidizing system) so as to efficiently deplete triplet states and to quickly recover oxidized and reduced states, thereby suppressing multiple photobleaching reactions at the same time.

Finally, as an important parameter for selecting a fluorescent label, one has to consider the targeting of the fluorophore to the biomolecule of choice. The label has to be equipped with a functional group that allows binding to the target. Here, target properties come into play, such as available chemical functionalities by which to attach the label, concentration, and localization. These functional groups also determine the labeling specificity. Questions to address are the number of putative attachment sites and the location of these sites on the target, both in view of conserving functional properties of the target and permitting label accessibility and specificity of the conjugation chemistry. Accessibility of the target molecule and its concentration are also important points for the proper choice of labeling strategy.

4.3 Synthetic Fluorophores

4.3.1 Organic Dyes

Organic dyes are generally planar molecules with conjugated π-electron systems, with typical extensions of 1–2 nm and molecular masses of ∼500–1500 g mol^{-1}. Arguably the most popular representative of organic dyes is fluorescein, which was synthesized for the first time in 1871. Since then, the number of synthetic dyes has increased enormously [2], so that by now a large variety of small organic fluorophores have become commercially available for covalent labeling of macromolecules [3, 4].

These dyes can be assigned to different families based on their core scaffolds. Figure 4.2 shows examples of the most common families: coumarins, BODIPY dyes, fluoresceins, rhodamines, and cyanines. The dye scaffolds are shaded according to their excitation/emission wavelength. When searching for a suitable dye, one will come across "ATTO dyes" or "AlexaFluor dyes." These apparent classifications are brand names (ATTO dyes: ATTO-TEC GmbH, Siegen, Germany; AlexaFluor dyes: Thermo Fisher Scientific, Waltham, MA) and do not give any information on the structure of the dye.

As can be seen from the structures in Figure 4.2, excitation and emission wavelengths increase with the extension of the delocalized π-electron system of the fluorophore and, hence, its size. For example, coumarins consist of two rings and absorb and emit in the blue and fluoresceins with three rings in the green range of the spectrum.

A key advantage of all these synthetic fluorophores is that one can use chemistry to control their properties. Optimization with respect to wavelength range, brightness, photostability, and self-quenching has been achieved by

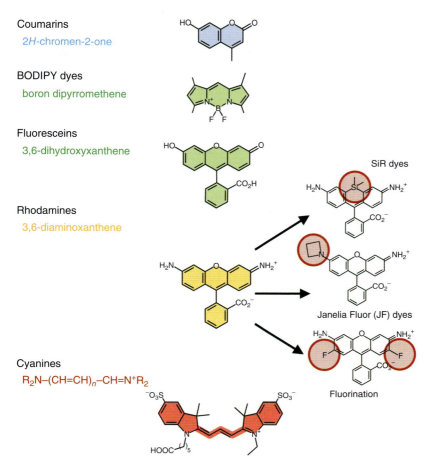

Figure 4.2 Examples of synthetic fluorescent dyes. The core scaffolds are colored according to the wavelengths of their emission maxima.

extending double-bond conjugations, rigidification through extra rings, and decoration with electron-withdrawing or obligatorily charged substituents such as fluorines or sulfonates.

The rhodamine scaffold is arguably the most preferred template for dye development. For example, SiR dyes are derivatives of rhodamines in which the oxygen atom at the 10-position is replaced by a silicon atom (Figure 4.2). Compared to those of the original dyes, their excitation and emission maxima are shifted markedly to the red. The so-called Janelia Fluor (JF) dyes were obtained by replacing the N,N-dimethylamino substituents in tetramethylrhodamine with four-membered azetidine rings. They show a large increase in brightness and photostability. Fluorination of rhodamine dyes was shown to shift the emission to the red and to endow the dye with the capability to penetrate into live cells. Further modifications will follow, driven by the need for the optimal fluorophore for a particular application.

Each class of dyes has its specific advantages and disadvantages (Table 4.1). In general, organic dyes are characterized by high extinction coefficients (10^4–10^5 M^{-1} cm^{-1}), Stokes shifts of typically 20–40 nm, and moderate to high QYs. Typically, the absorption and emission spectra are mirror images of each other. Given the considerable widths of the spectral lines, only a small number of dye molecules are sufficient to cover the entire visible and NIR regions of the spectrum. Therefore, the use of multiple dyes to label different targets in the same sample ("multiplexing") is limited because the poor separation of the optical bands introduces cross-talk between different dye molecules. If the excitation spectra of two dyes partially overlap, it may still be possible to select a wavelength for excitation at which only one of them is excited. In fact, cross-excitation is not a problem as long as the emission bands from the two dyes are well separated. However, partially overlapping excitation spectra typically go along with overlapping emission bands. These will give rise to emission cross-talk ("emission bleed-through"), so that the emission cannot be assigned to one or the other dye unless they have no cross-excitation.

To facilitate specific targeting of these labels to biomolecules both *in vitro* and *in vivo*, they have to be functionalized with suitable groups, as described in Section 4.3.3. Fortunately, in recent years, a large toolbox of functionalized dyes has become commercially available. It is constantly being expanded by new and advanced variants, optimized for a specific application. However, more sophisticated dyes are in general very expensive (>\$200 mg^{-1}). More details can be found in the "Molecular Probes Handbook" [2], arguably the most popular reference for commercially available dyes.

4.3.2 Fluorescent Nanoparticles

Semiconductor QDs are nanocrystals, often made from a core of CdSe or CdSe and a shell of ZnS [5]. For these QDs, core radii are typically 2–5 nm (Figure 4.3a,b), with their emission wavelengths depending on their size (Figure 4.3a) and the core material. They absorb from short wavelengths in the UV up to just below the emission wavelength (Figure 4.3c), so that a single-wavelength excitation light source readily excites QDs of different sizes and, consequently, emission wavelengths. They feature high extinction coefficients, especially in the UV ($\varepsilon > 10^6$ M^{-1} cm^{-1}) and high QYs. The emission bands are rather narrow (FWHM 30–90 nm) and symmetric (Figure 4.3d). Furthermore, QDs are remarkably photostable and, therefore, particularly suited for experiments that require prolonged illumination at high light intensity. An interesting phenomenon observed for individual QDs (but for organic dyes and FPs as well) is fluorescent intermittency, also called *flickering* or *blinking*, which can be disadvantageous for applications such as single-particle tracking (SPT).

Semiconductor core–shell QDs are typically synthesized in an organic solvent, and are initially insoluble in aqueous solutions. Biological applications require additional (hydrophilic) coatings that render QDs water-soluble [7]. These coatings may also provide functional groups for the conjugation to biomolecules (Figure 4.3b). Various approaches have been reported for water solubilization

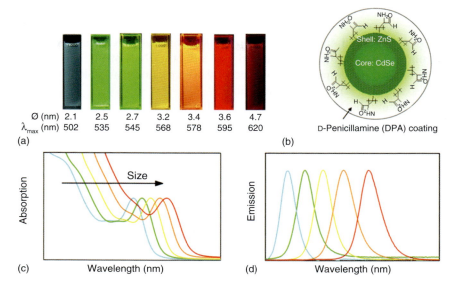

Figure 4.3 (a) Fluorescence emission of quantum dots. Sizes are given below. (b) Schematic depiction of the assembly of a water-soluble semiconductor quantum dot (QD): CdSe core, ZnS shell, coating for solubilization. (c) Absorption and (d) emission spectra of QDs of various sizes. (Breus et al. 2009 [6]. Reproduced with permission of American Chemical Society.)

of QDs, but all of them have their specific advantages and disadvantages and compromise one or more of the essential properties required for optimal biological labels, including small size, chemical stability, and fluorescence QY. Frequently, the application of water-soluble QDs in the life sciences is limited by their poor colloidal stability in physiological media and their nonspecific interaction with biomaterials, particularly cell membranes. Owing to their size, it is difficult to control the number and specific spatial arrangement of biomolecules attached to a QD. Because of an additional polymer shell, some commercially available QDs have overall radii of 10–20 nm, which can be a disadvantage compared to organic dyes. For example, access to receptor binding sites or trafficking of proteins may be hindered by the large label. Recently, however, smaller QDs with monovalent conjugation have also been engineered.

Metal NPs also display size- and shape-dependent optical and electronic features that make them useful as biolabels [8]. Gold NPs (AuNPs) of a few 10 nm diameter scatter light very strongly due to plasmonic excitations of bulk electrons. Very small fluorescent gold nanoclusters (< 2 nm diameter) consisting of only several to tens of atoms show discrete and size-tunable electronic transitions and reasonably high fluorescence QY (~10%). AuNP surfaces can be readily modified with ligands containing functional groups such as thiols, phosphines, and amines to endow them with colloidal stability in water. Additional moieties such as oligonucleotides, proteins, and antibodies can also be attached to the AuNPs to impart specific functionalities. However, problems associated with chemical and fluorescence instability still limit the application of fluorescent nanoclusters to cell or tissue imaging.

Carbon dots constitute another fascinating class of recently discovered NPs with sizes below 10 nm [9], which also feature size- and excitation-wavelength-dependent photoluminescence. They could eventually replace the toxic semiconductor-based QDs currently in use. Still other NPs made from silicon, silicon carbide, and germanium have recently been proposed as biological labels [10], also showing exquisite biocompatibility.

From their photophysical properties, fluorescent NPs appear superior to organic dyes, having much higher brightness and better photostability. Furthermore, their emission wavelength can easily be tuned just by changing the NP size. However, their routine use is restricted by the small number of commercially available marker systems and their limited reproducibility. At present, it is advisable to use a single NP batch within a set of experiments, and to characterize the properties of NPs for each batch prior to use. A key problem that limits their application remains the lack of simple and easy-to-use tools for targeted delivery of NPs to specific biomolecules within cells or into cells, tissues, and organisms (Section 4.3.5).

4.3.3 Conjugation Strategies for Synthetic Fluorophores

Ideally, modification of a biomolecule with a fluorescent label should result in a conjugate in which the biomolecule performs exactly the way it would in its unmodified form. However, it is quite obvious that this may not be true in reality, and thorough control experiments are required to ensure that labeling does not interfere with the functional processes under study.

Each conjugation process involves the reaction of a functional group (on the biomolecule) with another one (on the fluorophore), resulting in the formation of a chemical bond [11]. Depending on the application, one might aim for a 1 : 1 stoichiometric ratio of fluorophore and biomolecule. In certain instances, one may prefer maximum fluorescence emission from the construct. However, that is not necessarily achieved by attaching the largest possible number of fluorophores to the biomolecule because the dyes are prone to self-quenching at high densities. In addition, fluorophores might also be quenched by photoinduced electron transfer from oxidizable donors such as guanosine bases and tryptophan residues to the excited fluorophore. Even though electron transfer is a short-range process, attachment of a fluorophore close to these groups will likely lead to reduced fluorescence QY.

Commercially available organic dyes are typically furnished with suitable functional groups, for example, amine-reactive N-hydroxysuccinimide (NHS) esters, isothiocyanates, or thiol-reactive maleimides and iodoacetyl groups. NHS ester derivatives (Figure 4.4) enable cross-linking to primary and secondary amines, yielding stable amide and imide linkages, respectively [12]. In proteins, NHS ester reagents couple to the α-amine at the N-terminal end and the ε-amines of lysine side chains. A protein can be efficiently labeled with this approach, but the degree and specificity of labeling is unpredictable. This is because a protein molecule typically has several lysine residues with different accessibilities to the dye; hence, both the exact attachment site(s) and the number of labels per molecules will vary throughout the sample. In general, published labeling

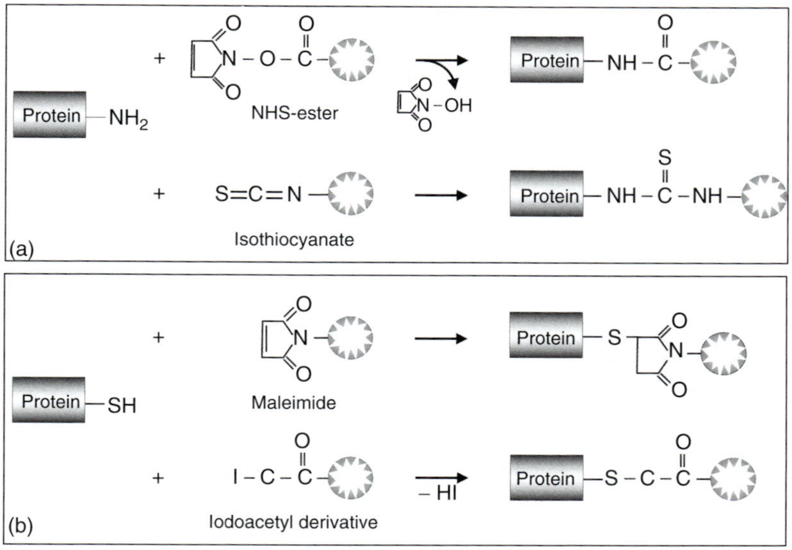

Figure 4.4 Coupling chemistry for the attachment of fluorescent labels: (a) amine- and (b) thiol-reactive functional groups.

protocols require some degree of protein-specific optimization with respect to the dye-to-protein ratio and the reaction conditions (time, temperature, and pH).

Maleimides (Figure 4.4) undergo an alkylation reaction with thiol (−SH) groups to form stable thioether bonds. At pH 7, the maleimide reaction with thiols is ~1000 times faster than with amines. However, some cross-reactivity may occur at higher pH. Because the amino acid cysteine, which carries a side chain with a thiol group, is much less abundant than lysine in a typical protein, thiol derivatization is likely to permit labeling at a single site. In fact, site-directed mutagenesis is often used to exchange the few cysteines already present in the polypeptide chain for other amino acids and to introduce an additional cysteine residue for site-specific labeling of the protein. Because of the ubiquitous presence of lysine and cysteine residues in all proteins, these techniques are restricted to *in vitro* labeling of purified biomolecules (Box 4.1).

Box 4.1 Practical Considerations for Conjugation with NHS Ester and Maleimide-Derivatized Dyes

Once dissolved in aqueous solution, NHS esters are prone to hydrolysis. The reaction buffers should be free of amines/imines (no Tris, no glycine, and no imidazole); the pH should be in the range 7–9. We note that the reaction kinetics are strongly pH dependent. A pH of 8.5–9.5 is usually optimal for modifying lysine residues. In contrast, the α-amino group at a protein's N-terminus normally has a pK_a of ~7, and so it can sometimes be selectively modified by carrying out the reaction at or close to neutral pH.

(Continued)

> **Box 4.1 (Continued)**
>
> Once dissolved in aqueous solution, maleimides are prone to hydrolysis as well. For maleimide coupling, the pH should be in the range 6.5–7.5. At higher pH values, cross-reactivity with amines might occur. To reduce possible disulfide bonds that would prevent the labeling of the cysteine thiol groups, TCEP (tris(2-carboxyethyl)phosphine) should be added before the maleimide-derivatized dye. Note that TCEP, if used in high concentrations, might considerably decrease the sample pH. Alternatively, dithiothreitol (DDT) or β-mercaptoethanol can be used for reduction; however, unlike TCEP, these compounds also carry thiol groups so that any excess has to be removed from the solution before adding the maleimide reagent.

To achieve high conjugation specificity both *in vitro* and *in vivo*, recombinant proteins are often expressed as fusion proteins with a so-called tag, an additional polypeptide moiety that mediates subsequent labeling with a fluorophore. Initially, such tags were introduced for the purification of the protein of interest, the most prominent example being the hexahistidine tag. This sequence of six histidines, when added to the C- or N-terminal end of the protein, endows the protein with high affinity toward affinity resins containing bivalent nickel or cobalt ions complexed via nitrilotriacetic acid (NTA).

The FLAG tag consists of eight amino acids (AspTyrLysAspAspAspAspLys). The Asp residues of the FLAG tag coordinate to a Zn^{2+}–imidazole complex, which is prelabeled with the fluorophore. The Strep tag is another synthetic octapeptide (Trp-Ser-His-Pro-Gln-Phe-Glu-Lys), with an intrinsic affinity toward a specifically engineered streptavidin.

The tetracysteine (TC) tag (Cys-Cys-Pro-Gly-Cys-Cys) specifically binds bis-arsenical compounds such as FlAsH or ReAsH (Figure 4.5), both of which are commercially available as Lumio Green and Lumio Red (life Technologies; http://www.lifetechnologies.com). Blue-fluorescent compounds such as CHoX-AsH are also available. In these compounds, two arsenic atoms are kept at a set distance from each other. Initially, it was reported that the bis-arsenical compound recognized a motif consisting of four cysteine side chains spanning a single α-helical turn. Later on it was realized that the TC tag can also be added at the N- or C-terminal end of a protein under study as long as the spacing and relative orientation of the two Cys-Cys pairs matches the distance of the arsenic atoms. This requirement is best met by Pro-Gly. If the tetracysteine motif is "complementary" to the initially nonfluorescent bis-arsenical dye, the two moieties can form a fluorescent complex in which each arsenic atom is bound to a pair of cysteine sulfur atoms. The main advantage of this tag is its relatively small size; it can be as small as six amino acids, with CCPGCC (in single-letter amino acid code) being the preferred motif. A limitation of this method is that the dye can bind to the target cysteines only if they are in the reduced form.

Other tags are derived from enzymes, and are therefore significantly larger. The SNAP tag is a variant of a small enzyme O^6-alkylguanine-DNA alkyltransferase (AGT), which can be specifically labeled with benzylguanine (BG) derivatives. In

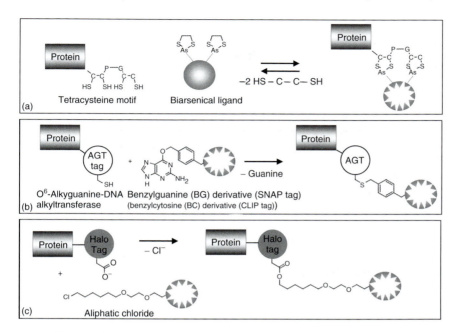

Figure 4.5 (a–c) Coupling chemistry for the attachment of fluorescent labels: protein tags and fusion approaches.

its regular function, AGT repairs DNA lesions resulting from the O^6-alkylation of guanine by irreversibly transferring the alkyl group to a reactive cysteine of AGT. Another AGT-derived tag is the CLIP tag; it reacts specifically with O^2-benzylcytosine (BC) derivatives. Both constructs are commercially available (New England Biolabs; http://www.neb.com). They can be fused N-terminally or C-terminally to the protein of interest at the DNA level, so that the fusion protein contains an extra AGT domain for dye attachment. A variety of fluorescent BG and BC derivatives with different emission wavelengths are commercially available for multicolor imaging applications, both *in vitro* and *in vivo*. Recently, NIR silicon–rhodamine dye derivatives (SiR-SNAP, SiR-CLIP) have been introduced. The specificity of SNAP-/CLIP-tag labeling and the broad range of available substrates are attractive features. Some substrates can even cross cell membranes, and therefore can be used for live-cell applications. However, the size of the AGT-based tags (~200 amino acids) may be an intrinsic disadvantage in some applications.

The HaloTag, a modified haloalkane dehalogenase from *Rhodococcus rhodochrous*, is the largest of the widely used tags (~290 amino acids). It is engineered to covalently bind synthetic ligands (Halo-Tag ligands), comprised of a chloroalkane linker attached to a variety of fluorescent dyes. Covalent bond formation between the protein tag and the chloroalkane linker is highly specific. The essentially irreversible reaction proceeds rapidly under physiological conditions (Figure 4.5).

An alternative approach to labeling fusion constructs involves post-translational modification by appropriate enzymes. The general principle is simple: a

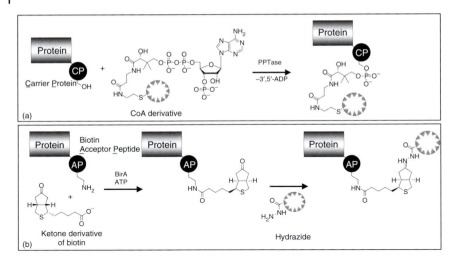

Figure 4.6 (a,b) Coupling chemistry for the attachment of fluorescent labels: labeling of fusion constructs through posttranslational modifications.

recognition peptide is fused to the protein of interest, and a natural or engineered enzyme ligates a small-molecule probe (which itself may be fluorescent or can be specifically labeled) to the recognition peptide. Phosphopantetheinyl transferase (PPTase)-catalyzed protein labeling is depicted in Figure 4.6. The basic function of PPTases is to transfer the phosphopantetheinyl group from Coenzyme A (CoA) to a conserved serine residue of a carrier protein (CP) domain. In the experiment, the CP domain acts as the recognition peptide. Fluorescently marked CoA is added as a substrate; PPT recognizes the substrate and transfers the fluorescently labeled phosphopantetheinyl group to the fusion protein.

Another general method for labeling fusion proteins through post-translational modification is based on biotinylation of a lysine residue within a 15-residue acceptor peptide (AP) by the biotin ligase BirA, from *Escherichia coli*. The AP peptide is fused to the protein of interest on the DNA level. Importantly, BirA can also recognize a ketone-containing analog of biotin, namely ketobiotin, as a substrate (Figure 4.6). The introduced ketobiotin function can be covalently labeled with hydrazide compounds linked to fluorescent probes.

4.3.4 Non-natural Amino Acids

Conjugation of organic dyes functionalized with NHS ester or maleimide groups to a protein has certain drawbacks (see Box 4.1). Importantly, the degree and specificity of labeling are unpredictable. Therefore, researchers have aimed at introducing other chemically orthogonal functional groups into proteins that allow precise control of the conjugation site and stoichiometry. A very elegant approach to reach this goal has been advanced by Schultz and coworkers [13]. They developed a recombinant method to incorporate unnatural amino acids (UAAs) into proteins. These UAAs have modified side chains and can be incorporated at a desired position in any protein. The structure of the protein is in general only minimally perturbed by UAA incorporation, and the site-specifically

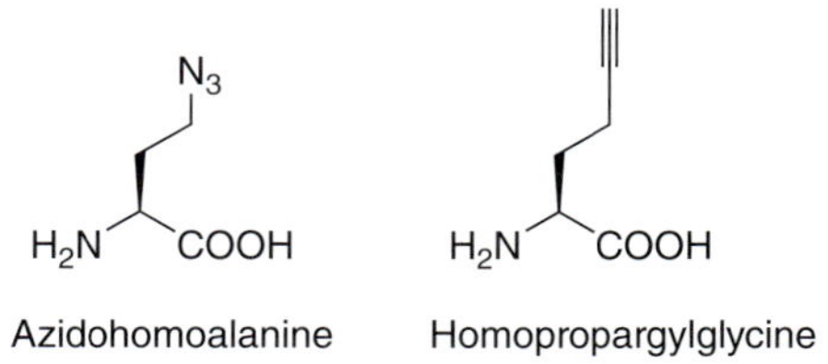

Figure 4.7 Non-natural amino acids with azide and alkyne functionality.

Azidohomoalanine   Homopropargylglycine

modified proteins are typically expressed in high yields in bacteria, yeast, or mammalian cells. In the meantime, a wide variety of UAAs, including those containing ketone, azide, alkyne, alkene, and tetrazine side chains, have been genetically encoded (Figure 4.7). Introduction of an azide group, for instance, enables its chemoselective modification with triarylphosphine reagents via a Staudinger ligation or the copper(I) catalyzed [3 + 2]-cycloaddition of an alkyne in a so-called click reaction. For *in vivo* applications, cyclooctynes can be used to react with the azide groups through a strain-promoted cycloaddition that does not require copper(I) as a catalyst. Copper ions are toxic to living cells and thus should be avoided. Azide-containing non-natural amino acids frequently suffer from reduction of the azido group during expression and purification. Therefore, amino acids with alkyne functionality, such as propargyllysine and propargylphenylalanine, are preferable. Various azide-functionalized organic dyes are commercially available (Life Technologies; http://www.lifetechnologies.com).

4.3.5 Bringing the Fluorophore to Its Target

Conjugation chemistry determines, at least in part, the labeling specificity. Techniques such as NHS ester or maleimide coupling are not specific to one kind of biomolecule owing to the ubiquitous presence of lysines and cysteines, and are therefore not recommended for cellular applications. The more specific labeling methods for live-cell applications rely on the specific "lock-and-key" principle, for example, the interaction between a tag and its complementary functionality [11, 14].

After selecting fluorophore and target molecule, the two reaction partners have to be brought in close proximity for the conjugation reaction to take place. For *in vitro* labeling of purified biomolecules, the dye and target molecule are dissolved in aqueous solution (pH 7–9) and mixed in appropriate concentrations. The required time for the reaction to complete depends, among other parameters, on the reaction partners, pH, and temperature. Subsequently, the protein–dye conjugate may be separated from unreacted dye, for example, by gel filtration chromatography. Ion-exchange columns allow separation of protein molecules according to the number of attached labels.

Labeling a particular target within a cell or within certain tissue specimens is more challenging, however. The procedure is relatively straightforward for fixed specimens with permeabilized cell membranes. In this case, immunostaining is the most popular technique for specific labeling. Immunofluorescence uses the specificity of immunoglobulins to their antigens to target fluorophores to specific biomolecular targets within a cell. Primary, or direct, immunofluorescence uses a single immunoglobulin G (IgG) antibody, which is chemically linked to the fluorophore, typically via NHS ester coupling. Secondary, or indirect, immunofluorescence uses two IgGs; the primary antibody recognizes the target molecule and binds to it. The secondary antibody carrying the fluorophore recognizes the primary antibody and binds to it. The fluorophore might be a dye, a QD, or an FP. However, antibodies cannot cross the cell membrane, so immunofluorescence is limited to fixed and permeabilized cells (for intracellular targets) or to the extracellular side of the cell membrane (for live cells).

IgGs are comparatively large proteins of 150 kDa, so that the fluorescent label is significantly displaced from the site to be marked, which can be detrimental for certain applications, for example, super-resolution imaging. In such cases, immunostaining using nanobodies [15] is preferable. A nanobody is the variable domain (VHH, 12–15 kDa) of a camelid heavy chain-only antibody. This smallest functional antigen-binding fragment can be produced recombinantly and easily coupled to fluorescent dyes via NHS-ester chemistry, and then used for fixed-cell staining.

Live cells and tissues present additional challenges for label delivery. Neutral, mono-anionic, and mono-cationic dye molecules with molecular mass $< 1000 \,\text{g mol}^{-1}$ such as LysoTracker Red, JC-1, MitoTracker Red CMXRos, or BODIPY FL ceramide penetrate the cell membrane fairly easily. They might also be internalized via ATP-gated cation channels if these are present in the particular cell type. An acetate ester derivative of a polyanionic dye might also diffuse across the plasma membrane. Once inside the cell, the dye is released by endogenous esterase activity. Organic dyes and QDs can also be conjugated with peptides for their intracellular delivery [16]. Polycationic peptides, in particular, are known to penetrate the cell membrane and to translocate to the cytosol. The efficiency and the exact mechanism of peptide-mediated uptake depend on the details of the peptide, including the charge and amino acid sequence of the peptide, size of the labeled construct, and the particular cell type under study.

The compounds SiR-tubulin and SiR-actin are cell-permeable and emit in the far-red region of the spectrum. In these compounds, SiR has been ligated to docetaxel and desbromo-desmethyl-jasplakinolide, which are small cytotoxic compounds that are known to specifically target tubulin and actin, respectively. Therefore, they enable specific labeling of actin and tubulin in live cells. These probes combine minimal cytotoxicity with excellent brightness and photostability. For *in vivo* studies, the labeling efficiency can be enhanced by adding Verapamil, a broad-spectrum efflux-pump inhibitor (tebu-bio; http://www.tebu-bio.com).

FlAsH and ReAsH labeling reagents are membrane-permeable and readily cross the cell membrane, allowing labeling and detection of recombinant proteins fused to the TC tag in live cells. The SNAP and CLIP tag systems and the HaloTag share this crucial advantage; the majority of their fluorescent derivatives can cross the membrane also fairly easily. Notably, nanobodies were also shown to penetrate the cell wall of live yeast cells.

Alternatively, it is possible to transiently render the cell membrane permeable by using low doses of bacterial toxins, by osmotic permeabilization or by electroporation. Finally, (micro-)injection procedures can be employed. They are highly invasive, but allow extremely precise control. Such a forced delivery of fluorophores may, for instance, be used to inject quantitative amounts of QDs into the cytosol.

4.4 Genetically Encoded Labels

4.4.1 Phycobiliproteins

The first FP labels used in cell biology were phycobiliproteins, that is, photosynthetic antenna pigments extracted from cyanobacteria. The most commonly used phycobiliprotein for biomolecular labeling is R-phycoerythrin, a red fluorescent, multisubunit protein. R-phycoerythrin has 34 bilin groups (Figure 4.8), which endow the reagent with a high net extinction coefficient. Strong absorption bands in the visible region of the spectrum extend from green to far-red wavelengths, with maximum absorption at 565 nm. R-phycoerythrin can be efficiently excited with 488 nm light ($\varepsilon_{488} \sim 1.1 \times 10^6$ M^{-1} cm^{-1}); its QY is 0.82. However, its considerable size (molecular mass 240 kDa) limits its applicability as an intracellular fluorescent label.

Therefore, R-phycoerythrins are mainly used in immunoassays. Antibody labeling kits are commercially available from various companies. These kits also provide suitable bifunctional cross-linkers to prepare phycobiliprotein conjugates. For example, cross-linkers with NHS-ester and pyridyldithiol reactive groups react with primary amines on the surface of the phycobiliprotein and introduce pyridyldisulfide group(s) that can be reacted with another protein (typically an antibody) that contains a free sulfhydryl. Alternatively, R-phycoerythrins may be attached to their target protein via avidin/biotin coupling. The avidin/biotin system is an elegant system to link proteins in immunoassays. It exploits the very high affinity ($K_d \sim 10^{-15}$ M) of biotin (vitamin H) molecules toward avidin, streptavidin, and the engineered derivatives NeutrAvidin and CaptAvidin. Avidin is a glycoprotein found in the egg white and tissues of birds, reptiles, and amphibians. Because of its high isoelectric point of 10.5, it is prone to nonspecific adsorption to negatively charged surfaces such as cell membranes or silica substrates. Streptavidin is a non-glycosylated protein with a near-neutral isoelectric point, isolated from the

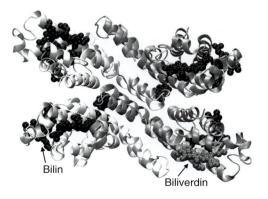

Figure 4.8 Cartoon representation of the R-phycoerythrin multimer (PDB code 1EYX). The biliverdin and bilin prosthetic group are shown in gray and black, respectively.

bacterium *Streptomyces avidinii*. NeutrAvidin is a deglycosylated form of avidin (pI = 6.3). CaptAvidin shows reduced affinity for biotinylated molecules above pH 9. All avidins can bind up to four biotin ligands. To couple two proteins, both are biotinylated and then linked via avidin.

In 2009, a bacteriophytochrome from *Deinococcus radiodurans* was engineered into two different monomeric infrared-fluorescent proteins (IFPs) of only ~320 amino acids, with biliverdin as the chromophore. Biliverdin is the initial intermediate during heme degradation by heme oxygenase (HO-1) in all aerobic organisms. In solution, biliverdin itself is nonfluorescent; even high concentrations (normal adult humans endogenously generate and metabolize 300–500 mg biliverdin/day) do not contribute to the fluorescence background. As soon as the cofactor is irreversibly incorporated into an IFP, its conformational flexibility is reduced by the protein matrix and it turns fluorescent. In the meantime, several NIR fluorescent probes derived from different bacterial phytochrome photoreceptors have become available [17]. The excitation and emission peaks of the IFPs are far in the red region of the spectrum, above 680 nm. Thus, they enable cell and tissue imaging with low scattering and minimal contributions of cellular autofluorescence. IFPs can be excited by cheap laser diodes, offer new wavelengths for multicolor labeling, and may serve as acceptor dyes in FRET (Förster resonance energy transfer) pairs.

Most of the NIR fluorescent probes derived from bacterial phytochrome photoreceptors are multimeric complexes, predominantly dimers. However, an FP should be monomeric in fusion protein applications. Recently, researchers have succeeded in engineering a truncated form of a bacteriophytochrome into a monomeric infrared fluorescent protein (mIFP).

4.4.2 GFP-Like Proteins

The green fluorescent protein (GFP) from the hydromedusa *Aequorea victoria* and the related FPs of the GFP family have found widespread application as fluorescence labels in the life sciences [18–22]. GFP's polypeptide chain consists of 238 amino acids; it folds into a rigid, 11-stranded β-can, with a central helix running along its axis (Figure 4.9). The β-can is capped at either end by short helical sections and loops. The fluorescent chromophore interrupts the central helix and resides close to the geometric center of the protein. The GFP chromophore 4-(*p*-hydroxybenzylidene)-5-imidazolinone (*p*-HBI) forms autocatalytically from a tripeptide (for GFP: Ser65-Tyr66-Gly67), requiring only molecular oxygen (Figure 4.9). Advanced variants of *A. victoria* GFP were

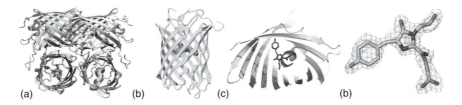

Figure 4.9 Molecular structure of GFP-like proteins: (a) quaternary arrangement, (b) monomer, (c) view into a monomer, and (d) the chromophore itself.

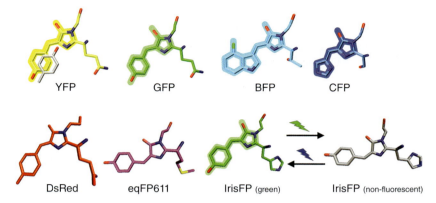

Figure 4.10 Variations in the chromophore structure of GFP-like proteins. For abbreviations, see Box 4.2.

developed by mutagenesis, with emission peaks ranging from blue (blue fluorescent protein; BFP to yellow (yellow fluorescent protein; YFP [18] (Figure 4.10). Homologs of GFP were found in anthozoa, most importantly, the long-sought orange and red FPs. In recent years, the so-called photoactivatable (PA) or optical highlighter FPs have emerged as powerful new tools for cellular imaging [20, 23, 24]. On irradiation with light of specific wavelengths, these FPs either switch reversibly between a fluorescent and a nonfluorescent state (photoswitching) or change their fluorescence properties irreversibly (photoconversion).

In principle, FPs could be functionalized with any of the mentioned groups/tags and conjugated to the complementary functional group on the target biomolecule (Sections 4.4.3 and 4.4.5), with all the advantages and disadvantages already discussed. For example, the fusion protein of a nanobody with an FP results in a so-called chromobody, which can be used to trace an antigen in various compartments of a living cells expressing this chromobody. In practice, however, functionalization with a tag is mainly pursued for immobilization of FPs on surfaces.

Much more prevalent and unique to this class of labels is their use in fusion proteins in which the FP gene is fused to the gene of the protein of interest at the DNA level, separated by a sequence coding for a short amino acid linker. Upon expression of the fusion construct in the cell, the additional FP domain ensures perfect labeling of the target protein (Figure 4.11); no further conjugation chemistry is necessary. However, although this approach appears straightforward, it still requires some considerations to be successful.

FPs obtained from natural sources often show a tendency to aggregate or oligomerize. For some applications, the oligomeric nature of an FP is entirely irrelevant, such as tissue imaging, monitoring of gene activity, and highlighting of cells and cellular compartments. However, oligomerization may be detrimental in studies involving FP fusions with other proteins because it may interfere with the function of the fusion partner. FP monomerization is essential to prevent artifacts because the formation of dimers and higher order oligomers induced by the FP moiety of a fusion construct may lead to improper targeting and

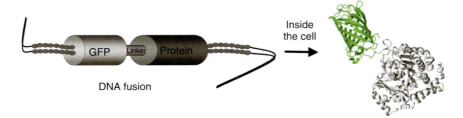

Figure 4.11 The gene coding for GFP is fused to the gene of the protein of interest on the DNA level. The fusion construct is subsequently expressed by the cell, eliminating the need for any additional conjugation chemistry.

atypical localization, disrupt normal function, alter subcellular dynamics of the tagged protein, or lead to aggregation and cytotoxicity. The basic strategy for overcoming oligomerization is to modify the FP amino acid sequence to include residues that disrupt the binding interfaces between the protomers.

For live-cell applications that involve mammalian cell cultures, it is necessary that the FP expresses and functions well at 37 °C. Most native FPs fold fairly efficiently when expressed at or below room temperature, but their folding efficiency declines steeply at higher temperatures. Codon optimization of GFP has led to an increase in the levels of protein expression by fourfold at 37 °C. For other FPs, a few amino acid replacements were sufficient to obtain a thermostable FP, for example, to enhance protein expression and maturation at elevated temperatures. For studying dynamic gene expression patterns with high temporal resolution or for monitoring short-lived proteins under these conditions, rapidly maturing and thermostable FPs are preferable.

Box 4.2 FPs with Modified Chromophores: A Detailed Look

A variety of spontaneous chemical alterations of the *p*-HBI chromophore have been found in GFP-like proteins from natural sources that affect their optical properties (Figure 4.10). Modifications have also been artificially introduced by site-directed mutagenesis. Substitution of Trp for Tyr at position 66 in the chromophore tripeptide produces a new chromophore with an indole instead of a phenol or phenolate. These proteins are called *cyan fluorescent proteins* or *CFPs* because of their blue-green emission. Substitution of His for Tyr shifts the emission maximum to even shorter wavelengths than Trp66, generating BFPs. In the YFP variant of *A. victoria* GFP, Thr203 is replaced by tyrosine; residue 65 is Gly or Thr instead of Ser so as to promote ionization of the chromophore. The Tyr203 hydroxyphenyl side chain is stacked on top of the phenol ring of the chromophore, thereby extending the aromatic system and shifting the fluorescence to the yellow/red region of the spectrum (Figure 4.10). eqFP611 is the only naturally occurring FP with its chromophore in the trans conformation (Figure 4.10). Red-shifted variants of this protein with enhanced fractions of cis isomers, RFP630 and RFP639, have been created, and presently the most red-shifted FP variants have emission maxima of ~670 nm.

Photoinduced modifications of the chromophore can also occur. The first PA protein known – PA-GFP – was derived from wild-type GFP by a single point mutation, Thr203His. This variant is devoid of any fluorescence until activated by irradiation with intense violet light. Crystal structures of PA-GFP determined before and after irradiation revealed that photoactivation is the result of a UV-induced decarboxylation of the Glu222 side chain which causes a shift of the chromophore equilibrium to the anionic form.

In another class of photoconvertible FPs, which includes Kaede, EosFP and its variant IrisFP, and a few other proteins, irradiation into the absorption band of the neutral green chromophore (~400 nm) shifts the emission maximum from the green to the red. This irreversible process is associated with a backbone cleavage between the Nα and Cα atoms of the first amino acid in the chromophore-forming tripeptide, which is always a histidine in this class (Figure 4.10).

A new generation of improved optical highlighters, which include Dronpa, kindling fluorescent protein 1 (KFP1), and mTFP0.7, among others, can toggle between a bright fluorescent state and a dark state in response to irradiation with light of specific wavelengths. The change in emission intensity is based on light-induced cis–trans (or trans–cis) isomerization of the chromophore, accompanied by its protonation (or deprotonation). In most photoswitchers, the anionic cis chromophore is the thermodynamically stable and fluorescent species. Illumination into its absorption band not only excites fluorescence but also switches the chromophore into its trans state, albeit with a probability of <1%. Cis–trans isomerization is accompanied by chromophore protonation. The trans species is nonfluorescent and has a strongly blue-shifted absorption. Irradiation into its absorption band switches the chromophore back into the fluorescent, deprotonated cis state. Alternatively, this relaxation process occurs within minutes to hours in the dark by thermal activation. Because fluorescence excitation turns off the emission, these photoswitchers are called *negative switchers* [20]. Accordingly, positive photoswitchers show an increase in the emission when illuminated at the peak excitation wavelength.

(Continued)

Box 4.2 (Continued)

Reversible and irreversible phototransformations in IrisFP. The wavelength of the respective activating light is indicated.

The Phe173Ser mutant of the tetrameric variant of EosFP, called *IrisFP*, shows both reversible and irreversible phototransformations. In its green fluorescent state, IrisFP displays reversible photoswitching between a bright state and a dark state, which involves cis–trans isomerization of the chromophore. Similar to its parent protein EosFP, IrisFP also photoconverts irreversibly to a red-emitting state under violet light because of an extension of the conjugated π-electron system of the chromophore, accompanied by a cleavage of the polypeptide backbone. The red form of IrisFP exhibits a second reversible photoswitching process, also involving cis–trans isomerization of the chromophore. This peculiar photoactivation behavior enables pulse–chase experiments with super-resolution.

Especially when expression of the target protein is low, bright FPs increase the signal-to-noise ratio and thus the detection sensitivity, which facilitates quantification. Red-emitting FPs are preferable because (i) cells tolerate red light better than green or blue light, (ii) the autofluorescence of a cell is in the green region of the spectrum, and (iii) red light is scattered less and thus penetrates deeper into the tissue. Yet, high-intensity light of any wavelength is inherently damaging to live cells because it may lead to accumulation of ROS. Long-term experiments of protein migration and translocation within cells require photostable FPs to minimize undesired photobleaching during image acquisition. Proteins targeted to cellular compartments with non-neutral pH should be tagged with pH-insensitive FPs ($pK_a < 5.0$). Another important issue is the construction of the fusion protein itself. Experiments are required to find out which fusion site (N-terminal, C-terminal, or in loops) is best tolerated so that functionality and fluorescence are preserved. PA FPs are frequently employed for optical pulse–chase experiments in live cells, tissues, and organisms. The FPs are activated in specific locations, and their migration can be followed without interference from newly expressed proteins. They can be tagged by light, with no interference from new protein synthesis. Localization-based super-resolution imaging techniques rely on the presence of PA fluorophores, and PA FPs are also most convenient for live-cell imaging applications using these methods. Another type of super-resolution microscopy, which is based on stimulated emission depletion (STED), is preferentially performed with organic dye labeling because very photostable markers are required; however, applications using FPs have also been reported. The optical properties of selected FPs are listed in Table 4.3.

4.5 Label Selection for Particular Applications

In the preceding sections, we have discussed the chemical and physical properties of small organic dyes, nanocrystals, and genetically encodable FPs (Table 4.4). Obviously, there is no perfect "one kind fits all" fluorescent label. If size matters most, small organic dyes are the best choice. For high photostability, QDs are unmatched. And only FP labels can be genetically introduced into their targets, without requiring additional conjugation chemistry. Therefore, one has to consider all pros and cons to select the most appropriate label for a particular application, as discussed in the following sections [25].

4.5.1 FRET to Monitor Intramolecular Conformational Dynamics

Proteins and other biomolecules are highly complex systems, exhibiting a substantial degree of structural variability in their folded state. In the presence of denaturants, the heterogeneity is further enhanced, and fluctuations among vast numbers of folded and unfolded conformations occur via many different pathways. Using single-molecule fluorescence microscopy, the structure and dynamics of an individual protein or nucleic acid can be measured via FRET between two dye labels denoted as donor and acceptor. FRET is a nonradiative mechanism of acceptor excitation via radiative donor excitation. For details, we refer the reader to Chapter 13.

For FRET labeling, two different fluorophores are attached at specific sites of the biomolecule, typically ~30–70 Å apart. The dye pair is selected such that donor emission and acceptor excitation spectrum have an appropriate spectral overlap (Figure 4.12). This quantity (among others) controls the Förster radius, the distance between the two dyes at which the FRET efficiency is 0.5.

Protein dynamics can also be measured within live cells. To ensure complete and site-specific labeling, it is recommended to use a fusion construct, with the

Table 4.3 Fluorescent proteins.

Name	Excitation (nm)	Emission (nm)	Brightness[a]	Photostability[a]	Oligomerization state[a]
Fluorescent proteins					
GFP	395/475	508/503	×	×	D
EBFP	383	447	××	×	D
mECFP	433/452	475/505	××	×××	M
mTFP1	462	492	××××	××××	M
EGFP	488	507	×××	××××	D
EYFP	514	527	××××	××	D
Venus	515	528	××××	×	D
DsRed	558	583	××××	××××	T
mRFP1	584	607	××	×	M
tdTomato	554	581	××××	×××	tD
mOrange	548	562	××××	×	M
mKO	548	559	×××	×××	M
mRuby	558	605	×××	×××	M
eqFP611	559	611	××××	×××	T
mStrawberry	574	596	×××	×	M
mCherry	587	610	××	×××	M
mKate	588	635	×××	××××	M
mPlum	590	649	×	××	M
mGarnet	598	670	×	××	M
mCardinal	604	659	×	××	M
eqFP670	605	670	×	××	D
Photoactivatable proteins			Conversion states		
Irreversibly photoactivatable proteins					
PA-GFP	504	517	Dark/green		D
Kaede	508/572	518/580	Green/red		T
mEosFP	505/569	516/581	Green/red		M
Dendra	486/558	505/575	Green/red		M
mIrisFP	486/546	516/578	Green/red		M
Reversibly photoactivatable proteins					
asFP595	572	595	Dark/red		T
KFP	580	600	Dark/red		T
Dronpa	503	518	Dark/green		M
mIrisFP	486/546	516/578	Dark/green, dark/red		M
NijiFP	469/507	526/569	Dark/green, dark/red		M

a) The number of '×' symbols encodes a coarse scale for the shown quantity. M, monomer; D, weak dimer; tD, tandem dimer; and T, tetramer.

Table 4.4 Comparison of fluorescence markers.

Property	Organic dye	Quantum dot	FP
Optical properties			
Absorption spectra	Discrete band(s)	Continuous absorption from UV almost to emission	Discrete band(s)
Extinction coefficient, ε (M^{-1} cm^{-1})	$2.5 \times 10^4 - 2.5 \times 10^5$	$10^5 - 10^6$ [a]	$10^4 - 2 \times 10^5$
Emission spectra	Asymmetric band with tail to the red	Symmetric band, narrow	Asymmetric band with tail to the red
Emission maximum	UV–NIR	Vis–NIR	Vis
Stokes shift (nm)	Typically <50	Up to >100	10–50
Quantum yield (QY)	0.5–1.0 (vis)	0.1–0.8 (vis)	0.05–0.8
	0.05–0.25 (NIR)	0.2–0.7 (NIR)	
Photostability[b]	××	×××××	××
Fluorescence lifetime (ns)	1–10	10–100	1–10
Possible environmental effects on emission properties	pH, polarity	—	pH, solvent
Biocompatibility			
Solubility in aqueous solution	Achieved by functional groups	Requires surface modification	Yes
Largest diameter (nm)	~0.5–1	~5–60	~3
Conjugation to biomolecules	Conjugation chemistry	Conjugation chemistry	Fusion on DNA level (conjugation chemistry)
Toxicity[b]	××	×××××	×
In vitro conjugation	Yes	Yes	Yes
In vivo conjugation	Yes	Yes	Yes

a) For most red-shifted absorption peak.
b) The number of '×' symbols encodes a coarse scale for the shown quantity.

protein gene sandwiched between the genes of two FPs. Suitable FP-based FRET pairs can be found in recent reviews [26–29]).

The FRET technique can also be used to detect protein–protein interactions, both *in vitro* and in live cells. This application requires that each interacting partner is labeled with one fluorophore of a FRET pair (Figure 4.13a). A rather elegant method for observing interactions between two biomolecules relies on bimolecular fluorescence complementation. Here, the FP is split at a suitable site and the two halves are fused to the supposed interaction partners via flexible linkers. As soon as the two partners interact and, concomitantly, the two FP halves combine to reconstitute the complete FP, the fluorescence signal is restored as well.

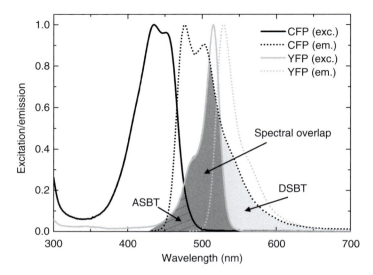

Figure 4.12 Excitation (solid lines; exc) and emission (dotted lines; em) spectra of a donor (CFP) and an acceptor (YFP). The FRET pair CFP/YFP has sufficient spectral overlap, although there is considerable spectral bleed-through (ASBT, acceptor spectral bleed-through and DSBT, donor spectral bleed-through).

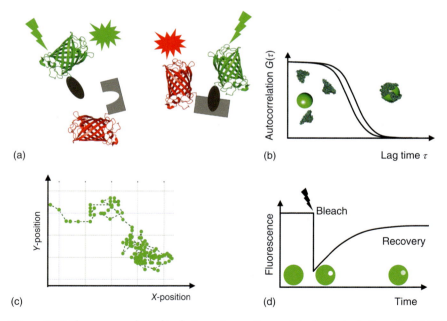

Figure 4.13 Fluorescence-based techniques to monitor protein dynamics in live cells: (a) FRET (Förster resonance energy transfer), (b) FCS (fluorescence correlation spectroscopy), (c) SPT (single-particle tracking), and (d) FRAP (fluorescence recovery after photobleaching). For details, see the text.

4.5.2 Protein Expression in Cells

Expression of a particular protein in the cell can be observed *in vivo* if the protein has been fused to a suitable FP and the gene construct has been transferred into the cell. The procedure that introduces foreign DNA into cells is called *transfection*. Transiently transfected genes are not integrated into the genome of the cell and only expressed for a limited period. In contrast, upon stable transfection, the foreign genes become integrated into the host genome so that descendants of these transfected cells will also express the new gene.

There are several techniques for transfection. Conventional methods include virus-mediated transport of the DNA, complexation of the DNA with cationic polymers or cationic lipids to favor transport across the cell membrane, direct microinjection of the DNA, electroporation of the cell membrane, and laser-based transfection.

It must be recognized, however, that any amount of fusion protein expressed in such transfected cells is, by definition, overexpression relative to the endogenous protein. While transient transfection provides the greatest flexibility for analyzing protein function, this approach may yield very high levels of fusion protein expression in the target cells, resulting in improper subcellular distribution and protein dysfunction. Eventually, this could lead to erroneous interpretations of protein localization and activity.

Recent advances in genome engineering technologies have enabled researchers to directly edit or modulate the genome in its endogenous context in virtually any organism of choice; that is, the foreign gene can be inserted directly into the genome [30]. This so-called targeted genome editing uses site-specific nucleases, such as meganucleases, zinc finger (ZF) nucleases, transcription activator-like effectors (TALEs), and, most importantly, RNA-guided endonucleases known as *Cas9* from the microbial adaptive immune system clustered regularly interspaced short palindromic repeats (CRISPR). Briefly, Cas9 has two active sites each of which can cleave one strand of a double-stranded DNA. The enzyme is guided to the target DNA by an RNA molecule that contains a sequence that matches the sequence to be cleaved. These site-specific double-stranded DNA breaks are then repaired by the cell repair machinery. During homologous recombination, a foreign gene can be inserted at the break site, and, therefore, it will be expressed at endogenous levels.

For optimum detection efficiency of the fluorophore and minimal phototoxicity due to the excitation light, red-emitting FPs are preferred. For target proteins with low expression yields or with widely distributed expression sites, the FP should be as bright as possible.

A quantitative determination of the expression yield requires that newly expressed proteins can be distinguished from old copies. Fusion constructs of a target protein with an FP can be bleached or photoactivated to select proteins expressed before or after illumination. By photoactivation, it is also possible to label all proteins present at a particular time with one color and those synthesized afterward with another color.

4.5.3 Fluorophores as Sensors Inside the Cell

A wide array of fluorescent probes exist that are sensitive to metal ions such as Ca^{2+}. Calcium-sensitive fluorophores change either their emission intensity or wavelength as a function of Ca^{2+} concentration. However, essentially all metal ion probes feature side reactions with other metal ions than the targeted one. Among the most studied ion sensors are H^+-selective probes. Many fluorophores such as fluorescein have intrinsic pH sensitivity. Certain FPs can reflect pH changes via the emission intensity or spectral region. Other FPs can detect halides, and fluorophores have also been reported that detect ROS and reactive nitrogen species. The key challenge in the development of sensors is to achieve a high specificity for only a single compound [31, 32].

4.5.4 Live-Cell Dynamics

Translocation of fluorescently labeled proteins within a live cell can be directly imaged by using a variety of approaches [33]. For example, fluorescence correlation spectroscopy (FCS) is a technique that monitors the fluorescence signal emanating from a very small (~1 fl) optically defined volume (Chapter 5). The fluorescence intensity fluctuates as the labeled molecules diffuse in and out, and the residence time in the detection volume yields the diffusion coefficient for the FP-labeled proteins in the volume (Figure 4.13b).

In SPT experiments (Chapter 12), individual molecules have to be brightly labeled and sparse enough to be tracked individually (Figure 4.13c). For long-term observation, labels have to be particularly photostable. Here, the unique optical properties of QDs can be especially beneficial. Because of their large extinction coefficients, narrow emission spectra, and resistance to photodegradation, QDs may be individually detected with high signal-to-noise ratio for many minutes under continuous irradiation, which is not possible when using conventional organic dyes.

Within a cell, fluorescent labels can also be photoselected, for example, by bleaching or photoactivation, with a short and intense focused laser pulse. Fluorescence recovery after photobleaching (FRAP; Chapter 11) refers to an optical technique in which a certain region of the sample is photobleached (Figure 4.13d). The subsequent recovery of the fluorescence due to diffusion within this area is followed as a function of time and yields information on the dynamics of the fluorophore.

4.5.5 Super-Resolution Imaging

In STED microscopy, the sample is raster-scanned by an excitation spot generated by a tightly focused laser. A second high-power laser beam with a doughnut-shaped intensity profile is overlaid and efficiently de-excites all fluorophores in the periphery of the excitation spot via stimulated emission. As the two laser beams are moved in small steps across the sample, a particular fluorophore will be excited and de-excited many times. Therefore, robust organic dyes are preferable as labels rather than FPs (see Chapter 10).

Instead of de-exciting with high-power lasers, one may also exploit the photoswitching properties of fluorophores between bright and dark states to achieve de-excitation. This generalization is known as *reversible saturable optical fluorescence transitions* (RESOLFT) microscopy. Typically, very low laser powers are sufficient for the depletion beam because the lifetime in the on-state is typically orders of magnitude greater than the fluorescence lifetime. As for STED, however, the markers still have to undergo many transitions between their fluorescent and nonfluorescent states during image acquisition.

Structured illumination microscopy (SIM) uses patterned illumination to excite the sample and generate fluorescence with a corresponding emission pattern (see Chapter 9). Super-resolution is achieved by using high excitation intensity so that the fluorescence emission becomes saturated and, therefore, is no longer proportional to the excitation power. The spatial resolution of saturated structured illumination microscopy (SSIM) increases with the degree of saturation. Accordingly, this technique requires very photostable fluorophores to avoid photobleaching. As for RESOLFT microscopy, the intensities can dramatically be reduced when employing photoinduced on–off switching of fluorophores.

Super-resolution localization microscopy utilizes photoactivatable fluorophores, so that individual fluorophores can be randomly photoactivated, that is, switched to the fluorescent state (on), while the surrounding molecules remain in the dark (off) state (see Chapter 8). Stochastic optical reconstruction microscopy (STORM) and direct stochastic optical reconstruction microscopy (dSTORM) use photoswitchable organic dyes. They should feature very high brightness and contrast levels to yield the maximum number of photons per molecule before photobleaching or return to the dark, nonfluorescent state. Various dyes are commercially available and have been characterized for this application [34].

For localization microscopy of live cells, PA FPs have become very popular. Presently, green-to-red photoconvertible FPs such as Kaede or EosFP are the optimal choice. They are expressed in their deactivated state, which is a green-emitting species. Consequently, they can be located in the cell prior to photoactivation. The light used for photoconversion (~400 nm) is well separated in wavelength from that exciting the green (~480 nm) and red (532 nm) fluorophores, so the risk of unintended photoconversion is greatly reduced. A key advantage is that red-emitting fluorophores form only after photoconversion, so the dynamic range (intensity ratio between the activated and non-activated forms) is very high. Photoswitchable FPs have less dynamic range but can also be used. Furthermore, caged dyes and photoswitchable dyes that are initially nonfluorescent and can be converted to a fluorescent state are also commercially available (Abberior; http://www.abberior.com).

4.6 Summary

Fluorescence-based microscopy has matured into a high-resolution technique that gives unprecedented insight into the machinery of living cells and organisms. This success story would not have been possible without the parallel advancement of fluorescent labels featuring a wide variety of key physical and chemical properties. Likewise, progress in the specific conjugation of biomolecules with fluorophores has been crucial as well. The diversity of labels and the numerous parameters that may affect their usefulness in a specific application pose serious challenges to the experimenter; there is no "one-fits-all" fluorophore. One has to consider whether a key parameter such as size, conjugation chemistry, selectivity, (photo)stability, or any other property tips the scale toward a particular fluorophore.

In this chapter, we have summarized the properties of various fluorescent labels, including small organic dyes, NPs such as semiconductor QDs and metal nanoclusters, and genetically encoded FPs, to facilitate the critical choice of the most suitable marker. QDs are arguably the brightest and also the most photostable labels. However, they are rather large, not always biocompatible, and oftentimes difficult to attach to a specific target moiety. Organic dyes are also bright and photostable, and their optical properties can easily be tuned by chemistry. If size matters most, small organic dyes are the best choice, although their conjugation to a biomolecule of interest may further add to the overall size of the label. If size does not matter, the expression of proteins fused with genetically encoded chemical tags (SNAP, etc.) can provide very bright, specific labeling. However, many dyes cannot penetrate cell membranes, so labeling of live samples can be problematic. FPs are comparatively poor performers with respect to brightness and photostability. However, in live-cell and especially live-organism applications, these disadvantages are completely outweighed by the fact that they are genetically endodable. As a result, they ensure perfect labeling of the target protein also in live samples; no further conjugation chemistry is necessary. For a fluorescence imaging experiment to be successful, judicious choice of a suitable marker is of utmost relevance. The concise overview given here may serve the reader as an entry point in the process of finding the optimal labeling strategy.

References

1 Hilderbrand, S.A. (2010) Labels and probes for live cell imaging: overview and selection guide. *Methods Mol. Biol.*, **591**, 17–45.
2 Haugland, R.P. (2005) *The Handbook—A Guide to Fluorescent Probes and Labeling Technologies*, 11th edn, Molecular Probes, Eugene, OR.
3 Lavis, L.D. and Raines, R.T. (2008) Bright ideas for chemical biology. *ACS Chem. Biol.*, **3**, 142–155.
4 Lavis, L.D. and Raines, R.T. (2014) Bright building blocks for chemical biology. *ACS Chem. Biol.*, **9**, 855–866.
5 Michalet, X., Pinaud, F.F., Bentolila, L.A., Tsay, J.M., Doose, S., Li, J.J., Sundaresan, G., Wu, A.M., Gambhir, S.S., and Weiss, S. (2005) Quantum dots for live cells, in vivo imaging, and diagnostics. *Science*, **307**, 538–544.
6 Breus, V.V., Heyes, C.D., Tron, K., and Nienhaus, G.U. (2009) Zwitterionic biocompatible quantum dots for wide pH stability and weak nonspecific binding to cells. *ACS Nano*, **3**, 2573–2580.
7 Hotz, C.Z. (2005) Applications of quantum dots in biology: an overview. *Methods Mol. Biol.*, **303**, 1–17.

8 Giljohann, D.A., Seferos, D.S., Daniel, W.L., Massich, M.D., Patel, P.C., and Mirkin, C.A. (2010) Gold nanoparticles for biology and medicine. *Angew. Chem. Int. Ed.*, **49**, 3280–3294.
9 Baker, S.N. and Baker, G.A. (2010) Luminescent carbon nanodots: emergent nanolights. *Angew. Chem. Int. Ed.*, **49**, 6726–6744.
10 Fan, J. and Chu, P.K. (2010) Group IV nanoparticles: synthesis, properties, and biological applications. *Small*, **6**, 2080–2098.
11 Sahoo, H. (2012) Fluorescent labeling techniques in biomolecules: a flashback. *RSC Adv.*, **2**, 7017–7029.
12 Hermanson, G.T. (1996) *Bioconjugate Techniques*, Academic Press, San Diego, CA.
13 Wang, L., Xie, J., and Schultz, P.G. (2006) Expanding the genetic code. *Annu. Rev. Biophys. Biomol. Struct.*, **35**, 225–249.
14 Sletten, E.M. and Bertozzi, C.R. (2009) Bioorthogonal chemistry: fishing for selectivity in a sea of functionality. *Angew. Chem. Int. Ed.*, **48**, 6974–6998.
15 Muyldermans, S. (2013) Nanobodies: natural single-domain antibodies. *Annu. Rev. Biochem.*, **82**, 775–797.
16 Biju, V., Itoh, T., and Ishikawa, M. (2010) Delivering quantum dots to cells: bioconjugated quantum dots for targeted and nonspecific extracellular and intracellular imaging. *Chem. Soc. Rev.*, **39**, 3031–3056.
17 Shcherbakova, D.M., Baloban, M., and Verkhusha, V.V. (2015) Near-infrared fluorescent proteins engineered from bacterial phytochromes. *Curr. Opin. Chem. Biol.*, **27**, 52–63.
18 Tsien, R.Y. (1998) The green fluorescent protein. *Annu. Rev. Biochem.*, **67**, 509–544.
19 Frommer, W.B., Davidson, M.W., and Campbell, R.E. (2009) Genetically encoded biosensors based on engineered fluorescent proteins. *Chem. Soc. Rev.*, **38**, 2833–2841.
20 Nienhaus, K. and Nienhaus, G.U. (2014) Fluorescent proteins for live-cell imaging with super-resolution. *Chem. Soc. Rev.*, **43**, 1088–1106.
21 Day, R.N. and Davidson, M.W. (2009) The fluorescent protein palette: tools for cellular imaging. *Chem. Soc. Rev.*, **38**, 2887–2921.
22 Chudakov, D.M., Matz, M.V., Lukyanov, S., and Lukyanov, K.A. (2010) Fluorescent proteins and their applications in imaging living cells and tissues. *Physiol. Rev.*, **90**, 1103–1163.
23 Wiedenmann, J. and Nienhaus, G.U. (2006) Live-cell imaging with EosFP and other photoactivatable marker proteins of the GFP family. *Expert Rev. Proteomics*, **3**, 361–374.
24 Lukyanov, K.A., Chudakov, D.M., Lukyanov, S., and Verkhusha, V.V. (2005) Innovation: photoactivatable fluorescent proteins. *Nat. Rev. Mol. Cell Biol.*, **6**, 885–891.
25 Correa, I.R. (2014) Live-cell reporters for fluorescence imaging. *Curr. Opin. Chem. Biol.*, **20**, 36–45.
26 Shaner, N.C., Patterson, G.H., and Davidson, M.W. (2007) Advances in fluorescent protein technology. *J. Cell Sci.*, **120**, 4247–4260.
27 Shaner, N.C., Steinbach, P.A., and Tsien, R.Y. (2005) A guide to choosing fluorescent proteins. *Nat. Methods*, **2**, 905–909.

28 Wang, Y., Shyy, J.Y., and Chien, S. (2008) Fluorescence proteins, live-cell imaging, and mechanobiology: seeing is believing. *Annu. Rev. Biomed. Eng.*, **10**, 1–38.

29 Miyawaki, A. and Niino, Y. (2015) Molecular spies for bioimaging-fluorescent protein-based probes. *Mol. Cell.*, **58**, 632–643.

30 Sander, J.D. and Joung, J.K. (2014) CRIRISPR-Cas systems for editing, regulating and targeting genomes. *Nat. Biotechnol.*, **32**, 347–355.

31 Yao, J., Yang, M., and Duan, Y.X. (2014) Chemistry, biology, and medicine of fluorescent nanomaterials and related systems: new insights into biosensing, bioimaging, genomics, diagnostics, and therapy. *Chem. Rev.*, **114**, 6130–6178.

32 Guo, Z.Q., Park, S., Yoon, J., and Shin, I. (2014) Recent progress in the development of near-infrared fluorescent probes for bioimaging applications. *Chem. Soc. Rev.*, **43**, 16–29.

33 Liu, Z., Lavis, L.D., and Betzig, E. (2015) Imaging live-cell dynamics and structure at the single-molecule level. *Mol. Cell.*, **58**, 644–659.

34 Dempsey, G.T., Vaughan, J.C., Chen, K.H., Bates, M., and Zhuang, X. (2011) Evaluation of fluorophores for optimal performance in localization-based super-resolution imaging. *Nat. Methods*, **8**, 1027–1036.

5

Confocal Microscopy

Nikolaus Naredi-Rainer, Jens Prescher, Achim Hartschuh, and Don C. Lamb

Department of Chemistry, Ludwig-Maximilians-Universität München, Butenandtstr. 5–13, München 81377, Germany

5.1 Evolution and Limits of Conventional Widefield Microscopy

Conventional widefield fluorescence microscopy (cf. Chapters 1 and 2) captivates the observer with details that it can reveal about the microscopic world with comparatively low technical effort. Scientists have always been striving to find ways to enhance this powerful tool, and significant advances have been achieved. In its long history, Robert Hooke was one of the early pioneers, publishing the seminal book *Micrographia* in 1665, and is considered to be the father of optical microscopy. However, the capacity of Hooke's microscope was limited technically by the quality of the lenses, which was rather poor. A contemporary of Hooke, Antoni van Leeuwenhoek, was an excellent lens grinder and was able to make outstanding lenses with short focal lengths, greatly improving the resolving power of microscopes. The microscopes built by van Leeuwenhoek were constructed using a single lens, and hence they were technically only simple magnifying glasses, but were capable of magnifications of 200–300-fold, allowing some of the first detailed observations of the microscopic world. The investigations of van Leeuwenhoek are still considered exceptional, and from his descriptions and drawings, it is easy to recognize the bacteria (referred to as *animalcules*) or algae that he observed.

The next breakthrough in microscopy occurred 200 years later in 1846 when Carl Zeiss set out to build optical microscopes whose spatial resolution was no longer limited by technology. He joined up with Ernst Abbe, who presented a theoretical derivation of the resolution capacity of an optical microscope. Expanding on an earlier theory by George Airy [1], he showed that diffraction of light on the finite opening of the microscope objective is the most noticeable factor that limits resolution [2]. The diffraction pattern of light focused through an ideal lens with a circular aperture is described by the Airy pattern. Its size depends on the angular aperture α and the refractive index n of the surrounding medium. In this context, he also coined the term *numerical aperture* (NA) defined as $NA = n \sin \alpha$ (cf. Section 2.2.4). Together with Hermann von

---

Fluorescence Microscopy: From Principles to Biological Applications, Second Edition.
Edited by Ulrich Kubitscheck.
© 2017 Wiley-VCH Verlag GmbH & Co. KGaA. Published 2017 by Wiley-VCH Verlag GmbH & Co. KGaA.

Helmholtz, Abbe was able to describe the smallest distance, Δx, at which objects could be separated and still resolved:

$$\Delta x = \frac{\lambda}{2\mathrm{NA}} \tag{5.1}$$

where λ is the wavelength of light. Lord Rayleigh later took into account how an ideal point object would be imaged in three-dimensional space. According to Rayleigh's criterion [3], the distance Δx at which two points can just be resolved corresponds to the distance from the center of the Airy pattern to its first minimum (Figure 2.9b):

$$\Delta x = \frac{0.61\lambda}{\mathrm{NA}} \tag{5.2}$$

A formal and more detailed approach to the Rayleigh criterion is given in Chapter 2.

Although the Rayleigh criterion is often used in microscopy for estimating the diffraction-limited resolution, the real resolving power of a microscope depends on the image contrast and photon statistics, or, in general, on the signal-to-noise ratio of the measurement. This led the astrophysicist C. M. Sparrow to present another specification for the *resolution limit* (the Sparrow criterion). Two signals are still resolvable until the dip between the two point-spread functions (PSFs), as shown in Figure 2.10c, is no longer detectable [4].

There are several reasons why contrast can be disadvantageously low in widefield microscopy, thus limiting its resolving power. First, assuming that absorption of the sample is negligible, excitation in widefield microscopy is not limited to the focal plane. The resulting image on the detector is not restricted to signals originating from the focal plane. However, points that do not lie in the focal plane of the detector and the objective will be imaged out of focus and their intensity will be dispersed in space, increasing the background signal. Furthermore, especially for biological samples, scattering effects such as the Tyndall effect are very common. This effect describes the scattering of light from particles whose size is comparable to the wavelength of the light. Because of the scattering of light, photons that were originally coming from out-of-focus areas of the sample may be redirected into the focal plane and subsequently contribute to a diffuse background. It is often the low image contrast that limits the spatial resolution rather than pure wave-optics criteria. For this reason, it was obvious that new ways to increase image contrast had to be found in order to broaden the hitherto-existing applications of optical microscopy.

5.2 Theory of Confocal Microscopy

5.2.1 Principle of Confocal Microscopy

In the late 1950s, Marvin Minsky developed confocal microscopy with the aim of being able to perform imaging in dense tissue such as the brain (see Box 5.1). The trick of confocal microscopy is to use two pinholes to restrict the excitation and detection beams and thereby suppress the detection of out-of-focus light.

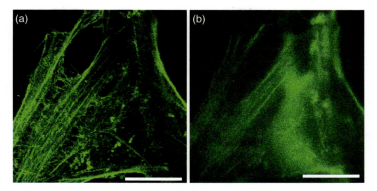

Figure 5.1 Confocal and widefield image of HeLa cell actin filaments. These images demonstrate the capability of confocal microscopy to provide detailed images of cellular structures through the improved axial resolution. The actin filaments of a HeLa cell were stained with Alexa Fluor 647-phalloidin and imaged. (a) On the scanning confocal microscope, actin filaments are clearly visible without blurring from out-of-focus fluorescence. A three-dimensional reconstruction from a confocal z-stack of the cell can be viewed in the online material. (b) The corresponding conventional widefield microscope image is blurred and a high fluorescent background is visible. Scale bars: 20 μm.

The two pinholes are both mounted in image planes of the microscope and are thereby "con"-focal. Figure 5.1 illustrates the sectioning capabilities of confocal microscopy. Actin filaments in HeLa cells were stained with Alexa Fluor 647-labeled phalloidin. The confocal fluorescence image reveals details of complex structures in three dimensions (Figure 5.1a and online supplementary material) that would not be accessible by widefield microscopy (Figure 5.1b). A schematic of the beam paths of a modern confocal microscope is shown in Figure 5.2a. For comparison, we have also included a schematic of one of our home-built confocal microscope setups in Figure 5.2b. Typically, a coherent laser beam is chosen as the excitation source (shown in Figure 5.2a as a dotted

Figure 5.2 Principles of confocal microscopy. (a) Schematic of a simple confocal microscope illustrating the function of the confocal detection pinhole. The excitation beam (dotted line, black) is focused onto a sample, and light originating from the focus will pass through the pinhole and reach the detector (solid line, green), whereas light originating from positions adjacent to the focal spot (dotted line, red) or from a different focal plane will be cut out by the pinhole (dotted line, blue). (b) Example for experimental realization of an advanced laser scanning confocal microscope. In this case, pulsed lasers were chosen as excitation sources rather than the continuous wave lasers that are typically used in confocal microscopy. The lasers are combined using dichroic mirrors (DMs). In this setup, one of the lasers also serves as a master clock to synchronize the entire system. The excitation light is guided through a single-mode fiber toward the microscope objective, with an XY galvo mirror system being used for laser scanning. The fluorescence is collected by the same objective, de-scanned by the galvo mirrors, and separated from excitation light by a polychroic mirror (PM). After passing through the confocal pinhole, the fluorescence is separated by color using DMs. The spectra are subsequently cleaned up by the appropriate emission filters (EFs). Avalanche photodiodes (APDs) were chosen as detectors to collect fluorescence photons. Photon counts are registered subsequently by time-correlated single-photon counting (TCSPC) electronics.

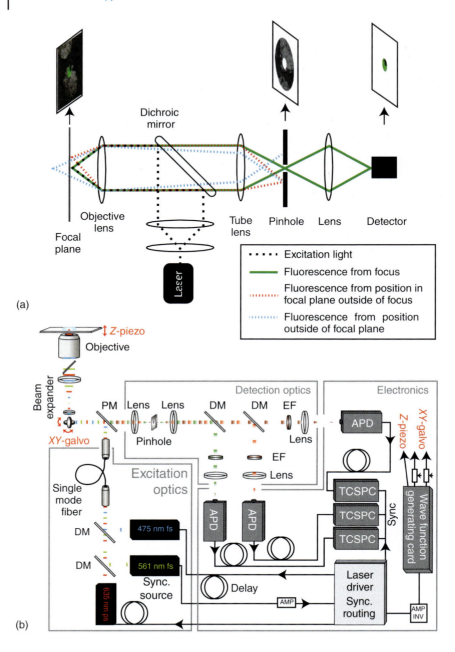

black line), in which case an excitation pinhole is no longer necessary. The excitation beam, expanded by a set of lenses, is subsequently reflected by a dichroic mirror that splits the excitation light from the fluorescence signal (shown as a solid green line) based upon their different wavelengths (cf. Section 3.4.3). Fluorescence light is collected with the same objective lens and transmitted through the dichroic mirror. To ensure that only light from a small region of the sample reaches the detector, a pinhole is placed in the image plane of the sample.

The confocal detection pinhole defines what light reaches the detector: Fluorescence photons originating from the focus of the excitation beam pass through the pinhole and are detected (green line, Figure 5.2a). Out-of-focus light (or in other words, light that is not confocal with the pinhole) is blocked by the pinhole. For example, the blue line in Figure 5.2a depicts the path of photons coming from a different axial plane of the sample. This light is focused at a different position within the microscope and is out of focus with respect to the plane of the pinhole. Hence, the vast majority of light will be blocked by the pinhole. Likewise, background light coming from the focal plane but displaced laterally from the position of the pinhole in the sample plane (red line, Figure 5.2a) will also be blocked. Although less light reaches the detector, the confocal pinhole leads to a significant improvement in the signal-to-noise ratio of the collected image compared to bright-field methods. As only photons emitted from the focal plane are detected, it is possible to detect an optical slice of the sample, as illustrated by the example of a confocal image of a cell depicted above the beam path in Figure 5.2a. As we will see in the more mathematical discussion in the following section, the major impact of the pinhole under normal conditions is in generating z-resolution with little influence on the lateral resolution.

After the pinhole, the remaining light beam can be directed immediately onto a detector, which might be a photomultiplier tube (PMT), an avalanche photodiode (APD), or, in some cases, also a charge-coupled device (CCD) camera as is the case for a spinning disk confocal microscope (SDCM). Further information about detectors for confocal microscopy can be found in the information box about instrumentation. It is also possible to introduce further optical elements into the beam path after the pinhole. For example, a single lens can be used, as depicted in Figure 5.2a, to image the pinhole on the detector surface. It is also possible to include additional optics such as a polarizing beam splitter or a dichroic mirror. In these cases, it is recommended to first recollimate the light transmitted by the pinhole before mounting the additional optics and then focusing the detection beam on the respective detectors.

A second aspect of confocal microscopy that is important for imaging is scanning. The detection volume in confocal microscopy is typically diffraction-limited. Hence, an entire image cannot be recorded at the same time, as is the case in widefield microscopy. To gain spatial information, either the sample is moved with respect to a stationary optical path or the laser beam is scanned over the sample. More details regarding different confocal laser scanning methods are given in Section 5.2.3.

Box 5.1 A Brief History of the Development of Confocal Microscopy

A new breakthrough to enhance the capability of optical microscopes occurred in the late 1950s with the development of a new type of microscope by the American mathematician Marvin Minsky, which is nowadays known as the

(Continued)

Box 5.1 (Continued)

confocal microscope. Minsky is more known for his numerous and seminal works on artificial intelligence, which led him to think about ways of improving optical microscopy. His original intention was to study the functioning of the human brain. He realized that optical measurements in this relatively dense tissue were very difficult with conventional light microscopes owing to the high background in the images. In order to solve this problem, he developed a new microscope that used a pinhole to block all light that did not come from the focal region. The sample was then scanned, point by point, instead of illuminating the whole sample at once as in widefield microscopy. In 1958, Minsky was able to present a first prototype for a confocal microscope. A sketch of this prototype is shown in Figure 5.3, taken from the patent that was granted to him for his invention in 1961 [5].

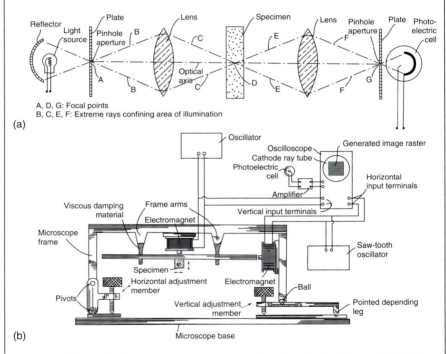

Figure 5.3 The confocal microscope prototype developed by Marvin Minsky. (a) Optical beam path of a confocal microscope. (b) Schematic of the scanner proposed for sample scanning with the prototype confocal microscope. (Source: After Minsky 1961 [5].)

Although Minsky receives all the credit for the development of the confocal microscope, it should be mentioned that a non-imaging scanning confocal microscope was built by the Japanese scientist Naora in 1951 [6] to perform spectroscopic studies on nucleic acids. In contrast to Minsky's version, this system was only designed to measure spectra and did not provide any images of the sample.

In the years shortly after the invention of the confocal microscope, the device had minimal success, which can be largely attributed to the limited technical equipment available at that time and forced the early pioneers to accept several compromises. First, a high-intensity light source was needed, and lasers were not known in 1958. The best light sources that were available were carbon arc lamps, but these lamps required high maintenance. Therefore, Minsky decided to use zirconium-arc lamps that represented the second best alternative. The second compromise was between scanning performance and the pixel dwell time. Minsky decided to move the sample itself in order to achieve higher optical stability, with the disadvantage of a relatively slow scan speed. Detection of the resulting signals was done using PMTs whose output was made visible on the screen of a long-persistence radio oscilloscope. This method had the advantage that a complete image of the sample could be visualized at once, although it could be kept on the screen of the oscilloscope for only about 10 s. Furthermore, to obtain the entire image on the screen, very short dwell times had to be accepted owing to the relatively slow scanning process, which consequently meant a decrease in image quality. When looking back, Minsky said, *…it occurs to me that this concern for real-time speed may have been what delayed the use of that scheme for almost thirty years* [7]. Had he accepted slower data acquisitions rates and recorded the resulting picture point by point on a film, his new method may have been more successful.

In the decades that followed, several technical advancements helped confocal microscopy to obtain more and more acceptance in the scientific community. Here, especially the invention of the lasers in the 1960s has to be mentioned, which offered a bundled, intense, monochromatic, and coherent light source that was able to efficiently excite fluorescent samples. Another important advancement in this context was the development of the first dichroic mirrors in 1967 [8]. With the help of this technical masterpiece, it became much easier to effectively separate excitation and detection pathways, which allowed confocal fluorescence microscopy to be performed with a single objective.

The next challenge was improving the scanning process (cf. also Section 5.2.3.2). Using the so-called galvanic mirrors (a variation of classic galvanometer where a mirror can be moved by an electric current), the excitation and detection beam could be scanned over the sample very accurately. A second approach, the so-called SDCM (cf. Section 5.2.3.3), was first presented by Egger and Petran in 1967 [9]. They used an array of pinholes placed on a spinning disk, making it possible to scan several points of the image simultaneously. Thus, they were able to scan in real time. After this first practical application of confocal microscopy, another 20 years of steady progress was necessary before the technique became accepted as the gold standard for microscopy. Two key publications in 1987 finally led to the acceptance of confocal microscopy. In the first work, White *et al.* [10] used confocal microscopy to image various elements of cells and larger tissues. They were able to demonstrate the considerable improvement in contrast and resolution compared to conventional microscopy. In the second work, Meer *et al.* [11] showed that it was not only possible to study structures in fixed cells,

(Continued)

> **Box 5.1 (Continued)**
>
> as in White's work, but that confocal microscopy is also a suitable tool to observe dynamic processes in living cells, which they demonstrated by following the pathway of sphingolipids in epithelial cells.

> **Box 5.2 Instrumentation**
>
> In general, both continuous wave (CW) and pulsed lasers can be used in a confocal setup. CW lasers are used in most conventional confocal microscopes. However, the use of sub-nanosecond pulsed lasers brings an additional dimension into play and makes it possible, for example, to determine the fluorescence lifetime and thereby perform fluorescence lifetime imaging microscopy (FLIM) (see Chapter 13) in addition to performing confocal imaging. The disadvantage of pulsed lasers is that they may lead to enhanced photophysics in fluorescent dyes, as the photon flux and thereby the energy density during the laser pulse is higher than for CW lasers. In contrast, CW lasers are more suited for applications where an extremely stable flux of photons over a wide range of timescales is needed (e.g., nanosecond fluorescence correlation spectroscopy (FCS) and rotational FCS).
>
> Detectors should be chosen based on the details of the measurement. First, one has to decide whether a widefield detector or a point detector is desired. Widefield detectors such as scientific complementary metal-oxide-semiconductor (sCMOS) or electron multiplying CCD (EMCCD) cameras offer high sensitivity and are suitable for applications such as spinning disk confocal microscopy (see Section 5.2.3.3). For laser scanning applications, a single point detector such as PMTs or APDs are typically the better option (see Section 5.2.3.2). PMTs have the advantage of a very short response time – the so-called instrument response function (IRF) – when operated in single-photon counting mode and hence are well suited for fluorescence lifetime measurements. They also have a large dynamic range and can be used to measure bright samples. However, they typically have low sensitivity and suffer from high dark count when they are not cooled. APDs, on the other hand, offer very low dark counts (down to <10 Hz) and high sensitivity, but they usually have a broader IRF, and the maximum count rates are limited to ~10 MHz. Hence, they are well suited for low-light-level experiments where high sensitivity is desired.
>
> Detectors can be run in either analog or digital mode. The technical difference occurs after the photosensitive electronics. The electrical pulses are either integrated over a certain time by an *integrator* in analog mode or digitalized by a *discriminator*. In the first case, the output voltage is proportional to the number of incoming photons. In the second case, each detected photon is converted into a transistor-to-transistor logic (TTL) pulse, which can be recorded by a counting card or a time-correlated single photon counting (TCSPC) card. Analog detection is advantageous for bright samples, whereas digital detection is better when more information from a limited number of photons is needed.

5.2.2 Radial and Axial Resolution and the Impact of the Pinhole Size

In confocal microscopy, the pinhole is the key element that provides the improved resolution and sectioning capabilities of the instrument. For a deeper understanding of the method, it is important to briefly discuss the details of image formation and the influence of the pinhole.

The intensity distribution for a point-like object imaged using ideal lenses (i.e., only the effect of diffraction is considered, the lenses are assumed to produce no chromatic or spherical aberrations) is given by the Airy pattern. Mathematically, the Airy pattern is the square modulus of the Fourier transform of an aperture:

$$I(r) = I_0 \left[\frac{2J_1(2\pi r \, \text{NA}/\lambda_0)}{2\pi r \, \text{NA}/\lambda_0} \right]^2 \tag{5.3}$$

where J_1 is the first-order Bessel function of the first kind. The *resolution of a system* is defined as the minimum distance at which two point-like objects can still be resolved. Using Rayleigh's criterion, the minimum lateral resolution r_{lateral} obtainable with ideal lenses is given by

$$r_{\text{lateral}} = 1.22 \frac{\lambda_0}{\text{NA}_{\text{Objective}} + \text{NA}_{\text{Condensor}}} \tag{5.4}$$

where the value of 1.22 comes from the first zero of the Bessel function in the Airy pattern, λ_0 is the wavelength of light in vacuum, and $\text{NA}_{\text{Objective}}$ and $\text{NA}_{\text{Condensor}}$ are the numerical apertures of the objective and the condenser, respectively. Often, the NA of the objective and the NA of condenser lenses are matched, or measurements are performed using epi-fluorescence (i.e., the same objective is used for illumination and detection). In these cases, Eq. (5.4) reduces to

$$r_{\text{lateral}} = 0.61 \frac{\lambda_0}{\text{NA}_{\text{Objective}}} \tag{5.5}$$

where r_{lateral} is the distance of the maximum of the Airy pattern to the first minimum. This radius of the Airy disk in the plane of the pinhole can be calculated by multiplying the result from Eq. (5.5) with the total magnification of the instrument. The diameter d ($=2r_{\text{lateral}}$) is referred to as 1 Airy unit (AU). The relationship between Airy unit and the full-width at half-maximum (FWHM) of the Airy pattern is given by

$$\text{FWHM}_{\text{lateral}} = \frac{0.51}{1.22} \text{AU}, \quad \text{where} \quad \text{AU} = 2\, r_{\text{lateral}} = 1.22 \frac{\lambda_0}{\text{NA}_{\text{Objective}}} \tag{5.6}$$

It is important to note that Eq. (5.4) and its derivatives are based on the paraxial approximation (i.e., the parts of the electromagnetic field propagating in x–y plane are assumed to be small compared to those propagating in the z-direction). Fortunately, Novotny and Hecht [12] could show that the paraxial approximation delivers surprisingly good results even for high-NA objectives. Otherwise, a full vectorial description of the electromagnetic field would be required for the discussion of high-NA optics.

Until now, we have only considered the effect of diffraction from the aperture of lenses on the formed image. In a confocal microscope, the excitation light is first focused on the sample, and the resulting fluorescence is collected via the objective and guided through the pinhole to the detector. To calculate the optical resolution of the confocal microscope, the effect of diffraction on both the excitation and emission beam paths needs to be considered. In terms of the PSF introduced in Chapter 2, the total PSF in confocal microscopy can be approximated for small pinholes ($d \leq 0.5$ AU) by the product of the excitation and detection PSFs:

$$\text{PSF}_{\text{confocal}} \approx \text{PSF}_{\text{excitation}} \cdot \text{PSF}_{\text{detection}} \tag{5.7}$$

For large pinholes, the Airy pattern of the detection PSF has to be convoluted with the transfer function of the pinhole. This means that the total confocal PSF is the product of the detection PSF, integrated over the transfer function of the pinhole, with the excitation PSF. This is mathematically equivalent to convoluting the excitation and detection PSFs and multiplying the convolution with the transfer function of the pinhole [13]. The effect of the pinhole is illustrated in one dimension in Figure 5.4.

The radial dependence of the Airy pattern can be well approximated by a Gaussian function. The intensity distribution $I(r)$ of a Gaussian in one dimension is given by

$$I(r) = I_0 * e^{-\frac{(r-r_0)^2}{2\sigma^2}} \tag{5.8}$$

where r_0 is the center of the Gaussian peak, σ its standard deviation, and I_0 the amplitude. When using epi-fluorescence, the size of the excitation and detection

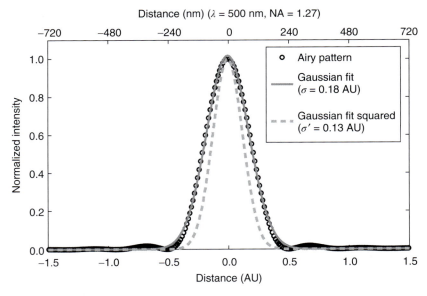

Figure 5.4 PSF and Gaussian approximation. The PSF is represented by black dots, and the Gaussian fit by the solid gray line. The dashed gray line represents the squared Gaussian function and depicts the increased resolution of $\sqrt{2}$ of a confocal microscope when compared to a conventional widefield microscope.

PSFs are equal ($I_{Exc}(r) = I_{Det}(r) = I(r)$). The total PSF is given by the product of the two Gaussian functions, which is itself a Gaussian function:

$$I'(r) = I_{Exc}(r) * I_{Det}(r) = I_0^2 * e^{-\frac{(r-r_0)^2}{2(\sigma')^2}} \text{ with } \sigma' = \frac{\sigma}{\sqrt{2}} \quad (5.9)$$

The standard deviation (and also the FWHM) of the resulting Gaussian is reduced by a factor $\sqrt{2}$, leading to an increased lateral resolution of a confocal microscope by $\sqrt{2}$:

$$\Delta r_{confocal} = \frac{1}{\sqrt{2}} * \Delta r_{widefield} = 0.43 \frac{\lambda_0}{NA_{Objective}} \quad (5.10)$$

Assuming equal excitation and detection PSFs, the full three-dimensional confocal PSF is shown in Figure 5.5.

The size of the pinhole can be varied to modify the resolution or increase the light throughput. The axial and lateral FWHM, as well as the fraction of light detected from a point source in the focus of the microscope, are plotted as a function of the pinhole size in Figure 5.6. The excitation and detection PSFs were calculated for an objective with an NA of 0.5, magnification of 60×, and an excitation wavelength of 500 nm using Eq. (4.44) from Novotny and Hecht [12]. For a pinhole size of diameter ~1 AU, the detection yield reaches a plateau. A plateau is observed when the size of the pinhole is equal to a minimum of the Airy pattern. As there is no intensity at the minima of the PSF, a small variation in the size of the pinhole does not lead to a change in the measured intensity. For a pinhole size of ~1 AU, 86% of the light collected from a point-like source passes through the pinhole. The *lateral* resolution determined by the detection PSF then approaches that achieved by widefield detection, but the pinhole still has an impact on the axial resolution. For most applications in confocal microscopy, a pinhole size between 0.8 and 1.0 AU is optimal.

Selecting a pinhole size significantly smaller than 1 AU (typically with diameter ≤0.25 AU) leads to an increase in both *lateral* and *axial* resolution. In this case, the *lateral* resolution can be improved by a factor of 1.4 compared to Eq. (5.6):

$$FWHM_{lateral} = 0.37 \frac{\lambda_0}{NA_{Objective}} \quad (5.11)$$

However, the increase in resolution comes at a significant cost in the intensity. For a pinhole size of 0.25 AU, only ~20% of the collected photons pass through the pinhole. Increasing the size of the pinhole beyond ~1 AU does not significantly change the *lateral* resolution, as it becomes dominated by the excitation PSF.

In contrast, the optical thickness is sensitive to pinhole size and becomes very important when measuring real objects. The optical thickness in widefield microscopy is not defined by the optical elements but by the thickness of the sample itself. Sharp images of biological objects can be generated if the specimen itself is sufficiently thin; otherwise, the sharp image of the focal plane is blurred due to out-of-focus light. In confocal microscopy, the optical thickness is determined by the size of the pinhole and the product of the *excitation* and *detection* PSFs. The depth of the *detection* PSF is a result of the diffraction pattern for the emitted light and the geometric-optical impact of the pinhole

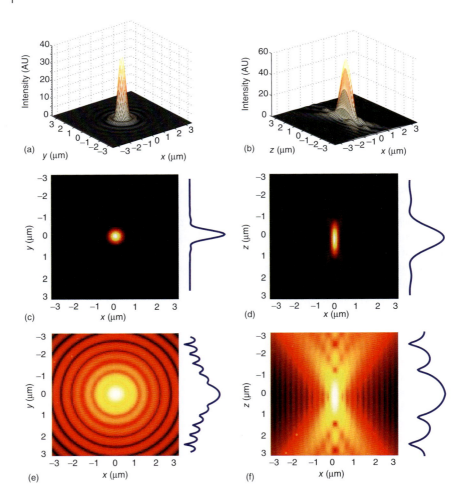

Figure 5.5 Airy pattern resulting from a luminescent point light source. (a, b) 3D representation of the (a) lateral and (b) axial dimensions of the PSF. (c–f) Projections of the intensities of the whole focus onto the x–y and x–z planes, respectively, (c, d) in linear scale, and (e, f) in logarithmic scale. Cross sections through the middle of the displayed PSF are shown on the plots to the right of the images.

[14]. The axial FWHM of the detection PSF is given by:

$$\text{FWHM}_{\text{det;axial}} = \sqrt{\left(\frac{0.88\lambda_0}{n - \sqrt{n^2 - \text{NA}^2}}\right)^2 + \left(\frac{\sqrt{2}n\text{PH}}{\text{NA}}\right)^2} \quad (5.12)$$

where PH is the absolute size of the pinhole in micrometers in the object space. The first squared term under the square root, often called the *wave-optical term*, is constant as long as the objective and the excitation wavelength remain the same. The second squared term under the square root, the *geometric-optical term*, depends on the pinhole size. As the size of the pinhole increases, the

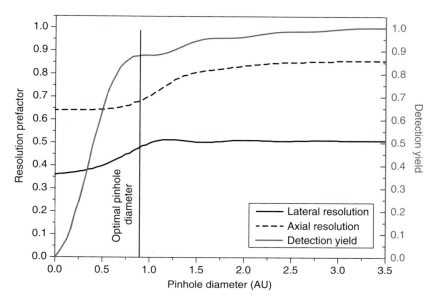

Figure 5.6 Resolution and signal throughput as a function of pinhole size. The solid and dashed black lines show how the resolution prefactor (i.e., resolution in units of $\lambda_0/\mathrm{NA}_{\mathrm{Objective}}$) changes for lateral and axial resolution, respectively (left scale). This depicts the transition from Eqs (5.5) to (5.11) for lateral resolution and the analog transition from Eqs (5.13) to (5.15) for axial resolution. The gray line shows the detection yield as a function of the pinhole diameter (right scale).

second term becomes prominent and the axial resolution of the detection PSF varies almost linearly with the pinhole size. In the case of small pinholes (diameter <1 AU), the width of the detection PSF is dominated by the first term in Eq. (5.12) and can be approximated by

$$\mathrm{FWHM}_{\mathrm{det;axial,\ PH\sim 1AU}} = \frac{0.88\lambda_0}{n - \sqrt{n^2 - \mathrm{NA}^2}} \tag{5.13}$$

The *total* PSF depends on the relative size of the excitation and detection PSFs. When the excitation PSF is significantly larger than the detection PSF, the FWHM of the total PSF will be determined by the detection PSF and, therefore, increases with the pinhole size. When the excitation PSF is smaller than the detection PSF, the total PSF will be given by the excitation PSF. This can be seen in Figure 5.6. The lateral and axial resolution are shown as a function of pinhole size. The prefactors plotted represent the number before $\lambda_0/\mathrm{NA}_{\mathrm{Objective}}$ in Eqs (5.6) and (5.11) for the lateral resolution and before $\lambda_0/(n - \sqrt{n^2 - \mathrm{NA}^2})$ in Eqs (5.12) and (5.13) for the axial resolution. The prefactors saturate for larger pinholes.

Box 5.3 Wavelength Corrections

In the text, we have not discriminated between excitation (λ_{exc}) and emission wavelength (λ_{em}) for the calculations. Because of the Stokes shift, the excitation and emission wavelengths are usually separated by 20–50 nm. To obtain the

(Continued)

> **Box 5.3 (Continued)**
>
> closest possible result, λ_0 has to be replaced by an average wavelength:
>
> $$\lambda_{AV} = \sqrt{2}\frac{\lambda_{em}\lambda_{exc}}{\sqrt{\lambda_{exc}^2 + \lambda_{em}^2}} \qquad (5.14)$$
>
> While defining the excitation wavelength λ_{exc} is easy, as modern lasers are monochromic and even diode lasers have a narrow bandwidth of only a few nanometers, the emission wavelength λ_{em} should be weighted by the emission spectrum to obtain the correct average wavelength.

If the size of the pinhole becomes extremely small (diameter ≤ 0.25 AU), not only the *geometric-optical* term in Eq. (5.12) vanishes but also diffraction effects of the pinhole play a significant role. The product with the excitation PSF further decreases the width of the total axial PSF, as shown in Figure 5.6:

$$\text{FWHM}_{\text{tot;axial, PH}\sim 0.25\text{AU}} = \frac{0.64\lambda_0}{n - \sqrt{n^2 - \text{NA}^2}} \qquad (5.15)$$

Jonkman and Stelzer [15] and Pawley [16] describe the axial resolution of a confocal microscope with the following formula:

$$r_{\text{axial}} = \frac{F\lambda_0 n}{\text{NA}^2} \qquad (5.16)$$

where F is a normalization factor that varies from 1.4 to 2.0 depending on the size of the pinhole and is also referred to as the *focal depth*. As a practical guideline, the confocal pinhole is typically used to control the optical section thickness of the microscope rather than to gain the best possible resolution in the lateral dimension. The axial resolution is at least 3–4 times worse than the resolution in the lateral dimension.

Table 5.1 compares the theoretically achievable resolution in widefield and confocal microscopy for two different objectives – an air and a water objective – assuming an excitation wavelength of 500 nm, a detection wavelength of 520 nm, and the usage of a pinhole size of 1 AU. For high-NA objectives, the increase in axial resolution is larger (about a factor of 1.6) than the improvement in lateral resolution (about a factor of 1.4).

> **Box 5.4 Increasing the Confocal PSF Size**
>
> For single-molecule studies in solution, the photons detected from a single molecule diffusing through the focus of a confocal microscope are collected and analyzed. To allow more photons to be collected, it is often useful to work with a larger focus size, as the amount of time a single molecule spends in the observation volume can be increased by a factor of ~ 10. This is achieved by not fully filling the back aperture of the objective with the laser beam. By reducing the size of the laser beam, the effective excitation NA of the system is decreased, leading to a larger diffraction-limited volume in both the lateral and

Table 5.1 Exemplary comparisons of the axial and lateral resolution limit for widefield and confocal microscopy for two different objectives by means of the FWHM of the respective PSF.

FWHM	Widefield (nm)	Confocal (nm)
NA = 0.5, n = 1.00		
Lateral	530	374
Axial	4896	3348
NA = 1.27, n = 1.33		
Lateral	208	147
Axial	759	480

First two rows: $\lambda_{exc} = 500$ nm, $\lambda_{det} = 520$ nm, PH = 1 AU, NA = 0.5, and $n = 1$. These values were used for the simulation of Figure 5.7. The lower two rows represent the maximum achievable resolution with high NA water-immersion objectives. Here, the values of the NA and n change to 1.27 and 1.33, respectively.

---

axial dimensions. In this case, the ratio of axial to radial size will increase from ~3 to 4, as discussed previously, to a value between 5 and 8, depending on what fraction of the back aperture of the objective is filled with the laser beam.

5.2.3 Scanning Confocal Imaging

The light passing through the pinhole in confocal microscopy is typically detected either on a point detector or a small number of pixels of a camera. Therefore, to record an image, it is necessary to perform a scan of the sample. It is the scanning, in addition to the pinhole, that differentiates a confocal microscope from other optical instruments. Confocal microscopes can be classified by the technique with which the laser and sample are scanned with respect to each other. The most typical scanning approaches are discussed in the following.

5.2.3.1 Stage Scanning

The simplest and most straightforward approach for scanning a sample is stage scanning. This approach was successfully implemented by Minsky in his first confocal microscope. The optics remains stationary while the sample is scanned, and the intensity of each point is measured before moving the sample to the next position. This technique is still widely used today, especially in the material sciences, and offers several advantages: a confocal microscope with stage scanning is straightforward to build with a minimum of optical elements, is very stable, and can be easily aligned. The scanning can be done very precisely using modern piezo scanners that can move the sample in three dimensions with a resolution on the subnanometer scale. Scanning in the z-direction is important when one wishes to exploit the optical sectioning capability of confocal microscopy to collect 3D information. However, stage scanning also suffers from one drawback: the sample has to be moved. Biological samples are often sensitive to movement, and the recorded image may be distorted if the specimen shifts during scanning. It is more difficult to interact with the sample, for example, to manipulate the sample with micropipettes unless they are mounted on the scanning stage. The maximum scan speed for piezo scanners is limited, and there is typically a significant relaxation time of several milliseconds or longer for the piezo scanner to move to a new position. This can make the application of piezo scanners difficult for applications where quick and precise control of the sample and laser beam with respect to each other is crucial.

5.2.3.2 Laser Scanning

A second approach to recording confocal images is raster-scanning of the light beam over the sample while leaving the sample stationary. Raster scanning is older than the idea of the confocal microscope [17]. However, it was first in 1987 that White et al. successfully combined the techniques of laser scanning and confocal microscopy [10]. The method of laser scanning typically uses a pair of galvanometric mirrors that are controlled by the acquisition software in order to move the beam in x- and y-dimensions. Figure 5.7a shows one possible setup for a laser scanning confocal microscope (LSCM), where two mirrors that scan the beam in perpendicular directions are closely spaced. To avoid clipping of the beam by the objective, a telescope is built into the optical path after the scan mirrors, which images an intermediate position between the two mirrors on the back aperture of the objective. This means that the position of the excitation beam with respect to the objective is relatively stationary while the angle of the beam changes. Hence, the position of the beam on the sample will change with mirror positions, but the position of the light path entering the back aperture of the objective will be stationary. Although the configuration in Figure 5.7a is sufficient for most imaging purposes, it is not ideal and can be improved as shown in Figure 5.7b. Here, a telescope is installed between both mirrors to project the image of the first mirror onto the second mirror. An

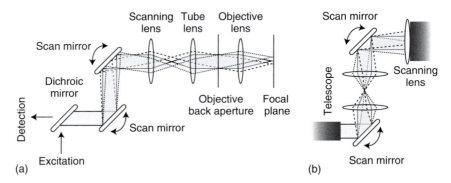

Figure 5.7 Two approaches to laser scanning. (a) A straightforward approach is to position the two mirrors close together. A telescope, usually with the tube lens as one of the elements, is then used to image the plane in between the two mirrors onto the back aperture of the objective. The telescope ensures that the excitation beam is not clipped by the back aperture of the objective during scanning. (b) A second, more elegant approach is to also insert a telescope between the two scanning mirrors. Hence, the first mirror is imaged onto the second, which is then imaged onto the back aperture of the objective.

additional telescope is used to image the second mirror on the back aperture of the objective. When correctly aligned, the excitation beam is stationary at the back aperture of the objective, passes through the objective, and excites a given position of the specimen depending on the angles of the two mirrors. Using galvanometer mirrors, scanning can be performed at almost video rates. When one is willing to give up the flexibility in scan speed, a resonance scanner can be used, which allows higher frame rates to be achieved.

Laser scanning also has its price. One has to live with the fact that the beam is almost always off-axis. Hence, the alignment of the optics is more critical, and excellent optics is needed to minimize off-axis spherical and chromatic aberrations.

5.2.3.3 Spinning Disk Confocal Microscope

Although modern technology has significantly improved the performance of LSCMs, it is still necessary to find a good compromise between the pixel dwell time for good photon statistics and a reasonable scanning speed for every measurement. In general, one is able to obtain acquisition speeds in the range of seconds (depending on the scanner type, the size of the scan region, etc.). Nevertheless, this is often too slow for observing many dynamic biological processes, particularly when one is interested in real-time dynamics. An alternative is the SDCM.

SDCM combines the advantages of confocal microscopy with real-time data acquisition rates known from widefield microscopy. The scan speed is increased by simultaneously scanning multiple pinholes across the sample. The scanning technique used for spinning disk microscopy was developed already in 1884 by the German engineer Nipkow for a machine he called an *electric telescope* [18]. The core element of his invention was a disk (now known as the *Nipkow disk*) with a series of holes arranged in an Archimedean spiral (shown in Figure 5.8). When this disk is placed in front of an image, only sections of the image that are in front of a hole are allowed to pass. The holes are arranged such that each section of an image is scanned for the same amount of time during one revolution of the disk. By spinning the disk, it is possible to translate the image into a series of analog signals which can subsequently be reassembled into an image by passing light through another disk with the same arrangement of holes. This method of mechanical signal transmission was used for the first television broadcasts in the 1920s, but could not compete with the completely electronic method using cathode ray tubes known as *Braun tubes* . Thus, Nipkow disks slid into oblivion for almost 40 years until 1967 when Egger and Petran recognized the relevance of this tool for confocal microscopy [9].

By placing a Nipkow disk with its array of pinholes in a conjugate image plane of the objective, as shown in Figure 5.8a, it is possible to scan every point of the sample successively in a very short time (down to 1 ms) compared to an LSCM. An image of the pinhole structure from a nonrotating Nipkow disk is shown in Figure 5.8b. The excitation beam is split by the disk into an array of beams that create individual focal volumes. To produce the confocal detection and reassemble the distinct fluorescence signals from the different pinholes in the correct order, it is possible to use either conjugated pinholes

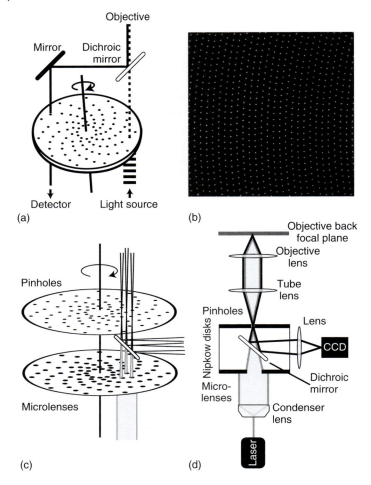

Figure 5.8 Spinning disk confocal microscopy. (a) Schematic of a Nipkow disk and sketch of scanning head of a tandem scanning disk microscope, which uses different regions of the same Nipkow disk for excitation and detection, respectively. (b) Image of the pinhole structure taken from a nonrotating Nipkow disk. (c) Schematic of a SDCM setup with two Nipkow disks, one containing the pinholes and a second one filled with microlenses for increasing the excitation intensity through the pinholes. Gray, excitation light; white with black borders, detection light. (d) Schematic of the excitation (gray) and detection (white with black borders) pathways in an SDCM.

located on diametrically opposed spirals for excitation and detection, as shown in Figure 5.8c (tandem disk scanning microscope), or the same set of pinholes for excitation and detection pathway as shown in the experimental setup in Figure 5.8d (single-sided disk scanning microscope). The latter method is preferred owing to the lower required maintenance efforts. As multiple detection volumes are measured simultaneously, a camera is used for the detection of the confocal image rather than an APD or a PMT. Thus, the microscope can benefit from the high quantum efficiency of CCD cameras. For correct image reproduction, the frame rate of the camera has to be synchronized with the rotation frequency of the Nipkow disk. Otherwise, the disk rotations will be incomplete, leading to unequal illumination of the image, which appears as a striped pattern on the image. This is particularly important when the collection times are short.

There are several parameters that have to be considered in determining the optical configuration of a Nipkow disk. Because of the design of the Nipkow disk, a considerable amount of the excitation light is blocked. The transmission factor or fill factor T for an SDCM is given by

$$T_{\text{pinhole}} = \left(\frac{D}{S}\right)^2 \tag{5.17}$$

where D is the diameter of the pinhole and S is the distance between the distinct apertures. T is generally much lower for SDCMs compared to LSCMs. In order to provide some rough estimate, about 50% of the excitation light passes through the pinhole of a confocal point-scanning microscope, whereas only about 5% is achievable with a pinhole-based SDCM. The transmission factor can be improved by decreasing the spacing between pinholes or by replacing the pinhole of a confocal microscope with a slit-shaped aperture. For a slit-based confocal microscope

$$T_{\text{slit}} = \left(\frac{D}{S}\right) \tag{5.18}$$

where D is the width of the slit and the transmission is typically ~10%. However, the use of slit-shaped pinholes or closely spaced apertures leads to the problem of cross-talk between the pinholes. This means that light coming from out-of-focus regions, which would be blocked by the respective pinhole, is able to pass through the neighboring pinholes and thereby decreases the axial resolution (see below). For this reason, most commercially available systems are based on pinholes with a low fill factor. To compensate for the low throughput of the excitation beam, a microlens system, first presented by the Yokogawa Company, is typically used. In the Yokogawa design, a disk of microlenses is arranged in the same pattern as the pinholes on the Nipkow disk and rotated synchronically, as shown in Figure 5.8c,d. With the microlenses, the fill factor can be increased up to a value of about 40%. However, to reach this value, it is also important to use the optimal pinhole size. On the one hand, a significant amount of excitation light is blocked when a small D or a relatively large S is chosen. On the other hand, when D is too large or S too small, pinhole cross-talk increases. Thus, the spacing S is chosen to find the best balance between minimum pinhole cross-talk, optimum scanning speed, and sample exposure.

While the lateral resolution of the SDCM is similar to that of other confocal imaging methods and the choice of 1 AU is typically a good compromise, the axial resolution is notably influenced by pinhole cross-talk, which can significantly reduce the optical sectioning capacity of the microscope. For very thin samples with a small pinhole (~0.25 AU), Eq. (5.15) can be used in good approximation to estimate the optical sectioning capacity of an SDCM using a prefactor of 0.67 for a disk of pinholes and 0.95 for a slit aperture. Otherwise, Eq. (5.12) can be used for the detection PSF, but neither Eq. (5.12) nor Eq. (5.15) takes into account potential pinhole cross-talk. As this effect also depends on nonquantifiable

Table 5.2 Comparison of SDCM with LSCM.

Advantages of the SDCM	Disadvantages of the SDCM
Faster scanning capabilities	Low transmission factors
Can be combined with higher quantum-yield detectors	Lower flexibility (e.g. fixed pinhole size)
Higher biocompatibility as a result of a reduction in excitation energy per unit area	Lower axial resolution

parameters such as the labeling density of the respective sample, it is difficult to give a specific formula for the actual z-resolution. In short, pinhole cross-talk is small when the fluorescence labeling is confined to a thin layer. Otherwise, signal from out-of-focus regions will be able to pass through adjacent pinholes. Of course, it is possible to increase the axial resolution by reducing the pinhole size or increasing the distance between the pinholes, but, as explained above, this goes at cost of signal quality.

In addition to the SDCM, there are multiple ways of generating and scanning multiple points on the sample. One possibility is to use two-photon excitation with a holographic beam splitter to generate multiple excitation beams or by scanning a two-dimensional (2D) array of pinholes rather than a disk in the appropriate conjugate image plane. While in all cases the scanning speed can be improved, this improvement is not possible without making compromises (cf. Table 5.2). Nevertheless, as suggested by the importance Minsky placed on real-time imaging, SDCM and similar approaches have become an extremely valuable tool, as they combine the high contrast of confocal microscopy with the fast data acquisition available from widefield microscopy. By scanning with multiple pinholes simultaneously, the excitation power is more distributed both spatially and temporally over the sample, which can lead to a lower excitation power per unit area and thus higher biocompatibility compared to LSCM.

5.2.4 Confocal Deconvolution

Despite the improvement in image contrast, confocal microscopy images are still diffraction-limited. In order to circumvent this diffraction limit, it is possible to apply super-resolution microscopy methods such as those described in Chapters 8–10. However, it is also possible to improve the resolution of confocal images by performing a deconvolution. The decrease in optical resolution in confocal microscopy is due to the convolution of the actual image with the diffraction-limited PSF of the microscope. Mathematically speaking, the detected signal $S(r)$ at a spatial coordinate r is given by the convolution of the signal $O(r)$ from the original object with the point spread function PSF(r):

$$S(r) = O(r) \otimes \text{PSF}(r) = \int_{-\infty}^{\infty} O(r') \, \text{PSF}(r - r') \, dr' \tag{5.19}$$

When two of the functions are known – as is the case for the observed signal $S(r)$ and the point spread function PSF(r) – the third function – in this

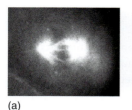 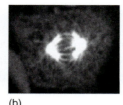 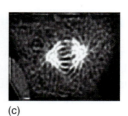

(a)   (b)   (c)

Figure 5.9 Improving resolution by confocal microscopy and confocal deconvolution. (a) A standard widefield image of a tubulin-labeled cell undergoing mitosis. (b) A confocal image of the same cell as shown in (a). The improvement in contrast is clearly visible. (c) The same image as shown in (b) deconvoluted using the Richardson–Lucy algorithm.

case the actual spatial distribution $O(r)$ of the object – can be determined. In practice, noise or limited information regarding one or both of the known functions can make deconvolution difficult to impossible. Thus, $O(r)$ cannot be straightforwardly obtained by using Eq. (5.19), but more complex deconvolution algorithms have to be applied.

Figure 5.9 shows the image of a tubulin-labeled cell undergoing mitosis. Compared with the widefield image (Figure 5.9a), the confocal image (Figure 5.9b) already shows a significant improvement in contrast. An additional improvement is seen in the deconvoluted image (Figure 5.9c) where the complete structure of the spindle apparatus is resolved.

One prominent deconvolution approach is the Richardson–Lucy algorithm, which was used in Figure 5.9 [19]. The signal detected in an image depends on the original distribution of signal from the object and the probability that emitted signal through the PSF of the system ends up in the respective pixel. When u_{kl} represents the intensity originating from pixel k,l and $\text{PSF}_{ij;kl}$ is the probability of light that originated from location k,l, being detected at location i,j, the detected signal in pixel i,j is given by

$$d_{ij} = \sum_{k,l} \text{PSF}_{ij;kl} u_{kl} \qquad (5.20)$$

If u_{kl} is Poisson-distributed, which is the case when the dominant noise source is the photon noise, one can iteratively calculate the most probable value for $u_{kl}^{(t+1)}$ at iteration step $t+1$ when all other parameters from Eq. (5.20) are known:

$$u_{kl}^{(t+1)} = u_{kl}^{t} \sum_{i,j} \text{PSF}_{ij;kl} \frac{d_{ij}}{\sum_{kl} \text{PSF}_{ij;kl} u_{kl}^{t}} \qquad (5.21)$$

It has been shown that, when these iterations converge, the result will be the maximum likelihood for u_{kl}.

For a more accurate deconvolution, detailed knowledge of the PSF is required. Close inspection of the lower left corner of Figure 5.9c shows that the PSF might have changed in this region owing to aberrations in the objective, and the deconvolution does not yield the correct value for u_{kl} in this part of the image. When the PSF is not known, the quality of the image can still be enhanced by using *blind deconvolution*, where the PSF is estimated in an iterative process.

5.3 Applications of Confocal Microscopy

For the biological sciences, the main application of confocal microscopy is imaging. It has become routine to record multicolor 3D images using a confocal microscope. An example of a 3D confocal image recorded using an SDCM is shown in Figure 5.10 and also given in the online material. Here, the plasma membrane of an HMEC-1 cell has been labeled in red and the mitochondria have been labeled in green. The structure of the complex 3D network of the mitochondria is clearly visible throughout the cell. Also early endosomes, which are labeled by internalization of the membrane marker, can be distinguished from the plasma membrane via the position and morphology of the measured signal.

Although imaging is the most common application of confocal microscopy, there are many important applications in the biochemical and biophysical sciences that go beyond imaging. In the following sections, we discuss a number of fluorescence fluctuation spectroscopy techniques that utilize confocal microscopy.

5.3.1 Nonscanning Applications

5.3.1.1 Fluorescence Correlation Spectroscopy

FCS is a versatile technique to extract information out of fluctuations in fluorescence intensity. Several detailed papers and reviews can be found in the literature (e.g., [20–22]). FCS was first published in the beginning of the 1970s by Magde, Elson, and Webb [23, 24]. The method analyzes the temporal fluctuations of the fluorescence intensity using a correlation analysis to gain information regarding the processes responsible for the fluctuations. Along with the development of ultrasensitive detectors and stable laser light sources, it was the use of a small confocal detection volume (~1 fl) that made it possible

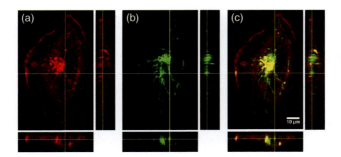

Figure 5.10 Three-dimensional (3D) confocal imaging. Images of a HMEC-1 cell are shown, where different organelles have been labeled with different colors. Slices of the three-dimensional image in the x–y, x–z, and y–z planes are shown for (a) membrane labeling using CellMask™ (deep red, Life Technologies) (red) and (b) the mitochondria labeled with MitoTracker® (Green FM, Life Technologies) (green). (c) A merged image is shown where early endosomes, indicated by the internalized membrane marker, are shown in yellow. The various structures of the plasma membrane, mitochondria network, and endosomes are clearly distinguishable via the improved 3D resolution of confocal microscopy. A 3D representation of these images can be found in the online material.

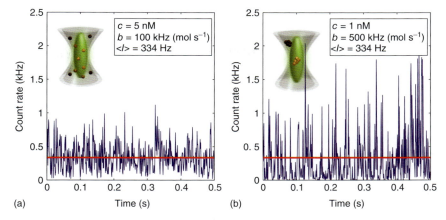

Figure 5.11 Fluorescence intensity fluctuations at low concentration. (a) Simulation of a sample of molecules with a concentration, c, of 5 nM and a molecular brightness, b, at the center of the PSF of 100 kHz per molecule. (b) Simulation of a sample at 1 nM concentration with a molecular brightness of 500 kHz per particle at the center of the PSF. The average intensity $\langle I \rangle$ is equal in both cases.

to detect signals down to a single fluorophore [25]. This sensitivity has made FCS easily applicable to a broad range of applications. Figure 5.11 shows the intensity versus time trace for two species with different concentrations and molecular brightnesses freely diffusing in solution. The two samples were chosen to have the same average fluorescence intensity (solid line). However, the fluctuations in the fluorescence intensity are very different for the two samples, with the fluctuations from the brighter species at lower concentration being larger. The variations in fluorescence intensity arise from thermodynamic fluctuations in the number of particles within the confocal detection volume due to Brownian motion. By analyzing the temporal fluctuations in fluorescence intensity, information can be extracted regarding the average number of particles in the volume as well as the average time it takes for them to travel through the confocal volume. When the shape and size of the confocal volume is known, it is possible to convert the average number of molecules and the time they spend in the confocal volume into an average concentration and a translational diffusion coefficient. FCS is not limited to diffusion; any process that leads to fluctuations in fluorescence intensity can be analyzed with FCS. Hence, it is possible to measure, for example, singlet–triplet state dynamics and the rate of chemical reactions (when the reaction leads to a change in molecular brightness) even though the measurements are performed in equilibrium.

To extract the information buried in the fluctuations of the fluorescence signal, a correlation analysis is used. The photons detected from the same molecule will lead to a correlation signal, whereas randomly detected photons from noncorrelated processes do not correlate. The correlation process is highly sensitive and can be used to determine the timescale over which two signals remain similar, or, in the case of the autocorrelation analysis, how long the fluctuations of a single signal persist on average. Consider the fluctuations due to molecules diffusing in and out of a confocal focal volume. The fluorescence signal will fluctuate as molecules enter and leave the focus. The average timescale of the fluctuation depends on the time a molecule needs to cross the confocal spot, referred to as the *diffusion time* τ_D. τ_D depends on the size of confocal volume and on the diffusion coefficient D. On timescales shorter than or equal to τ_D, the position of molecules within the volume does not change significantly and the detected signal remains similar. On longer timescales, molecules can enter or leave the confocal volume, the fluorescence signal changes, and the correlation drops. Hence, for times longer than τ_D, the correlation function decays.

The derivation of the autocorrelation for FCS can be found, for example, in Elson and Magde [23]. For a measurement with freely diffusing molecules in three dimensions, the intensity of the fluorescence signal F at a certain time t is given as

$$F(t) = \varepsilon \int dr \overline{W(r)} C(r, t) \tag{5.22}$$

where ε is the molecular brightness, $C(r, t)$ is the number density of molecules at a certain position r at time t, and $\overline{W(r)} = W(r)/W(0)$ is the normalized PSF, with $W(0)$ being the laser intensity at the center of the PSF. The molecular brightness ε is defined as the fluorescence intensity of a single fluorophore at the center of the PSF and is given by

$$\varepsilon = \kappa \sigma_{Abs} \phi W(0) \tag{5.23}$$

where κ is the total detection efficiency of the microscope, σ_{Abs} is the absorption cross section of the fluorophore at the wavelength of the laser, and ϕ is the fluorescence quantum yield of the fluorophore. The autocorrelation function (ACF) is given by

$$G(\tau) = \frac{\langle \delta F(t) \delta F(t+\tau) \rangle}{\langle F(t) \rangle^2} = \frac{\langle F(t) F(t+\tau) \rangle - \langle F(t) \rangle^2}{\langle F(t) \rangle^2}$$
$$= \frac{\langle F(t) F(t+\tau) \rangle}{\langle F(t) \rangle^2} - 1 \quad \text{with} \quad \delta F(t) = F(t) - \langle F(t) \rangle \tag{5.24}$$

where $\langle \, \rangle$ represents the time-averaged value. A graphical interpretation of the ACF is shown in Figure 5.12. During the correlation analysis, the signal is shifted by a time interval τ, multiplied with the original curve, and integrated over the fluorescence time trace. In the end, the correlation analysis is normalized by the square of the average fluorescence intensity and 1 is subtracted.

The PSF of a confocal microscope can be approximated as a 3D Gaussian:

$$W(r) = W(0) \exp\left[-\left(\frac{(x^2+y^2)}{2\omega_r^2}\right) - \left(\frac{z^2}{2\omega_z^2}\right)\right] \tag{5.25}$$

where ω_r and ω_z are the dimensions of the PSF from the center to the position where the intensity has decreased by a factor $1/e^2$ in the lateral and axial directions, respectively. Substituting Eqs (5.22) and (5.25) into Eq. (5.24), we obtain an analytical form for the ACF of a freely diffusing particle:

$$G_D(N, D, \tau) = \frac{\gamma_{FCS}}{N} \left(\frac{1}{1+(\tau/\tau_D)}\right) \left[\frac{1}{1+(\omega_r/\omega_z)^2 \cdot (\tau/\tau_D)}\right]^{\frac{1}{2}} \tag{5.26}$$

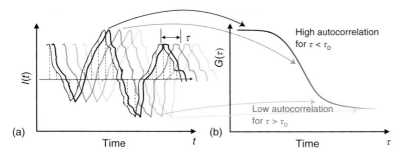

Figure 5.12 Determination of the autocorrelation function. (a) A schematic of a fluorescence signal (black curve) fluctuating about a mean value (black horizontal line) and (b) the resulting autocorrelation function. The trace is duplicated and shifted by different time delays τ (different gray values). The shifted curve is multiplied with the original curve and integrated to determine the residual self-similarity of the signal. The similarity decays from black to light gray, which can be seen by the drop in the correlation amplitude.

where $\langle N \rangle$ is the average number of particles in the confocal volume, γ_{FCS} is a geometrical factor that depends on the shape of the volume ($\gamma_{\text{FCS}} = 2^{-3/2}$ for a 3D Gaussian PSF), and $\tau_D = \omega_r^2/4D$ is the diffusion time.

One can also define the ACF as

$$g(\tau) = \frac{\langle F(t)F(t+\tau) \rangle}{\langle F(t) \rangle^2} = G(\tau) + 1 \tag{5.27}$$

The expression for the ACF shown in Eq. (5.27) is proportional to the probability of detecting a photon at a delay time τ when a photon was detected at the time $t = 0$. For timescales much smaller than the diffusion time τ_D, the molecule still remains in the confocal volume, and so the probability to detect a second photon is high. On longer timescales, this probability decreases owing to the possibility that the particles leave the confocal focal volume via Brownian motion and no further photon is detected. A correlation analysis is powerful, as it is capable of detecting any type of fluctuation in fluorescence intensity that is not stochastic. Examples of processes that can be detected using FCS are shown in Figure 5.13.

The full correlation function here consists not only of the term arising from diffusion but also terms due to photon anti-bunching, rotational motion, and blinking of the fluorophore when it enters and escapes the triplet state. Fortunately, these influences take place on different timescales and are distinguishable. While FCS analyzes the temporal behavior of the fluorescence intensity, there are other methods available that make use of the amplitudes of the fluctuations to gather information regarding the average number of molecules and their molecular brightness such as the photon counting histogram (PCH) analysis, fluorescence intensity distribution analysis (FIDA), and cumulant analysis. For an excellent introduction to the different brightness analysis methods, we refer the reader to Macdonald *et al.* [26].

The amplitude of the ACF is inversely proportional to the number of particles in the confocal volume. For a single species, the amplitude of the ACF is independent of the molecular brightness, whereas the signal-to-noise ratio

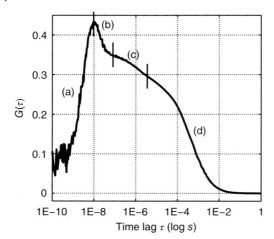

Figure 5.13 The measured ACF of green fluorescent protein (GFP) in aqueous solution. A number of processes can be detected by FCS as highlighted in the ACF: (a) anti-bunching, which emerges because the fluorophore has to be re-excited before it can emit a second photon (to detect anti-bunching, two detectors are necessary to avoid the dead time of detectors and data collection card), (b) rotational diffusion, (c) intramolecular reactions such as transitions between the singlet to the triplet states, and (d) translational diffusion.

depends on the molecular brightness but is independent of sample concentration, at least in the range where FCS is most sensitive (from ~100 pM to ~100 nM).

5.3.1.2 Fluorescence Cross-Correlation Spectroscopy

It is also possible to correlate two signals with each other, a method known as *cross-correlation*. Typically, two detection channels sensitive to different wavelengths are used, and the fluorescence signals from the two channels are cross-correlated. This method of dual-color fluorescence cross-correlation spectroscopy (FCCS) is well suited for investigating the interaction of molecules [27]. Figure 5.14 shows the autocorrelation and cross-correlation of double-stranded DNA that is labeled on one strand with Atto532 and on the second strand with Atto647. A strong correlation is expected, as the DNA is double-labeled with high efficiency and a fluctuation in green channel due to a DNA molecule entering the confocal volume will be correlated with an increase in signal in the red channel.

The number of double-labeled molecules detectable in both channels, N_{12}, can be extracted from the amplitude of the cross-correlation function (CCF). Assuming two overlapping 3D Gaussian confocal volumes and that the system is cross-talk-free, the CCF is given by

$$G_D(N, D, \tau) = \frac{\gamma_{FCS} \langle N_{12} \rangle}{\langle N_1 + N_{12} \rangle \langle N_2 + N_{12} \rangle} \left(\frac{1}{1 + (\tau/\tau_D)} \right) \left(\frac{1}{1 + (\omega_r/\omega_z)^2 \cdot (\tau/\tau_D)} \right) \tag{5.28}$$

where γ_{FCS} is a geometrical factor equal to $2^{-3/2}$ when the PSF is approximated with a 3D Gaussian, N_i is the number of molecules of type i (where i can be single-labeled molecules visible in channel 1 or 2, or double-labeled molecules), and $\tau_D = \omega_r^2/4D_{12}$ is the diffusion time of the double-labeled molecules where D_{12} is the diffusion coefficient for the double-labeled species. Although the CCF is often used as a digital answer to a problem – whether two molecules interact – the amplitude of the CCF itself and its relative height with respect to the autocorrelation function contain information regarding the percentage of complexes carrying both fluorophores. A detailed description of how to perform

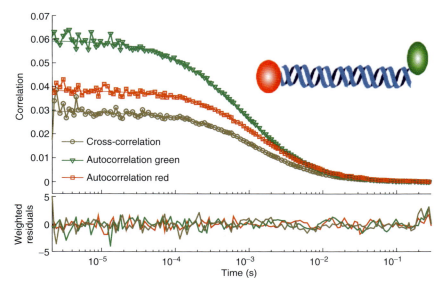

Figure 5.14 Cross-correlation of double-labeled DNA. Double-stranded 60 bp DNA was labeled at the ends with Atto532 and Atto647. The autocorrelation functions for Atto532 (green), Atto647 (red), and corresponding cross-correlation function (gold) are shown. The fits are shown as solid lines of the respective color, and the weighted residuals are plotted in the lower panel.

quantitative analysis of FCCS data, in particular with fluorescent proteins, is given in Foo *et al.* [28].

5.3.1.3 Pulsed Interleaved Excitation

The sensitivity of FCCS can be further enhanced with the use of pulsed interleaved excitation (PIE) [29]. PIE is based upon alternating laser excitation (ALEX) [30], developed in the laboratory of Shimon Weiss, with the modifications that subnanosecond pulsed lasers are used for excitation and they are alternated (or interleaved) on the nanosecond timescale. To achieve this, the excitation lasers are synchronized and delayed with respect to each other by 12–20 ns (4–5 times the fluorescence lifetime of the dyes used). For detection, photons can be recorded using a counting board that is synchronized to the excitation lasers. It is then possible to assign every detected photon to its excitation source (Figure 5.15a). A more elegant possibility for data collection is the use of TCSPC detection, which has the additional advantage that lifetime histograms of the fluorophores can be generated and analyzed. With the lifetime information, the presence of quenching, for example, due to Förster resonance energy transfer (FRET, Chapter 13), can be determined.

Figure 5.15b shows the ACFs and CCFs for two noninteracting fluorophores in solution. As correlation spectroscopy is very sensitive to correlated events, the cross-talk signal of the Atto532 fluorophore into the Atto665 channel gives rise to a residual cross-correlation amplitude in the CCF. Spectral cross-talk arises from the tail of the fluorescence emission spectrum of the green fluorophore being detected in the red channel. Spectral cross-talk cannot be totally suppressed

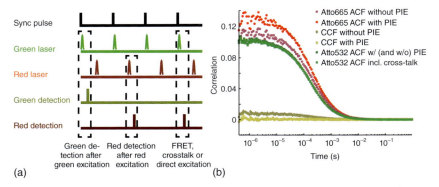

Figure 5.15 Pulsed interleaved excitation (PIE). (a) A schematic showing the principle of PIE. (b) The ACF and CCF for Atto532 and Atto665 freely diffusing in solution determined with and without PIE. The influence of cross-talk and direct excitation on the red ACF and CCF can be observed, which is completely removed when using PIE.

in the red channel by simply using the appropriate emission filter without significantly decreasing the sensitivity of the system to the red fluorophore. Cross-talk affects not only the amplitude of the CCF but also the amplitude of the red ACF as the green fluorophore acts as a second species in the red detection channel. Spectral cross-talk can be totally avoided when using PIE by cross-correlating the photons detected in the green channel after green excitation with photons detected in the red channel after red excitation (light gold data, Figure 5.15b). With the appropriate choice of an emission filter in the green channel, no fluorescence from the red fluorophore will be present in the green channel. Furthermore, with selection of the correct wavelength for red excitation, it is possible to avoid excitation of the green fluorophore. Hence, the PIE CCF is cross-talk-free. In addition, the ACF determined from photons detected in the red channel after red excitation has no spectral cross-talk. Removal of spectral cross-talk results in a quantitatively correct amplitude for the red ACF, as demonstrated by the increase in the amplitude of the red ACF when using PIE (red data, Figure 5.15b). The amplitude of the green ACF is not affected by PIE (light green data, Figure 5.15b). Reassignment of the cross-talk photons to the correct detection channel will increase the detected molecular brightness and can be done when the direct excitation of the red dye (here: Atto665) with the green laser (530 nm) is negligible. An increase in molecular brightness will lead to an improvement of the signal-to-noise ratio of the ACF but will not alter the shape or amplitude of the function when only a single diffusing species is present.

This improvement in FCCS is very important for interaction studies on biological samples, as PIE makes it possible to distinguish between a weak interaction of relevant biomolecules and no interaction at all. As PIE records the detection channel and excitation source for each photon, different channels can be defined *ex facto* and "misplaced" photons can be reassigned to the correct channel. For example, cross-correlating all the photons detected after green excitation with photons detected after red excitation yields CCFs that can be evaluated quantitatively even in the presence of FRET [29].

5.3.1.4 Burst Analysis with Multiparameter Fluorescence Detection

With the small detection volume of a confocal microscope, it is possible to detect single molecules in a dilute sample [31, 32]. When a single fluorescent molecule diffuses through the focus of the microscope, a burst of photons is detected. All of the photons collected from a burst due to a single particle are accumulated and analyzed. To maximize the information that can be extracted from each photon, multiparameter fluorescence detection (MFD) is used (shown schematically in Figure 5.16) [33]. In MFD, the fluorescence emission signal is first split with a polarizing beamsplitter into parallel and perpendicular polarizations, and then both polarization channels are split with a dichroic beamsplitter to separate the photons spectrally.

From the MFD data, it is possible to determine the fluorescence lifetime τ and anisotropy r for the different channels (Figure 5.16b), the molecular brightness η for each species, the fluorescence intensity of each channel, and the arrival time of each photon with respect to the start of the measurement (with 100 ns resolution or better). PIE can be combined with MFD, as shown in Figure 5.17, to add stoichiometry information to the measurement [34].

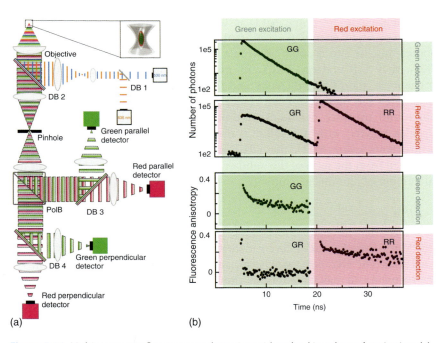

Figure 5.16 Multiparameter fluorescence detection with pulsed interleaved excitation. (a) Schematic drawing of an multiparameter fluorescence detection-pulsed interleaved excitation (MFD-PIE) setup and (b) experimental data of double-labeled DnaK, a bacterial heat-shock protein 70, recorded with MFD-PIE. MFD-PIE allows segregation of photons based on their spectral range, polarization, and the excitation source. The excitation source is assigned according to the arrival time of the detected photon with respect to the synchronization pulse as in PIE. From the MFD-PIE measurements, the stoichiometry, the fluorescence lifetimes of the various channels (upper panel), or the time-resolved anisotropy (lower panel) of the various channels can be determined.

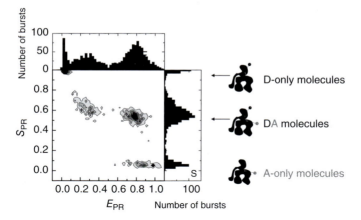

Figure 5.17 FRET efficiency versus stoichiometry plot of double-labeled DnaK molecules. From the stoichiometry values, the donor- and acceptor-only species can be separated from the double-labeled molecules or complexes. Hence, single-pair FRET histograms can be determined that are not contaminated with donor-only molecules.

With dual-color detection, MFD is well suited for performing single-pair Förster resonance energy transfer (spFRET) experiments. When two fluorophores attached to a single molecule or complex are within a distance of ~100 Å, they will undergo FRET. By measuring the relative intensity of the two channels, the FRET efficiency and thereby the distance between the two fluorophores can be determined. Hence, it is possible to determine structural features and measure dynamics of proteins and other biologically relevant molecules *in vivo* with a resolution down to a few ångstroms. In spFRET measurements using MFD, the FRET efficiency for every burst is typically determined from the ratio of the number of transferred photons (I_{FRET}) over the total number of photons in a burst ($I_{FRET} + I_{Donor}$):

$$E = \frac{I_{FRET}}{I_{FRET} + \gamma I_{Donor}} \tag{5.29}$$

where γ accounts for the different detection efficiencies of the donor (D) and the acceptor (A) channels, which can either be determined in a different set of experiments or calculated from the respective quantum yields ϕ of the fluorophores and the efficiencies η of the different detection pathways:

$$\gamma = \frac{\eta_A \phi_A}{\eta_D \phi_D} \tag{5.30}$$

The FRET values calculated for the individual molecules are typically plotted in a 1D histogram. When PIE is used, the stoichiometry for every burst can also be determined. The stoichiometry describes the ratio of photons collected after green excitation to the total photons detected:

$$S = \frac{I_{FRET} + I_{Donor}}{I_{FRET} + I_{Donor} + I_{Acceptor}} \tag{5.31}$$

where I_{Acceptor} is the number of photons collected in the acceptor channel after red excitation. A 2D histogram of FRET efficiency versus stoichiometry is shown in Figure 5.17 for the bacterial heat-shock protein 70 DnaK. With the help of the stoichiometry value (Eq. (5.31)), it is easy to distinguish proteins labeled with only a single donor or acceptor fluorophore from double-labeled molecules. From the FRET efficiency versus stoichiometry histogram, bursts from double-labeled molecules can be selected and further analyses performed with the data available from MFD. The details that can be extracted from a burst analysis experiment regarding dynamics, subpopulations, distances and so on are impressive. The main limitations in burst analysis experiments are the number of collected photons and the signal-to-noise ratio of the experiment. Hence, the advantages of confocal microscopy have played an important role in the advancement of solution-based single-molecule methods.

5.3.2 Scanning Applications beyond Imaging

There are a number of other methods that analyze the information available in the fluctuations of fluorescent signals. Many of these methods can be performed on image data collected using a confocal microscope. We highlight two of these methods here: the number and brightness (N&B) analysis and raster image correlation spectroscopy (RICS).

5.3.2.1 Number and Brightness Analysis

The effect of particle number and molecular brightness on the average intensity and variance is illustrated in Figure 5.11. Experiments were simulated for two species, A and B, where the molecular brightness of species A is 5 times lower than that of species B, but A is present at 5 times higher concentration. The mean intensity $\langle I \rangle$ is the same for both solutions, but the fluctuations in solution B (Figure 5.11b) – and therefore the variance σ^2 – are significantly larger than in solution A (Figure 5.11a). N&B analysis [35] is the most direct approach for analyzing the fluctuations in an intensity trace or series of fluorescence images. Two assumptions are made for an N&B analysis: only one species with a particular brightness is present within the detection volume, and the detected mean fluorescence intensity $\langle I \rangle$ is directly proportional to the number of molecules n in the detection volume and the apparent brightness b of the fluorescent species:

$$\langle I \rangle = nb \qquad (5.32)$$

where b has units of photons per molecule per unit time. The variance of the fluorescence signal will depend upon the fluctuations in the number of fluorescent molecules in the detection volume and the fluctuations due to shot noise of the detected signal. The variance in the fluorescence signal arising from fluctuations in the number of particles within the detection volume due to diffusion, $\sigma^2_{\text{Particles}}$, is proportional to the square of the molecular brightness. The fluctuations in particle number follow a Poissonian distribution, which has the property that the variance of the distribution is equal to the mean value. Hence, the variance in the number of particles in the detection volume is equal to the average number of molecules:

$$\sigma^2_{\text{Particles}} = nb^2 = \langle I \rangle b \tag{5.33}$$

Shot noise comes from the fact that fluorescence emission does not occur steadily, but also follows a Poissonian probability distribution:

$$\text{Poi}(k, \lambda) = \frac{\lambda^k e^{-\lambda}}{k!} \tag{5.34}$$

where k is the number of occurrences, and λ is the average number of occurrences (e.g., emitted photons). Hence, the variance from the shot noise is given by the average intensity, $\sigma^2_{\text{Shot Noise}} = \langle I \rangle$. The total variance in the signal is the sum of the two processes:

$$\sigma^2 = \sigma^2_{\text{Particles}} + \sigma^2_{\text{Shot Noise}} = b\langle I \rangle + \langle I \rangle = (b+1)\langle I \rangle \tag{5.35}$$

Including the shot-noise corrections into Eqs (5.32) and (5.33) yields

$$b = \frac{\sigma^2 - \langle I \rangle}{\langle I \rangle} \tag{5.36}$$

$$n = \frac{\langle I \rangle^2}{\sigma^2 - \langle I \rangle} \tag{5.37}$$

In order to obtain a more meaningful physical property than the number of photons emitted in an arbitrary time interval, the *molecular brightness* ε is introduced [36]. Molecular brightness is defined as the number of detected photons per second per molecule when the fluorophore is at the center of the PSF. There is a linear interrelation between the apparent brightness b and the molecular brightness ε via the binning time T_{bin} of the fluorescence trace:

$$\varepsilon = \frac{b}{T_{\text{bin}}} \tag{5.38}$$

One should keep in mind that the molecular brightness for the same fluorophore may be different when measured with different setups, as it depends on the number of laser photons per second $W(0)$ in the confocal spot and the detection efficiency κ of the entire system (Eq. (5.23)).

The method of N&B provides a straightforward approach to analyze the fluctuations in a given intensity trace or individual pixels from a movie. As it assumes a single species and makes no assumptions regarding the shape of the PSF, it is not as accurate as other brightness-based approaches such as the PCH analysis or FIDA. However, it is quick and easy to calculate and often provides a sufficient amount of information. An example of the N&B analysis is shown in Figure 5.18 [37]. Figure 5.18a shows the fluorescence intensity of DNA methyltransferase 1 in a buffer as a function of time. The overall fluorescence intensity decays as a function of time. From the N&B analysis (Figure 5.18b), we observe that the number of particles decreases with time whereas the molecular brightness remains constant. This suggests that the concentration of DNA methyltransferase 1 is decreasing due to adsorption of the protein to the surface of the sample holder. In Figure 5.18c, the fluorescence intensity of a sample of Atto532 in solution is shown as the laser power is increased in a stepwise

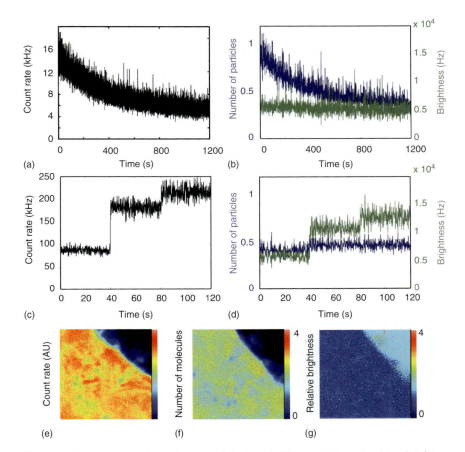

Figure 5.18 Fluorescence intensity traces (black) and N&B analysis (number blue, brightness green) for two different measurements. (a) Measurements of DNA methyltransferase 1 (Dnmt1) in solution. The fluorescence intensity is shown as a function of time. (b) An N&B analysis was performed with a sliding window on the intensity trace shown in (a). The number (blue) and brightness (green) of the molecules are plotted as a function of time. As more and more molecules adsorb to the sample chamber surface, the number of molecules in the solution decreases while the brightness remains constant. (c) The fluorescent intensity of Atto532 in solution. During the measurement, the laser power was increased stepwise. (d) The N&B analysis of the intensity trace shown in (c) using a sliding window. The N&B analysis shows a stepwise increase in the brightness (green) correlating to the changes in laser power, whereas the number of molecules (blue) remains constant. (e) Fluorescence measurements of cytosolic GFP in HeLa cells. An image from the movie is shown. Regions of intermediate and high count rates are visible. (f) The number of molecules and (g) the brightness determined using the N&B analysis for each pixel of the image are shown. (Courtesy of Dr Höller.)

manner during the measurement. As the laser power is increased, the molecular brightness also increases (Eq. (5.22)). This is reflected in the N&B analysis, where the increase in brightness follows the laser power while the number of molecules remains the same (Figure 5.18d).

The strength of N&B analysis is that it can be performed on movies. Figure 5.18e shows the average fluorescence detected from a HeLa cell expressing the green fluorescent protein (GFP). Variations in the fluorescence intensity are observed. By performing an N&B analysis on each pixel of the image (Figure 5.18f,g), it is clear that the concentration of protein varies within the cytosol whereas the molecular brightness remains the same. While this is a trivial result for GFP alone, it demonstrates the capabilities of the method. When GFP is fused to other proteins, N&B analysis can provide insights into the oligomerization state and stoichiometry of the protein or complex.

5.3.2.2 Raster Image Correlation Spectroscopy

RICS [38] combines LSCM with correlation spectroscopy. One difficulty with point FCS measurements in live cells is photobleaching due to the slow diffusion of biomolecules in the crowded environment of the cell. Photobleaching can be reduced in slowly diffusing media by scanning the laser beam. One possibility is to scan in a circle as in scanning FCS or, alternatively, to raster-scan the sample as is done for imaging and for RICS. In a raster-scanned image (Figure 5.19a), there is a relationship between the spatial separation of pixels in the image and the time delay between when the pixels were measured. As in FCS, the fluctuations are analyzed, but, in RICS, the temporal information is encoded in the spatial information within the image and analyzed using image correlation spectroscopy. As the excitation and detection beams are scanned relative to the sample, it is possible to gather more statistics while minimizing the exposure of each position within the sample to the laser light.

The 2D correlation pattern offers information about the diffusion process along the respective axes, which are scanned on different timescales. The two neighboring pixels in a line (x-axis) are measured consecutively with a time delay given by the pixel clock, whereas two neighboring pixels in adjacent lines (y-axis)

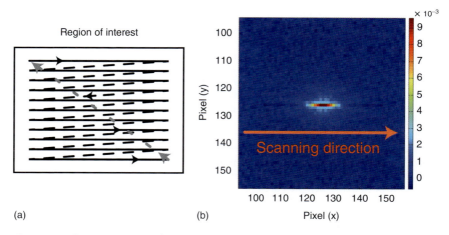

Figure 5.19 Raster image correlation spectroscopy (RICS). (a) The principle of raster-scan imaging with a laser scanning confocal microscope. A line is scanned linearly in time after which the beam is directed to the beginning of the next line and the following line scanned linearly in time. (b) RICS correlation function from of series of 50 images (total length 50 μm) of labeled DNA diffusing in solution with a recording time of 500 ms per frame.

are separated by the time it takes to scan a line. In the case of free Brownian diffusion, the RICS ACF assuming one-photon excitation and a 3D Gaussian PSF (Eq. (5.25)) is given by

$$G(\xi, \psi) = \frac{2^{-\frac{3}{2}}}{N}\left(1 + \frac{4D(\tau_x\xi + \tau_y\psi)}{\omega_r^2}\right)^{-1}\left(1 + \frac{4D(\tau_x\xi + \tau_y\psi)}{\omega_z^2}\right)^{-\frac{1}{2}}$$

$$\times \exp\left\{\frac{\left[\left(\frac{s\xi}{\omega_r}\right)^2 + \left(\frac{s\psi}{\omega_r}\right)^2\right]}{\left[1 + \frac{4D(\tau_x\xi + \tau_y\psi)}{\omega_r^2}\right]}\right\} \quad (5.39)$$

where $\langle N \rangle$ is the average number of fluorescent molecules in the PSF, D is the diffusion coefficient of the molecules, τ_x and τ_y are the pixel sampling times in the x- and y-axes, respectively, ω_x and ω_z are the lateral and axial dimensions of the confocal volume, ξ and ψ are the x and y spatial lags in pixels, respectively,

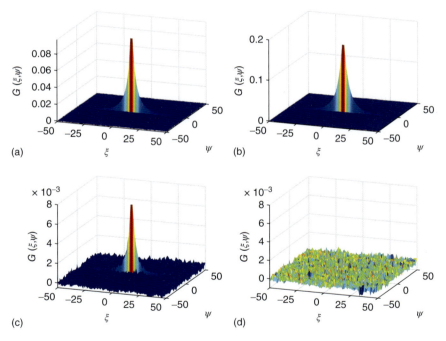

Figure 5.20 Influence of PIE on cross-correlation RICS. Raster-scanning images were collected for a solution of Atto488 and Atto565. The RICS autocorrelation function is shown for (a) Atto488 and (b) Atto565. As these dye molecules do not interact, there should be no residual cross-correlation. (c) The cross-correlation function is shown where the information available from PIE is not incorporated and all photons detected in the green channel are correlated with all photons detected in the red channel. (d) The same dataset is analyzed, but this time utilizing the PIE information to eliminate cross-talk. Here, the photons detected in the green channel after green excitation are correlated with photons detected in the red channel after red excitation. The residual cross-correlation due to spectral cross-talk is completely removed with PIE.

and s is the distance between contiguous pixels. The RICS ACF of DNA labeled with Atto565 freely diffusing in buffered solution is shown in Figure 5.19b. Here, the two different timescales become apparent: along the x-axis, the laser is scanned quickly and gives the width of the PSF unless the molecules diffuse more rapidly than the laser is scanned. Along the y-axis, more time passes before the laser beam returns to the same area. Fast moving particles will diffuse away, leading to a lower correlation amplitude at short distances but a higher probability of the particle being detected further away. For slowly diffusing molecules, the amplitude will be higher at short distances but decays to zero at distances larger than the PSF. For immobilized molecules that are much smaller than the size of the PSF, RICS returns the shape of the PSF. By fitting the RICS data with the theoretical expression for the ACF (Eq. (5.39)), the diffusion coefficient and average number of particles in the PSF can be determined. It is also possible to measure interactions using cross-correlation spectroscopy by performing RICS with two colors. Figure 5.20 shows the RICS ACF and CCF for two noninteracting fluorophores, Atto488 (Figure 5.20a) and Atto565 (Figure 5.20b), in solution. Figure 5.20c shows the RICS CCF without segregating the photons based on PIE. The spectral cross-talk of Atto448 into the Atto565 channel leads to a residual correlation, even though the particles do not interact. When taking the PIE information into account and correlating the photons detected in the green channel after green excitation with the photons detected in the red channel after red excitation, the residual cross-correlation completely disappears as shown in Figure 5.20d. Hence, as with FCCS, the sensitivity of two-color RICS experiments is enhanced with PIE. The applicability of RICS to cellular measurements can also be improved by a generalization of the ICS algorithm referred to as arbitrary region RICS (ARICS) [39]. The improved algorithm makes it possible to perform RICS on regions of interest of arbitrary shape that can selected manually, automatically via thresholding or based on parameters measured in a second channel. Thus, ARICS enables measurements of diffusional motion within subcellular compartments.

Acknowledgments

We thank Dr Gregor Heiss, Dr Martin Sikor, Dr Adriano A. Torrano, Anders Barth, Nader Danaf, and Ivo Glück for measurements presented in this work. We gratefully acknowledge the financial support of the Deutsche Forschungsgemeinschaft through the Excellence Cluster Nanosystems Initiative Munich (NIM) and the Ludwig Maximilian University of Munich through the LMU Innovativ BioImaging Network (BIN) and the Center for NanoScience (CeNS).

References

1 Airy, G.B. (1835) On the diffraction of an object-glass with circular aperture. *Trans. Cambridge Philos. Soc.*, **5**, 283–291.
2 Abbé, E. (1873) Beiträge zur Theorie des Mikroskops und der mikroskopischen Wahrnehmung. *Arch. Mikrosk. Anat.*, **9**, 413–468.
3 Strutt, J.W. (1879) Investigations in optics, with special reference to the spectroscope. *Philos. Mag.*, **5**, 261–274.

4 Sparrow, C.M. (1916) On spectroscopic resolving power. *Astrophys. J.*, **44**, 76–86.
5 Minsky, M. (1961) Microscopy apparatus. US Patent 3013467, Dec. 19, 1961.
6 Naora, H. (1951) Microspectrophotometry and cytochemical analysis of nucleic acids. *Science*, **114** (2959), 279–280.
7 Minsky, M. (1988) Memoir on inventing the confocal scanning microscope. *Scanning*, **10**, 128–138.
8 Ploem, J.S. (1967) The use of a vertical illuminator with interchangeable dichroic mirrors for fluorescence microscopy with incident light. *Z. Wiss. Mikrosk.*, **68**, 129–142.
9 Egger, M.D. and Petran, M. (1967) New reflected-light microscope for viewing unstained brain and ganglion cells. *Science*, **157** (3786), 305–307.
10 White, J.G., Amos, W.B., and Fordham, M. (1987) An evaluation of confocal versus conventional imaging of biological structures by fluorescence light microscopy. *J. Cell Biol.*, **105** (1), 41–48.
11 van Meer, G., Stelzer, E.H.K., Wijnaendts-van-Resandt, R.W., and Simons, K. (1987) Sorting of sphingolipids in epithelial (Madin–Darby canine kidney) cells. *J. Cell Biol.*, **105** (4), 1623–1635.
12 Novotny, L. and Hecht, B. (2006) *Principles of Nano-Optics*, 1st edn, Cambridge University Press, Cambridge.
13 Webb, R.H. (1996) Confocal optical microscopy. *Rep. Prog. Phys.*, **59**, 427–471.
14 Wilhelm, S. (2008) *Confocal Laser Scanning Microscopy*, Carl Zeiss MicroImaging GmbH, Jena.
15 Jonkman, J.E.N. and Stelzer, E.H.K. (2001) in *Confocal and Two-Photon Microscopy: Foundations, Applications and Advances* (ed. A. Diaspro), Wiley-Liss Inc., New York, pp. 101–125.
16 Pawley, J.B. (2006) *Handbook of Biological Confocal Microscopy*, 3rd edn, Springer, New York.
17 Young, J.Z. and Roberts, F. (1951) A flying-spot microscope. *Nature*, **167** (4241), 231.
18 Nipkow, P. (1884) Elektrisches Teleskop. Germany Patent 30105, Jan. 15, 1885.
19 Richardson, W.H. (1972) Bayesian-based iterative method of image restoration. *J. Opt. Soc. Am.*, **62** (1), 55–59.
20 Haustein, E. and Schwille, P. (2007) Fluorescence correlation spectroscopy: novel variations of an established technique. *Annu. Rev. Biophys. Biomol. Struct.*, **36** (1), 151–169.
21 Lakowicz, J.R. (2006) *Principles of Fluorescence Spectroscopy*, 3rd edn, Springer, Berlin.
22 Lamb, D.C. (2009) in *Single Particle Tracking and Single Molecule Energy Transfer* (eds C. Bräuchle, D.C. Lamb, and J. Michaelis), Wiley-VCH Verlag GmbH, Weinheim, pp. 99–129.
23 Elson, E.L. and Magde, D. (1974) Fluorescence correlation spectroscopy. I. Conceptual basis and theory. *Biopolymers*, **13** (1), 1–27.
24 Magde, D., Elson, E., and Webb, W.W. (1972) Thermodynamic fluctuations in a reacting system: measurement by fluorescence correlation spectroscopy. *Phys. Rev. Lett.*, **29** (11), 705.

25 Rigler, R., Mets, Ü., Widengren, J., and Kask, P. (1993) Fluorescence correlation spectroscopy with high count rate and low background: analysis of translational diffusion. *Eur. Biophys. J.*, **22** (3), 169–175.
26 Macdonald, P., Johnson, J., Smith, E., Chen, Y., and Mueller, J.D. (2013) in *Methods in Enzymology: Fluorescence Fluctuation Spectroscopy (FFS)*, Part A, vol. **518** (ed. S.Y. Tetin), Elsevier, Oxford, pp. 71–98.
27 Schwille, P., Meyer-Almes, F.J., and Rigler, R. (1997) Dual-color fluorescence cross-correlation spectroscopy for multicomponent diffusional analysis in solution. *Biophys. J.*, **72** (4), 1878–1886.
28 Foo, Y.H., Naredi-Rainer, N., Lamb Don, C., Ahmed, S., and Wohland, T. (2012) Factors affecting the quantification of biomolecular interactions by fluorescence cross-correlation spectroscopy. *Biophys. J.*, **102** (5), 1174–1183.
29 Muller, B.K., Zaychikov, E., Brauchle, C., and Lamb, D.C. (2005) Pulsed interleaved excitation. *Biophys. J.*, **89** (5), 3508–3522.
30 Kapanidis, A.N., Lee, N.K., Laurence, T.A., Doose, S., Margeat, E., and Weiss, S. (2004) Fluorescence-aided molecule sorting: analysis of structure and interactions by alternating-laser excitation of single molecules. *Proc. Natl. Acad. Sci. U.S.A.*, **101** (24), 8936–8941.
31 Shera, E.B., Seitzinger, N.K., Davis, L.M., Keller, R.A., and Soper, S.A. (1990) Detection of single fluorescent molecules. *Chem. Phys. Lett.*, **174** (6), 553–557.
32 Zander, C., Sauer, M., Drexhage, K.H., Ko, D.S., Schulz, A., Wolfrum, J. *et al.* (1996) Detection and characterization of single molecules in aqueous solution. *Appl. Phys. B*, **63** (5), 517–523.
33 Widengren, J., Kudryavtsev, V., Antonik, M., Berger, S., Gerken, M., and Seidel, C.A.M. (2006) Single-molecule detection and identification of multiple species by multiparameter fluorescence detection. *Anal. Chem.*, **78** (6), 2039–2050.
34 Kudryavtsev, V., Sikor, M., Kalinin, S., Mokranjac, D., Seidel, C.A.M., and Lamb, D.C. (2012) Combining MFD and PIE for accurate single-pair Förster resonance energy transfer measurements. *ChemPhysChem*, **13** (4), 1060–1078.
35 Digman, M.A., Dalal, R., Horwitz, A.F., and Gratton, E. (2008) Mapping the number of molecules and brightness in the laser scanning microscope. *Biophys. J.*, **94** (6), 2320–2332.
36 Chen, Y., Müller, J.D., Ruan, Q., and Gratton, E. (2002) Molecular brightness characterization of EGFP in vivo by fluorescence fluctuation spectroscopy. *Biophys. J.*, **82** (1), 133–144.
37 Höller, M. (2011) *Advanced Fluorescence Fluctuation Spectroscopy with Pulsed Interleaved Excitation*, Ludwig-Maximilians Universität, Munich.
38 Digman, M.A., Sengupta, P., Wiseman, P.W., Brown, C.M., Horwitz, A.R., and Gratton, E. (2005) Fluctuation correlation spectroscopy with a laser-scanning microscope: exploiting the hidden time structure. *Biophys. J.*, **88** (5), L33–L36.
39 Hendrix, J., Dekens, T., Schrimpf, W., and Lamb, D.C. (2016). Arbitrary-Region Raster Image Correlation Spectroscopy. *Biophys. J.*, **111** (8), 1785–1796.

6

Two-Photon Excitation Microscopy for Three-Dimensional Imaging of Living Intact Tissues

David W. Piston

Washington University in St. Louis, Department of Cell Biology and Physiology, School of Medicine, 660 S. Euclid Avenue, St. Louis, MO 63110-1093, USA

6.1 Introduction

Two-photon excitation microscopy is an alternative to confocal microscopy that provides attractive features especially for optical sectioning deep into intact tissues or whole animals. Two-photon excitation microscopy is the most commonly used type of a wider class of imaging approaches called *multiphoton* or *nonlinear* microscopies. These nonlinear microscopies include two-photon excitation (2PM) and three-photon excitation (3PM) microscopies, second-harmonic generation (SHG) and third-harmonic generation (THG) microscopy, as well as coherent anti-Stokes Raman spectroscopy (CARS) and stimulated Raman spectroscopy (SRS) imaging. All these approaches offer similar advantages in terms of inherent optical sectioning, but two-photon excitation yields the most efficient signal generation and is therefore the most common. So this chapter will focus on the 2PM technique. As detailed throughout this chapter, the advantages of 2PM over linear optical sectioning techniques are greater imaging depths and reduced overall phototoxicity. As is true for all fluorescence microscopy approaches, 2PM enables dynamic cellular measurements, so the extra depth penetration opens up a broad range of experiments that rely on noninvasive intravital imaging of cellular and subcellular processes. While 2PM provides the capability to perform quantitative imaging in tissues at depths well beyond those that can be reached with confocal microscopy, inherent difficulties with the nonlinear optics underlying 2PM limit its main usefulness to deep-tissue imaging and a few other specialized applications.

For investigations of samples thicker than a couple of micrometers – that is, anything thicker than a monolayer cell culture – out-of-focus background limits the usefulness of fluorescence microscopy, especially for quantitative measurements. For many years, researchers overcame this limitation by embedding and sectioning cells, typically in conjunction with immunofluorescent labeling. The advent of commercially available optical sectioning microscopes, including confocal, deconvolution, and more recently light sheet (or selective plane illumination) microscopy, allowed diffraction-limited imaging studies to move from single cells to the tissue level, and even to whole animals. These optical sectioning methods allow investigators to acquire three-dimensional (3D) image data from intact, live samples without the need for fixation and mechanical slicing.

The most widespread optical sectioning microscopy used is confocal microscopy. As detailed in Chapter 5, the confocal microscopy concept relies on the fluorescence signal generated everywhere throughout the sample (whether the fluorescence arises from above, below, or within the plane of focus), and then uses a conjugate focus (i.e., confocal) pinhole to act as a spatial filter and reject the out-of-focus background fluorescence. As such, clean optical sections of the region near the focus are produced with minimal background contamination. However, imaging live samples with confocal microscopy can create deleterious effects in the sample, in particular due to photobleaching of the probes and the phototoxicity of the excitation light. The fluorescence signal is detected from only a single optical plane, but background fluorescence is excited throughout the sample. Most of the background fluorescence is rejected by the pinhole. This out-of-focus excitation can be particularly damaging in causing rapid photodestruction of the fluorophore in question as well as exciting autofluorescent compounds that can contribute to the killing of the live sample. As described below, the physics underlying 2PM allows much deeper imaging with 2PM since it is much less susceptible to degradation by out-of-focus absorption and scattering. A good rule of thumb is that 2PM can provide useful image data about 6 times deeper into tissue than can confocal microscopy.

While 2PM was initially demonstrated over 25 years ago [1], for the first ~15 years, its use was limited to specialist laboratories due to the expertise needed to control the sophisticated ultrafast laser technology required for 2PM. Nowadays, reliable "turn-key" laser systems have been developed, allowing a hands-off approach to 2PM and rendering it as easy to use as commercial confocal microscopes. This has allowed 2PM to transition from a novelty of the laser laboratory into a useful tool in a true biological research setting, where investigators can concentrate on studying the biological question at hand without worrying about aligning, operating, and maintaining the microscope system. Beyond the instrumentation advances, though, greater understanding of 2PM and its advantages has allowed this technique to be effectively utilized in biological research.

This chapter will cover the basics of 2PM and how the underlying photophysics leads to both its advantages and limitations. To take full advantage of its strengths, the instrumentation design needs to be optimized for 2PM, so these points will also be described in detail. The other nonlinear microscopies will be briefly described, with emphasis on the particular advantages of each. Finally, some recent examples of 2PM will be presented to highlight the wide range of physiological studies enabled in fields such as neuroscience, developmental biology, immunology, cancer, and endocrinology.

6.2 What is Two-Photon Excitation?

Given the plethora of optical sectioning methods, such as confocal, deconvolution, and light sheet microscopies, a good question to ask is why is yet another approach needed. To answer this question, we can consider the limitations of the other approaches, which are all based on linear absorption. That is, fluorescence can occur when a single excitation photon (typically of ultraviolet (UV), blue, or green wavelengths) is absorbed by a fluorescent molecule (fluorophore), which leads to a transition of the molecule into an excited state. The molecule then relaxes back to the ground state by emitting a single photon with less energy (or longer wavelength than the excitation wavelength), see Figure 6.1. In this case, which holds true under low to moderate excitation intensities, one photon is absorbed by the fluorophore, which can in turn emit a single fluorescent photon. If the excitation power is doubled, then twice the fluorescence is generated. This linearity is useful for quantitation but also leads to significant limitations, especially in confocal microscopy. This limitation is exacerbated in thick samples, as schematized in Figure 6.2. Because of focusing, the laser intensity is highest at the focus (focal plane) so a fluorophore there (f_1) is strongly excited. Fluorophores

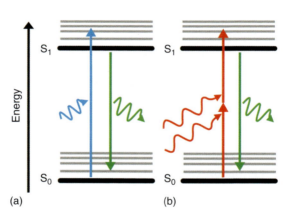

Figure 6.1 Jablonski diagram illustrating (a) one-photon versus (b) two-photon excitation of a molecule, and the subsequent fluorescence emission (shown in green). Two-photon-excited fluorescence results from the simultaneous absorption of two photons (red), each of half the energy of that from single-photon absorption (blue).

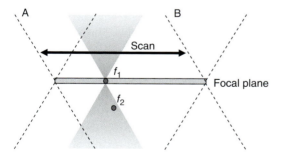

Figure 6.2 Photobleaching in confocal microscopy occurs above and below the plane of focus as shown in this axial photobleaching diagram. In confocal microscopy, as the beam is scanned from A to B to form an image, fluorophore f_2 is photobleached equally as f_1. Data is only acquired from f_1, which is in the focal plane, while data from f_2 is excluded from detection by the confocal pinhole.

above or below the plane of focus are illuminated less strongly at any given time, but as the laser beam is scanned across the field of view from A to B to form an image, out-of-focuses fluorophores are illuminated for a concomitantly longer time. During collection of an image, the entire sample is irradiated equally by the excitation light. That is, the regions above and below the focal plane see just as much excitation as does the focal plane. As a result, the entire thickness of the sample is subjected to photobleaching and associated photodamage similar to what is seen at the focus, but given the confocal pinhole, data is collected only from the focal plane. The fluorescence that arises from out-of-focus fluorophores is rejected by the pinhole, but photobleaching and photodamage still occur despite the lack of data collected from those regions. Deconvolution microscopy uses the out-of-focus fluorescence to increase the signal-to-noise ratio, and though this mathematical analysis approach works best for samples with low background, it has not been shown to function well for imaging of intact tissue. The more recently developed optical sectioning microscopies based on light sheet or selective plane illumination microscopy (SPIM) are methods where only the observed section of the sample is illuminated. Thus, SPIM approaches not limited by out-of-focus photobleaching and photodamage, and they may evolve as a replacement for confocal microscopy. However, both the deconvolution and SPIM imaging techniques are limited in their imaging depth by out-of-focus absorption of the excitation light and scattering of both the excitation and emission light. For SPIM, the excitation light is absorbed as the light sheet passes through the sample, so the penetration depth is limited by the concentration of fluorophores, which is usually quite high for intact tissue imaging.

6.2.1 Nonlinear Optics and 2PM

In contrast, two-photon excitation microscopy relies on nonlinear interactions between light and matter. Under very high excitation intensities, it is possible that a fluorophore can simultaneously absorb two excitation photons, each at half the excitation energy required for a single-photon absorption event (typically red to infrared wavelengths, see Figure 6.1). Because of the inverse relationship between photon energy and wavelength, a photon with half of the energy will have twice the wavelength. For example, a photon in the blue range (~450 nm) will have twice the energy as one in the near-infrared range (~900 nm). Thus, fluorescent probes that are normally excited in the UV (~350 nm) can be excited in 2PM with red light (~700 nm), while probes that are normally excited by blue, green, or orange light (450–550 nm) can be excited in 2PM with near-infrared light (900–1100 nm). Probes that normally absorb red light (>600 nm) require deeper infrared excitation (>1200 nm) where the photophysical properties of water can become a limitation. As detailed below, this nonlinear absorption means that in a microscope, two-photon excitation occurs only in the region of the focus. Since the excitation photons do not have sufficient energy by themselves to excite fluorescence, they pass through the sample with minimal interactions. This has two direct major implications. First, fluorophores above and below the microscope's focal plane are not excited by the illumination laser, and thus they are not photobleached, nor do they cause photodamage. It is important to realize that fluorophores in the plane of focus are still subject to photobleaching and the associated photodamage – in fact, these deleterious events are often accelerated under the high laser intensities needed for two-photon excitation. Second, the excitation light is not attenuated by out-of-focus absorption as it travels through the sample. This allows stronger excitation deep into thick samples and more uniform illumination as a function of the focus depth. As detailed below, however, an additional practical consequence of this localized excitation causes 2PM images to be degraded less by light scattering in the sample, and as a result 2PM is much better than other optical sectioning approaches for deep-tissue imaging.

6.2.2 History and Theory of 2PM

The possibility of two-photon absorption was initially predicted by Maria Goppert-Mayer in her Ph.D. thesis in 1931 [2], but the high photon fluxes needed to achieve such nonlinear effects were not available at that time (Dr Goppert-Mayer went on to win the 1963 Nobel Prize in physics for her development of the nuclear shell model). Even the brightest arc lamps focused through the best objective lenses available at that time could only have created a single two-photon event every few minutes, well below the detection limits. Soon after the invention of the laser, however, Kaiser and Garrett first observed two-photon excitation in 1961, but the number of two-photon excitation events generated by those early lasers was still too low to permit imaging. It was not until the invention of ultrashort mode-locked lasers (see Box 6.1) that 2PM became practical, as was first demonstrated by Denk *et al.* [1] in 1990. Over the last 25 years, continuous improvements in instrumentation have made 2PM a practical alternative for optical sectioning microscopy, especially for deep-tissue imaging.

We now know that Goppert-Mayer correctly predicted the simultaneous absorption of multiple photons, and set the level of photon density (light intensity) that would be needed for this effect to occur. According to her original theory, which has been verified by many measurements over the last 50+ years, the intensity needed for efficient two-photon excitation is *about one million fold greater* than what is needed for the same number of one-photon excitations. Reaching such high intensities would not have been possible for biological experiments but for the advent of ultrashort pulsed lasers that provided high peak power (well suited for two-photon excitation) but have sufficiently low average power (which permits imaging samples without undue heating or other effects that come from linear absorption). In fact, development of ultrafast pulsed lasers is what initially enabled 2PM. These details are presented in Box 6.1.

> **Box 6.1 Practical Excitation Parameters for 2PM**
>
> To define the requirements of appropriate laser excitation sources for two-photon excitation microscopy, it is useful to understand the underlying photophysical parameters. In any fluorescence experiment, the number of excitation events depends on the strength of absorption and the intensity of the incoming light. As shown in the first row of Table 6.1, for the linear absorption used in confocal

(Continued)

Box 6.1 (Continued)

microscopy this number (N_{ex}) depends on the absorption cross section (σ) times the exciting laser intensity (I). The cross section is directly related to the extinction coefficient, which is typically measured in an absorption spectrometer, and for typical fluorophores it has a value of $\sim 10^{-16}$ cm² (row 2). For a typical image size of 512 × 512 pixels and fluorophore concentration in the submillimolar range, the excitation intensity needs to be in the range of 1 MW cm⁻² (row 3). While this may seem like a large intensity, the focusing down to a submicrometer spot allows this level to be easily reached with a continuous wave (CW) laser with powers <1 mW, which is less than the intensity of a standard laser pointer. In fact, only 1 mW of laser power (P_{sat}) is sufficient to saturate the fluorescence (i.e., to excite all of the fluorophores) in a typical confocal microscope. For two-photon excitation (right-hand column in Table 6.1), the number of excitations (row 1) depends on the square of the excitation intensity and the two-photon cross section (δ). Since both photons must interact with the fluorophore at "the same time," δ has both a spatial component (as does σ) and a temporal overlap component (row 2). If we assume that the spatial components are similar to the 10^{-16} cm² measured in linear absorption, and factor out that amount for each photon, we are left with 10^{-18} s/photon, which gives a good estimate for what is meant by "the same time." That is, both photons must arrive within 10^{-18} s. Given the speed of light, the photons must be closer than 3 Å, which is consistent with the $\sim$10 Å size of typical fluorophores. In this case, the excitation intensity needed for imaging is roughly a million-fold higher than what is needed in linear absorption (row 3). Using a CW laser, this would require kilowatts of power, which is comparable to what is used in laser welding. To reduce the total power needed, ultrashort-pulse, mode-locked lasers are used (row 4). These lasers output pulses with a duration of $\sim 10^{-13}$ s at a repetition rate of 100 MHz, which means that the lasers are actually off 100 000-fold longer than they are on. While these pulses seem short to us, the 10^{-13} s is also 100 000 times longer than the simultaneous absorption time (10^{-18} s), so two-photon excitation has plenty of time to occur during each pulse. The use of lasers with a duty cycle of 10^{-5} means that, instead of kilowatt powers, fluorophore saturation can be reached with about 10 mW, or only 10-fold more laser power than is needed in linear absorption microscopy (row 5).

Table 6.1 Parameters for confocal and two-photon excitation microscopy.

	One-photon (linear) absorption	Two-photon (non-linear) absorption
1	$N_{ex} = \sigma I$	$N_{ex} = \delta I^2$
2	$\sigma = 10^{-16}$ cm²	$\delta = 10^{-50}$ cm⁴ s/photon
3	$I_{ex} = 1$ MW cm⁻²	$I_{ex} = 10^6$ MW cm⁻²
4	CW laser	$t = 10^{-13}$ s
		Freq. = 100 MHz
5	$P_{sat} = 1$ mW	$P_{sat} = 10$ mW

For laser scanning imaging, around one million pixels need to be acquired in about 1 s for each image, and to achieve a good signal-to-noise ratio in the image, over one thousand photons need to be detected in each pixel. To obtain powers at the sample sufficient for 2PM imaging, the excitation photons must be crowded not only in time with pulsed lasers but also in space using the focusing of a microscope objective lens. This critical point of crowding can be understood from the basic physical chemistry of a reaction that goes as A + 2B → C (where A is the fluorophore in the ground state, B is a photon, and C is the fluorophore in the excited state). This reaction will proceed with a probability that depends on the square of the concentration of B. As the excitation light is focused in the microscope, the photons are increasingly crowded so that their concentration increases, which in turn greatly increases the probability of two-photon excitation. Using a high numerical aperture (NA) lens, the laser beam is focused to a spot roughly 300 nm in diameter, which yields a photon density 10 million-fold greater than what is seen in the typical 1-mm laser beam diameter (or about 100 000-fold greater than the 100-μm diameter beam near the surface of the sample). As the two-photon excitation rates increase with the square of the photon concentration, this focusing enhances the probability several trillion-fold. Thus, through the combined crowding of photons in space by focusing and in time using pulsed lasers, it is possible to generate billions of the normally rare two-photon excitation events each second.

This requirement of photon crowding in space leads to a major advantage of 2PM – the localization of fluorescence excitation. This is schematized in Figure 6.3. As laser pulses pass through the sample, photons are crowded in time, but for most of their path they are not sufficiently crowded for photon pairs to interact simultaneously with single fluorophores. Only near the focus

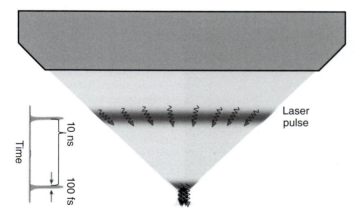

Figure 6.3 Excitation localization by photon crowding in time and space. Schematic representation of the fluorescence distribution in the focus resulting from single- and two-photon excitation. Under single-photon excitation, fluorescence is generated throughout the sample; in a confocal microscope only fluorescence generated at the focal plane is detected. Two-photon-excited fluorescence results in fluorescence being generated solely at the focal plane. The lack of out-of-focus excited fluorescence reduces overall phototoxicity in the sample and enhances contrast in deep-tissue imaging.

are photons sufficiently crowded to generate a significant number of two-photon absorption events. This region near the focus the two-photon excitation volume (often called the *focal volume*), and for a high NA objective lens it is ~0.1 fl. Two-photon excitation events fall off rapidly above or below the focal volume because of their dependence on the square of the excitation intensity, as shown in Figure 6.4. As can be seen from this figure, the only place where two-photon absorption occurs on a practical level is within the focal volume. This means that all the fluorescence comes from this small focal region and not from other depths through which the excitation light passes. This localization of excitation results in inherent optical sectioning, which is roughly equivalent to that achieved in confocal microscopy. However, in 2PM there is no need for a confocal pinhole to obtain optical sections, rather these are generated directly by the localization of the fluorescence. As an analogy, conventional fluorescence microscopy is equivalent to looking inside a house by shining a powerful spotlight from outside, while 2PM is shining a flashlight inside the house to examine its contents. The light needed is generated inside the sample of interest. This difference is illustrated in Figure 6.5, where two parallel laser beams are focused into a cuvette filled with a fluorescein solution. While the conventional one-photon absorption occurs throughout the sample, forming the hour-glass figure characteristic of a focused beam, the two-photon excitation is observed just at the focal point. In a confocal microscope, the pinhole is used to reject background fluorescence from outside this focal spot and permit detection of only the in-focus signal, but in 2PM there is no background created by the excitation, so there is no need for a pinhole to reject it. Since the pinhole is not needed to obtain optical sections, the detection geometry can be simplified, which also leads to further advantages in signal collection.

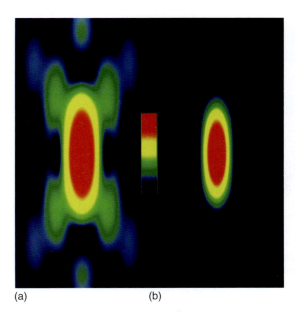

Figure 6.4 Axial image (X–Z) of the calculated point-spread functions for (a) one-photon and (b) two-photon excitation. For one-photon excitation, fluorescence is generated both above and below the focal spot, seen as the wings of the Airy disk profile. In two-photon excitation, the out-of-focus background created by these wings is not observed. (Sandison and Webb 1994 [3]. Reproduced with permission of The Optical Society.)

(a) (b)

Figure 6.5 Experimental illustration of the difference between linear and two-photon excitation: 380 nm (confocal) and 760 nm (two-photon) excitation wavelengths were used to generate fluorescence in a 50 µM fluorescein solution. Two-photon excitation generates fluorescence in the focal volume exclusively, while conventional single-photon excitation generates fluorescence all along the light path. (Image used by permission, Kevin Belfield Research Group, University of Central Florida; Zhen-Li Huang and Ciceron Yanez.)

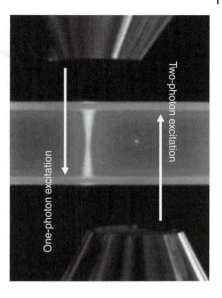

6.3 How Does Two-Photon Excitation Microscopy Work in Practice?

As described previously, the photophysical theory shows that 2PM excitation is limited to the focus. The strengths of 2PM for deep tissue imaging are all consequences of this effect [4]. The first consequence is that, as the laser focus is raster-scanned to form an image, two-photon excitation of the fluorophores as well as any photobleaching and photodamage associated with the probe excitation are also limited to the focal plane. This is demonstrated in Figure 6.6, which shows an axial (X–Z) scan through a polymer film that is uniformly stained with fluorescein. To create this image, a 20 µm × 20 µm square X–Y region was imaged repeatedly for an extended period, first using 488 nm and then, after moving the region 30 µm to the right, again using 780 nm 2PM. After these two regions were

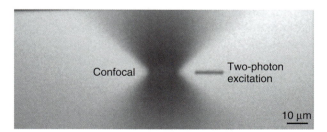

Figure 6.6 2PM limits photobleaching and associated photodamage. Axial photobleaching patterns produced by conventional confocal and two-photon excitation after imaging a single plane inside a polymer film containing sulforhodamine B. On the left, the effect of conventional confocal illumination (514 nm); on the right the effect of two-photon illumination (820 nm). Scale bar 10 µm.

photobleached effectively to zero, an axial $X-Z$ scan was acquired to examine the 3D photobleaching profile. In the first case of linear excitation, as is used in confocal microscopy, photobleaching is seen uniformly throughout the sample: above, below, and at the focal plane. If the data from only the focal plane was acquired using a confocal pinhole to reject the out-of-focus background, all the other photobleaching (and associated photodamage) would be nil. That is, out-of-focus molecules would be photobleached and out-of-focus cells would be subjected to photodamage, but there would be no data collected in return. In the second case of 2PM, the photobleaching (and associated photodamage) is limited only to the focal plane. Thus, out-of-focus fluorophores are not damaged, and out-of-focus cells are not subjected to photodamage. It is important to note, though, that photobleaching and associated photodamage still occur at the focal plane, and as detailed in the following, the rates of bleaching and damage within the focus can actually be accelerated during 2PM.

6.3.1 The Role of Light Absorption in 2PM

The second consequence of 2PM excitation being limited to the focus is that the incoming excitation light is not absorbed as it passes through the sample, and therefore more excitation can reach deeper into the tissue. This effect is demonstrated in Figure 6.7, which shows axial $(X-Z)$ scans through 250 μm of a colored, but otherwise clear and essentially nonscattering glass slide. This glass slide also has an index of refraction equal to that of the immersion oil used for imaging, so that there is no degradation of imaging quality from aberrations due to index of refraction mismatch. The first axial $X-Z$ scan image was acquired using one-photon 488-nm excitation. As can be seen in the image, but especially in the accompanying intensity profile along the Z-axis, the intensity drops off as the imaging depth increases so that only 25% of the signal is observed at the 250 μm level. Since this glass is clear, there is minimal contribution from scattering. This means that the resulting signal loss at depth is entirely due to the absorption of the incoming light by out-of-focus fluorophores. In other words, 75% of the excitation photons are absorbed before reaching the focus 250 μm into the sample. This glass slide is labeled at a fairly high concentration of fluorophores, but losses due to out-of-focus absorption always place a critical limitation on confocal microscopy, and, in fact, also on the efficacy of light sheet microscopies, which also rely on penetration into the sample (see below). The second image shows an identical axial scan, except that this one was created using 780-nm two-photon excitation. In contrast to what is seen in the first case, the 2PM scan shows uniform brightness all the way down to 250 μm. The reason for this uniformity is that there is no absorption of the incoming light until the focus: the individual photons do not have sufficient energy to excite the fluorophores, and the photons are not crowded enough to generate two-photon excitation events until they reach the focus. Thus, one main reason for the increased imaging depth provided by 2PM is the lack of out-of-focus absorption of the incoming excitation light. The practical advantage of 2PM depends on the fact that the red and infrared excitation light is generally not absorbed by biological tissue. This is easily observed by covering a flashlight with a hand – it is the red light that comes

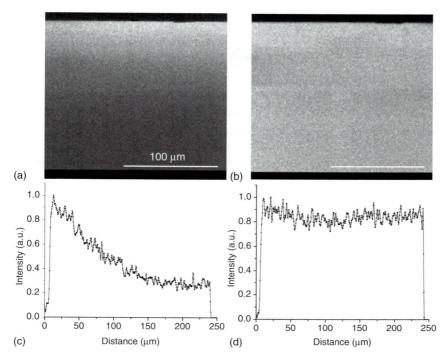

Figure 6.7 Out-of-focus absorption effects on fluorescence imaging with one- and two-photon excitation demonstrated using fluorescent glass (with negligible scattering). In one-photon excitation (a, c), excitation photons are absorbed throughout the sample, so fewer photons reach the focal plane. This can be seen as a decrease of the final fluorescence collected as a function of imaging depth. On the other hand, two-photon excitation light is not absorbed until the photons are sufficiently crowded at the focal spot (b, d). Thus, there is no loss of fluorescence signal as a function of imaging depth.

through – the other colors are absorbed by naturally occurring molecules, most strongly by flavins, lipo-pigments, and porphyrins (heme proteins) in the blue, green, and orange regions of the spectrum. While this assumption holds for most animal cells, it should be noted that many species, especially plants, can have very strong absorption in the red and near-infrared regions and that this absorption may limit or preclude the use of 2PM in such specimen. Finally, even though the excitation photons may not be absorbed as they pass through the sample, there is usually significant absorption of the resulting fluorescence as it comes out through the sample, and this contributes to limiting the maximum imaging depth of 2PM.

6.3.2 The Role of Light Scattering in 2PM

The third consequence of 2PM excitation being limited to the focus is that 2PM is less sensitive to degradation by light scattering in the sample than confocal microscopy. Even though 2PM is less sensitive to this, light scattering defines the ultimate limit of the imaging depth. The relative roles of absorption and scattering can be demonstrated by a simple experiment using red and green laser pointers.

Covering the red laser pointer with a finger, the red light is observed coming through the finger (similar to what is seen using a flashlight as discussed above). The red light is not absorbed by the molecules and cells in the finger. However, the laser beam does not come through – the light is all heavily scattered as it passes through the finger and it comes out in all directions. In contrast to the red laser pointer, doing the same thing with the green laser pointer shows a much different result. None of the green light comes through the finger – it is all absorbed. A small amount of scattered green light may be observed close to where the laser beam impinges on the finger, but none traverses the finger and comes out on the other side. Still, even though the red light penetrates the tissue, it is highly scattered, which limits its usefulness for imaging.

The depth penetration of 2PM is limited by scattering in the sample of both the incoming excitation light and the outgoing fluorescence. Still, 2PM is less sensitive than confocal microscopy to degradation by light scattering, as demonstrated in Figure 6.8, which shows the ways light scattering affects both confocal microscopy and 2PM. Figure 6.8a depicts confocal microscopy with blue excitation and green fluorescence. The microscope focuses the blue laser to the focus (1), and the resulting fluorescence is collected and directed through the confocal pinhole to the detector (2). However, some of the fluorescence coming back through the sample can be scattered and miss the pinhole (3). This is fluorescence

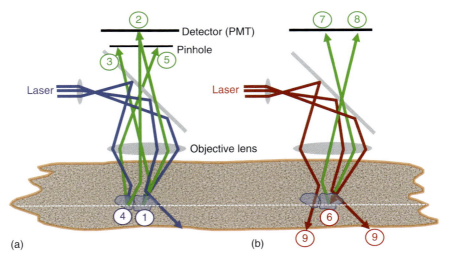

Figure 6.8 Scattering effects on fluorescence imaging with one- and two-photon excitation. In one-photon excitation (a), excitation photons may reach the focus (1) and generate fluorescence. Only the unscattered photons pass through the pinhole (2) and reach the detector (photomultiplier tube, PMT). However, some of the photons generated at the focus are scattered in the sample, and miss the pinhole (3). Excitation can also be scattered, which creates extra out-of-focus background (4). This background is largely rejected by the pinhole (5), but some of it can be scattered in the sample such that it passes cleanly through the pinhole and reaches the detector (2). In two-photon excitation (b), excitation light can also be scattered before it reaches the focus (6), but these scattered photons (9) are too few to generate two-photon excitation. For the fluorescence, both scattered (7) and unscattered (8) photons are detected since the pinhole is not needed to reject out-of-focus background.

that comes from the focus, and is thus signal that should be detected, but is lost due to scattering. In the same manner, the excitation light can also be scattered as it passes through the sample on the way to the focal spot (4). This light can interact with fluorophores outside the focal spot and generate extraneous background fluorescence. In a weakly scattering sample, most of this background will be excluded from collection by the pinhole, so only negligible amounts of it will be detected. In a more strongly scattering sample, or when imaging deeper into a sample, the increased amount of excitation scattering will result in significant amounts of background fluorescence being scattered into the pinhole (2). The detection of this extra background will be more or less constant over the image, causing a haze that reduces the image contrast (as shown in Figure 6.11). Thus, the scattering of both excitation and fluorescence reduces the detected signal, and scattering of background fluorescence reduces image contrast.

Figure 6.8b depicts 2PM with red (two-photon) excitation and the same green fluorescence. The laser is focused (6) to a point where the excitation photons are sufficiently crowded to generate two-photon excitation. As in a confocal microscope, the fluorescence is collected and directed to the detector. However, since we know that for 2PM the fluorescence is coming only from the focus, there is no need for a confocal pinhole to reject out-of-focus fluorescence. Therefore, fluorescence that passes cleanly back through the sample (7) as well as that which is scattered (8) can all be efficiently detected, which can greatly increase the signal. As described in the instrumentation section below, such a non-descanned detector can be used immediately after the objective lens, which further increases the efficiency of signal collection. As detailed in the next paragraph, the red and infrared excitation light used in 2PM is still scattered by the sample similar to what is seen in confocal microscopy. However, in 2PM these scattered excitation photons do not create additional background fluorescence since the probability of two photons scattering to *the same place at the same time* is essentially zero. Hence, both collection efficiency and imaging depth are increased over confocal microscopy by using 2PM.

Much has been written about how the longer wavelengths used in 2PM greatly reduce this scattering, but that is not really the reason behind the improved depth penetration. The reason for this confusion is that light scattering is best known in the atmosphere, which causes the sky to look blue. This effect is based on the Rayleigh scattering theory, which assumes that the scattering particles are much smaller than the wavelength of the light. This results in scattering strength that is inversely proportional to the fourth power of the light's wavelength. Thus, Rayleigh scattering is much stronger for violet and blue light than for the longer wavelengths toward the red end of the spectrum. However, in cells and tissue, the scattering particles (organelles, vesicles, chromosomes) are not much smaller than the wavelength of light; in fact, they can be larger than this wavelength. Scattering of light in tissue has been extensively studied in association with tissue imaging by diffuse optical tomography, which is used to measure tissue oxygenation and detect tumors. These studies show that tissue scattering can be much better predicted using the Mie scattering theory, which describes the scattering of light by spherical particles of any diameter. For particles on the order of the

6.4 Instrumentation

wavelength of light or larger, the amount of Mie scattering is not strongly dependent on the wavelength but is proportional to the square of the particle diameter.

A 2PM instrument is generally built similar to a confocal microscope, consisting of a laser, a scanning system, and a detector. In fact, most confocal microscope manufacturers offer an option to use their confocal microscope as a 2PM instrument. Two key components, though, differ significantly between confocal microscopy and 2PM: the laser and the location of the detector, as previously mentioned previously. A schematic of a typical two-photon microscope setup is shown in Figure 6.9, which is a basic laser-scanning microscope arrangement. The laser comes in and is scanned in a raster pattern to form an imaging field of view, which is focused onto the sample with a scan lens/objective lens system. The laser induces fluorescence, which is collected by the objective lens and directed back up the optical path.

6.4.1 Lasers for 2PM

The first major difference between confocal microscopy and 2PM is the laser. As described previously, for 2PM the laser must be "ultrafast," which is typically defined as subpicosecond pulses of laser light. Using these short laser pulses, it is

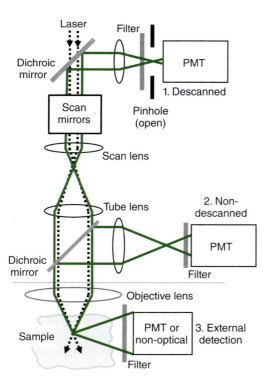

Figure 6.9 Schematic of a typical two-photon microscope setup with the different possible detection strategies. Most of the apparatus is similar to a confocal microscope with the exception of the laser source, the lack of detection pinhole, and the various possible detection geometries.

possible to deliver high laser intensities to the specimen, without simultaneously vaporizing it (see Box 6.1). 2PM is now a well-established technique largely due to the availability of easy-to-use, ultrafast, mode-locked laser systems that are appropriate for this approach. While 2PM first became possible with the invention of subpicosecond-pulse mode-locked lasers, these early versions were far from turn-key systems. The expertise required for daily laser operation was generally not available in cell biology or physiology laboratories. Over the last 15 years, however, laser manufacturers have developed Ti:sapphire and other laser systems in which the alignment and wavelength-tuning is automatic, and the laser system is essentially as much a turn-key system as a laser pointer. In addition to facilitating the use of 2PM in a nonspecialist laboratory, these lasers permit easy excitation wavelength scanning, which is needed for 2PM imaging of different fluorophores.

The laser source most commonly used is the Ti:sapphire laser oscillator, which typically provides tuning over the wavelength range 680–1050 nm (approximately equivalent to 340 and 525 nm single-photon absorption wavelength range). This wide tunable wavelength range brings with it a further advantage: a continuum of excitation wavelengths is possible to select in 2PM, whereas confocal microscope excitation is typically limited to several discrete wavelength lines. These mode-locked Ti:sapphire lasers generate ultrashort light pulses (~100 fs) with a repetition rate of ~100 MHz (which is equal to one pulse every 10 ns). Since the laser is essentially off 10^5-fold more time than it is on, the average power is kept low while providing high photon fluxes during pulsed operation. As described previously, this crowding of the laser photons in time is combined with focusing to a diffraction-limited focal volume in the microscope to achieve the high photon fluxes needed to generate sufficient nonlinear excitation events for 2PM imaging. A newer development is the broadly tunable optical parametric oscillator (OPO) laser, which is now commercially available in a hands-free automated package. These lasers provide pulse and power characteristics similar to those of the Ti:sapphire laser but are tunable from 680 up to 1300 nm. This broader tuning range opens up the possibility of using 2PM with fluorophores that are normally one-photon-excited in the red region (~650 nm). Fluorescence from these fluorophores is very weakly absorbed in tissue, which should allow imaging at greater depths than with blue, green, or orange fluorophores. At this point, though, the use of red fluorophores for 2PM has only begun to be explored. There are some practical limitations expected with using these probes, but none of these has been experimentally verified to date. First, we know that the absorption of water greatly increases throughout the infrared spectrum. Since this absorption is linear, the high powers needed to generate two-photon excitation can lead to heating (described in more detail below). Using 1300 nm light, a simple calculation suggests that heating could be as high as 5 °C during imaging. This would be sufficient to denature lipid membranes and greatly perturb cellular function. If this turns out to be the case, then care will be needed to keep laser powers low, which may force decreased spatial and temporal resolution. Other potential limitations may arise from the scarcity of red fluorescent proteins and the relatively poor quality of red fluorescence probes. There is considerable interest in developing such probes for *in vivo* imaging, but still the quantum yield of even the best red and near-infrared fluorophores is ~10% (as compared to near 100% for green fluorophores). Finally, the use of deeper infrared wavelengths will require dedicated microscope optics. Fortunately, microscope manufacturers have developed objective lenses that are optimized for two-photon microscopy. These account for the need to efficiently transmit both infrared excitation and visible emission wavelengths and to image deep into a variety of tissues and other specimen without introducing significant image aberrations. While the Ti:sapphire laser wavelengths can be used to obtain reasonable results with regular microscope optics, the 2PM-optimized objective lenses will be crucial for extension of 2PM into the infrared region using the newest turn-key OPO lasers.

A major consideration regarding appropriate lasers for 2PM is just how short the pulses need to be. There is no hard and fast rule for this, but in general the shorter the pulse, the better. The original 2PM experiments were done using a colliding-pulse mode-locked (CPM) dye laser that provided pulses <100 fs ($<10^{-13}$ s), and similar pulses are available from the typically used Ti:sapphire and OPO lasers. Thus, the standard pulse width is ~100 fs, and this is known to work well in a wide range of applications. Pulses longer than 1 ps (10^{-12} s) have also been used effectively for 2PM. The trade-offs associated with different laser pulses are detailed in Box 6.2.

Box 6.2 Laser Pulse Widths for 2PM

The laser pulse width is a key parameter that defines the efficiency of two-photon excitation at a given laser power level. That is, for the identical laser power, more two-photon excitation will occur for a narrower pulse width. However, the dependence of two-photon excitation efficiency on pulse width is linear, so it is not as critical as might be initially thought.

Consider a 1 mW laser with 100 fs pulses coming at 100 MHz. In this case, the laser is "on" for 10^{-13} s and the pulses are spaced 10^{-8} s apart. This factor of 10^5 in duty cycle (the portion of time that the laser is "on") means that the peak laser power during each pulse will be 10 kW, or 10^5-fold greater than the average power of 1 mW. The rate of two-photon excitation will be proportional to the square of this peak power (100 kW^2), and the number of events per pulse will be this rate times the pulse duration (100 fs). For these parameters, the number of two-photon excitation events per pulse will be proportional to 100×100 fs kW^2 = 10 000 fs kW^2. If we now consider a laser with identical parameters except for a longer pulse width of 1 ps (1000 fs), the laser is "on" for 10^{-12} s, which gives a 10^4 duty cycle and a peak laser power during each pulse of 1 kW. In this case, the rate of two-photon excitation will be proportional to 1 kW^2, but number of events per pulse will be proportional to 1×1000 fs kW^2 = 1000 fs kW^2. This is only a factor of 10 less than with the 100-fs pulses. While the efficiency drops with the square of the peak laser power, the longer pulse duration partially compensates for this lost efficiency. Using the 1-ps pulse, it is possible to obtain the same number of two-photon excitation events generated by the 100-fs pulses by simply increasing the average laser power by a factor 10. As the lasers used in 2PM these days mostly provide average powers of >1 W, it is generally no problem to deliver the extra power needed for 2PM with a longer pulse width. However, some potential photodamage (such as laser heating, especially when using longer infrared wavelengths) depends linearly on the average laser power. This type of damage will be minimized by using shorter laser pulses. Another motivation for the use of shorter pulses is the loss of power due to scattering in deep-tissue imaging. In cases where the amount of input laser power is limited to a certain level, it is possible to image deeper by making the pulse widths narrower, thus making the two-photon excitation probability more efficient. For the deepest penetration depths (see Box 6.3), >1 W is needed at the surface of the tissue to generate sufficient two-photon absorption events at the focus. In this case, it is generally not possible to increase the laser power, but deeper penetration of imaging may be achieved by using shorter laser pulses.

It is best to use shorter pulses, but physical limits dictate the minimum practical pulse width. Light is governed by the Heisenberg uncertainty principle, which limits the precision of the product of energy and time. This means that photons confined within a short time window cannot have a precise energy (wavelength) but instead must have a spread of energies. The shorter the pulse, the more spread out this energy range. Light of different energies (colors) interacts with matter differently, so these ultrashort pulses with their color spread are affected not only by the focusing optics by chromatic aberrations but also by traveling through air where the speed of light varies as a function of wavelength. This variation of photon speeds leads to group velocity dispersion (GVD), which spreads the pulse width. While strategies exist to compensate for chromatic aberrations and GVD, in practice these factors are not a limitation on 2PM unless the laser pulses durations are <80 fs.

6.4.2 Detection Strategies for 2PM

The second major difference between 2PM and the confocal microscopy design is the location of the detectors. In a confocal microscope, the fluorescence passes back through the scanning mirrors, so that it is "descanned" and then focused onto the confocal pinhole to reject out-of-focus background fluorescence that would otherwise blur the image. While the pinhole leads to the desired optical sectioning in a confocal system, it does so at the cost of rejecting some in-focus fluorescence together with the out-of-focus background. This loss of "good" photons becomes more significant for deeper imaging into the specimen. In addition to losses at the pinhole, the fluorescence must also pass back through the scanning mirror optics before reaching the confocal pinhole. Passing back through this optics inevitably leads to further signal losses. In contrast, 2PM does not require the use of a pinhole since we know that the fluorescence originates solely from the focus. Thus, all the fluorescence emission can be collected without the use of a pinhole, which opens up a number of new possibilities for detector locations.

In 2PM, it is possible to simply open the pinhole (#1 in Figure 6.9), and thus capture some of the scattered fluorescence photons (Figure 6.8, photon (8)). While this will increase the collection efficiency compared to confocal microscopy, it is still subject to the losses as the fluorescence passes back through the scanning system and is refocused onto the detector. Also, because of the optical system design, even a fully open pinhole still acts as a spatial filter and excludes many of the scattered fluorescence photons that should be collected in 2PM. Indeed, there is a simpler detector arrangement that be used, where all the photons collected by the objective lens are directed immediately to a photomultiplier detector. This strategy is called *non-descanned detection* (NDD, #2 in Figure 6.9) and is the most common detection arrangement for 2PM [4]. For NDD, the only selection applied is based on wavelength by using the appropriate optical filter to reject the excitation light and detect only the fluorescence of interest. Using an NDD, all the scattered fluorescence that is collected by the objective lens can be measured, which greatly increases the signal level. As for confocal microscopy, new photomultiplier detectors using GaAsP photocathodes can further improve the collection efficiency, and these should be used for 2PM if at all possible. Finally, it is possible to use an external detector that does not even rely on the objective lens (#3 in Figure 6.9). Such a detector could be a large-area photon detector, or a non-optical detector, such as the cell electrical response measured by electrophysiology (as given in the applications of 2PM example below).

6.4.3 The Advantages of 2PM for Deep-Tissue Imaging

Combining optimized NDD with the inherent depth penetration of 2PM permits maximal depth imaging. The actual maximum depth depends on the sample, but a good rule of thumb is that 2PM using an NDD can image 6–10-fold deeper than with confocal microscopy. As demonstrated in Figures 6.10 and 6.11, an increase in depth of 2–3-fold comes from the lack of out-of-focus absorption, and another factor of 3 in depth comes from improved collection of the scattered fluorescence with the NDD. Figure 6.10 shows a series of X–Z scans of a mouse kidney section that is heavily stained with eosin. The laser power was adjusted for each X–Z scan so that the signal level at the tissue surface was the same in each image. Using confocal microscopy, useful signal can be obtained only within the first ~10 µm from the surface. It is possible to collect fluorescence from deeper levels, but the contrast will be lost (as demonstrated in Figure 6.11). Switching from confocal to 2PM, but using descanned detection (detector strategy #1 in Figure 6.9) with the pinhole open, it is possible to collect useful data down to ~20–30 µm from the surface. This improvement is largely due to the lack of out-of-focus absorption in the case of 2PM, which allows more excitation light to reach the focal spot deep in the sample. Some scattered fluorescence is collected through the open pinhole, but many of these useful photons are lost as they stray from the optical path as it returns through the scan system and is focused on the pinhole. The NDD (detector strategy #2 in Figure 6.9) can be used immediately after the objective, which thus greatly increases the collection efficiency of the scattered fluorescence. Using the NDD, useful data can be acquired from >50 µm into the sample even without increasing the laser power. By increasing the laser power, it would be possible to obtain useful data down to 100 µm below the surface of this specimen. As was described for Figure 6.8, in the case of 2PM,

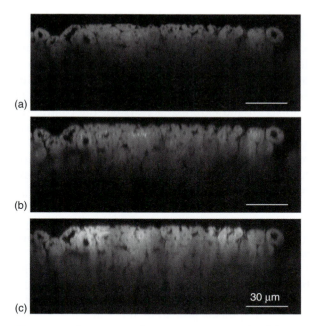

Figure 6.10 Comparison of imaging penetration depth between one- and two-photon excitation microscopy. X–Z profiles through an eosin-stained mouse kidney sample imaged through a depth of 70 μm with (a) confocal, (b) 2PM descanned, and (c) 2-PM with non-descanned detection microscopy. The imaging penetration depth with 2PM (descanned, b) is improved approximately twofold relative to confocal microscopy (a). Non-descanned detection (c) yields increased fluorescence collection, which further increases the effective depth of imaging. Scale bars = 30 μm.

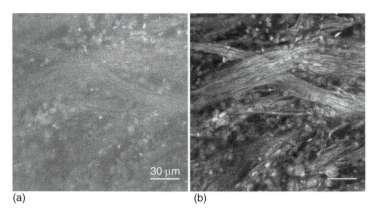

Figure 6.11 Comparison of imaging contrast between one- and two-photon excitation imaging at deep focal planes within a scattering sample. The two panels show the same X–Y plane imaged 50 μm into a rhodamine-stained slice of mouse brain with (a) confocal microscopy and (b) 2PM. In confocal microscopy, the laser power can be increased to obtain sufficient signal at depth, but this results in increased scattered background (see Figure 6.8), which appears as haze and reduces the image contrast. In 2PM, on the other hand, increasing the laser power creates more signal without decreasing contrast, so that fine structures can still be resolved and measured. Scale bars = 30 μm.

increasing the excitation power does not generate additional background fluorescence, which would create a haze and decrease the image contrast. This effect is demonstrated in Figure 6.11, which shows two comparable X–Y planes 50 µm into a rhodamine-stained slice of mouse brain. The first was taken using confocal microscopy, where the laser power was increased in an attempt to raise the fluorescence signal. Indeed, the fluorescence signal level is increased, but most of this fluorescence is seen as a background haze that smears out the contrast of the biological structure. On the other hand, with 2PM it is possible to increase the excitation laser power without creating additional background haze. In this case, the signal level is increased, and the fine biological structures are still visible. Importantly, areas in the image where no structure is present remain black, which indicates that the image contrast is a maximum.

In summary, there are several practical advantages to 2PM that derive from the photophysics of two-photon excitation in a microscope, where the excitation is limited to the focus. This leads to the inherent production of optical sections (often called *confocal for free* because no pinhole is required), and the lack of out-of-focus absorption of the excitation light allows more of it to reach the focus. Importantly, photobleaching and photodamage are also limited to the focal plane, and thus in a thick sample these deleterious effects can be greatly reduced. Because of the details of 2PM image formation, this approach is much less sensitive to degradation by light scattering in the tissue than with other optical sectioning approaches. Further, the red and infrared laser light used in 2PM is less absorbed by many biological samples, and so is less biologically damaging. Together, these factors give 2PM tremendous advantages for deep-tissue imaging with sub-cellular resolution.

6.5 Limitations of Two-Photon Excitation Microscopy

While all of the strengths described previously give 2PM significant advantages over other approaches for diffraction-limited imaging deep into intact tissue, 2PM is by no means a panacea. There are limitations and challenges to 2PM that must be appreciated to take full advantage of this approach. These include limits to spatial resolution, possible heating of the sample by the high laser powers used in 2PM, difficult to predict or measure excitation spectra, accelerated photobleaching (and associated photodamage) in the plane of focus, and the extra expense of appropriate laser systems, which limit the applicability of 2PM in spectral imaging. Despite these challenges, however, it is important to realize that many important studies have been enabled by 2PM, which would not have been possible using other imaging techniques.

6.5.1 Limits of Spatial Resolution in 2PM

Because of the longer wavelengths used in 2PM, in practice this approach yields somewhat poorer spatial resolution than confocal microscopy. In 2PM, the quadratic dependence on the excitation intensity reduces the size of the excitation volume by $\sqrt{2}$, but the use of light of double the wavelength broadens the focal spot by a factor of 2. Use of a detection pinhole can improve the spatial resolution of 2PM, but this would negate most of the advantages of the approach (see Figure 6.8). Because of its use of shorter wavelengths, the resolution attained with a confocal microscope can be as small as half of that from a two-photon microscope. In biological fluorescence confocal microscopy, however, signal collection is typically optimized by opening the pinhole, which degrades spatial resolution. Thus, in practice two-photon microscopy provides spatial resolution that is only slightly worse than in confocal microscopy.

6.5.2 Potential Sample Heating by the High Laser Powers in 2PM

Red and infrared excitation light can cause localized heating effects in the sample when excessive powers are used. This heating is largely due to one-photon absorption of this excitation light by water. Since water is present in high concentrations in all biological systems, even a very small direct absorption can lead to significant numbers of excitation events. Water is not fluorescent, so most, if not all, of the absorbed energy is dissipated in the sample as heat. The absorption profile of water in the red and near-infrared region is shown in Figure 6.12 [5]. This is plotted on semilog scale, so the absorption at 980 nm is actually 100-fold greater than at 680 nm. Calculations of the localized heating during an imaging experiment suggest that the maximum heating even at 980 nm is <1 °C. For single-point experiments, though, where the laser beam is parked and not scanned, heating could be >1 °C s^{-1}. Using other wavelengths where the water absorption is lower (e.g., 800 or 1060 nm) can minimize any potential artifact from sample heating. Thus, for the laser intensities required for imaging using a Ti:sapphire laser, heating rarely causes any problem as long as one avoids parking the 980 nm beam. Use of the new OPO systems, which are tunable out to 1300 nm, though, could lead to the risk of more significant heating artifacts. Water absorption continues to rise throughout the infrared spectrum such that the absorption near 1300 nm can be almost 100-fold greater than at 980 nm. This means that a temperature rise of >10 °C might be possible for high laser intensities. As discussed previously, the spectral region from 1050 to 1300 nm has not been thoroughly explored so far, but it is clear that sample heating will be an important challenge in this spectral window.

Figure 6.12 Semilog plot of the absorption coefficient for water in the near-infrared spectral region. Note that the water's absorption of light at 980 nm is over 100-fold greater than it is at 680 nm. (Data from Hale and Querry 1973 [5].)

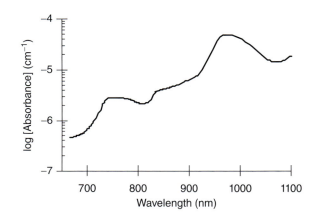

6.5.3 Difficulties in Predicting and Measuring Two-Photon Excitation Spectra

Another challenge when using 2PM is that the two-photon excitation spectrum of a fluorophore can often be very different from the one-photon excitation spectrum due to the nature of the two-photon absorption process. This can be critical since choosing the wrong excitation wavelength can exacerbate other problems in 2PM, such as the sample heating and accelerated photobleaching. An example of the differences between the one-photon and two-photon absorption spectra is shown for the fluorophore diphenylhexatriene (DPH) in Figure 6.13. In this figure, the two-photon excitation spectrum has been plotted at half the wavelength for comparison (i.e., the two-photon absorption peak at 400 nm is actually at 800 nm). Knowing the one-photon spectrum, which is easy to measure, and doubling the absorption peak, an initial guess for the two-photon absorption peak would be 770 nm (2 × 385 nm). However, 770 nm is at a valley of the two-photon absorption. A much better choice would be 800 nm. This example, however, suggests a reasonable approach for optimizing the two-photon excitation wavelength. Beginning by doubling the one-photon peak, and then tuning the two-photon excitation wavelength above and below that initial wavelength, the maximum two-photon excitation can be found by maximizing the amount of fluorescence observed. In this case, the beginning wavelength would be 770 nm, but tuning every 10 nm in each direction would yield increased fluorescence signals at 750 and 800 nm.

One of the main reasons for the differences between the one-photon and two-photon absorption spectra is the differences in the strength of interactions with higher level excited states [7]. Figure 6.14 shows the relative absorptions in the fluorescent protein TagRFP by the lowest level singlet state (S_1) and higher order states (S_n) for both one-photon and two-photon excitation (in this case, the one-photon values are plotted at twice their wavelength for comparison – that is, the 550-nm one-photon absorption peak is plotted at 1100 nm). While the two-photon excitation efficiency of the S_1 state is greatly reduced in comparison with that of one-photon absorption, the two-photon excitation of the S_n states is greatly enhanced. This opens the possibility to use S_n excitation wavelengths (e.g., 750 nm) with 2PM, even though these transitions are too weak for efficient one-photon imaging. In this case, the transitions are well separated, which

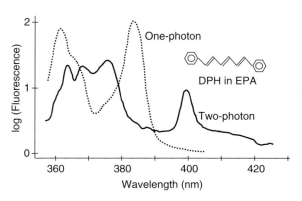

Figure 6.13 Two-photon excitation spectrum (solid line) of the membrane probe diphenylhexatriene (DPH) in an ether–isopentane–ethanol random glass matrix at 77 K. A portion of the one-photon absorption spectrum is shown by the dashed curve. (Adapted from Fang et al. 1978 [6]. Reproduced with permission of Elsevier.)

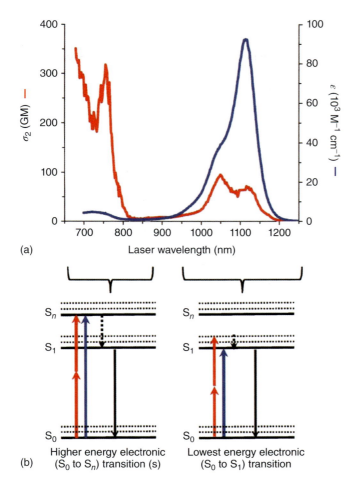

Figure 6.14 Structure of the two-photon absorption spectrum of a fluorescent protein. (a) One-photon absorption (blue) and two-photon absorption (red) spectra of TagRFP. (b) Jablonski diagram of 1PA and 2PA transitions. (Drobizhev et al. 2011 [7]. Reproduced with permission of Nature Publishing Group.)

makes sorting out these differences straightforward. However, even among the vibrational levels within the S_1 transition, there are differences between the one-photon and two-photon absorptions. This can be seen where the energy levels that form a shoulder in the one-photon spectrum at 525 nm are enhanced in the two-photon excitation spectrum. This leads to enhanced two-photon efficiency at 1050 over 1100 nm.

One common theme regarding the differences between one-photon and two-photon absorption spectra is the effect of the molecular symmetry of the fluorophore on the differences. The general rule of thumb is that symmetric fluorophores will exhibit a blue shift of the two-photon absorption peak. This means that the one-photon peak of 400 nm will appear in the two-photon spectrum at 700–750 nm rather than 800 nm (2 × 400 nm). Asymmetric fluorescent

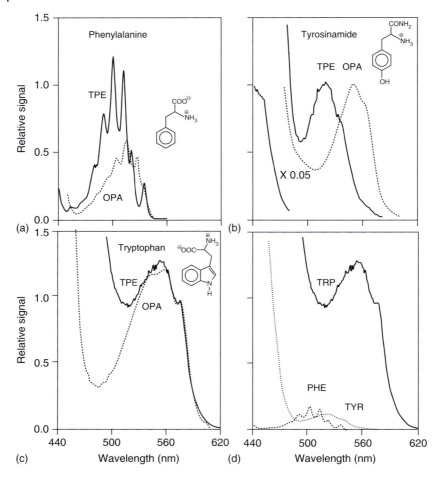

Figure 6.15 Two-photon excitation spectra (solid lines) with the corresponding one-photon UV absorption plotted at twice their actual wavelengths (dashed lines) for the aromatic amino acids (a) phenylalanine, (b) tyrosinamide, and (c) tryptophan in aqueous solutions at room temperature. (d) Comparison of relative two-photon signals of phenylalanine (Phe), tyrosinamide (Tyr), and tryptophan (Trp) observed at each compound's fluorescence maximum so that the relative signals are approximately in proportion to the relative product of the two-photon absorptivity and fluorescence quantum yield. (Reprinted with permission from Chemical Physics Letters, ©1993 [8].)

molecules are much more likely to show two-photon absorption peaks at exactly twice the one-photon peak wavelength. A good example of this is the different absorption spectra of the aromatic amino acids, especially tyrosine (symmetric) and tryptophan (asymmetric), as shown in Figure 6.15, where the one-photon wavelengths are plotted at twice their actual values [8]. The two-photon absorption spectrum of tryptophan almost perfectly overlays twice the one-photon absorption spectrum, while the two-photon absorption spectrum of tyrosine is blue-shifted by ~40 nm from the "expected" twice the one-photon absorption spectrum. This symmetry rule for two-photon excitation is very common among fluorophores, and generally provides the best first guess at what an unknown two-photon absorption spectrum might look like.

Fluorophores are typically complex molecules, and it is not easy to predict their two-photon absorption spectra. Likewise, it is difficult to measure two-photon absorption, since this process utilizes a very small portion of the incident light. Thus, as described for the DPH example in Figure 6.13, excitation maxima are usually determined by tuning the laser and measuring the fluorescence of the fluorophore of interest normalized to a known two-photon excitation spectrum. Nowadays, the two-photon excitation spectra have been characterized for many commonly used fluorophores and fluorescent proteins, which can greatly facilitate their use with 2PM.

6.5.4 Accelerated Photobleaching (and Associated Photodamage) in the Focal Plane

One of the major advantages of 2PM is that photobleaching and any associated photodamage are limited to the focal plane. Over a thick specimen, this can greatly reduce the total photobleaching, especially in the acquisition of 3D data stacks. However, even if the total bleaching is reduced, there is a significant risk of actually increasing the local photobleaching within the focus. Because of the nonlinear absorption and high laser powers used in 2PM, there is an increased probability of photobleaching *at the focal plane.* The origin of this increased photobleaching remains unclear, but it is consistent with additional photons interacting with the excited state of the fluorophore and causing a transition to a state with a higher probability of photobleaching. Because of the high photon concentrations in the focus, it is easy to see how such transitions could be greater in 2PM. The practical manifestation of this accelerated photobleaching is demonstrated for fluorescein fluorescence and photobleaching in Figures 6.16 and 6.17. As described in Box 6.1, the amount of two-photon excited fluorescence is proportional to the square of the laser intensity ($F = I^2$). This intensity-squared dependence can be demonstrated clearly by a log–log plot ($\log(F) = 2\log(I)$), which should yield a straight line with a slope of 2 for two-photon excitation, as shown in Figure 6.16. It is possible to analyze the photobleaching rate in the same

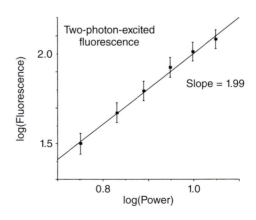

Figure 6.16 Log–log plot of the excitation dependence of fluorescein dextran fluorescence intensity for six levels of two-photon excitation with 710 nm laser light. The dependence exhibits a slope of ~2 indicating two-photon excitation. Data are displayed as mean ± SE ($n = 5$). (Data from Patterson *et al.* 2000 [9].)

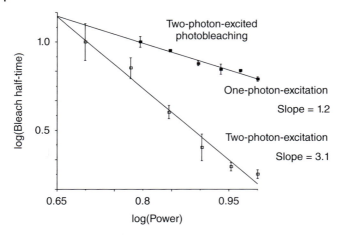

Figure 6.17 Log–log plot of the excitation dependence of fluorescein dextran average photobleaching rate as a function of laser power for one-photon excitation with 488 nm light (filled squares) and two-photon excitation with 710 nm light (open squares). The one-photon photobleaching slope is close to 1, indicating a near-linear photobleaching process, but the two-photon dependence is >3 (rather than the expected slope of 2). This higher slope generates accelerated photobleaching in 2PM. Data are displayed as mean ± SE ($n = 5$). (Data from Patterson and Piston 2000 [10].)

manner. If photobleaching is dependent only on the amount of fluorescence (as is thought to be the case for one-photon excitation), then the amount of photobleaching should be proportional to the square of the laser intensity ($P = I^2$). In turn, the log–log plot ($\log(P) = 2\log(I)$) should yield a straight line with a slope of 2 for two-photon excitation. However, measuring the amount of photobleaching for various laser intensities yields a different result, as shown in Figure 6.17. For one-photon excitation, the photobleaching dependence shown in this log–log plot is close to 1, as expected. However, the bleaching dependence on intensity for 2PM is not 2 but rather >3. This means that for fluorescein, the photobleaching is dependent on $I^{3.1}$. In other words, using twice the excitation power will yield fourfold more fluorescence but eightfold greater photobleaching. For this reason, minimization of excitation power is essential in 2PM to minimize not only any direct damage to the sample but also photobleaching (and any photodamage associated with the photobleaching).

It is also due to this accelerated photobleaching that 2PM will not always lead to significant improvements over confocal microscopy when imaging thin samples. Because of the accelerated photobleaching and photodamage, 2PM is poorly suited for most single-plane time-lapse imaging experiments. Of course, there are some cases (e.g., a single plane deep inside a tissue or use of UV-excitable fluorophores) where the experiments are just not feasible with other approaches, and this is where 2PM adds a novel approach.

6.5.5 Expensive Lasers Create a Practical Limitation for Some Experiments

A final practical limitation of 2PM is that the lasers needed for 2PM are expensive. Currently a state-of-the-art OPO system can cost as much as an entire entry-level confocal microscope system. This can limit the access to the technology and also prevent some approaches that might be useful. For instance, because of the high cost of the lasers used, simultaneous multiple excitation wavelengths are rarely available. This makes multicolor imaging somewhat more difficult, although the tunability of most lasers used in 2PM allows one to find a single wavelength that comparably excites two fluorophores at the same time. So while two-color imaging is straightforward to implement in 2PM, hyperspectral imaging is more difficult to achieve, as compared to the multitracking (strobing on single excitations in sequence and detecting one fluorophore at a time) used in confocal and widefield fluorescence microscopy.

6.6 When is 2PM the Best Option?

The strengths and limitations of 2PM have been laid out, but it is useful to consider different scenarios regarding when it is better to use 2PM versus another approach, such as confocal microscopy. Since confocal microscopes are widely available, easy to use, and capable of delivering high-quality data, it makes good sense to use confocal microscopy first, and then try 2PM if the confocal approach is incapable of providing the needed data. However, there are three major situations where 2PM is almost always the better choice, and those are listed here along with a brief rationale explaining the advantages of 2PM for each situation.

6.6.1 Thick Specimen including *In Vivo* Imaging

The fact the 2PM is good for imaging deep into tissue has been stressed throughout this chapter, and the key advantages of 2PM for this work have been mentioned many times. Still, imaging structures that extend into the specimen or are buried in the specimen absolutely requires optical sectioning. Confocal microscopy can often accomplish the needed experiments by providing high-quality images at depths up to 100 µm or more into the sample. As detailed in Box 6.3, 2PM can extend this limit up to 1 mm. The reasons for this have been detailed previously. Similarly, 2PM is also the best option for *in vivo* imaging of structures within a living animal. First, it allows deeper imaging compared to confocal microscopy, which can be a major advantage, especially when excision of the specimen is not possible. 2PM also offers the advantage that it uses infrared light which is not seen by the animal. More importantly, though, reduced photobleaching and photodamage both contribute to reduced phototoxicity in the animal and thus make 2PM the preferred method for *in vivo* imaging. However, confocal microscopy is the better (and also cheaper) option when imaging thinner specimens, unless some other factors such as those listed below need to be considered. Further, for single cell layers, a conventional widefield fluorescence microscope is often an even better solution than confocal microscopy, since there might not be a need for optical sectioning in such a thin sample.

Box 6.3 Limits on the Depth of Imaging in 2PM

As discussed in the text, the potential limiting factors on the penetration depth in fluorescence imaging are the absorption of excitation light, absorption of fluorescence returning to the surface, and scattering of the excitation and fluorescence light. In the case of 2PM, the absorption of excitation light is quite small, so it does not play a limiting role. Similarly, scattering of the fluorescence is not a major source of signal loss, especially if NDD detection is used. Thus the two main sources of signal degradation are losses due to scattering of the excitation light and absorption of the fluorescence returning through the sample. The intensity of excitation will fall off as a result of light scattering in the sample as an exponential decay, e^{-ax}, where a is constant based on the amount of scattering and x is the distance the light travels through the sample. Similarly, the fluorescence intensity will also fall off exponentially, e^{-bx}, as it returns through the sample back to the microscope. Both these decreases can be described by a characteristic decay length related to the constants a and b. These constants vary between different tissues. To compensate for these losses and equalize the amount of signal collected at depth to that attained near the surface, the input laser power can be increased. However, there is a physical limit to how high this can be raised: at some point, the input laser power becomes high enough (even without the photon crowding at the focus) to generate two-photon excitation events at the surface of the sample. As described, the losses in the sample are proportional to e^{-x}, but the laser beam waist outside the focus is proportional to x^2. If we assume that the scattering coefficient a and absorption coefficient b are similar, then the loss of excitation due to scattering of the excitation light into the sample and the loss due to absorbance of fluorescence coming back out of the sample will be equal (this is a simplification, as the characteristic length of light scattering is likely much shorter than that for absorption in tissue). Imaging at a depth of 10 scattering lengths ($x = 10/a$) will attenuate the laser power by 450 000-fold, which will decrease the amount of two-photon excitation by a factor of $\sim 9 \times 10^9$. Under the assumption that $a = b$, the fluorescence will decrease by 450 000-fold as it returns through the sample, for a total signal decrease of $\sim 10^{15}$. To compensate for this loss, the excitation power would need to be raised by $\sqrt{10^{15}}$ or $\sim 3 \times 10^7$. At the same time, the area of the laser beam at the surface will have increased from the focus by only 10^2 or 100-fold. Thus, the laser intensity (power/area) at the surface would now be increased by 3×10^5 over what is needed to image through a nonscattering, nonabsorbing sample. This is comparable to the factor of 100 000-fold crowding by the objective lens focusing (see above). Since the tissue scattering length is typically $\sim 100\,\mu m$, this calculation is consistent with practical experience that the maximum depth of imaging is $\sim 1\,mm$ using green and orange fluorophores. Imaging red or near-infrared fluorophores using infrared 2PM may allow significantly greater imaging depths, but that has yet to be demonstrated.

6.6.2 Imaging Fluorophores with Excitation Peaks in the Ultraviolet (UV) Spectrum

Because of the large number of biomolecules that absorb UV light, it is often difficult to image UV fluorophores in live cells without compromising specimen integrity. UV photons have high energy, and they can easily produce cellular damage, especially in time-lapse imaging where the same region is imaged repeatedly over time. For optical sectioning, UV excitation becomes more problematic, as special objectives are necessary to properly transmit and focus UV light. 2PM using 680–800 nm light can excite fluorophores that are normally excited by UV light in the 340–400 nm range. The use of red and near-infrared light reduces scattering, and 2PM limits any photodamage to the focal plane, which greatly reduces it overall. Thus, 2PM can permit increased number of scans in a time-lapse experiment without incurring the amount of photodamage that UV light produces.

6.6.3 Localized Photochemistry

Two-photon excitation can stimulate any kind of optical transition, not just the excitation of fluorescence. In fact, 2PM is extremely useful for localizing photochemical reactions to selected regions of a specimen, since its nonlinear nature constrains excitation events to the focal volume ($\sim 1\,\mu m^3$). Using 2PM, it is possible to elicit a reaction inside a cell without any activation outside the cell or on the cell surface. The nature of the reaction depends on the photoactive chemicals used, but two-photon photoactivation is commonly used for uncaging photolabile compounds and photoswitching of fluorescent molecules (including fluorescent proteins). In all cases, the reactions could be triggered using one-photon excitation, but not with the same spatial confinement provided by 2PM. Similar to what is seen in one-photon photobleaching (Figure 6.6), photoactivation in confocal microscopy would occur throughout the sample – both above and below the target region. A further concern is that most photoactivation processes utilize UV light, so 2PM is even more preferred for these experiments as described previously. Several examples of 2PM photoactivation are given in the applications section.

6.7 Applications of Two-Photon Microscopy

Since the initial development of 2PM in 1990, the technique has expanded rapidly and impacted many fields of biological research. As described previously, the biggest application areas have been thick tissues, such as brain slices, and *in vivo* imaging, mainly in mice. A few of these applications are detailed in this section to highlight the role of 2PM in enabling them.

6.7.1 Imaging UV-Excited Fluorophores, such as NADH for Metabolic Activity

As discussed earlier, 2PM offers a convenient alternative to UV excitation, where single-photon excitation in the 340–400 nm range can be replaced by two-photon excitation in the 680–800 nm range. This range is provided by commercial turn-key Ti:sapphire systems. There are many useful fluorescent probes that require UV excitation, but high phototoxicity associated with UV excitation can be problematic in live samples where relatively low fluorophore concentrations are necessary. One of the most successful applications of 2PM has been imaging metabolic activity via the endogenous fluorescent cofactor β-nicotinamide adenine dinucleotide (phosphate), abbreviated to NAD(P)H. NAD(P)H can be efficiently two-photon-excited at ~710 nm with much reduced phototoxicity compared with 355 nm UV excitation. In its oxidized form, NAD(P)H is fluorescent (it is one of the main components of UV-excited autofluorescence), but the reduced form, $NAD(P)^+$, is not fluorescent, so measuring NAD(P)H levels provides an assay of the *in situ* redox state. This assay has been used to elucidate the metabolic pathways in the islet of Langerhans, as highlighted in Figure 6.18. NAD(P)H is produced from NAD^+ in many metabolic processes such as glycolysis (in the cytoplasm) and citric acid cycle metabolism (in the mitochondria) [9]. Thus, as glucose is metabolized in the islet cells, the level of metabolic activity can be observed through NAD(P)H fluorescence. 2PM imaging of NADH levels has also been used to classify tissues as normal, precancerous, or invasive cancer, to determine how redox state regulates transcription, and to connect astrocyte metabolic activity and electrical activity *in vivo*.

Similar to imaging NAD(P)H, 2PM can be used to image other UV-excitable fluorophores with excellent 3D spatial resolution and overall reduced phototoxicity. Another useful UV-excitable fluorophore is Laurdan, which is a marker for membrane fluidity. 2PM of Laurdan has been used to investigate the importance of membrane order in cellular adhesion to the extracellular matrix, and how membrane order promotes acetylcholine receptor clustering in postsynaptic membranes.

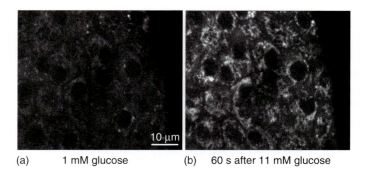

(a)   1 mM glucose   (b)   60 s after 11 mM glucose

Figure 6.18 Imaging NAD(P)H with subcellular resolution in β-cells within a cultured pancreatic islet. Panel (a) shows the basal 1 level at 1 mM glucose and panel (b) shows the brighter autofluorescence reflecting the change in NAD(P)H-to-NAD+ ratio due to increased glucose metabolism. Both the cytoplasmic and mitochondrial NAD(P)H can be quantified representing cytoplasmic glycolysis metabolism and mitochondrial TCA metabolism, respectively. (Image adapted from data described in [9].)

6.7.2 Localized Photoactivation of "Caged" Compounds

2PM was originally conceived as a method for highly localized photoactivation of "caged" compounds such as caged calcium or the neurotransmitter caged glutamate. These caged compounds are initially inactive but can be activated by irradiation with light, typically in the UV region. The limitations of UV excitation were discussed earlier, but using UV light for uncaging cannot release the caged compound solely in a well-defined location because out-of-focus light will photoactivate the compound throughout the sample. Using 2PM, the uncaging is localized to the focus, and the use of red and infrared excitation light greatly reduces phototoxicity.

There are several general areas of photoactivation where 2PM has proven its worth. One of the simplest applications is the photoactivation of caged fluorophores. These molecules have a photolabile chemical moiety attached that quenches the fluorescence, but after stimulation of that moiety, the chemical dissociates and frees the fluorophore. Generally, these caged compounds are built around strong fluorescent molecules such as fluorescein and rhodamine, so the photoreleased signals are bright and robust. Fluorophore uncaging can be used for many purposes, such as cell coupling and flow measurements, intracellular diffusion measurements via an inverse FRAP (fluorescence recovery after photobleaching) approach where fluorescence is initially photoreleased in one spot and then its spread is measured, and cellular lineage tracing in cell growth and development. An example of this latter use is given in Figure 6.19, which shows a 3D reconstruction of an image stack acquired through a pre-gastrulation sea urchin (*Lytechinus variegatus*) embryo. This embryo arose from fertilization of an egg that had been injected with a caged fluorescein-10 kDa dextran, where one cell of the eight-cell embryo was scanned with the two-photon excitation beam. In the developing embryo, all descendants from the single photolabeled cell are fluorescent but the other cells are not. As seen in the two individual slices, cells in just one quadrant of the top half of the embryo are fluorescent, while no cells in the bottom half are fluorescent. The specificity of two-photon excitation along with its ability to penetrate deep within tissue allows this approach to be performed with any cell or group of cells anywhere within the specimen in an entirely noninvasive manner [11].

2PM has also found significant applications using photoswitchable fluorescent proteins, such as PA-green fluorescent protein (PA-GFP), which is efficiently converted to bright green fluorescence using two-photon excitation at 800 nm, and PAmCherry, which exhibits complicated photoactivation kinetics. For these genetically encoded labels, photoactivation is generally produced by photoisomerization between a cis and a trans state of the fluorophore within the fluorescent protein. It is often the case that the process of two-photon activation is complicated, as the laser light can cause transitions between multiple carboxylated and protonated states in addition to the photoisomerization. Further, the activation laser can trigger deprotonation within the decarboxylated states, while excitation of the photoactivated fluorescence simultaneously switches the fluorophores between their cis and trans forms and triggers protonation.

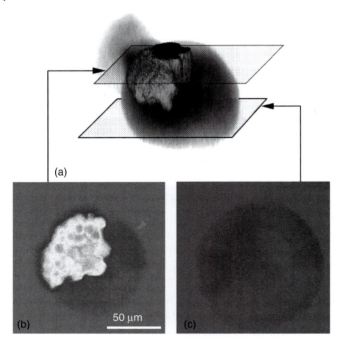

Figure 6.19 Two-photon uncaging for lineage tracing in embryo development. A caged fluorophore attached to dextran was released in a single cell of an eight-cell sea urchin embryo (*Lytechinus variegatus*). In the later pre-gastrulation embryo, shown as a full 3D reconstruction in (a), one-quarter of one hemisphere is brightly fluorescent (one-eighth of the total embryo). Two single optical sections of the 3D dataset are shown from the top hemisphere (b) which shows brightly labeled cells in one quandrant, and from the bottom hemisphere (c) which shows no fluorescence.

In contrast to photoactivation of fluorescence, 2PM photoactivation was first used to uncage carbamoylcholine for mapping of cell surface ligand-gated ion channel distribution in combination with electrophysiology, and the neuroscience field has continued to exploit this approach. One example combining two-photon photorelease of caged glutamate with 2PM imaging of neuronal structures is shown in Figure 6.20 [12]. This work used 2PM to activate excitatory potentials in well-defined locations along single dendritic spines in a brain slice culture. The precise photoactivation given by 2PM allowed the quantitative correlation of action potential strength and the spine length. This work expanded upon numerous previous studies using 2PM imaging and uncaging that defined how the network of individual dendritic spines and shafts processes excitatory inputs in terms of membrane potential summation and filtering. The reader is directed to reviews such as [13] covering the application of two-photon uncaging in neurosciences.

One area where the unique capabilities of 2PM are useful is optogenetics, where photoactivatable ion channels are used to manipulate cellular membrane potential. An elegant example of this approach is shown in Figure 6.21 [14]. Pyramidal cells in a brain slice from an optogenetic transgenic mouse can be patched and

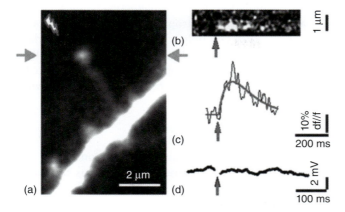

Figure 6.20 Inverse correlation between action potential amplitude and spine neck length in brain slice neurons stimulated with two-photon photoactivated caged glutamate. Synaptic activation by two-photon uncaging of glutamate near a single long-necked spine (a) generated a clear calcium response at its head (b,c), but no voltage deflection for uncaging at the soma (d). Arrows on either side of (a) mark the line scan intersecting the spine head for uncaging. Arrows on (b)–(d) denote the time of uncaging. (Reprinted with permission from Proceedings of the National Academy of Sciences, USA, ©2014 [12].)

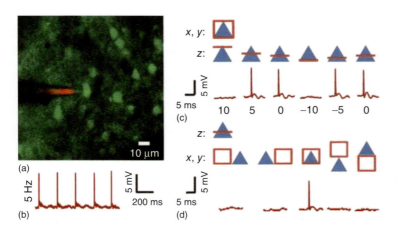

Figure 6.21 Optogenetics with 2PM. (a) *In vivo* two-photon image of layer 2/3 pyramidal cells (~150 μm from the surface of the brain) transduced with C1V1$_T$-p2A-EYFP (imaged during loose patch). (b) A trace of precise spike-train control with 5 Hz 1040 nm raster-scanning illumination; the amplitude and waveform of these evoked spikes recorded in cell-attached mode matched the spontaneous spikes in each cell. (c) The axial resolution of two-photon optogenetic control of spiking *in vivo*. Blue triangles indicate pyramidal neurons, and red boxes illustrate region-of-interest (ROI) positioning. Spiking of layer 2/3 cells as a function of raster scan position is shown (20×/0.5-NA objective; λ: 1040 nm; dwell time per pixel: 3.2 μs; scan resolution: 0.6 μm per pixel; line scan speed: 0.19 μm μs^{-1}; laser intensity: 20 mW). (d) Lateral resolution of two-photon optogenetic control of spiking *in vivo*. (Prakash *et al*. 2012 [14]. Reproduced with permission of Nature Publishing Group.)

imaged *in situ*. By two-photon stimulation of the optogenetic channels, action potentials can be triggered and observed by electrophysiology. As demonstrated in the figure, action potentials are observed only when the 2PM scan overlaps with the cell in all three dimensions, X, Y, and Z. No other optical approach offers this degree of precise specificity. Thus, two-photon optogenetics is expected to be a robust growth field over the coming years. More details are available in review articles, such as [15].

6.7.3 Imaging Electrical Activity in Deep Tissue

The principal advantage of 2PM is the increased penetration depth for thick-tissue imaging. One of the earliest applications of 2PM was in neuroscience, where this approach was used to map neuronal electrical activity in brain slices through imaging of cellular calcium dynamics. 2PM has become an indispensable tool in neuroscience for studying neuronal electrical activity and mapping neuronal architecture in brain slices as well as the brains of live animals. Imaging depths greater than 1 mm have been achieved in live mouse brains, which allows access to large swaths of the brain's information processing regions. An example of this superb depth penetration is shown in Figure 6.22, where a set of genetically labeled mouse neurons are imaged down to >800 µm into the brain [16]. To achieve useful images throughout this depth, the laser intensity was increased several times as the microscope was focused deeper into the brain. As described previously, however, this increase in intensity does

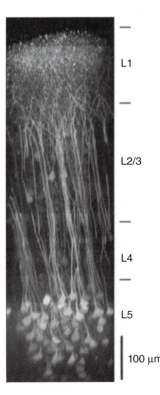

Figure 6.22 Deep 2PM in the brain of a live mouse expressing a genetically encoded indicator demonstrated with a maximum intensity X–Z projection of a fluorescence image stack. Nearly the entire depth of the neocortex can be imaged, but the laser excitation intensity needed to be increased several times (seen as lines of intensity change near the bottom of L2/3, and in L4 and L5). (Reprinted by permission from Macmillan Publishers Ltd, copyright (2005) [16].)

not compromise resolution and can be used to compensate for losses due to scattering of the excitation light and absorption of the resulting fluorescence.

2PM has also been applied to study electrical activity in the heart, whose tissue is denser than brain tissue, and whose movement also makes imaging more challenging. Studies of electrical activity through [Ca^{2+}] imaging have been performed in excised hearts for the study of electrical synchronization. 2PM can enable subcellular studies of functions deep into excised hearts, for example, to measure mitochondrial responses to ischemia using a fluorescent indicator of membrane potential.

6.7.4 Light Sheet Microscopy Using Two-Photon Excitation

As described in the introduction, light sheet microscopy is a promising alternative to confocal microscopy, but it can be limited by penetration depth restrictions that arise from the use of single-photon excitation. This problem can be partially ameliorated by the use of two-photon excitation light sheets, where the combination of nonlinear excitation with the orthogonal illumination of light sheet microscopy can yield excellent performance for 3D imaging of thick biological samples. This is demonstrated in Figure 6.23, where equivalent images show that two-photon light sheets can image up to twice as deep as one-photon light sheet microscopy [17]. The geometry of light sheet microscopy does not permit NDD to be used, thus the depth increases only by a factor of ~2. Still, the hallmark advantages of 2PM are clear in the improved contrast and lack of background haze that degrades the one-photon light sheet image. Importantly, the light sheet approach was shown to collect images up to 10 times faster than point-scanning 2PM without compromising biological viability.

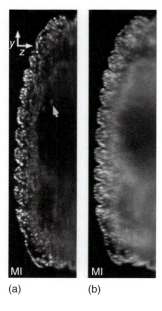

Figure 6.23 (a) Two- and (b) one-photon light sheet images of stage-13 embryos shown as Y–Z image slices of at $X = 90$ μm from embryo surface. The light sheet imaging was done with monodirectional illumination. Arrow in (a) indicates a deep cell at $X \approx 90$ μm and $Z \approx 70$ μm, which is clearly visible in the two-photon light sheet image, but not in the one-photon light sheet image. (Truong *et al.* 2011 [17]. Reproduced with permission of Nature Publishing Group.)

6.7.5 Other Applications of 2PM

There are numerous applications of 2PM, and the list grows longer each day. Here, a few of the other successful *in vivo* 2PM imaging experiments are briefly described. These include measurements of blood flow, kidney function, immune cell dynamics, and embryo development.

Functional intravital 2PM has been used extensively to study blood flow. These studies have been complementary to functional magnetic resonance imaging (fMRI), which can detect localized increases in blood oxygenation levels but does not have sufficient temporal and spatial resolution to resolve individual cellular events. 2PM allows resolution of individual capillaries in live mice to noninvasively study blood flow and quantitatively measure flow velocity with fluorophore-labeled red blood cells and blood plasma. As examples, this approach has been used to measure diverse physiological parameters such as blood flow variations in the olfactory bulb glomeruli as different smells activated specific neuronal units and in the islet of Langerhans as a function of glycemic state. 3D blood flow architecture has also been elucidated through 2PM-enabled targeted ablation of blood vessels, where single capillary ablation was possible with minimal collateral damage through the unique intrinsic confinement of two-photon excitation.

An obvious extension of 2PM blood flow measurements is for the investigation of renal tissue function. The kidney contains a complex network of vasculature, particularly in the glomerulus in which filtration occurs. 2PM imaging with fluorophores of different sizes and net charges can be used to quantify glomerular permeability, which is especially important in studying renal diseases.

The enhanced imaging depth provided by two-photon microscopy also allows for substantial improvements in 4D imaging (x, y, z, t) for observing *in vivo* dynamic processes. There has been a large body of work using two-photon microscopy for dynamically imaging immune cell motility, interactions, and clustering *in vivo*. These studies, many involving intravital imaging of lymph nodes and bone marrow, have all required high temporal and spatial resolution, 100 μm imaging depths, and with the minimal invasiveness to image for periods of hours, all afforded by two-photon microscopy.

Two-photon microscopy has many advantages when imaging live embryo development. One important demonstration of these improvements over confocal microscopy used time-lapse 3D imaging of hamster embryo development [18]. Since mammalian embryos are highly sensitive to culture conditions as well as exposure to light, the survival of these embryos to birth and normal development clearly showed the noninvasiveness of two-photon microscopy. Another strategy used with 2PM to study embryo development is the use of targeted ablations. For example, this approach has been used in Drosophila embryos to generate targeted deformation of specific tissue patterns. Quantifying these morphological changes yielded insight into the mechanisms of mechano-sensitive tissue development and gene expression.

Finally, intravital 2PM has been used to provide new insights into tumor pathology and physiology, since it can reveal gene expression and physiological function deep within regions of tumors that are not accessible with confocal microscopy.

This approach can be used to quantitatively resolve the vascular architecture, giving an insight into the mechanisms of angiogenesis in tumors, and to measure growth and localization of mutations in the tumor cells, such as those resistant to hypoxia-induced apoptosis. 2PM utilizing fluorescent semiconductor nanocrystals, called *quantum dots*, can also be used to image tumors. Quantum dots are very bright and resistant to photodamage, and, in combination with two-photon microscopy, have been shown to enhance imaging through highly scattering tissue such as skin and adipose tissue at 100 μm depth.

6.8 Other Nonlinear Microscopies

As mentioned at the outset, 2PM is a special case of a larger family of nonlinear microscopies. This family can be further divided into three sets of approaches: multiphoton absorption, harmonic generation, and Raman scattering. 2PM is the most prominent example of the multiphoton absorption approach, and the other utilized approach from this set is three-photon excitation. As the name implies, three photons can be absorbed simultaneously, and each of these provides one-third of the energy needed to reach the excited state. For example, infrared light at 1050 nm can be used to excite UV fluorescence that would normally be excited with 350 nm light. However, the higher order of nonlinearity means even weaker signals. Because this consequently requires higher photon fluxes, and does not add any additional advantages over 2PM, this technique has not been used as extensively as two-photon excitation.

Harmonic generation imaging is a fundamentally different nonlinear optical process. As for multiphoton absorption, the most common harmonic-generation-based approach is SHG imaging [19]. SHG imaging requires high photon densities just as 2PM, so the laser requirements for harmonic imaging are similar. The difference is in how the signal is created. In fluorescence, a photon is absorbed and then another is re-emitted. On the other hand, for optical harmonic generation, photons are not absorbed, but instead they interact with the sample via scattering. In SHG, incoming photons are simultaneously scattered and the two combine to generate a single photon that has double the energy of the incident photons (i.e., the scattered light has a wavelength that is half of the incident wavelength). SHG requires that the molecules in the specimen lack inversion symmetry and that they are spatially ordered, so this approach is useful for imaging ordered structures such as collagen fibers or microtubules, with the advantage that it does not require any labeling of the structures. In particular, the molecular structure of fibrillar collagen produces a strong SHG signal with minimal background from non-fibrillar components. Since collagen is an important structural protein in mammal, SHG imaging has become a useful tool for imaging its structural organization. THG, a three-photon analog of SHG, can also be observed, and used for imaging in some cases. Since the third harmonic signal does not require the lack of inversion symmetry in the specimen, this approach can be used to image more disordered molecules. For example, lipid bodies provide a major source of THG, allowing lipid metabolism to be studied without the use of exogenous probes.

The third set of nonlinear microscopies fall into the category of light scattering approaches, based on Raman scattering. Two implementations that have proven feasible are CARS and SRS. The underlying photophysics, strengths, and limitations of these approaches are eloquently detailed in [20]. CARS is a third-order nonlinear optical process that is generally performed using three laser beams, which interact with the sample and generate a coherent optical signal at the anti-Stokes difference frequency. This anti-Stokes light is at a well-defined color that can be easily separated from the incident beams, and its intensity is resonantly enhanced at intrinsic vibrational modes of molecules in the sample. Leveraging the specific vibrational modes of lipid molecules, CARS can be used for noninvasive imaging of biological membranes, and implementations of CARS microscopy have been used to image myelin structures. The second Raman imaging approach uses SRS. In SRS, two laser beams are used to efficiently excite a specific molecular vibrational level, where for each quantum of the vibrational excitation being excited, one photon is annihilated from the pump beam and simultaneously a photon is created in the probe beam. The resulting intensity gain in the probe beam is called the *stimulated Raman signal*. Much like CARS, this signal can be used to detect intrinsic molecules within biological samples by targeting various vibrational bands. As an example, SRS imaging could reveal changes in the regulation of lipid storage in *Caenorhabditis elegans* resulting from genetic manipulation.

6.9 Future Outlook for 2PM

The outlook for 2PM is promising for further studies of *in vivo* physiology. Many important advances are currently under way that will enhance the utility of this approach. On the instrumentation front, the development of 2PM endoscopes has enabled 3D imaging with lateral resolutions of ~1 µm and axial resolutions of ~10 µm. Using these devices, 2PM imaging is no longer limited to 1 mm from the surface tissue, but may instead be used to image structures deep in living animals by introducing an endoscope via the esophagus, blood vessels, ear, or other entry points. Further developments in two-photon excitation light sheet or line-scanning microscopy will also expand the usefulness of 2PM by enhancing the speed of image acquisition. Finally, intravital 2PM can be improved through the use of adaptive wavefront corrections, which can compensate for image distortions caused by heterogeneous biological material. A second area of development is the optimization of fluorophores that are brighter and more red-shifted, including genetically encoded red and infrared fluorescent proteins. As described earlier, red-shifting the fluorescence reduces emission absorption and allows greater imaging depths.

6.10 Summary

Advances in our understanding of two-photon excitation microscopy (2PM), coupled with modern digital imaging instrumentation, have facilitated the use of this physical phenomenon whose existence was predicted 80 years ago.

Turnkey mode-locked laser systems have made 2PM straightforward for all researchers to use. Because of the nonlinear nature of two-photon excitation, it is confined to the focal plane, which results in inherent optical sectioning capability without the need for a confocal pinhole. Thus, photon collection can be more efficient, especially through the use of NDD. Further, confinement of the excitation reduces photobleaching and associated photodamage to the specimen. Because of the unique photophysics of two-photon excitation, 2PM imaging is not significantly degraded by light scattering in the sample, so it can be used to image thick specimens down to depths >1 mm, up to 10-fold deeper than with confocal microscopy.

Despite its strengths, it is important to keep in mind that 2PM is not a magical do-all replacement to the confocal microscope. Complications having to do with possible sample heating, unknown two-photon absorption spectra, and accelerated photobleaching and photodamage at the focal plane mean that, for many experiments, the confocal microscope is still a better instrument. 2PM provides much improved depth penetration for imaging, so it is usually the best option for *in vivo* fluorescence microscopy. Leveraging this advantage, many studies have used 2PM to determine 3D structure and morphology as well as to gain functional information on 3D motility, electrical activity, blood flow, and metabolic activity. 2PM also permits spatially localized perturbations to be introduced into systems through using targeted photo-uncaging or photoablations or optogenetics. Finally, endogenous UV-excited molecules, such as NAD(P)H or elastin that are difficult to image in live cells with conventional fluorescence microscopy, can be imaged by 2PM with minimal loss of cell viability.

As 2PM has become more accessible to biological research, it has been used for a wide range of challenging physiological studies. Successful 2PM applications span the gamut from neuroscience to immunology to cancer to developmental biology to endocrinology, all demonstrating the advantages that 2PM can bring to these and other fields of study.

Acknowledgment

I am grateful to Zeno Lavagnino for his collection of data for some of the demonstration images.

References

1 Denk, W., Strickler, J.H., and Webb, W.W. (1990) Two-photon laser scanning fluorescence microscopy. *Science*, **248** (4951), 73–76.
2 Goppert-Mayer, M. (1931) Elementary file with two quantum fissures. *Ann. Phys.*, **9** (3), 273–294.
3 Sandison, D.R. and Webb, W.W. (1994) Background rejection and signal-to-noise optimization in confocal and alternative fluorescence microscopes. *Appl. Opt.*, **33** (4), 603–615.
4 Denk, W. and Svoboda, K. (1997) Photon upmanship: why multiphoton imaging is more than a gimmick. *Neuron*, **18** (3), 351–357.

5 Hale, G.M. and Querry, M.R. (1973) Optical-constants of water in 200-nm to 200-μm wavelength region. *Appl. Opt.*, **12** (3), 555–563.
6 Fang, H.L.B., Thrash, R.J., and Leroi, G.E. (1978) Observation of low-energy Ag-1 state of diphenylhexatriene by 2-photon excitation spectroscopy. *Chem. Phys. Lett.*, **57** (1), 59–63.
7 Drobizhev, M., Makarov, N.S., Tillo, S.E., Hughes, T.E., and Rebane, A. (2011) Two-photon absorption properties of fluorescent proteins. *Nat. Methods*, **8** (5), 393–399.
8 Rehms, A.A. and Callis, P.R. (1993) Two-photon fluorescence excitation-spectra of aromatic-amino-acids. *Chem. Phys. Lett.*, **208** (3-4), 276–282.
9 Patterson, G.H., Knobel, S.M., Arkhammar, P., Thastrup, O., and Piston, D.W. (2000) Separation of the glucose-stimulated cytoplasmic and mitochondrial NAD(P)H responses in pancreatic islet beta cells. *Proc. Natl. Acad. Sci. U.S.A.*, **97** (10), 5203–5207.
10 Patterson, G.H. and Piston, D.W. (2000) Photobleaching in two-photon excitation microscopy. *Biophys. J.*, **78** (4), 2159–2162.
11 Piston, D.W., Summers, R.G., Knobel, S.M., and Morrill, J.B. (1998) Characterization of involution during sea urchin gastrulation using two-photon excited photorelease and confocal microscopy. *Microsc. Microanal.*, **4** (4), 404–414.
12 Araya, R., Vogels, T.P., and Yuste, R. (2014) Activity-dependent dendritic spine neck changes are correlated with synaptic strength. *Proc. Natl. Acad. Sci. U.S.A.*, **111** (28), E2895–E2904.
13 Svoboda, K. and Yasuda, R. (2006) Principles of two-photon excitation microscopy and its applications to neuroscience. *Neuron*, **50** (6), 823–839.
14 Prakash, R., Yizhar, O., Grewe, B., Ramakrishnan, C., Wang, N., Goshen, I. et al. (2012) Two-photon optogenetic toolbox for fast inhibition, excitation and bistable modulation. *Nat. Methods*, **9** (12), 1171–1179.
15 Oron, D., Papagiakoumou, E., Anselmi, F., and Emiliani, V. (2012) Two-photon optogenetics. *Prog. Brain Res.*, **196**, 119–143.
16 Helmchen, F. and Denk, W. (2005) Deep tissue two-photon microscopy. *Nat. Methods*, **2** (12), 932–940.
17 Truong, T.V., Supatto, W., Koos, D.S., Choi, J.M., and Fraser, S.E. (2011) Deep and fast live imaging with two-photon scanned light-sheet microscopy. *Nat. Methods*, **8** (9), 757–760.
18 Squirrell, J.M., Wokosin, D.L., White, J.G., and Bavister, B.D. (1999) Long-term two-photon fluorescence imaging of mammalian embryos without compromising viability. *Nat. Biotechnol.*, **17** (8), 763–767.
19 Campagnola, P.J., Wei, M.D., Lewis, A., and Loew, L.M. (1999) High-resolution nonlinear optical imaging of live cells by second harmonic generation. *Biophys. J.*, **77** (6), 3341–3349.
20 Min, W., Freudiger, C.W., Lu, S., and Xie, X.S. (2011) Coherent nonlinear optical imaging: beyond fluorescence microscopy. *Annu. Rev. Phys. Chem.*, **62**, 507–530.

7

Light Sheet Microscopy

Gopi Shah[1,2], *Michael Weber*[1,3], *and Jan Huisken*[1,4]

[1] Max Planck Institute of Molecular Cell Biology and Genetics, Pfotenhauerstr. 108, 01307 Dresden, Germany
[2] University of Cambridge, Cancer Research UK Cambridge Institute, Robinson Way, CB20RE Cambridge, UK
[3] Harvard Medical School, Department of Cell Biology, 200 Longwood Ave, LHRRB 113, Boston MA 02115, USA
[4] Morgridge Institute for Research, Department of Medical Engineering, 330 N Orchard Street, Madison WI 53715, USA

Live imaging of whole tissues and organs is becoming increasingly important in modern-day biology. In recent years, light sheet microscopy, with its high speed and low phototoxicity, has become the technique of choice for long-term live imaging of developing organs and organisms. The earliest instrument to employ the principle of light sheet microscopy was the slit-ultramicroscope for counting gold particles in solution in 1902 [1]. Light sheet microscopy found its application for biology after the development of genetically encoded fluorescent proteins in the 1990s [2]. In 1993, a technique called orthogonal-plane fluorescence optical sectioning (OPFOS) used light sheet illumination to image the internal architecture of fixed cochlea [3]. However, it was not until 2004 that the application of light sheet microscopy was demonstrated for fast four-dimensional (4D) (x, y, z, t) live imaging of millimeter-sized embryos [4].

Every *in vivo* fluorescence microscope has a certain range of sample sizes where it performs best, with no single instrument that covers the entire range from subcellular resolution all the way up to whole embryos. On the small scale, confocal microscopy (Chapter 5) has been used to image tissue sections and parts of small model organisms at high resolution. On the large end of the scale, tomographic techniques have been established for reconstructing centimeter-sized samples at lower resolution. The gap between these technologies is now bridged by light sheet microscopy, which offers cellular resolution in millimeter-sized samples. Phototoxicity has been reduced to a minimum even at high acquisition rates, proving the technique most suitable for imaging developing embryos at high speed and very good resolution over several hours or days. In the last decade, light sheet microscopy has been successfully used to image dynamic processes in various model organisms: zebrafish [5, 6], Drosophila [4, 7, 8], *Caenorhabditis elegans* [9], spheroids [10], cell cultures as well as single cells [11–13]. At the same time, it has become the preferred technique for visualizing the structure of large specimens such as fixed and cleared brain and other organs in mouse [14]. It has also been used for imaging rapid processes such as the beating zebrafish

Fluorescence Microscopy: From Principles to Biological Applications, Second Edition.
Edited by Ulrich Kubitscheck.
© 2017 Wiley-VCH Verlag GmbH & Co. KGaA. Published 2017 by Wiley-VCH Verlag GmbH & Co. KGaA.

heart and the blood flow [15, 16], as well as long-term slow events such as plant growth [17]. It is this versatile nature that makes light sheet microscopy a popular technology among cell and developmental biologists.

Although called by different names in the literature (selective plane illumination microscopy, SPIM, digitally scanned light sheet microscopy, DSLM, light sheet fluorescence microscopy, LSFM, ultramicroscopy, etc.), all implementations are based on the same fundamental principle described in this chapter. We will focus on the basics of the light sheet microscopy principle and construction, followed by an overview of its various implementations. In the end, we will discuss some of the novel data handling solutions that light sheet microscopy gave rise to, as well as future prospects in this area.

7.1 Principle of Light Sheet Microscopy

In light sheet microscopy, the sample is illuminated with a thin sheet of laser light to obtain optical sections. The microscope generally consists of two orthogonal optical axes: one for generating the light sheet for illumination, and the other for widefield detection of the emitted fluorescence. The two axes are aligned such that the illuminating light sheet is positioned in the focal plane of the detection unit (Figure 7.1).

As the specimen is illuminated with a sheet of laser light, the entire focal plane of the detection arm is illuminated providing instant optical sectioning as opposed to the slow point scanning used in confocal microscopy (Chapter 5).

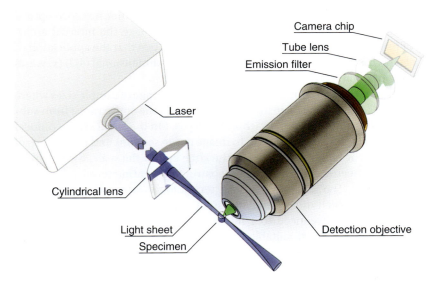

Figure 7.1 Principle of light sheet microscopy. Illumination and detection axes are oriented orthogonally, and the specimen is placed at their intersection. A sheet of laser light is produced, for example, using a cylindrical lens, and illuminates a thin slice of the sample in the focal plane of the detection objective. A camera, which is placed behind the detection lens, the fluorescence filter, and the tube lens, captures an image of the fluorescence.

The height and thickness of the light sheet can be adjusted to achieve the desired extent of optical sectioning and to illuminate the entire field of view (FOV). Fluorescence generated by the light sheet is collected very efficiently, as all the camera's pixels collect photons simultaneously during the entire exposure time. By using fast and sensitive cameras, the sample is imaged rapidly with low light exposure. Therefore, light sheet microscopy is an ideal technique to capture dynamic developmental processes without detrimental phototoxicity and to image large, cleared organs fast and efficiently.

Theoretically, a light sheet microscope's lateral resolution is equivalent to that of an epi-fluorescence microscope (with the same objective lens), given by the objective lens' numerical aperture (NA) and the fluorophore's wavelength λ_{em} (Chapter 2, Section 2.3.4). In practice, however, the contrast in light sheet microscopy is generally better because of the optical sectioning. The depth of the optical section is defined by the thickness of the light sheet. As a consequence, high axial resolution can be obtained even with low-NA, low-magnification detection lenses, which usually have a large working distance desirable for large sample imaging.

7.2 Light Sheet Microscopy: Key Advantages

Light sheet microscopy offers several advantages over conventional imaging modalities. The main reason for the success of light sheet microscopy in the biological sciences is certainly the heavily reduced phototoxicity. Traditional compound and confocal microscopes illuminate the entire volume of the sample, even when imaging only a single plane. Even worse, when a z-stack of N images is recorded, the entire sample is exposed to the excitation light N times. In contrast, in light sheet microscopy one image plane at a time is illuminated. Therefore, during a stack across the entire sample, every plane is exposed only once, minimizing the risk of photobleaching and phototoxicity.

In point scanning microscopes, each point is sequentially exposed and detected, requiring higher laser power and typical exposure times of several seconds to obtain a single image. In light sheet microscopy, the entire plane is illuminated simultaneously, and photons from the entire field of view are collected in parallel within an exposure time of a few milliseconds. This exposure time is about a thousand times longer than the pixel dwell time in a point scanning system, which is in the range of a few microseconds per pixel. Using fast and sensitive EM-CCD and scientific complementary metal-oxide-semiconductor (sCMOS) cameras, one can rapidly acquire images with excellent dynamic range and signal-to-noise ratio. Fluorophore saturation, unlike in confocal point scanning microscopy, is less of an issue in light sheet microscopy. The acquisition speed is limited only by the camera technology and available fluorescence signal and can be exploited in a number of ways. Most importantly, one can study very fast three-dimensional phenomena that are too complex to reconstruct with conventional widefield microscopy and too fast for confocal microscopy [15, 18].

The fast recoding in light sheet microscopy is also another key advantage for the imaging of large scattering objects. The ability to record an entire z-stack in a few seconds opens up the possibility to record additional views of the sample from other directions. The classical horizontal setup of the objective lenses is particularly well suited for this multiview imaging: since the sample hangs vertically at the intersection of the two optical axes, it can be rotated without deformation. While other imaging modalities are simply too slow, several views can be obtained within a few seconds on a light sheet microscope and fused to reconstruct the whole sample without artifacts [19]. Two orthogonal or more views are merged for improving the axial resolution, for example, by multiview deconvolution, or to simply increase the overall information content of the volume when imaging large specimen that cannot be optically penetrated from a single side [20–22].

Furthermore, the orthogonal geometry provides flexibility to tweak the illumination and the detection arms independently to adapt to the experimental needs. Consequently, there are several technical implementations of light sheet microscopy developed for a wide range of applications discussed later in this chapter.

7.3 Construction and Working of a Light Sheet Microscope

Two different ways of generating a light sheet are commonly used: in SPIM, the laser beam is expanded and focused using a cylindrical lens to form the sheet of light (Figure 7.2a) [4]. This method is simple to implement and widely used for fast and gentle low-light *in vivo* applications. Alternatively, a "virtual" light sheet can be generated by the DSLM technique, which uses a scanning mirror to rapidly sweep a beam across the FOV (Figure 7.2b) [5]. This method offers more flexibility, as the height of the light sheet can be easily adapted by changing the scanning amplitude, and its thickness by changing the diameter of the incoming laser beam. Here, because of the sequential line illumination, only a fraction of the final image is illuminated at a given point in time, exposing the sample to a local light intensity much higher than in SPIM in order to achieve the same fluorescence yield in the same amount of time. While this may result in fluorophore saturation and higher photobleaching, it is compatible with applications that require high laser power such as two-photon excitation [23], beam-shaping applications like Bessel beam illumination [12, 24], and structured illumination [25].

In DSLM, the light sheet has a uniform intensity profile along its height, whereas in SPIM the laser light is expanded and cropped with apertures to cover the FOV of the detection lens, resulting in a less uniform intensity profile. However, the simultaneous plane illumination obtained in SPIM is preferred for high-speed imaging applications, especially when using electrically tunable lenses (ETLs) and ultrafast cameras for image acquisition without motion artifacts. Therefore, when designing a light sheet microscope, it is crucial to determine the application for which it will be used in order to choose the appropriate means to produce the light sheet. The following section presents a detailed overview of designing a light sheet microscope setup with a static light sheet.

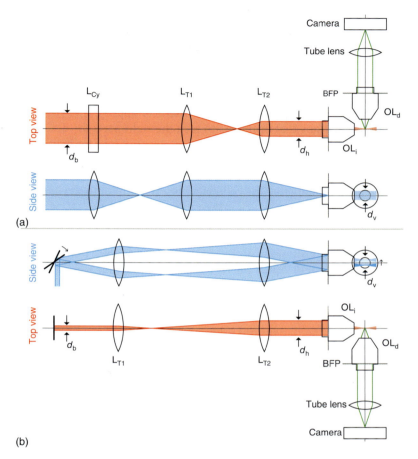

Figure 7.2 Beam paths of microscopes with static and scanned light sheet illumination. (a) Static light sheet illumination. An incoming laser beam of diameter d_b passes a cylindrical lens L_{Cy} and is focused to a line. A telescope comprising the lenses L_{T1} and L_{T2} adapts the beam height d_h and focuses it into the back focal plane (BFP) of the illumination objective OL_i, which projects a light sheet into the focal plane of the detection objective OL_d. The excited fluorescence signal is collected by OL_d and focused onto a camera by means of a tube lens. (b) Scanned light sheet illumination. The major difference from a microscope with static light sheet illumination is an incoming laser beam that is rapidly scanned along one axis in the BFP of L_{T1}, resulting in a sweeping laser beam that generates a "virtual" light sheet in the focal plane of OL_d.

7.4 Theory of Light Sheet Microscopy

A light sheet can be easily generated with a single cylindrical lens. Its focal length and the incoming beam diameter determine the thickness and extent of the light sheet. While this simple setup may be sufficient to generate a single-colored sheet in air, it does not fulfill the requirements in biological imaging. For example, it is desirable to generate several overlapping light sheets with different wavelengths and to work in an immersion medium such as water. It is therefore advisable to use a well-corrected objective lens, specifically designed for the desired immersion medium. Hence, many SPIM setups feature not only a water-dipping detection lens but also a matching water-dipping illumination lens. The optics need to be designed in such a way that the last element is an objective lens; the cylindrical lens is placed earlier in the beam path. In addition, it is desirable to have enough degrees of freedom to adjust the position, orientation, and thickness of the light sheet. Therefore, mirrors and slits are inserted wherever appropriate. Here, we will go through calculations that enable us to pick the right parts for designing a SPIM setup that features a cylindrical lens and an objective lens for light sheet formation.

The light path of the illumination arm consists of a laser and a beam expander, optionally a fiber and a collimator. The collimated beam is sent onto a cylindrical lens L_{Cy} to generate a light sheet, which is then imaged into the back focal plane (BFP) of the illumination objective lens. In order to get a vertical light sheet at the sample, the sheet in the BFP needs to be horizontal. Ideally, a mirror is placed in the BFP to allow precise positioning of the sheet (in order to place it in the focal plane of the detection plane). Since the BFP of the illumination lens is usually inside the housing of the lens and therefore not accessible, one uses a telescope of two lenses to image the back focal plane onto a mirror. This arrangement gives access not only to the BFP but also to the focal plane of the illumination lens. With mirrors and slits, the dimensions and position of the beam can be easily adjusted. The following considerations are important to obtain a light sheet that is thin yet uniform across the entire FOV. We consider a set of four lenses to form the light sheet in the sample chamber: a cylindrical lens L_{Cy}, two lenses L_{T1} and L_{T2}, and an objective lens OL_i. Further we assume a Gaussian beam of diameter d_b to hit L_{Cy} (Figure 7.2). The light sheet in the sample chamber has a shape as depicted in Figure 7.3. The beam converges to a waist of $2\omega_0$ and diverges again.

A convenient measure of the extent of the light sheet is given by the Rayleigh length x_R, which is defined as the distance from the waist to the plane where the beam has expanded to $\sqrt{2}\,\omega_0$, that is,

$$\omega(x_R) = \sqrt{2}\,\omega_0.$$

The Rayleigh length is given by

$$x_R = \frac{\pi n \omega_0^2}{\lambda} \quad \text{or} \quad \omega_0 = \sqrt{\frac{x_R \lambda}{\pi n}} \tag{7.1}$$

where ω_0 is half the thickness at the waist of the light sheet, λ is the wavelength of the beam, and n is the refractive index of the medium. We can further approximate (see Figure 7.3)

$$\phi/2 \approx \tan(\phi/2) = \frac{d_{BFP}/2}{f_{IO}} \quad \text{or} \quad \phi \approx \frac{d_{BFP}}{f_{IO}} \tag{7.2}$$

where f_{IO} is the focal length of the illumination objective and d_{BFP} is the width of the beam in the BFP of the objective lens. The total angular spread of a Gaussian beam in radians is related to the beam waist ω_0 by

$$\phi = \frac{2\lambda}{\pi n \omega_0} \tag{7.3}$$

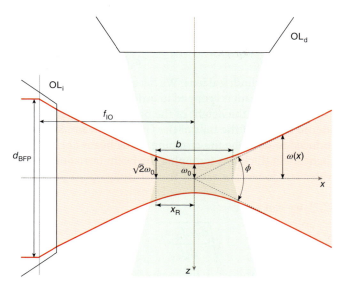

Figure 7.3 Relationship between light sheet dimensions and field of view. A light sheet is generated from a laser with Gaussian beam width $\omega(x)$ and total angular spread (ϕ). The waist of the light sheet is placed in the middle of the field of view and its properties are adjusted such that the distance (2× Rayleigh length x_R) equals the width of the field of view.

From Eqs (7.2) and (7.3), we get

$$\frac{d_{BFP}}{f_{IO}} = \frac{2\lambda}{\pi n \omega_0}.$$

Substituting the value for ω_0 from Eq. (7.1), we have

$$d_{BFP} = \frac{2\lambda f_{IO}}{\pi n}\sqrt{\frac{\pi n}{x_R \lambda}} = 2f_{IO}\sqrt{\frac{\lambda}{\pi n x_R}}. \tag{7.4}$$

From this equation, we can calculate how wide our beam needs to be at the BFP of our illumination lens d_{BFP} in order to get a sheet with a length of x_R. The width of the FOV (along x-axis) should be equal to $2x_R$, and d_{BFP} can be calculated. In other words, x_R can be expressed as

$$2x_R = \frac{N_x^{px} d^{px}}{M_{DO}} \tag{7.5}$$

where N_x^{px} is the number of pixels along the x-axis, d^{px} is the pixel size, and M_{DO} is the magnification of the detection objective.

From the top view of the light path in Figure 7.2, d_{BFP} can be expressed as

$$d_{BFP} = d_b \frac{f_{T2}}{f_{T1}} = d_b M_T \quad \text{or} \quad d_b = \frac{d_{BFP}}{M_T} \tag{7.6}$$

where d_b is the diameter of the incoming circular beam, M_T is the magnification, and f_{T1}, f_{T2} are the focal lengths of the telescope in front of the illumination lens. At the same time, from the side view of the light path in Figure 7.2, the height of the light sheet can be expressed as

$$d_h = \alpha_y d_b \frac{f_{T1} f_{IO}}{f_{Cy} f_{T2}} = \alpha_y d_b \frac{1}{M_T} \frac{f_{IO}}{f_{Cy}} \quad \text{or} \quad d_b = \frac{d_h M_T f_{Cy}}{\alpha_y f_{IO}} \quad (7.7)$$

where f_{Cy} is the focal length of the cylindrical lens. We have introduced the factor α_y, which describes how much the beam is cropped in order to achieve an almost uniform intensity distribution across the height of the beam. In the instrument, an iris or a slit is introduced to cut off the tails of the Gaussian beam. Similarly, a factor α_z may be introduced to describe the cropping of the beam horizontally (along z) to adjust the light sheet thickness (cropping the beam horizontally will make the light sheet wider and longer). α ranges from 1 (fully open) to 0 (fully closed).

From Eqs (7.6) and (7.7), we can write

$$\frac{d_{BFP}}{M_T} = \frac{d_h M_T f_{Cy}}{\alpha_y f_{IO}} \quad \text{or} \quad M_T = \sqrt{\frac{d_{BFP} f_{IO} \alpha_y}{d_h f_{Cy}}} \quad (7.8)$$

d_h is the height of the desired FOV, therefore the magnification of the telescope M_T can now be determined, by inserting the value for d_{BFP} from Eq. (7.4). The incoming beam diameter d_b can be obtained, and an appropriate beam expander or collimator can be included to obtain the desired diameter.

In the case of DSLM, the height of the light sheet is regulated by the scanning amplitude and the thickness by the diameter of the incoming beam. While the light path does not contain a cylindrical lens (Figure 7.2b), the remaining calculations are identical for both SPIM and DSLM.

Example 1: System Design for Imaging an Early Zebrafish Embryo that Fits Entirely in an 800 × 800 µm FOV

We know

- width of the FOV: $2x_R = 800$ µm $= 0.8$ mm
- height of the FOV (and light sheet): $d_h = 800$ µm $= 0.8$ mm
- wavelength of illumination $\lambda = 0.488$ µm
- refractive index of medium $n = 1.33$
- illumination objective: 10×, $f_{IO} = 19$ mm
- focal length of cylindrical lens commonly used $f_{Cy} = 50$ mm
- slit $\alpha_y = 0.25$.

Substituting values in Eq. (7.4), the beam diameter in the BFP of the illumination objective should be $d_{BFP} = 0.649$ mm. Substituting values in Eq. (7.8), the required magnification of the telescope is obtained as $M_T = 0.278$. Therefore, the lenses T1 and T2 need to be chosen such that $f_{T2}/f_{T1} = 0.278$.

From Eq. (7.6), the required incoming beam diameter d_b should be at least 2.34 mm. Using these components, the resulting light sheet thickness can be calculated from Eq. (7.2) to be $2\omega_0 = 13.7$ µm.

> As the FOV here is quite large, the light sheet is relatively thick at its waist. In the next example, we will look at a smaller FOV and see how the light sheet thickness and beam diameter in the BFP change.

> **Example 2: System Design for Imaging a Drosophila Embryo, Which Would Require 300 × 500 μm FOV**
>
> We use the same parameters as before, except for the following:
>
> - width of the FOV: $2x_R = 300$ μm $= 0.3$ mm
> - height of the FOV (and light sheet): $d_h = 500$ μm $= 0.5$ mm.
>
> Substituting values in Eq. (7.4), the beam diameter in the BFP of the illumination objective should be $d_{BFP} = 1.06$ mm. Substituting the values in Eq. (7.8), the required magnification of the telescope is obtained as $M_T = 0.449$.
>
> Therefore, from Eq. (7.6), the required incoming beam diameter d_b should be at least 2.36 mm. Using these components, the resulting light sheet thickness can be calculated from Eq. (7.2) to be $2\omega_0 = 8.37$ μm.

7.5 Light Sheet Interaction with Tissue

All light microscopes suffer from artifacts by the interaction of the light with the sample. Scattering and attenuation of excitation and emission light are major issues. While traveling through dense tissue, the excitation light is attenuated and scattered: scattering within the plane of the light sheet is not a problem, but scattering along the z-axis leads to an increase in light sheet thickness and loss of z-sectioning (Figure 7.4a). Choosing the right polarization of the light sheet is crucial to minimize this effect: the laser beams in light sheet microscopy are polarized, and the scattering in tissue is polarization dependent. This effect can be easily observed by using a $\lambda/2$-plate to rotate the polarization. Turning the polarization and maximizing the scattering within the plane, that is, toward an observer looking at a SPIM chamber from the top, yields the correct polarization.

Attenuation of the excitation light and the emitted fluorescence results in stripy and patchy images. Especially in large samples, the part of the sample facing the incoming light sheet yields better contrast and signal-to-noise ratio, as the signal degrades going deeper into the sample (Figure 7.4a). To overcome this issue, the sample can be illuminated from two opposite sides, obtaining good image quality in both halves of the sample (Figure 7.4b). It is important to note that the double-sided illumination is ideally performed sequentially, recording one image for each illumination. The two images are merged either directly or later during post-processing. Thereby, the aberration-free parts facing the illumination are preserved and contribute to the final image. Illumination from both sides simultaneously is advisable only when the sample is sufficiently transparent and the

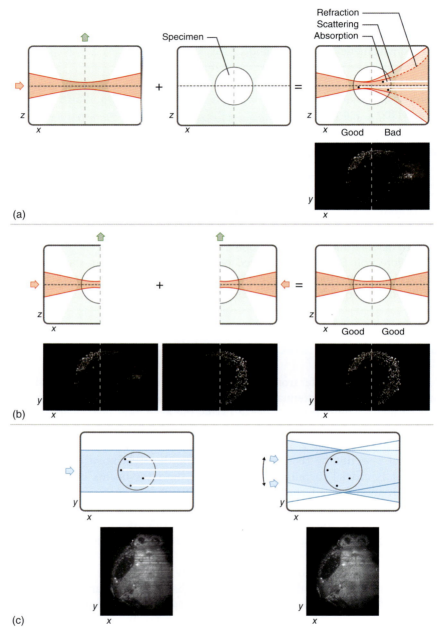

Figure 7.4 Correction of sample-induced effects on the light sheet. (a) Single-sided illumination. (b) Double-sided illumination. (c) Pivoting light sheet. Obstacles in the field of view potentially result in stripes along the propagation direction of the light sheet illumination. By using a light sheet that pivots around the center of the field of view within one exposure, the propagation direction is constantly altered. Thereby, obstacles are homogeneously illuminated from a range of angles, and the sum of the excited fluorescence results in an image with a minimum number of visible stripes.

scattered thicker light sheet from one side does not deteriorate the other thinner light sheet.

The absorption of the excitation light produces artifacts such as stripes of bright and shadowed regions across the entire FOV (Figure 7.4c). In light sheet microscopy, this effect is especially pronounced and visible as a result of the illumination from the side. To overcome this problem, multidirectional SPIM (mSPIM) has been developed [26], which includes a combination of double-sided sequential illumination and light sheet pivoting. More even illumination of the FOV is achieved by pivoting the light sheet, that is, scanning the beam over an angle of ca. 10 deg at a frequency of about 1 kHz with a resonant mirror during a single exposure of the camera. By doing so, the stripes and shadows in the image are greatly reduced, thereby achieving a more homogenous image quality across the whole sample (Figure 7.4c).

7.6 3D Imaging

So far we have discussed how to acquire 2D images of a sample using light sheet microscopy. However, most biological applications require 3D imaging. Given its excellent optical sectioning capability, speed, and low phototoxicity, light sheet microscopy is ideally suited for fast 3D imaging. For a variety of samples, different modes of 3D image acquisition have been developed.

I. A widely used approach is to acquire a z-stack by translating the sample along the detection axis, through a static light sheet, as the camera continuously records images (Figure 7.5a). Only the sample is moved, and the rest of the

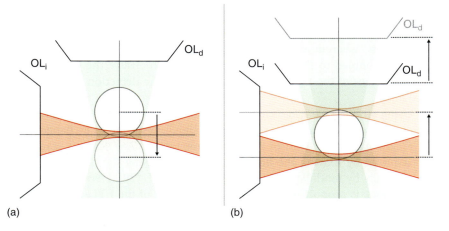

Figure 7.5 Two strategies to record z-stacks in a light sheet microscope. (a) Translating the sample. The specimen is moved through the static focal plane of the detection objective OL_d, which is continuously illuminated with a light sheet. (b) Translating illumination and detection. The focal plane is moved through the static specimen, for example, by translating OL_d. Simultaneously, a scan mirror maintains a light sheet in the focal plane of OL_d.

optics remains fixed. This method ensures that the light sheet stays in the focal plane of the detection lens after alignment of the system, making it robust and straightforward to implement. Moving the sample continuously rather than stepwise ensures fast and smooth data acquisition with minimum artifacts. The illumination time of the sample needs to be kept short, or the motors need to move slowly, to minimize bluring and thickening of the light sheet as the sample is moved during the exposure. The imaging speed is ultimately limited by the speed of the motors and the speed at which a given sample can be moved without affecting its physiology.

II. A more recent approach is to keep the sample static while scanning the light sheet through it. A major advantage of such a system is that it is independent of the sample, and even fragile samples can be imaged contactless. Such a system is especially beneficial for imaging samples that cannot be embedded in a solid medium such as agarose, and need to be in a water-like medium for their proper growth and movement. However, when continuously moving the light sheet, the focal plane of the detection system needs to be synchronized with the light sheet movement to keep the fluorescence in focus. Alternatively, one could also expand the depth of focus to cover the entire scan range, which generally requires deconvolution or sacrificing lateral resolution. The two main remote focusing approaches are as follows:

a) Using a motorized objective lens: In such a system, fast and precise short-range stages or piezo-driven stages are used to rapidly move the objective lens back and forth. When synchronized with a galvanometric mirror used to scan the light sheet, fast z-stacks can be obtained without moving the sample (Figure 7.5b). While many light sheet microscopes use water-dipping objective lenses, air lenses are better suited for this setup to prevent any leakage from the sample chamber and avoid pressure waves in the medium. Nevertheless, the use of motorized water-dipping lenses has been demonstrated for brain and whole-animal imaging of Drosophila larvae [27]. An alternate approach had been developed earlier to image neurons, wherein an optical fiber producing a light sheet was mechanically coupled to the moving detection lens to ensure the synchronized movement of the light sheet with respect to the detection objective lens [28].

b) Using an ETL: An ETL is a liquid-filled, deformable lens that changes its curvature in response to an electrical signal, resulting in a change of its focal length and a shift of the focal plane (Figure 7.6). Rapid z-stacks are acquired by moving the light sheet and refocusing the detection optics remotely by modulating the current in the ETL [29]. Here, the depth of the stack is proportional to the focusing range of the ETL. Large volume scanning at moderate speed can be obtained. Ultimately, the speed of such a system is limited by the fluorescence yield; cameras with 10 000 frames per second are available and principally applicable to achieve hundreds of volumes per second. High-speed scanning of smaller volumes, necessary for imaging highly dynamic processes, such as a beating zebrafish heart and blood flow, have been demonstrated with a fast CMOS camera [15].

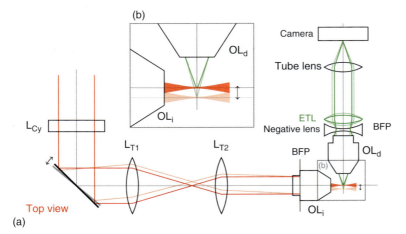

Figure 7.6 Beam path of a light sheet microscope with an electrical tunable lens (ETL). (a) An ETL in the microscope's detection path oscillates the focal plane of the detection objective OL_d along the detection axis, z. A scan mirror simultaneously maintains a light sheet in the focal plane of OL_d. (b) The intersection of illumination and detection axes with the oscillating focal plane and the scanned light sheet.

7.7 Multiview Imaging

The penetration depth of a light microscope is limited by attenuation and scattering of excitation and emission light. Thus, images recorded in areas facing the illumination and detection objectives provide better contrast and signal-to-noise ratio than those acquired in deeper regions. It would therefore be advantageous if the arrangement of the sample with respect to illumination and detection axes could be adapted. In a light sheet microscope, z-stacks can be obtained from different viewing angles by rotating the sample. Importantly, this is not necessarily a unique feature of light sheet microscopy. However, a prerequisite for successful multiview imaging in living samples is high acquisition speed. Multiple views need to be acquired in quick succession to avoid developmental changes in the sample, making the datasets incompatible. Multiple views of the sample are then fused to reconstruct the entire sample in three dimensions (Figure 7.7a). This method provides two major advantages: multidirectional illumination and detection of entire samples, resulting in increased image quality and information content throughout the sample. Moreover, different views of the same area of the sample can be fused to improve the axial resolution of the data: for example, ideally the poor axial resolution of one dataset is replaced by the good lateral resolution of a dataset recorded perpendicular to the first.

Several computational methods have been developed to merge z-stacks recorded from multiple viewing angles. Precise registration of the datasets is crucial for successful fusion. Ideally, the micromotors and rotational motors used to orient and position the sample are precise and reproducible so that an initial calibration is sufficient to register subsequent datasets. Otherwise, fluorescent beads need to be added as fiducial markers, or nuclear markers

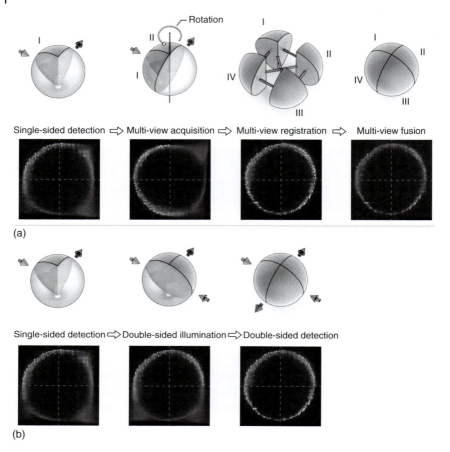

Figure 7.7 Multiview acquisition strategies. (a) Stepwise rotation of the sample. A specimen is rotated around its center in steps of, for example, 90°, and z-stacks are acquired from each angle (multiview acquisition). Single-sided illumination and detection result in about 25% coverage of the sample. Multiview registration and fusion combines the well-covered areas of each recording into a homogeneous dataset. (b) Multi-sided detection of a static sample. The limited coverage of single-sided illumination and detection is overcome by illuminating the specimen from two sides (double-sided illumination) while recording consecutive z-stacks from two sides (double-sided detection).

inside the sample can serve as landmarks to register the different views to each other. These reference points in different views are matched with each other to determine the transformation between adjacent views. The different views are then fused accordingly to obtain a 3D reconstruction of the entire sample (Figure 7.7a). Deconvolution based on point spread functions recorded from multiple views can also be used to improve the axial resolution and contrast of images [21]. Real-time deconvolution by re-slicing the acquired data and processing cross-sectional planes individually on the graphics processing unit (GPU) has been demonstrated and offers the efficient application of multiview acquisitions also in extended time lapse experiments [22]. An alternate and efficient method is to selectively acquire the well-resolved parts from each view (quadrants that are well illuminated and detected) and stitch the "good" parts together to reconstruct the entire embryo with good signal-to-noise ratio [6]. Oftentimes, it is not clear at the beginning of a timelapse recording which view may yield the best data. Multiview acquisitions can then be performed to defer the decision and simply delete inferior data after acquisition.

7.8 Different Lens Configurations

A typical two-lens SPIM setup consists of an illumination and a detection arm, with a water-filled sample chamber at their intersection (Figure 7.8a). Water-dipping lenses are preferred, as they minimize the number of interfaces for illumination and detection; however, the orthogonal arrangement of illumination and detection objectives limits the choice of lenses. Often, high-NA

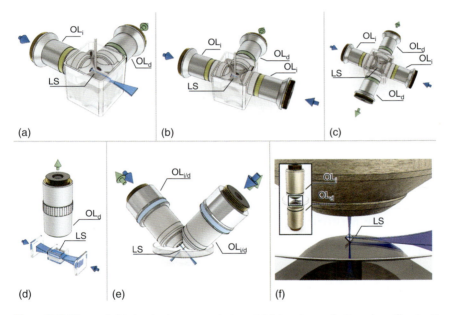

Figure 7.8 Diverse light sheet microscope designs. (a) A two-lens selective plane illumination microscope (SPIM) features single-sided illumination and detection and typically requires a vertically mounted sample. (b) In a three-lens or multidirectional SPIM (mSPIM), a second illumination path generates and additional light sheet to illuminate the sample simultaneously or alternately from two sides (double-sided detection). (c) A four-lens SPIM adds a second detection path to simultaneously or successively record images from two opposite sides. (d) Ultramicroscopes are typically designed around larger samples and facilitate a low-magnification objective in an upright detection path. The sample is illuminated with horizontally oriented light sheets from one or two sides. (e) The dual inverted SPIM (diSPIM) is an adaption of a two-lens configuration to accommodate specimens mounted in conventional dishes or multiwell plates. To provide a level of multiview imaging, illumination and detection can be alternated between the two objectives. (f) Reflected light sheet microscopy describes setups in which the orthogonal light sheet illumination is provided by reflecting the beam next to the specimen by means of a small mirror.

detection objectives are too big to be combined with an orthogonally placed water-dipping illumination objective. Therefore, in many cases, air illumination objectives need to be used. Fortunately, low NA illumination objectives such as 10×/0.3 lenses are sufficient to generate a light sheet with a thickness of only a few micrometers.

One of the beauties of light sheet microscopy is the flexibility of its design and the ease with which it can be customized for desired applications. Recent years have seen several implementations for imaging samples from single cells to multicellular vertebrate embryos, and developmental processes ranging from a plant growing over days to a zebrafish heart beating several times a second. The orthogonal optical arrangement of the illumination and detection paths can be set up horizontally for imaging most samples (Figure 7.8a). In a three-lens SPIM, a second identical illumination arm illuminates the sample from the opposite side (Figure 7.8b) [26]. The design of light sheet microscopes inspired the concept of multiview imaging. At the same time, the need for multiview imaging of large biological samples has inspired more advanced implementations of light sheet microscopy. Adding a second detection lens such that both light sheets are aligned with the shared focal plane of the two detection lenses gives two opposite views of the sample simultaneously (Figures 7.7b and 7.8c) [6–8].

A vertical configuration may be preferred for brain imaging [30] and imaging of cleared tissues when the sample is mounted horizontally such as in ultramicroscopy (Figure 7.8d) [14] or in single-molecule tracking [31]. For cells in culture or any sample that needs to be mounted on a coverslip, an inverted configuration like the diSPIM (Figure 7.8e) [9] or the reflected light sheet microscope, where a light beam is reflected from a scanning mirror to image through the same or oppositely placed detection lens, is more appropriate (Figure 7.8f) [32].

7.9 Sample Mounting

As is also true for other fields of light microscopy, sample mounting for light sheet microscopy has to fulfill two major tasks. The first one is to keep the specimen stable over the course of the recording to avoid blur and distortions. The second task is to minimize light attenuation and scattering, and maximizing resolution and contrast by using refractive index-matched and optically clear mounting materials. If living samples are examined, an additional task for sample mounting is to ensure the best possible conditions for the unaffected survival and well-being of the specimen over the course of the experiment.

The large variety of specimens imaged with light sheet microscopy, as well as the multitude of light sheet microscope designs, requires entirely different sample mounting strategies. Crucially, whether the specimen is alive or fixed (and potentially optically cleared) determines how to embed it. In case of living samples, such as fruit fly or zebrafish embryos, cultured tissue or cells, the mounting materials are typically matched to the refractive index of water (1.33). A frequently used strategy is to embed the specimen in a solid or viscous gel such as agarose, phytogel, or methylcellulose. If the specimens are fixed, mounting materials with higher refractive indices are ideal (1.4–1.5). The optional use of clearing solutions increases the transparency of thicker fixed samples, but also has an impact on the refractive index [33, 34]. Some of those solutions are harmful and must not come into contact with optics (or the user), which requires an enclosed sample chamber and air lenses.

Sample mounting for light sheet microscopy goes hand in hand with the respective technical implementation. A requirement in all designs is the access from at least two orthogonal directions: the illumination and the detection axes. Some light sheet microscopes resemble traditional compound microscopes by using common microscope bodies (iSPIM) [9] and some of them even by using the same objective for illumination and detection (reflected light sheet microscopy, swept confocally-aligned planar excitation microscopy) [32, 35]. This permits the use of established protocols such as mounting the specimen in a Petri dish. However, the design of many light sheet microscopes is inspired by the early horizontal design, which differs significantly from that of common compound microscopes [4]. Having the specimen hanging inside a medium-filled chamber required rethinking, and initiated the development of novel mounting strategies to accommodate a variety of specimen. Widely used were gel columns made from agarose or phytogel [17, 36], or medium-filled polymer tubes or bags [37]. Many of these protocols do not simply reproduce existing techniques, but rather enable new types of experiments. When embedded in an agarose column or polymer tube, the specimen can be imaged from the best possible angle [15] or reconstructed from multiple angles [6]. Dedicated mounting protocols also enable light sheet microscopy of cell cultures, spheroids and organoids [38], as well as developing plants [17].

7.10 Recent Advances in Light Sheet Microscopy

Despite the fast multiview capabilities of SPIM, it still remains challenging to image millimeter-sized embryos fast and efficiently. Internal organs like the pancreas, liver, and gut in the zebrafish have been difficult to resolve with satisfactory resolution. Custom microscope setups that are tailored to the specific needs of the experiment and provide better penetration without additional phototoxicity are now needed.

A number of methods have been developed to suppress the scattered light in DSLM. While this makes the images crisper, it eliminates information, albeit blurry. The most popular solution is to use the rolling shutter of an sCMOS camera [39]. In this mode, the beam's position on the chip is synchronized with the rolling shutter of the camera. By selecting the correct "slit" width, scattered light can be eliminated.

Several attempts have been made in recent years to increase the penetration of DSLM. Two-photon excitation had been successfully applied in confocal scanning microscopy to increase penetration. The nonlinearity of the excitation permits eliminating the pinhole, the major bottleneck in deep-tissue imaging in confocal microscopy. In addition, the longer excitation wavelength scatters less and penetrates deeper into optically dense tissue. Hence, the implementation of two-photon light sheet microscopy was straightforward in a DSLM system [23]. While providing improved penetration, serious issues have been the low excitation efficiency and the abrupt loss in excitation as soon as the light sheet gets scattered and widened, making quantitative analysis difficult. In addition, the potentially increased phototoxicity needs to be carefully evaluated.

DSLM also offers the opportunity to modify the beam modes and explore alternative profiles such as Bessel or Airy beams. The Bessel beam offers a long, thin core and would be ideally suited for the generation of thin and long light sheets. In addition, the beam is relatively robust against scattering [24, 40]. Unfortunately, the Bessel beam's thin core is accompanied by a set of rings that cause a lot of unwanted out-of-focus excitation. Suppression of these by confocal slit detection and two-photon excitation has been demonstrated. By generating a linear array of Bessel beams and making them to interfere, the rings can be eliminated and an optical lattice is generated (lattice light sheet microscopy [41]). Such ultrathin light sheets are particularly suited for imaging single cells.

Fluorescence is a prerequisite for light sheet microscopy and provides selective staining of distinct tissues. In many not-so-well-established model organisms, the absence of transgenic tools often limits *in vivo* imaging studies to a few, sometimes poorly penetrating, vital dyes. Another limitation of fluorescence microscopy is the fact that it only shows what has been fluorescently labelled; other structures remain (literally) in the dark. Structures that cannot be labeled simply cannot be visualized. It is therefore desirable to include other modalities in a light sheet instrument. By using the brightfield illumination that is typically present in a light sheet microscope, other microscopy techniques such as optical projection tomography (OPT) have been demonstrated [42]. The complementary data provides valuable insight into the state and stage of the sample.

Ideally, one would like to watch the development of a single embryo and have access to all the individual tissues, cellular compartments, and so on. Unfortunately, we currently cannot image more than three or four fluorescent colors simultaneously in a single sample due to the strong overlap of the emission spectra. Spectral detection with nanometer precision has been used to unmix several components of the full spectrum and distinguish many colors [43]. In the future, one can expect labeling and imaging techniques that offer a variety of modalities to extract as much information out of a single sample as possible.

7.11 Outlook

7.11.1 Big Data

Light sheet microscopes can record images with high spatial and temporal resolution. Consequently, the rate and amount of data generated is about three orders of magnitude higher than in conventional confocal microscopes, often running into several terabytes for a single long-term imaging experiment [6]. Setups with multiple fast, high-resolution cameras [6–8, 27] may provide quick multiview acquisition and important new information, but several-folds more data are produced. Therefore, in such setups, the number of experiments is limited by the available storage and/or transfer speed, preventing statistical analysis of image data necessary for quantitative biology. The issue of long-term storage can be tackled by data compression [8]; however, data processing, visualization, and analysis remain a challenge. As a consequence, there is a need to process, condense, and analyze light sheet microscopy data in real time.

Typical image stacks of organically shaped biological samples are cuboidal and therefore contain a lot of pixels with no information in the corners of the dataset. One way to efficiently reduce the amount of data is to crop the images as they come from the camera or, even better, selectively acquire regions of the sample where signal is expected and mask the rest. An even more efficient approach is to utilize the shape of the sample to create projections such as radial projections for the spherical early zebrafish embryos [6], cylindrical projections for Drosophila embryos [7], and other surface projections for arbitrary shaped objects [44]. These methods transform the 3D image data into 2D projections, thereby drastically reducing the amount of data while extracting maximum information from it on the fly and providing a novel way of visualizing and analyzing the data.

7.11.2 Smart Microscope: Imaging Concept of the Future

The concept of a smart microscope advocates that the microscope should decide how to best image the sample, given the experimenter's needs. It incorporates the idea of adaptive image acquisition: reading out only the relevant pixels to reduce the data stream, choosing the ROI based on prior knowledge, and identification of cellular events while imaging to adapt the FOV accordingly. Such a learning-based approach will make imaging and interpreting data much easier and comprehensible [45].

Sample health is one of the most crucial factors, as the reliability of the finding depends heavily on it. Hence, microscopes today are designed around their application, an idea aptly illustrated by the various implementations of light sheet microscopy. Ideally, samples are kept in their most natural environment and undisturbed while imaging. Noncontact imaging methods such as remote focusing using motorized or electrically tunable lenses [29] provide the best imaging conditions for keeping the sample immobile or freely swimming as need be, eliminating the need to embed sample in a stiff agarose gel. As these approaches also provide high imaging speeds, imaging of several samples in parallel becomes possible.

7.11.3 High-Throughput Imaging

Developmental processes are complex and vary significantly within a population. Obtaining a quantitative understanding of this variability and how it is dealt with to develop each embryo into a healthy individual is an important question. While it is possible to have a high sample count in other biological studies, imaging of developmental processes has so far been limited to a few samples only, owing to the exceedingly large amount of time and processing power required to perform each imaging experiment. As shown in this chapter, by using light sheet illumination instead of conventional point scanning, imaging speed increases drastically, making light sheet microscopy the preferred method for *in vivo* imaging of developmental processes. With high-speed cameras, SPIM and its advanced implementations can, in fact, acquire 2D images and even 3D volumes at a rate much faster than most biological processes. Therefore, imaging several samples would be a way to utilize this time to increase experimental throughput.

Most industrial drug screens are limited to single cells due to the need for testing thousands of compounds, slow imaging speeds, and data handling issues. Genetic screens, on the other hand, are usually performed in research labs and involve a lot of manual labor, especially for injecting drugs or plasmids prior to screening. A light sheet microscopy-based high-throughput imaging platform will expedite the entire process and take drug and genetic screening to the next level with rapid high-resolution imaging of many cells, spheroids, and small embryos such as zebrafish exposed to different compounds.

Thus, a combination of the above-mentioned features will prevent us from drowning in data, yet capturing all relevant aspects of biological processes, possibly in a high-throughput fashion. Light sheet microscopy has the potential to address these current challenges and facilitate systematic and noninvasive quantitative biology.

7.12 Summary

Light sheet microscopy is a surprisingly simple yet very potent technology. The large collection of implementations demonstrates its power, versatility, and simplicity. The main advantages of low phototoxicity and high speed offer numerous applications that are simply out of reach of conventional fluorescence microscopy techniques. In the long run, light sheet techniques will benefit from the affordability and the ease of customization: the microscope can be built "around the sample" and may thereby offer the best imaging conditions and performance for many novel and demanding applications. In some cases, however, it may be difficult to implement the two-way access for illumination and detection optics. Single-lens solutions may then prove to be superior, particularly for conventional sample preparations. In the future, we will see numerous new applications, even in novel model organisms. One thing is for sure: we have come closer to noninvasive imaging of fragile biological samples, and this may already be the key to new discoveries.

References

1 Siedentopf, H. and Zsigmondy, R. (1902) Uber sichtbarmachung und größenbestimmung ultramikoskopischer teilchen, mit besonderer anwendung auf goldrubingläser. *Ann. Phys.*, **315** (1), 1–39.
2 Tsien, R.Y. (2010) Nobel lecture: constructing and exploiting the fluorescent protein paintbox. *Integr. Biol.*, **2** (2-3), 77–93.
3 Voie, A.H., Burns, D.H., and Spelman, F.A. (1993) Orthogonal-plane fluorescence optical sectioning: three-dimensional imaging of macroscopic biological specimens. *J. Microsc.*, **170** (3), 229–236.

4 Huisken, J., Swoger, J., Del Bene, F., Wittbrodt, J., and Stelzer, E.H.K. (2004) Optical sectioning deep inside live embryos by selective plane illumination microscopy. *Science*, **305** (5686), 1007–1009.
5 Keller, P.J., Schmidt, A.D., Wittbrodt, J., and Stelzer, E.H.K. (2008) Reconstruction of zebrafish early embryonic development by scanned light sheet microscopy. *Science*, **322** (5904), 1065–1069.
6 Schmid, B., Shah, G., Scherf, N., Weber, M., Thierbach, K., Campos, C.P., Roeder, I., Aanstad, P., and Huisken, J. (2013) High-speed panoramic light-sheet microscopy reveals global endodermal cell dynamics. *Nat. Commun.*, **4**, 2207.
7 Krzic, U., Gunther, S., Saunders, T.E., Streichan, S.J., and Hufnagel, L. (2012) Multiview light-sheet microscope for rapid in toto imaging. *Nat. Methods*, **9** (7), 730–733.
8 Tomer, R., Khairy, K., Amat, F., and Keller, P.J. (2012) Quantitative high-speed imaging of entire developing embryos with simultaneous multiview light-sheet microscopy. *Nat. Methods*, **9** (7), 755–763.
9 Wu, Y., Wawrzusin, P., Senseney, J., Fischer, R.S., Christensen, R., Santella, A., York, A.G., Winter, P.W., Waterman, C.M., Bao, Z., Colón-Ramos, D.A., McAuliffe, M., and Shroff, H. (2013) Spatially isotropic four-dimensional imaging with dual-view plane illumination microscopy. *Nat. Biotechnol.*, **31** (11), 1032–1038.
10 Lorenzo, C., Frongia, C., Jorand, R., Fehrenbach, J., Weiss, P., Maandhui, A., Gay, G., Ducommun, B., and Lobjois, V. (2011) Live cell division dynamics monitoring in 3D large spheroid tumor models using light sheet microscopy. *Cell Div.*, **6** (1), 22.
11 Ritter, J.G., Spille, J.H., Kaminski, T., and Kubitscheck, U. (2010) A cylindrical zoom lens unit for adjustable optical sectioning in light sheet microscopy. *Biomed. Opt. Express*, **2** (1), 185–193.
12 Planchon, T.A., Gao, L., Milkie, D.E., Davidson, M.W., Galbraith, J.A., Galbraith, C.G., and Betzig, E. (2011) Rapid three-dimensional isotropic imaging of living cells using Bessel beam plane illumination. *Nat. Methods*, **8** (5), 417–423.
13 Kumar, A., Wu, Y., Christensen, R., Chandris, P., Gandler, W., McCreedy, E., Bokinsky, A., Colón-Ramos, D.A., Bao, Z., McAuliffe, M., Rondeau, G., and Shroff, H. (2014) Dual-view plane illumination microscopy for rapid and spatially isotropic imaging. *Nat. Protoc.*, **9** (11), 2555–2573.
14 Dodt, H.U., Leischner, U., Schierloh, A., Jährling, N., Mauch, C.P., Deininger, K., Deussing, J.M., Eder, M., Zieglgänsberger, W., and Becker, K. (2007) Ultramicroscopy: three-dimensional visualization of neuronal networks in the whole mouse brain. *Nat. Methods*, **4** (4), 331–336.
15 Mickoleit, M., Schmid, B., Weber, M., Fahrbach, F.O., Hombach, S., Reischauer, S., and Huisken, J. (2014) High-resolution reconstruction of the beating zebrafish heart. *Nat. Methods*, **11** (9), 919–922.
16 Trivedi, V., Truong, T.V., Trinh, L.A., Holland, D.B., Liebling, M., and Fraser, S.E. (2015) Dynamic structure and protein expression of the live embryonic heart captured by 2-photon light sheet microscopy and retrospective registration. *Biomed. Opt. Express*, **6** (6), 2056–2066.

17 Maizel, A., von Wangenheim, D., Federici, F., Haseloff, J., and Stelzer, E.H.K. (2011) High-resolution live imaging of plant growth in near physiological bright conditions using light sheet fluorescence microscopy. *Plant J.*, **68** (2), 377–385.
18 Scherz, P.J., Huisken, J., Sahai-Hernandez, P., and Stainier, D.Y.R. (2008) High-speed imaging of developing heart valves reveals interplay of morphogenesis and function. *Development*, **135** (6), 1179–1187.
19 Swoger, J., Verveer, P., Greger, K., Huisken, J., and Stelzer, E.H.K. (2007) Multi-view image fusion improves resolution in three-dimensional microscopy. *Opt. Express*, **15** (13), 8029–8042.
20 Verveer, P.J., Swoger, J., Pampaloni, F., Greger, K., Marcello, M., and Stelzer, E.H.K. (2007) High-resolution three-dimensional imaging of large specimens with light sheet-based microscopy. *Nat. Methods*, **4** (4), 311–313.
21 Preibisch, S., Amat, F., Stamataki, E., Sarov, M., Singer, R.H., Myers, E., and Tomancak, P. (2014) Efficient Bayesian-based multiview deconvolution. *Nat. Methods*, **11** (6), 645–648.
22 Schmid, B. and Huisken, J. (2015) Real-time multi-view deconvolution. *Bioinformatics*, **31** (20), 3398–3400.
23 Truong, T.V., Supatto, W., Koos, D.S., Choi, J.M., and Fraser, S.E. (2011) Deep and fast live imaging with two-photon scanned light-sheet microscopy. *Nat. Methods*, **8** (9), 757–760.
24 Fahrbach, F.O. and Rohrbach, A. (2010) A line scanned light-sheet microscope with phase shaped self-reconstructing beams. *Opt. Express*, **18** (23), 24 229–24 244.
25 Breuninger, T., Greger, K., and Stelzer, E.H.K. (2007) Lateral modulation boosts image quality in single plane illumination fluorescence microscopy. *Opt. Lett.*, **32** (13), 1938–1940.
26 Huisken, J. and Stainier, D.Y.R. (2007) Even fluorescence excitation by multidirectional selective plane illumination microscopy (mSPIM). *Opt. Lett.*, **32** (17), 2608–2610.
27 Chhetri, R.K., Amat, F., Wan, Y., Höckendorf, B., Lemon, W.C., and Keller, P.J. (2015) Whole-animal functional and developmental imaging with isotropic spatial resolution. *Nat. Methods*, **12** (12), 1171–1178.
28 Holekamp, T.F., Turaga, D., and Holy, T.E. (2008) Fast three-dimensional fluorescence imaging of activity in neural populations by objective-coupled planar illumination microscopy. *Neuron*, **57** (5), 661–672.
29 Fahrbach, F.O., Gurchenkov, V., Alessandri, K., Nassoy, P., and Rohrbach, A. (2013) Light-sheet microscopy in thick media using scanned Bessel beams and two-photon fluorescence excitation. *Opt. Express*, **21** (11), 13 824–13 839.
30 Vladimirov, N., Mu, Y., Kawashima, T., Bennett, D.V., Yang, C.T., Looger, L.L., Keller, P.J., Freeman, J., and Ahrens, M.B. (2014) Light-sheet functional imaging in fictively behaving zebrafish. *Nat. Methods*, **11** (9), 883–884.
31 Ritter, J.G., Veith, R., Siebrasse, J.P., and Kubitscheck, U. (2008) High-contrast single-particle tracking by selective focal plane illumination microscopy. *Opt. Express*, **16** (10), 7142–7152.

32 Gebhardt, J.C.M., Suter, D.M., Roy, R., Zhao, Z.W., Chapman, A.R., Basu, S., Maniatis, T., and Xie, X.S. (2013) Single-molecule imaging of transcription factor binding to DNA in live mammalian cells. *Nat. Methods*, **10** (5), 421–426.

33 Hama, H., Hioki, H., Namiki, K., Hoshida, T., Kurokawa, H., Ishidate, F., Kaneko, T., Akagi, T., Saito, T., Saido, T., and Miyawaki, A. (2015) ScaleS: an optical clearing palette for biological imaging. *Nat. Neurosci.*, **18** (10), 1518–1529.

34 Chung, K. and Deisseroth, K. (2013) CLARITY for mapping the nervous system. *Nat. Methods*, **10** (6), 508–513.

35 Bouchard, M.B., Voleti, V., Mendes, C.S., Lacefield, C., Grueber, W.B., Mann, R.S., Bruno, R.M., and Hillman, E.M.C. (2015) Swept confocally-aligned planar excitation (SCAPE) microscopy for high speed volumetric imaging of behaving organisms. *Nat. Photonics*, **9** (2), 113–119.

36 Reynaud, E.G., Krzic, U., Greger, K., and Stelzer, E.H.K. (2008) Light sheet-based fluorescence microscopy: more dimensions, more photons, and less photodamage. *HFSP J.*, **2** (5), 266–275.

37 Kaufmann, A., Mickoleit, M., Weber, M., and Huisken, J. (2012) Multilayer mounting enables long-term imaging of zebrafish development in a light sheet microscope. *Development*, **139** (17), 3242–3247.

38 Pampaloni, F., Berge, U., Marmaras, A., Horvath, P., Kroschewski, R., and Stelzer, E.H.K. (2014) Tissue-culture light sheet fluorescence microscopy (TC-LSFM) allows long-term imaging of three-dimensional cell cultures under controlled conditions. *Integr. Biol.*, **6** (10), 988–998.

39 Baumgart, E. and Kubitscheck, U. (2012) Scanned light sheet microscopy with confocal slit detection. *Opt. Express*, **20** (19), 21 805–21 814.

40 Fahrbach, F.O. and Rohrbach, A. (2012) Propagation stability of self-reconstructing Bessel beams enables contrast-enhanced imaging in thick media. *Nat. Commun.*, **3**, 632.

41 Chen, B.C., Legant, W.R., Wang, K., Shao, L., Milkie, D.E., Davidson, M.W., Janetopoulos, C., Wu, X.S., Hammer, J.A. III, Liu, Z., English, B.P., Mimori-Kiyosue, Y., Romero, D.P., Ritter, A.T., Lippincott-Schwartz, J., Fritz-Laylin, L., Mullins, R.D., Mitchell, D.M., Bembenek, J.N., Reymann, A.C., Böhme, R., Grill, S.W., Wang, J.T., Seydoux, G., Tulu, U.S., Kiehart, D.P., and Betzig, E. (2014) Lattice light-sheet microscopy: imaging molecules to embryos at high spatiotemporal resolution. *Science*, **346** (6208), 1257 998.

42 Bassi, A., Schmid, B., and Huisken, J. (2015) Optical tomography complements light sheet microscopy for in toto imaging of zebrafish development. *Development*, **142** (5), 1016–1020.

43 Jahr, W., Schmid, B., Schmied, C., Fahrbach, F.O., and Huisken, J. (2015) Hyperspectral light sheet microscopy. *Nat. Commun.*, **6**, 7990.

44 Heemskerk, I. and Streichan, S.J. (2015) Tissue cartography: compressing bio-image data by dimensional reduction. *Nat. Methods*, **12** (12), 1139–1142.

45 Scherf, N. and Huisken, J. (2015) The smart and gentle microscope. *Nat. Biotechnol.*, **33** (8), 815–818.

8

Localization-Based Super-Resolution Microscopy

Markus Sauer[1] and Mike Heilemann[2]

[1] University Würzburg, Department of Biotechnology and Biophysics, Am Hubland, 97074 Würzburg, Germany
[2] Goethe-University Frankfurt, Single Molecule Biophysics, Institute of Physical and Theoretical Chemistry, Max-von-Laue-Str. 7, 60438 Frankfurt, Germany

8.1 Super-Resolution Microscopy: An Introduction

Since the knowledge of their existence, the detection, manipulation, and control of atoms and molecules have been a dream of many scientists. Near-field interactions of molecules or atoms with tips have been used in the 1980s, for example, by scanning tunneling microscopy (STM), which allowed imaging surfaces at the atomic level. The development of STM in 1981 earned its inventors Gerd Binnig and Heinrich Robert, that time at IBM in Zurich, the Nobel Prize in physics in 1986. In parallel to these efforts, the optical detection of single molecules in complex condensed matter attracted considerable scientific interest. The fascination of single-molecule detection and characterization has been triggered by the vision to study the interactions of single atoms or molecules with their native environment unperturbed by tips and unobscured by ensemble averaging. Today, fluorescence microscopes are essential in biological and biomedical sciences and serve for 3D noninvasive imaging of cells and tissues. Diverse confocal and widefield optical fluorescence microscopes are in operation at most major research institutes. In addition to its high selectivity, fluorescence microscopy shows extreme sensitivity, enabling the detection and identification of individual molecules as well as the monitoring of inter- and intramolecular interactions with high temporal resolution (see Chapters 12 and 13).

However, conventional ensemble as well as single-molecule fluorescence microscopy enables only the spatial resolution of two neighboring emitting objects when they are separated by about the wavelength of the used light. That is, the ability to spatially resolve a structure has a physical limit, which is caused by the wave nature of light. Because of diffraction, focusing of light always results in a blurred spot, whose size determines the resolution. Already at the end of the nineteenth century, Abbe showed that the diffraction limit is proportional to the wavelength and inversely proportional to the angular distribution of the light observed [1]. Therefore, any lens-based microscope cannot resolve objects that are closer than half of the wavelength of light in the imaging plane, that is, for visible light in the range of ∼200 nm.

Fluorescence Microscopy: From Principles to Biological Applications, Second Edition.
Edited by Ulrich Kubitscheck.
© 2017 Wiley-VCH Verlag GmbH & Co. KGaA. Published 2017 by Wiley-VCH Verlag GmbH & Co. KGaA.

The spatial resolution limit of optical microscopy can also be explained with the point spread function (PSF). Because of the diffraction of light passing through an aperture, the fluorescence signal of a single fluorophore produces an Airy pattern in the image plane of a fluorescence microscope (Chapter 2). The dimensions of this image are much larger than those of the fluorophore itself, and are determined by the wavelength of the light and the size of the aperture. By approximating the PSF of a single emitter with a two-dimensional (2D) Gaussian function, its center of mass (i.e., the x, y coordinates) can be precisely determined, and the emitter can be localized in space.

Since the localization precision depends mainly on the number of collected photons and on the standard deviation of the PSF for negligible background noise (see Chapter 12, Section 12.4.1), individual fluorophores emitting thousands of fluorescence photons can be localized with nanometer accuracy. Thus, fluorescence imaging with one nanometer accuracy (FIONA) is possible, and was used successfully to monitor the migration of single fluorophores conjugated to myosin proteins along actin filaments [2]. However, two adjacent fluorophores that are closer than $\lambda/2$ cannot be distinguished. Consequently, they cannot be localized as single emitters. The first concepts to bypass this "natural" diffraction barrier and to measure distances between individual fluorophores that are shorter than the diffraction limit were developed in the early 1990s. These so-called ultrahigh-resolution colocalization studies used orthogonal spectroscopic properties (such as fluorescence emission wavelength or fluorescence lifetime) to separate independent fluorophores. In a similar way, stepwise photobleaching of fluorophores can be exploited for distance determination between single fluorophores that are separated by distances shorter than the diffraction limit.

> **Box 8.1 History of Super-Resolution Microscopy**
>
> At first glance, in order to achieve super-resolution the overlapping fluorescence emission (PSFs) of all emitters has to be separated. The first successful ideas that demonstrated a spatial resolution below the resolution limit employed physical or chemical concepts to confine fluorescence emission in an additional dimension, that is, time. This can be achieved either deterministically, for example, by generation of a light pattern that prevents emission of fluorophores in specific regions by transferring fluorophores into a nonfluorescent state, or in a stochastic way, for example, by allowing only a sparse population of fluorophores to emit light at any time of the experiment.
>
> The deterministic far-field super-resolution imaging methods include stimulated emission depletion (STED) [3] and structured illumination microscopy (SIM) [4]. Stochastic super-resolution imaging methods rely on the precise position determination (localization) of single emitters and are therefore often grouped in the term *single-molecule localization microscopy* (SMLM). These techniques include photoactivated localization microscopy (PALM) [5], fluorescence photoactivation localization microscopy (FPALM) [6], stochastic optical reconstruction microscopy (STORM) [7], and *direct* stochastic optical reconstruction microscopy

(dSTORM) [8]. Super-resolution optical fluctuation imaging (SOFI) analyzes temporal fluorescence fluctuations of fluorophores and can handle also samples with higher signal densities than SMLM approaches [9].

For their seminal contributions to the development of super-resolution fluorescence microcopy, the Royal Swedish Academy of Sciences awarded Eric Betzig, Stefan W. Hell, and William. E. Moerner the Nobel Prize in chemistry 2014.

In this chapter we will focus on SMLM with photoswitchable fluorophores. We will describe the historic developments and basic principles of 2D and 3D super-resolution imaging, live-cell imaging, image reconstruction, and quantification. We will furthermore discuss the advantages and limitations of the technique.

8.2 The Principle of Single-Molecule Localization Microscopy

When performing an SMLM experiment, one is commonly confronted with on/off switching of fluorescence and photobleaching. Even though continuously excited, the emission of a single fluorophore behaves as a random telegraph, that is, it shows on/off switching at random times for random durations, often denoted as *blinking*. This observation characterizes a single quantum system: two or more independently emitting fluorophores cannot switch on and off simultaneously without any synchronization.

However, these fluorescence intermittencies can be exploited advantageously to separate the fluorescence emission of two nearby fluorophores that have overlapping PSFs in the time domain. This key element of resolution enhancement (see Box 8.1) was introduced in 2005 [10] exploiting the blinking of individual semiconductor nanocrystals (quantum dots). If the blinking processes allows the spatially isolated fitting (localization) of individual PSFs, an artificial "image" consisting of individually localized points and exhibiting sub-diffraction spatial resolution may be reconstructed, which led to the term *Pointillism* [10]. This finding laid the foundation for all SMLM concepts. For precise position determination, that is, fitting of ideal PSFs to the measured photon distributions, active (fluorescent) fluorophores have to be spaced further apart than the distance resolved by the microscope. That is, only a small subpopulation of fluorophores must be fluorescent at any time of the experiment, and the majority of fluorophores have to reside in a nonfluorescent off state.

Therefore, the fundamental principle of all SMLM methods is identical: the target of interest is labeled with fluorescent probes and imaged using a widefield fluorescence microscope (Chapter 3). Stochastic transitions between fluorescent "on" and nonfluorescent "off" states of the fluorescent probes are induced by irradiation with light of appropriate wavelength, and a sequence (stack) of typically tens of thousands of images is recorded. In each of these images of the same sample, only a very sparse subset of the total fluorophore population is activated

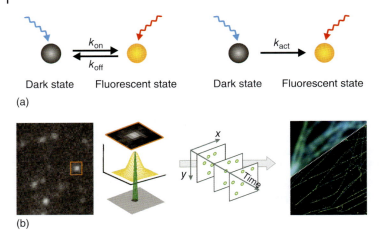

Figure 8.1 Principle of SMLM. (a) Fluorescent proteins or organic dyes can be operated as photoswitchable fluorophores. (b) The PSF of a single fluorophore can be approximated by a Gaussian function and the center of mass can be determined. From the ensemble of single-molecule localizations, an SMLM image with sub-diffraction resolution can be reconstructed.

(fluorescent) and imaged. It is therefore impossible to see any structure in any single image of the stack; however, by precisely localizing the position of each fluorophore in each image and superimposing the positions of all fluorophores, a high-resolution image can be reconstructed (Figure 8.1).

To achieve that only a sparse subset of fluorophores emits fluorescence at any time of the experiment, photoswitchable, photoactivatable, or photoconvertible synthetic fluorophores or fluorescent proteins are required. Ideally, the concentration of active fluorophores (i.e., fluorophores in the "on" state) is sufficiently low to ensure that the majority of them are spaced further apart than the resolution limit. The fluorescence of individual fluorophores is read out, and the fluorophores are photobleached or photoswitched to the "off" state before the next subset of fluorophores is activated. All localization microscopy methods share this modus operandi, differing mainly in the employed type of fluorophore. *d*STORM and related methods use standard organic fluorophores, whereas PALM employs photoactivatable fluorescent proteins (PA-FPs). The latter enable stoichiometric protein labeling with efficiencies of nearly 100%, but are limited by only a few hundred detectable photons emitted per FP before bleaching. The former are mostly used with chemical staining, which has a lower labeling efficiency, but exhibit a higher brightness of more than 1000 photons per cycle and a higher photostability, thereby permitting a higher localization precision [11].

The on/off switching of fluorescence emission can also be distinguished by the underlying physical or chemical process (e.g., photoswitching by light vs light together with chemical additives), by the reversibility, and by the nature of the involved dark state (Figure 8.2):

- *Photoactivatable/photoconvertible fluorophores*: irreversible, light-induced activation that typically involves a bond cleavage.

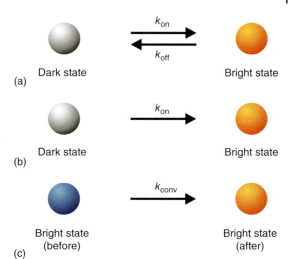

Figure 8.2 General photoswitching schemes. Fluorophores can be either (a) reversibly switched between a bright and a dark state (a, photoswitching), (b) irreversibly photoactivated from a nonfluorescent dark state to a bright state (photoactivation), or (c) converted between fluorescent states with different excitation and emission wavelengths (photoconversion). The interconversion rates (k_{on}, k_{off}, and k_{conv}) between these states usually depend on the light intensity and the chemical environment of the fluorophores.

- *Photoswitchable fluorophores*: reversible, light-induced switching usually arising from cis–trans isomerization steps or proton transfer.
- *Reaction-induced photoswitching*: non-switchable fluorophores can be operated as photoswitches in combination with chemical reactions involving extrinsic reactants (additives; e.g., through redox chemistry).

Notably, we wish to mention that a number of very similar super-resolution techniques were developed in the recent past, which also localize single emitters but achieve temporal separation by transient binding events of single fluorophores from solution (Box 8.2).

> **Box 8.2 Points Accumulation for Imaging in Nanoscale Topography (PAINT)**
>
> Points accumulation for imaging in nanoscale topography (PAINT) approaches rely on the fact that free diffusion of fluorophores in solution is too fast to be followed on a charge-coupled device (CCD) camera. Thus they appear as essentially nonfluorescent molecules. However, the fluorophores are detected once they bind or adhere to the target object and stay immobile for at least several milliseconds. This has several advantages. Because the binding process is stochastic, all fluorophores bind at different times and the fluorescence emission is separated in time. The binding rate can be adjusted by the concentration of the fluorophore in the solution. Moreover, because binding can ideally occur anywhere on the target object, the probe molecules will eventually map out the entire structure. A high-resolution image of the target object can then be reconstructed using the localized positions. The original PAINT idea [13] has recently be generalized under the acronym universal points accumulation for imaging in nanoscale topography (uPAINT) [14]. The following years have seen subsequent extensions of the principle, such as the use of transient binding of

(Continued)

> **Box 8.2 (Continued)**
>
> short fluorophore-labeled oligonucleotides to complementary oligonucleotides attached to antibodies that mark the structure of interest (DNA-PAINT) [15]. Multiplexing of the approach allows sequential imaging of multiple targets using only a single dye and a single laser source (exchange-PAINT).

8.3 Photoactivatable and Photoconvertible Probes

A large and relevant class of fluorophores for SMLM comprises fluorescent proteins, with photoactivatable green fluorescent protein (paGFP) as the first example of its kind [12]. Following maturation, paGFP exhibits a single, strong absorption band at ~400 nm in its native state but is only weakly fluorescent. Upon irradiation at 400 nm, the amplitude of the 400 nm absorption band decreases and a second and well-known absorption band of green fluorescent protein (GFP), peaking at 492 nm, appears. Mechanistically, irradiation with light at 400 nm induces the cleavage of a chemical bond and leads to decarboxylation of a glutamate residue in close vicinity to the chromophore inside paGFP. The changed nanoenvironment of the chromophore affects the photophysical properties of paGFP and transforms it into a fluorescent state. Other fluorescent proteins can undergo a light-induced photoconversion, which is an irreversible transition from one fluorescent state to another with each exhibiting different spectral properties (i.e., a different color). Examples of such proteins are Kaede, EosFP, and IrisFP (Chapter 4). To date, many other fluorescent proteins have been engineered to work as photoactivatable or photoconvertible probes.

Irreversible photoactivation is not limited to fluorescent proteins. Various groups have demonstrated efficient chemical quenching of fluorescence in organic fluorophores. In these caged fluorophores, the quencher moiety can be cleaved off by irradiation with light of a suitable wavelength (usually UV light).

8.4 Intrinsically Photoswitchable Probes

A second group of fluorophores exhibits intrinsic and reversible on/off switching by irradiation with light. Reversible photoswitching can be found both for fluorescent proteins, such as EYFP or Dronpa, and for the large group of photochromic molecules (Chapter 4). To date, intrinsically photoswitchable fluorescent proteins have been playing an important role in SMLM. Typically, two wavelengths are used for photoswitching a fluorophore between a fluorescent "on" and a nonfluorescent "off" state: the first wavelength to read out and for switching the fluorophore into a nonfluorescent state (488 nm for Dronpa, 514 nm for EYFP), and the second one for re-activation of the fluorophore (405 nm for EYFP and Dronpa). The underlying mechanisms of reversible photoswitching are intramolecular and include, for example, proton transfer and/or cis–trans isomerization.

8.5 Photoswitching of Organic Fluorophores by Chemical Reactions

A very general, and therefore attractive, approach is the use of standard organic fluorophores whose fluorescence can be modulated between an "on" and "off" state by a light-induced chemical reaction. As such, the fluorophore can be "programmed" and operated as a photoswitch, and the photoswitching properties can be tuned by the chemical additives and their concentration. A common concept to induce photoswitching of synthetic organic fluorophores is redox chemistry, typically realized by adding appropriate concentrations of reducing agents such as thiols (10–200 mM), ascorbic acid (50–100 µM), or phosphines (25 mM) to the sample buffer [11]. Here, photoswitching of organic fluorophores by millimolar concentrations of thiols in the presence (STORM) [7] or absence of a second shorter wavelength fluorophore (dSTORM) [8] is the most established technique.

A mechanistic picture of redox-induced photoswitching by thiols as basis for dSTORM was constructed from the investigation of various classes of synthetic fluorophores (Figure 8.3). It was observed that the reduction of a fluorophore in a first step leads to the formation of a radical anion (nonfluorescent), which itself is remarkably stable even in aqueous sample buffers and represents the nonfluorescent state of the photoswitch. Re-oxidation by oxygen present in the aqueous buffer may occur, restoring the fluorescent state of the photoswitch [11]. Depending on the chemical structure of the fluorophore, other nonfluorescent states are potentially involved in photoswitching. For example, carbocyanines can form adducts with thiols attacking the polymethine bridge upon prolonged irradiation, or oxazine fluorophores (e.g., ATTO655) can transit into a fully reduced state through the transfer of two electrons. Many rhodamine fluorophores (e.g., Alexa Fluor 488, Alexa Fluor 532, Alexa Fluor 568, ATTO488, ATTO532, ATTO565) form stable radical species (observed by electron paramagnetic resonance (EPR) and UV–vis spectroscopy). The stability of these radical ions in aqueous solution can reach minutes to hours. Similar to paGFP, the irradiation of the radical ion with a suitable wavelength returns the fluorophores into the fluorescent state.

8.6 Experimental Setup for Localization Microscopy

Precise position determination of single fluorophores requires high fluorescence signals and a very low background signal. Typically, an inverted fluorescence microscope is equipped with an objective with a high numerical aperture (Chapter 2). Because of their higher signal-to-noise ratio (SNR), oil-immersion objectives with NA ≥ 1.45 are often used for total internal reflection fluorescence (TIRF) microscopy (see Box 12.2 for details) that limits fluorophore excitation to a thin evanescent field (100–200 nm) close to the surface and thereby minimizes background light. The scheme can be extended to 3D or whole-cell imaging by using a widefield configuration, albeit at the cost of a lower SNR

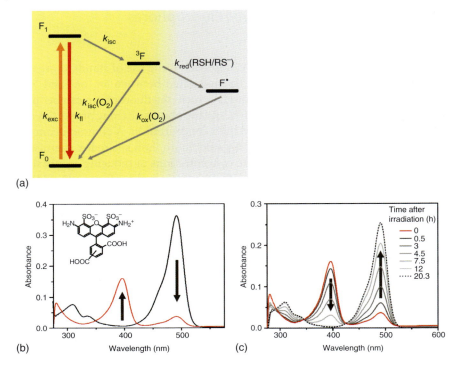

Figure 8.3 Chemically induced reversible photoswitching of organic dyes. (a) General photoswitching scheme of organic dyes. A dye that is excited with rate k_{exc} to its first singlet state F_1 can either emit fluorescence with rate k_{fl}, decay nonradiatively (not shown), or undergo intersystem crossing with rate k_{isc} to a triplet state ^3F. From the triplet state, the dye may react either with oxygen with rate $k'_{isc}(O_2)$ to return to the ground state F_0, or with a thiol with rate k_{red} (RSH/RS$^-$) to form a semireduced dye radical F$^\bullet$, which can in turn return to the ground state under the action of oxygen with rate $k_{ox}(O_2)$. (b) Absorption spectra of Alexa Fluor 488 in water in the absence (black) and in the presence of ß-mercaptoethylamine after irradiation at 488 nm for 5 min (red). (c) Absorption spectra of Alexa Fluor 488 in water in the presence of ß-mercaptoethylamine after irradiation at 488 nm for 5 min (red) and after various time points (gray curves). The dye radical exhibits a remarkable stability of many hours in aqueous solution. (Adapted from Linde et al. 2011 [16]. Reproduced with permission from Royal Society of Chemistry.).

and, consequently, lower localization precision and final image resolution. For intracellular super-resolution imaging, water-immersion objectives are often used.

The fluorophores are excited by lasers operating at a wavelength in the respective absorption spectra (e.g., at 488, 514, 532, 568, or 640 nm) to induce fluorescence and at 405 nm for photoswitching from the "off" to the "on" state. Excitation intensities vary in the range 0.5–50 kW cm^{-2} for excitation and are below 0.1 kW cm^{-2} for "off" to "on" photoswitching or photoactivation and photoconversion, respectively. The fluorescence light in the detection path is filtered using suitable bandpass filters and imaged with a sensitive camera, for example, an electron-multiplying charge-coupled device (EMCCD) camera with

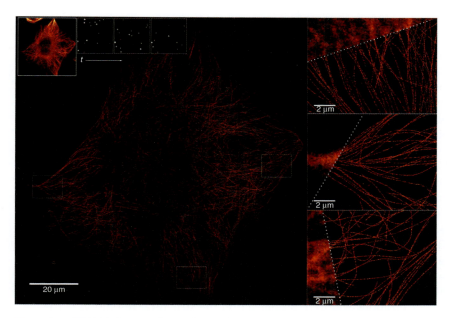

Figure 8.4 dSTORM of microtubulin obtained with an sCMOS camera. Fixed COS-7 cells were labeled with primary mouse antibodies against ß-tubulin and Alexa647-labeled goat anti-mouse secondary antibodies. dSTORM imaging was performed in 100 mM MEA (pH 8.3). Direct photoswitching of Alexa647 was performed by irradiating the fluorophores with 5 kW cm^{-2} at 641 nm. No additional activation at a shorter wavelength was used. The upper and left parts of the images show the corresponding widefield fluorescence images.

quantum yields of 80–90% in the visible range, or a scientific complementary metal–oxide–semiconductor (sCMOS) camera. To preserve most of the position information in the fluorescence signal data while reducing shot noise to a minimum, a pixel size of ∼0.4× of the Abbe resolution limit of the optical system is generally used. A typical image pixel is therefore ∼100 nm in size, which is achieved by the objective–tube lens combination and possibly a second magnification stage in the detection path. Figure 8.4 shows a dSTORM image of the microtubulin network in a mammalian COS-7 cell. This image was acquired with a sCMOS camera, which is the currently most promising detection device to image whole cells at high frame rates with high resolution.

Even though SMLM experiments using PA-FP or organic fluorophores share the same fundamental principles, the initial situation of the experiment is very different. PA-FPs are typically nonfluorescent at the beginning of the experiment, allowing the density of fluorescent molecules to be tightly controlled by the irradiation intensity of the laser used for photoactivation (often at 405 nm). In contrast, SMLM with organic fluorophores requires the efficient transfer of the fluorophores into a reversible, relatively stable "off" state at the beginning of the experiment while avoiding photodamage of the sample and photobleaching of the fluorophores. Thus, a highly reliable and, in the case of synthetic fluorophores, reversible photoswitching mechanism, a stable "off" state, and appropriate photoswitching rates are of utmost importance.

8.7 Optical Resolution and Imaging Artifacts

It is important to remember that the extractable structural information from SMLM data is determined not only by the localization precision but also by the labeling density. According to information theory, the required density of fluorescent probes has to be sufficiently high to satisfy the Nyquist–Shannon sampling theorem [17]. In essence, the theorem states that the mean distance between neighboring localized fluorophores (the sampling interval) must be at least twice as fine as the desired structural resolution. In order to resolve structural features of 20 nm in one dimension, a fluorophore must be localized at least every 10 nm. Considering a 2D structure, a labeling density of about 10^4 fluorophores per μm^2 or about 600 fluorophores within a diffraction-limited region is required. In order to detect individual fluorophores, only one fluorophore at a given time should reside in its fluorescent state (Figure 8.5).

This implies that the lifetime of the "off" state has to be substantially longer than the lifetime of the "on" state or, in other words, the photoswitching ratio $r = k_{off}/k_{on}$ has to be high enough to minimize false multiple-fluorophore localizations (Figure 8.5). If the fluorophores are arranged along a single filament, false localizations will have no impact on the structural resolution perpendicular to filament extension. In the case of two adjacent fluorophores on crossing filaments, the generation of a false localization will affect the ability to resolve both filaments independently. Consequently, false localizations due to inappropriate photoswitching rates or too high fluorophore densities can give rise to the appearance of fluorophore clusters instead of revealing, for example, crossing cytoskeletal filaments that are possibly loaded with cargo vesicles (Figure 8.5).

In practice, the density of fluorescent spots should be well below 1 emitter per square micrometer to enable reliable spot-finding and fitting. If, for any reason, the emitter density increases (e.g., to improve the temporal resolution in live-cell experiments), the number of overlapping PSFs increases as well. These

Figure 8.5 Effect of photoswitching kinetics and labeling densities on localization microscopy. (a) Crossing filaments are labeled with fluorophore-labeled primary antibodies. Every filament consists of a line labeled with antibodies. The circle surrounding the filaments has a diameter of 250 nm, corresponding approximately to the full-width at half-maximum (FWHM) of a PSF detected for a single emitting fluorophore. From all antibodies or rather fluorophores (assuming a degree of labeling of ∼1) residing in the PSF area, only one is allowed to be fluorescent at any time of the experiments in order to generate regular localizations, that is, localizations that represent the actual position of fluorophores on the structure (b). If two (c) or more (d) fluorophores reside simultaneously in the "on" state, the approximation of the PSF of multiple emitters yields a false position (red) that does not correspond to the regular physical position (blue). Such overlapping multiple-fluorophore PSFs result in false localizations and affect the ability to resolve independent filaments. (e–h) Simulation of the influence of different photoswitching ratios $r = k_{off}/k_{on}$ as possible source of false localizations on the ability to resolve crossing filaments (2 × 2 μm per image) labeled with a fluorophore every 40 nm. Only for a high photoswitching ratio of $r = 1000$, for example, the fluorophores stay on average for 1 ms in the "on" state and 1 s in the "off" state, the intersecting area is clearly resolved. For lower ratios, for example, $r = 10–30$, false localizations affect the presence of a fluorophore aggregate or cluster (f). Permission required from Sauer [41].

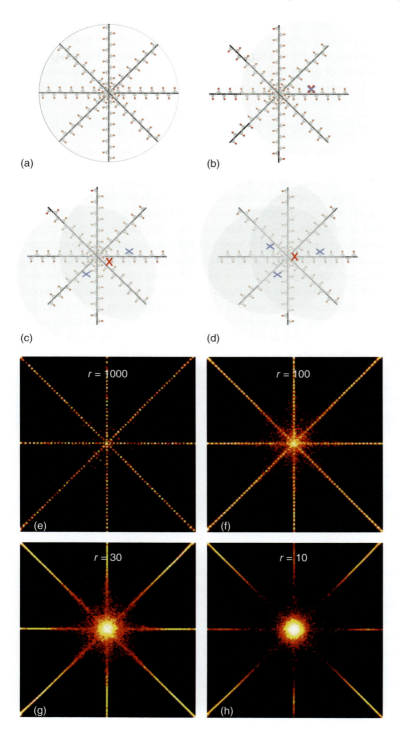

localization errors are included as artifacts in the reconstructed super-resolution image and can lead to unresolved features and misinterpretation. A possible solution are algorithms such as daoSTORM and compressed sensing that can process fluorophore densities of up to 10 emitters per square micrometer with tolerable error rates.

> **Box 8.3 Experimental Approaches to Determine the Spatial Resolution in SMLM Experiments**
>
> The spatial resolution of a fluorescence microscope depends on both optical and labeling parameters (by the Nyquist–Shannon sampling theorem [17]). The main optical parameter that determines the spatial resolution in an SMLM experiment is the number of photons detected per single molecule emission spot. The photon number is related to the precision at which the center of mass of the fluorophore can be determined, also termed *localization precision*, σ. Other optical parameters are the pixel size and background counts. The localization precision σ itself – given a sufficient spatial density of fluorescent labels that fulfills the sampling theorem – can be used for a rough estimate of the spatial resolution, res $= 2.35\sigma$. The localization precision of an SMLM experiment is determined using one of the following approaches: the first one is based on calculating the localization precision from the photon counts emitted by single fluorophores. However, this approach does not take into account specific features of the microscopic setup that might influence the localization precision. Localization precision can also be determined directly from single-molecule localizations and from the distribution of nearest neighbor positions in adjacent frames [18]. Another experimental approach is to calculate the Fourier ring correlation [19], an approach that was introduced first for electron microscopy.

8.8 Fluorescence Labeling for Super-Resolution Microscopy

8.8.1 Label Size versus Structural Resolution

Since fluorescence labeling is treated in detail in Chapter 4, the following considerations focus explicitly on aspects relevant for SMLM. Because the density of fluorophores controls the achievable structural resolution, efficient and specific labeling with small fluorescent probes is another decisive aspect of super-resolution imaging. PA-FPs are doubtless the labels of choice for live-cell SMLM because they can be genetically fused to the protein of interest. In the ideal case, each protein carries a single fluorophore and shows wild-type functionality. With a mass of ~27 kDa, a barrel-like structure, and a size of 2–5 nm (Figure 8.6), a fluorescent protein can, however, perturb protein functionality. Furthermore, besides a few exceptions, they do not have a direct extension to other classes of biomolecules, including lipids, nucleic acids, glycans, and secondary metabolites, and have less than optimal fluorescent properties.

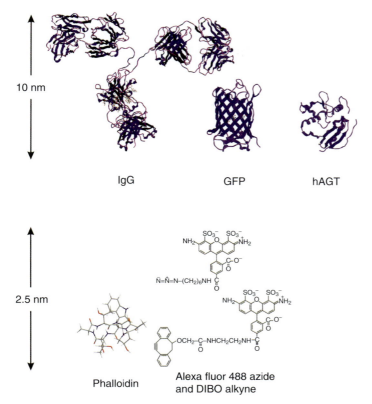

Figure 8.6 Different sizes of fluorescent labels. The green fluorescent protein (GFP) is a protein composed of 238 amino acid residues (2–5 nm, 27 kDa) and exhibits bright green fluorescence when exposed to light in the blue to ultraviolet range. Immunoglobulin G (IgG) antibodies are large molecules (~10 nm, 150 kDa) composed of four peptide chains. IgG are commonly used in immunocytochemistry applications to target specific peptide or protein antigens in cells. The SNAP tag uses the human DNA repair enzyme O6-alkylguanine-DNA alkyltransferase (hAGT). hAGT has a mass of 20 kDa and, similar to FPs, is genetically attached to a protein of interest. Labeling is achieved by using fluorophore-labeled O6-benzylguanine (BG) substrates that are processed by hAGT, which first cleaves off the BG unit and then covalently binds the fluorophore. Phalloidin, a toxin from the death cap (*Amanita phalloides*), binds F-actin and prevents its depolymerization. The bicyclic heptapeptide has a molecular mass of 789 g mol^{-1} and is ~4 times smaller than GFP and hAGT. The 1,2,3-triazole linkage (click chemistry) between a fluorophore modified as azide and reaction partner (e.g., protein) modified as alkyne, or vice versa, is extremely stable. Classic click reactions comprise a copper-catalyzed azide–alkyne cycloaddition to label molecules of interest. To avoid cytotoxic effects of Cu(I), DIBO (dibenzycyclooctyne) fluorophore derivatives have been developed. The strain in the eight-membered ring allows the reaction with azide-modified molecules in the absence of Cu(I). Permission required from Sauer [41].

Organic fluorophores such as rhodamine or carbocyanine dyes are substantially smaller (~1 nm) (Figure 8.6). They exhibit a higher fluorescence quantum yield and improved photostability, but require a chemical labeling procedure (see also Chapter 4). For fixed cells, immunofluorescence is the method of choice using primary and secondary antibodies. The drawback, however, is the size of the commonly used immunoglobulin G antibodies of ~10 nm (Figure 8.6), which can deteriorate epitope recognition and increase the apparent size of the actual structure. To reduce the size of the label, direct labeling of the primary antibody or the smaller fluorophore-labeled camelid antibodies (nanobodies) directed against GFP can be used. Some cellular structures can be targeted with small and specific labels, for example, phalloidin for actin and paclitaxel for microtubulin (Figure 8.6).

8.8.2 Live-Cell Labeling

Live-cell SMLM requires a method for specific and stoichiometric labeling of proteins in living cells with organic fluorophores that can be switched between an "on" and "off" state in the cellular redox environment. As living cells naturally contain the reducing glutathione disulfide–glutathione couple (GSH/GSSG) at millimolar concentrations in the cytosol, localization microscopy by dSTORM is feasible with many Alexa Fluor and ATTO dyes [11]. Post-translational labeling of proteins in living cells can be achieved using genetically encoded polypeptide tags in combination with organic fluorophore ligands [20]. However, most of these methods still use large protein tags comparable to the size of fluorescent proteins, which can sterically interfere with protein function, such as the HaloTag with a mass of 35 kDa, SNAP/CLIP tags with a mass of 20 kDa, and the DHFR/TMP tag with a mass of 18 kDa [20]. So far, different combinations of SNAP, Halo, TMP, CLIP tags, and organic fluorophores have also been successfully used in combination with PA-FPs for multicolor live-cell localization microscopy.

8.8.3 Click Chemistry

Another currently emerging labeling method that is ideally suited for high-density labeling of target structures involves the incorporation of a unique chemical functionality into a target molecule using the cell's own biosynthesis machinery [21]. This so-called bioorthogonal "click chemistry" has led to an explosion of interest in the selective covalent labeling of molecules in cells and living organisms. Originally, click chemistry relied on a copper-catalyzed azide–alkyne cycloaddition, whereas copper-free methods are nowadays also available (Figure 8.6). Owing to the small size of azide and alkyne tags, modified amino acids, monosaccharides, nucleotides, or fatty acids can be metabolically incorporated by living cells or organisms with high efficiency. It should be noted that click chemistry, just as any other synthetic labeling method, has the drawback that the labeling efficiency may not be known. Nevertheless, click chemistry has been successfully used for dSTORM imaging of cell surface glycans [22] (Figure 8.7) and nascent DNA and RNA [23].

For specific labeling of proteins by click chemistry, a method based on enzyme-catalyzed probe ligation, called *probe incorporation mediated by enzymes* (PRIME) [24], can be used. The central component of the PRIME method is an engineered lipoic acid ligase (LplA) from *Escherichia coli* that catalyzes the covalent tagging of the desired probe to a 13-amino-acid recognition sequence called the LplA acceptor peptide (LAP). In the first step, the genetically fused LAP is ligated site-specifically with 10-azidodecanoic

Figure 8.7 dSTORM imaging of cell membrane glycans. Membrane glycans of SK-N-MC neuroblastoma cells were stained via metabolic incorporation of azido-sugar analogs (Ac_4GalNAz) and copper-catalyzed azide–alkyne cycloaddition (CuAAC). At an Alexa Fluor 647 alkyne concentration of about 20 µM, the number of localizations per membrane area as determined by dSTORM saturates. Quantifications of basal and apical cell surface localizations estimates the presence of about 5–10 Mio glycans per cell membrane.

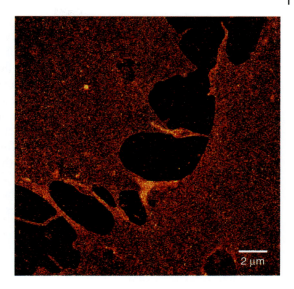

acid in cells, and in the second step fluorophore-modified cyclooctynes react with the azide and form the corresponding aza-dibenzocyclooctyne (ADIBO). Alternatively, intracellular proteins can be specifically labeled with, for example, fluorophore-tetrazine derivatives, following the incorporation of genetically encoded unnatural amino acids (UAAs) [25].

8.8.4 Three-Dimensional SMLM

To unravel 3D information about cellular structures and molecular architectures in cells, the axial symmetry of the PSF in SMLM has to be broken. In conventional microscopy (in the case of zero aberrations), the PSFs of a fluorophore slightly above and below the image plane appear equally blurred, preventing the extraction of the accurate axial fluorophore position. Various methods that tweak the signal in a way that the PSFs above and below the image plane look different have been developed. Some of the ideas have been already successfully used in single-particle tracking. With the ability to image only a few fluorophores defining a densely labeled structure at a time due to temporal separation of fluorophore emission, it was straightforward to use these established methods also for localization microscopy. Methods used to infer information about the axial origin of a signal can roughly be split into four groups: (i) astigmatism, (ii) biplane, (iii) PSF splitting into lobes, and (iv) interferometric approaches [26].

8.8.5 Astigmatic Imaging

The astigmatism approach introduces a cylindrical lens into the detection path of the microscope system. This leads to an elongation of the PSF in one lateral dimension because only one spatial direction is tightly focused while the other is defocused. In practice, there is one point of equal PSF widths (which can be set to be $z = 0$ nm), and PSFs originating from above or below are elongated in the x- and y-direction, respectively. The rough axial position can be extracted from the orientation of the PSF, and the widths in *x* and *y* can be calibrated to yield exact *z*-coordinates. After the initial use in single-particle tracking, the method has been broadly applied to SMLM. For calibration of the defocusing behavior, usually single fluorophores, quantum dots, or small fluorescent beads are adsorbed on a bare coverslip and moved in the *z*-direction while recording their PSF. The obtained PSF widths in *x* and *y* at different axial positions are fitted to a model polynomial of second order, which represents a physically derived model, or a fourth-order polynomial to account for imperfections in the optical system. To avoid fitting of a more or less physically derived function to the calibration data, a look-up table can be created for extraction of the actual axial position. In the open-source QuickPALM plugin for ImageJ, the standard deviations of the calibration PSF in *x*- and *y* are determined, and the known *z*-position is plotted against $\sigma_x - \sigma_y$. The obtained straight line serves to look up the *z*-coordinates during the course of a measurement.

8.8.6 Biplane Imaging

In the biplane imaging approach, the detected signal is split into two parts with an optical delay added to one of the two signals. Both signal paths can be either imaged on the same camera or, if a second camera is at hand, the two signals can be detected separately. In the second scheme, the full field of view is preserved. In both approaches, a relative offset in the axial position between both image planes has to be realized. The axial position can be extracted from the sharpness ratio of the respective planes. The PSF widths detected on detectors A and B encode the *z*-coordinate just as the ratio between the *x* and *y* PSF widths using the astigmatism approach.

8.8.7 Double Helix PSF

Another approach of 3D SMLM uses additional optics to split the PSF of a fluorophore into two lobes that rotate around each other with respect to the axial position of the emitter. It was first proposed to engineer the PSF to a double helix by the introduction of a phase mask, which is created by a polarization-sensitive spatial light modulator (SLM). The two resulting lobes rotate around each other in the form of a double-helix point spread function (DH-PSF), which baptized the technique. It was later proposed to use a simplified approach adding a phase ramp in the objective aperture to half of the signal instead of an SLM in the light path. In this method, termed *phase ramp imaging localization microscopy* (PRILM), in contrast to DH-PSF, one lobe remains fixed while the other rotates around it. Both techniques have in common that the available axial range is larger than in astigmatism or biplane imaging.

8.8.8 Interferometric Imaging

In the interferometric iPALM approach, the sample is placed between two objectives and thus the fluorescence light emitted from a single fluorophore travels along two different light paths. Exploiting the wave–particle duality, upon superposition of the respective signals with a three-way beam splitter, every single photon will interfere with itself. Depending on the axial position, that is, whether it is closer to the objective A or B, the fluorophore will show a distinct interference that gives information about the z-coordinate. In good agreement with the subwavelength sensitivity, interferometry so far yields the best axial localization precision but is limited in axial range to about half the emitted wavelength.

8.9 Measures for Improving Imaging Contrast

Most 3D SMLM approaches use high-NA oil-immersion objectives for surface-confined TIRF irradiation. Another very popular irradiation mode for intracellular imaging only available with oil-immersion is highly inclined and laminated optical (HILO) sheet microscopy [27]. In HILO, the sample is illuminated by a highly inclined and thin beam, allowing high irradiation intensities combined with high SNR even inside cells.

Using oil-immersion objectives for imaging while measuring in an aqueous buffer induces aberrations because of the different refractive indices. As a consequence, the volume image of a sample will appear stretched along the optical axis, an effect well described for both confocal microscopy and SMLM. The image stretching can be accounted for experimentally or analytically by rescaling with the factor $\eta_{buffer}/\eta_{immersion}$. Furthermore, the refractive index mismatch leads to spherical aberration along the z-axis, which is more difficult to correct. Both types of aberrations can be avoided by adjusting the refractive index of the imaging buffer [28] or by using objectives that use immersion media with matching refractive indices (glycerin immersion for tissue imaging or water immersion for adherent cells).

The time needed to fully image a large 3D volume is significantly longer than for one single 2D snapshot because several layers have to be imaged. With longer imaging times, setups are more prone to thermal or mechanical drift of the sample, which impacts data interpretation. Sample drift can be compensated, for example, by adding fiducial markers to the sample chamber. Their position can be recorded in each single image, and tracking over the time of the experiment allows correcting for sample drift. 3D drift correction with fiducial markers requires them to be immobilized within the 3D space of the sample. This can be achieved using a hydrogel with a refractive index close to that of water, where 100 nm fiducial markers were shown to be stably immobilized and homogeneously distributed in three dimensions even in close vicinity of cells [29]. Multicolor fiducial SMLM experiments additionally profit from the use of multispectral fiducial markers as reference signal for the registration of different spectral images.

8.10 SMLM Software

Different from other super-resolution techniques, SMLM requires processing of raw image data to generate an image. This first step of image processing aims to identify the emission patterns of single fluorophores and to determine their positions with high precision. The ease and speed of data acquisition afforded by the various SMLM methods increase the computational requirement concerning fast and reliable data processing and image reconstruction. The general problem of processing stochastic SMLM data is characterized by a huge volume of data, typically in the range of gigabytes, uncertainty in the exact size of the PSF, high background noise as common in single-molecule experiments, and stochastically occurring insufficient spatial separation between simultaneously emitting fluorophores. While efficient software for data analysis was a bottleneck in most laboratories in the early days of SMLM, different efficient software packages are currently available (Box 8.4).

> **Box 8.4 Software for SMLM**
>
> QuickPALM is an ImageJ-based plugin for super-resolution by Henriques *et al.* [32]. While fairly fast, it is based on center-of-mass computations, which are known to be quite suboptimal for particle localization [33]. However, practical results with QuickPALM are usually acceptable, and the method has the advantage of having only a few free parameters. Features include 2D and 3D localization and embedded laser control software. As it is based on ImageJ, it allows, on one hand, a wide range of image formats to be used and a good integration into a huge tool database but, on the other, is hard to automate and shares the harsh memory requirements of ImageJ, which can only process files that fully fit into the computer's main memory. QuickPALM is available as free open-source software (FOSS) from http://code.google.com/p/quickpalm/.
>
> Palm3d is a Python program [34] based on cross-correlating prerecorded instances of the PSF with the located image data and has the advantage of being able to work with a very aberrated PSF, but suffers from the distinct disadvantage of necessary lengthy calibration and of lower precision than full Gaussian fitting [33]. Palm3d can perform 2D and 3D localization. The computation time is acceptable for slow acquisition methods, and the program is very concise and sufficiently documented to allow easy extension. Palm3d is available as FOSS from http://code.google.com/p/palm3d/.
>
> LivePalm is a MATLAB suite [35] based on the fluoroBancroft algorithm, which is, according to the tests published with the program, only slightly less precise than Gaussian fitting. LivePalm can only perform 2D localization. The runtime seems to be acceptable, but the program depends on the proprietary and expensive MATLAB software suite and is itself closed source, thus making further analysis complicated. LivePalm is available from the supporting information of the article by Hedde *et al.*
>
> rapidSTORM is a standalone program [36] that employs Gaussian least-squares fitting and makes this computation (which has often been considered to be too expensive) feasible and even real-time-capable. From the mentioned programs, it seems to be the largest and by far most complete and configurable analysis suite; can perform 2D, 3D, and multicolor localization; and has a large number of additional features. However, its size, complexity of infrastructure, and build

process require in-depth knowledge for further extension or adaptation, and the learning curve is fairly steep. rapidSTORM is available as FOSS from http://www.super-resolution.biozentrum.uni-wuerzburg.de/home/rapidstorm/.

daoSTORM is a Python program [37] based on the Image Reduction and Analysis Facility (IRAF). It adapts the Daophot module of IRAF, which specializes in astronomical crowded-field photometry, and applies it to microscopy data. daoSTORM is specialized on such crowded-field acquisitions, that is, those with high fluorophore densities, and a prototype, and consequently not the best choice for more sparse images. However, the code is concise and documented and should be easily extensible. daoSTORM can only perform 2D localization and has quite a long runtime. daoSTORM is available as FOSS from the supporting information of the article by Holden *et al.* [37].

8.11 Reference Structures for SMLM

To judge the plausibility of localization microscopy data, robust test samples with well-defined molecular structure and composition have been introduced in the last years. Such test samples can also be used advantageously for calibration of super-resolution imaging instruments and optimization of experimental photoswitching conditions (e.g., irradiation wavelength and intensity, imaging buffer, and pH of the solvent) [30]. Among the various natural biological samples, the nuclear pore complex (NPC) with its well-defined symmetry and molecular composition as well as its dimensions and high density in the nuclear membrane takes a special position. Thus, the NPC can be used not only to calibrate microscopes but also to test the achievable resolution of different super-resolution imaging methods by resolving, for example, the central channel of the NPC with a diameter of ~40 nm.

Centrosomes, composed of two orthogonally arranged centrioles each consisting of nine triplet microtubules or pre- and postsynaptic proteins separated by the synaptic cleft, can also be used as natural biological calibration samples. Artificial calibration samples for 2D and 3D localization microscopy include DNA origami, single-molecule assembled patterns generated by cut-and-paste technology (SMCP), and DNA bricks.

Further cellular reference structures are microtubulin filaments, which are highly conserved in their structure and assemble into tubes with a diameter of ~25 nm. In addition, excellent and specific antibodies for tubulin labeling are commercially available. In the early days of super-resolution microscopy, it was the width of such cellular filaments that was used to demonstrate a resolution enhancement. However, this measure is prone to errors due to applied thresholds (for deterministic super-resolution methods such as SIM, STED), or a low localization statistics (for stochastic super-resolution methods, i.e., SMLM). A better measure is to visualize the projection of the labeled filament: because of its tubular structure, a microtubulin filament that is imaged perpendicular to the optical axis (within the imaging plane) will exhibit higher intensities

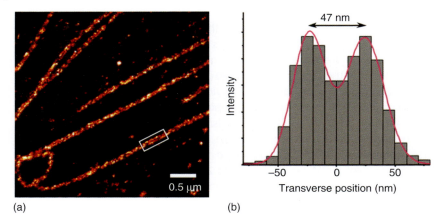

Figure 8.8 dSTORM imaging of microtubulin labeled with primary and secondary antibody. The super-resolution image (a) shows the appearance of two "lines" for some microtubulin filaments, which result from a projection effect in imaging. The intensity histogram (b) shows a distance between the peak maxima of 47 nm.

close to the walls of the filament (Figure 8.8). With a sufficiently high spatial resolution, the distance between these intensity maxima can be visualized, and was shown to vary between ∼40 nm (antibody labeling) [18] and <20 nm for direct labeling of tubulin monomers [31].

8.12 Quantification of SMLM Data

Since SMLM builds images from single-molecule localization events, it contains intrinsic information about the number of molecules present in a certain area. In the case where each target molecule is labeled with an intact fluorophore and the fluorophore is detected, the number of localizations measured per individual fluorophore is accessible. As such, SMLM is potentially ideally suited not only to resolve cellular structures with a near-molecular resolution but also to quantify by counting. Relevant applications are, for example, counting the number of subunits of receptor multimers [38], or determining the stoichiometry of protein clusters. Quantitative SMLM thereby makes use of the emission properties of a single fluorophore, which behaves as a single quantum system and, for example, exhibits digital, one-step photobleaching. In theory, each fluorescent spot appearing in an image sequence recorded during an SMLM experiment could be counted as a single fluorophore; however, and because of the photophysical properties such as reversible blinking, a careful correction of the emission pattern is needed, and calibration measurements are required.

In the case of PA-FP, the situation appears more intuitive and simpler: single-step photoactivation should highlight the fluorophore only once, and stoichiometric tagging of a protein of interest is possible with genetic tools. Using low photoactivation rates, ideally only one fluorescent protein is photoactivated, the fluorescence signal read out, and the molecule photobleached. Blinking occurs, however, to a much lower extent as observed for synthetic fluorophores. As blinking leads to multiple detection events and overcounting, a variety of algorithms were developed that correct for this phenomenon. Quantitative PALM allows protein counting and was, for example, used to quantify the number and cluster distribution of the RNA polymerase in *E. coli* cells [39].

Organic fluorophores are certainly only the second choice for protein quantification because of experimental challenges such as stoichiometric labeling and multiple detection events per fluorophore because of reversible photoswitching. On the other hand, they are typically much brighter and photostable than fluorescent proteins, so that the fraction of fluorophores that are detected is higher. Since the photophysical properties of fluorophores depend on the chemical nature of their nanoenvironment, reference experiments are required that allow correcting for multiple detection events [40].

8.13 Summary

SMLM achieves the highest spatial resolution in fixed and living cells at the cost of time, that is, SMLM is slow compared to STED and SIM.

The achievable resolution is mainly determined by photon statistics.

Various methods for 3D SMLM have been developed, achieving axial resolutions of a few to a few tens of nanometers.

To check the reliability of SMLM, different references structures such as the NPC have been introduced, which enable inspection of resolution, labeling efficiency, and chromatic aberration correction.

SMLM provides intrinsic single-molecule information and can provide quantitative data on protein copy numbers, clusters, and interactions.

References

1 Abbe, E. (1873) Beiträge zur Theorie des Mikroskops und der mikroskopischen Wahrnehmung. *Arch. Mikr. Anat.*, **9**, 413–420.
2 Yildiz, A. *et al.* (2003) Myosin V walks hand-over-hand: single fluorophore imaging with 1.5-nm localization. *Science*, **300** (5628), 2061–2065.
3 Hell, S.W. and Wichmann, J. (1994) Breaking the diffraction resolution limit by stimulated emission: stimulated-emission-depletion fluorescence microscopy. *Opt. Lett.*, **19** (11), 780–782.
4 Gustafsson, M.G. (2000) Surpassing the lateral resolution limit by a factor of two using structured illumination microscopy. *J. Microsc.*, **198**(Pt. 2), 82–87.
5 Betzig, E. *et al.* (2006) Imaging intracellular fluorescent proteins at nanometer resolution. *Science*, **313** (5793), 1642–1645.
6 Hess, S.T., Girirajan, T.P., and Mason, M.D. (2006) Ultra-high resolution imaging by fluorescence photoactivation localization microscopy. *Biophys. J.*, **91** (11), 4258–4272.
7 Rust, M.J., Bates, M., and Zhuang, X. (2006) Sub-diffraction-limit imaging by stochastic optical reconstruction microscopy (STORM). *Nat. Methods*, **3** (10), 793–795.

8 Heilemann, M. et al. (2008) Subdiffraction-resolution fluorescence imaging with conventional fluorescent probes. *Angew. Chem. Int. Ed.*, **47** (33), 6172–6176.
9 Dertinger, T. et al. (2009) Fast, background-free, 3D super-resolution optical fluctuation imaging (SOFI). *Proc. Natl. Acad. Sci. U.S.A.*, **106** (52), 22287–22292.
10 Lidke, K.A. et al. (2005) Superresolution with quantum dots: enhanced localization in fluorescence microscopy by exploitation of quantum dot blinking. *Biophys. J.*, **88** (1), 346a–346a.
11 van de Linde, S. et al. (2011) Direct stochastic optical reconstruction microscopy with standard fluorescent probes. *Nat. Protoc.*, **6** (7), 991–1009.
12 Patterson, G.H. and Lippincott-Schwartz, J. (2002) A photoactivatable GFP for selective photolabeling of proteins and cells. *Science*, **297** (5588), 1873–1877.
13 Sharonov, A. and Hochstrasser, R.M. (2006) Wide-field subdiffraction imaging by accumulated binding of diffusing probes. *Proc. Natl. Acad. Sci. U.S.A.*, **103** (50), 18911–18916.
14 Giannone, G. et al. (2010) Dynamic superresolution imaging of endogenous proteins on living cells at ultra-high density. *Biophys. J.*, **99** (4), 1303–1310.
15 Jungmann, R. et al. (2010) Single-molecule kinetics and super-resolution microscopy by fluorescence imaging of transient binding on DNA origami. *Nano Lett.*, **10** (11), 4756–4761.
16 van de Linde, S. et al. (2011) Photoinduced formation of reversible dye radicals and their impact on super-resolution imaging. *Photochem. Photobiol. Sci.*, **10** (4), 499–506.
17 Shannon, C.E. (1949) Communication in the presence of noise. *Proc. IRE*, **37** (1), 10–21.
18 Endesfelder, U. et al. (2014) A simple method to estimate the average localization precision of a single-molecule localization microscopy experiment. *Histochem. Cell Biol.*, **141** (6), 629–638.
19 Nieuwenhuizen, R.P. et al. (2013) Measuring image resolution in optical nanoscopy. *Nat. Methods*, **10** (6), 557–562.
20 van de Linde, S., Heilemann, M., and Sauer, M. (2012) Live-cell super-resolution imaging with synthetic fluorophores. *Annu. Rev. Phys. Chem.*, **63**, 519–540.
21 Prescher, J.A. and Bertozzi, C.R. (2005) Chemistry in living systems. *Nat. Chem. Biol.*, **1** (1), 13–21.
22 Letschert, S. et al. (2014) Super-resolution imaging of plasma membrane glycans. *Angew. Chem. Int. Ed.*, **53** (41), 10921–10924.
23 Raulf, A. et al. (2014) Click chemistry facilitates direct labelling and super-resolution imaging of nucleic acids and proteins. *RSC Adv.*, **4** (57), 30462–30466. doi: 10.1039/c4ra01027b
24 Uttamapinant, C. et al. (2010) A fluorophore ligase for site-specific protein labeling inside living cells. *Proc. Natl. Acad. Sci. U.S.A.*, **107** (24), 10914–10919.
25 Liu, C.C. and Schultz, P.G. (2010) Adding new chemistries to the genetic code. *Annu. Rev. Biochem.*, **79**, 413–444.

26 Klein, T., Proppert, S., and Sauer, M. (2014) Eight years of single-molecule localization microscopy. *Histochem. Cell Biol.*, **141** (6), 561–575.
27 Tokunaga, M., Imamoto, N., and Sakata-Sogawa, K. (2008) Highly inclined thin illumination enables clear single-molecule imaging in cells. *Nat. Methods*, **5** (2), 159–161.
28 Huang, B. *et al.* (2008) Whole-cell 3D STORM reveals interactions between cellular structures with nanometer-scale resolution. *Nat. Methods*, **5** (12), 1047–1052.
29 Zessin, P.J.M. *et al.* (2013) A hydrophilic gel matrix for single-molecule super-resolution microscopy. *Opt. Nanosc.*, **2** (1), 4.
30 Endesfelder, U. and Heilemann, M. (2014) Art and artifacts in single-molecule localization microscopy: beyond attractive images. *Nat. Methods*, **11** (3), 235–238.
31 Vaughan, J.C., Jia, S., and Zhuang, X. (2012) Ultrabright photoactivatable fluorophores created by reductive caging. *Nat. Methods*, **9** (12), 1181–1184.
32 Henriques, R. *et al.* (2010) QuickPALM: 3D real-time photoactivation nanoscopy image processing in ImageJ. *Nat. Methods*, **7** (5), 339–340.
33 Cheezum, M.K., Walker, W.F., and Guilford, W.H. (2001) Quantitative comparison of algorithms for tracking single fluorescent particles. *Biophys. J.*, **81** (4), 2378–2388.
34 York, A.G. *et al.* (2011) Confined activation and subdiffractive localization enables whole-cell PALM with genetically expressed probes. *Nat. Methods*, **8** (4), 327–333.
35 Hedde, P.N. *et al.* (2009) Online image analysis software for photoactivation localization microscopy. *Nat. Methods*, **6** (10), 689–690.
36 Wolter, S. *et al.* (2010) Real-time computation of subdiffraction-resolution fluorescence images. *J. Microsc.*, **237** (1), 12–22.
37 Holden, S.J., Uphoff, S., and Kapanidis, A.N. (2011) daoSTORM: an algorithm for high-density super-resolution microscopy. *Nat. Methods*, **8** (4), 279–280.
38 Fricke, F. *et al.* (2015) One, two or three? Probing the stoichiometry of membrane proteins by single-molecule localization microscopy. *Sci. Rep.*, **5**, 14072.
39 Endesfelder, U. *et al.* (2013) Multiscale spatial organization of RNA polymerase in Escherichia coli. *Biophys. J.*, **105** (1), 172–181.
40 Ehmann, N. *et al.* (2014) Quantitative super-resolution imaging of Bruchpilot distinguishes active zone states. *Nat. Commun.*, **5**, 4650.
41 Sauer, M. (2013) Localization microscopy coming of age: From concepts to biological impact. *J. Cell Sci.*, **126**, 3505–3513.

9

Super-Resolution Microscopy: Interference and Pattern Techniques

Udo Birk[1,2,3], Gerrit Best[3,4], Roman Amberger[3], and Christoph Cremer[1,2,3]

[1] Institute of Molecular Biology GmbH (IMB), Ackermannweg 4, 55128 Mainz, Germany
[2] Mainz University, Department of Physics, Mathematics and Computer Science, Institute of Physics, Staudingerweg 9, 55128 Mainz, Germany
[3] Heidelberg University, Applied Optics and Information Processing, Kirchhoff-Institute for Physics, Im Neuenheimer Feld 227, 69120 Heidelberg, Germany
[4] Heidelberg University Hospital, Department of Ophthalmology, Im Neuenheimer Feld 400, 69120 Heidelberg, Germany

9.1 Introduction

In many scientific fields such as biology, medicine, and material sciences, microscopy has become a major analytical tool. Electron microscopy (EM) with nanometer resolution, on one hand, and light microscopy with a broad applicability, on the other, have allowed groundbreaking scientific achievements. Even though EM delivers unmatched resolution, light microscopy has never lost its relevance. Because of the vast popularity of fluorescent labeling techniques in recent years, the method of fluorescence microscopy has actually become one of the most important imaging techniques in the life sciences.

However, the intrinsically limited resolution of standard fluorescence microscopy compared to EM and X-ray microscopy is still a major drawback, as many biological specimens are of a size in the nanometer to micrometer range. Therefore, the standard light-microscopic resolution defined by Rayleigh [1] (Chapter 2) is often not sufficient to resolve the objects of interest.

Over the last years, different techniques, such as super-resolution fluorescence microscopy, have been established to compensate for the deficiency of low spatial resolution. These approaches use fluorescence excitation because this allows – in combination with other techniques – access to high-resolution object information.

These super-resolution methods (i.e., 4Pi [2–5], stimulated emission depletion (STED) (Chapter 10), structured illumination microscopy/patterned excitation microscopy (SIM/PEM) [6, 7], and localization methods (Chapter 8)) seemingly break the conventional resolution limit. However, it should be noted that the fundamental resolution limit is not broken directly. The super-resolution methods are based on conditions that are different from the assumptions of Rayleigh. Rayleigh's conclusions for self-luminous (i.e., fluorescent) objects do not consider

Fluorescence Microscopy: From Principles to Biological Applications, Second Edition.
Edited by Ulrich Kubitscheck.
© 2017 Wiley-VCH Verlag GmbH & Co. KGaA. Published 2017 by Wiley-VCH Verlag GmbH & Co. KGaA.

spatial or temporal variations of the light intensity. Basically, all super-resolution techniques depend on juggling with fluorescence excitation or emission.

In this chapter, we describe in detail two widefield methods that apply interference of the excitation light to make high-resolution object information accessible. These are structured illumination microscopy (SIM) (also referred to as *patterned excitation microscopy*, PEM) and spatially modulated illumination (SMI).

9.1.1 Review: The Resolution Limit

Regarding imaging systems, it is obvious that the rendering power (i.e., the capability to transmit structural information) is always limited to some degree. To describe the rendering power of an imaging system, the term *resolution* is employed. However, this term is ambiguously used, and there exist different scientific definitions for the resolving power of a system.

A commonly used resolution measure is the definition provided by Lord Rayleigh, which is described in Chapter 2. Rayleigh's definition is based on the point spread function (PSF) of an optical system (Section 2.3.3).

The PSF resembles the image of an ideal point-like object. The Rayleigh distance defining the resolution is the lateral distance between the maximum and the closest zero of the PSF.

It is given by

$$d_{\text{Rayleigh}} = 0.61 \frac{\lambda}{n \sin \alpha} \tag{9.1}$$

When the PSF is the same for every position in the object plane, every point of the object is broadened by the same PSF. This is a reasonable approximation in many microscopy applications. It may be helpful to imagine the microscope as an artist who draws an image representation of the object with a brush; fine details of the object will be visible only if the artist uses a tiny brush, that is, a small PSF. Mathematically, this imaging process of drawing an image $A'(x,y)$ from an object $A(x,y)$ using the PSF as a brush can be represented by a convolution:

$$A'(x, y) = A(x, y) \otimes \text{PSF}(x, y) \tag{9.2}$$

In Chapter 2, the optical transfer function (OTF) was introduced. The OTF is the Fourier transform (FT) of the PSF:

$$\text{FT}[\text{PSF}(x, y)] = \text{OTF}(k_x, k_y) \tag{9.3}$$

The convolution theorem states that convolution in position space corresponds to multiplication in frequency space, and vice versa. Here and in the following, "frequency" denotes *spatial frequency*, that is, a modulation along the spatial coordinates. Hence, by applying the FT to the image $A'(x,y)$ and the object $A(x,y)$, we can describe the imaging process (Eq. (9.2)) as a multiplication:

$$\text{FT}[A'] = \text{FT}[A] \times \text{OTF} \tag{9.4}$$

The PSF is band-limited, which means that the OTF is zero for high frequencies beyond a cut-off frequency $k_{\text{cut-off}}$:

$$\text{OTF}(\mathbf{k}) = 0 \quad \text{for } |\mathbf{k}| \geq k_{\text{cut-off}} \tag{9.5}$$

All object information beyond this cut-off frequency is filtered out by the multiplication with the OTF and is therefore missing in the image. The lack of high

Figure 9.1 Cut-off frequency of an OTF. The supported frequency region of a conventional microscope is a toroid. For small lateral frequency components k_x and k_y, the axial component k_z of the cut-off frequency approaches zero. This represents the so-called missing cone.

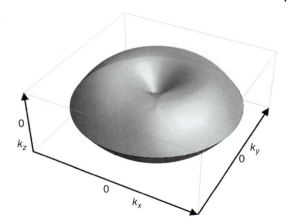

frequencies in the image is the reason for the limited resolution. In Figure 9.1, the cut-off frequency is shown in three dimensions.

The cut-off frequency can be used as a measure for the resolution of an optical system, as an alternative to the Rayleigh criterion.

A definition of resolution referring to the object's frequency is that of Abbe [8]. Abbe considered a fine grating placed on a microscope. The finest, yet resolvable pattern (with grating distance d_{Abbe}) would define the resolution of the microscope. Calculation yields

$$d_{Abbe} = \frac{\lambda}{2n \sin \alpha} \qquad (9.6)$$

The corresponding maximum transmittable frequency (i.e., cut-off frequency) is given by $k_{cut\text{-}off} = 2\pi/d_{Abbe}$.

In our case, both definitions ($d_{Rayleigh}$ and d_{Abbe}) are analogous and yield similar values. In the following paragraphs, ways to circumvent this resolution limit are discussed, which are based on manipulations of the illumination conditions.

9.2 Structured Illumination Microscopy (SIM)

In SIM, the object is illuminated with a periodic illumination pattern. This pattern is used in order to manipulate the object's spatial frequencies [6, 7], as described in the following sections. In SIM, the pattern is usually projected through the objective lens, and no additional devices or mechanisms are used to illuminate the object from the opposite side.

In practice, the pattern is generated by interference of a number of coherent beams proceeding from the objective lens and meeting in the object plane (Figure 9.2).

The periodic pattern can be of arbitrary shape, but typically the modulation of the pattern is only one dimensional in the focal plane (x–y plane), as is the case with the patterns shown in Figure 9.2. Two methods to spatially modulate the excitation intensity in SIM are common: two-beam interference, also called *fringe projection*, and three-beam interference, the so-called grid projection. Examples for optical setups generating the coherent beams are given in Section 9.2.6.

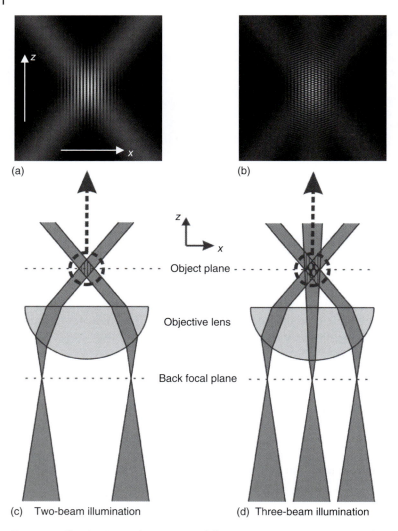

Figure 9.2 Illumination path in structured illumination microscopy. (a) Lateral interference pattern when two coherent beams are used for illumination, as shown in (c). (b) In case of interference of three beams, as shown in (d), the modulation is two dimensional (lateral and axial direction, e.g., along x- and z-axis).

In a typical three-beam interference SIM setup, the coherent beams are arranged in a symmetrical configuration as follows: one light beam propagates along the optical axis, and two beams span an angle with the first beam. One of the other beams propagates at an angle α with respect to the optical axis and the other one at the angle $-\alpha$. All three beams span a common plane (interferometer plane). The resulting interference pattern has a modulation along one direction in the focal plane and also a modulation along the optical axis.

The other method works with only two beams (Figure 9.2a) and corresponds to the three-beam case with a blocked central beam along the optical axis. The corresponding interference pattern exhibits only a modulation along the focal plane (x–y plane).

In the following, the image formation process and the consequent resolution improvement of SIM will be shown using the simpler method of two-beam interference.

9.2.1 Image Generation in Structured Illumination Microscopy

In this section, a mathematical description of the image generation process in SIM is developed.

As mentioned in Section 9.1.1 and in Chapter 2, the imaging process of a widefield microscope can be described by

$$A'(x, y) = A(x, y) \otimes \text{PSF}(x, y) \tag{9.7}$$

where A' is the image and A is the object. The object A is the distribution of the emitted fluorescence light in the object plane. The PSF is the impulse response, blurring the image.

When fluorescence saturation is insignificant (under usual fluorescence microscopy conditions), the object distribution is proportional to the illumination intensity $I_{\text{ill}}(x, y)$ and the fluorophore distribution $\rho(x, y)$:

$$A(x, y) = \rho(x, y) \times I_{\text{ill}}(x, y) \tag{9.8}$$

In conventional widefield microscopy, the illumination intensity is constant and the object is essentially the fluorophore density.

In SIM, however, the illumination is spatially varying and this variation is carried through to the image. For the case of two-beam illumination (Figure 9.2a), the illumination distribution is given by

$$I_{\text{ill}}(x, y) = I_0[1 + m \cos(k_{Gx} x + k_{Gy} y + \phi)] \tag{9.9}$$

or in vector notation with the position vector $\mathbf{r}$

$$I_{\text{ill}}(\mathbf{r}) = I_0[1 + m \cos(\mathbf{k}_G \mathbf{r} + \phi)] = (1 - m)I_0 + 2 m\, I_0 \cos^2(\mathbf{k}_G \mathbf{r}/2 + \phi/2) \tag{9.10}$$

The modulation strength m is a scalar value between 0 and 1. For $m = 1$, the distribution has a minimum of zero and a maximum of $2I_0$. The intensity modulation takes place in the x–y plane. To express the modulation in the 2D space, the $\mathbf{k}$-vector of the grating $\mathbf{k}_G$, with $|\mathbf{k}_G| = 2\pi/G_{\text{SIM}}$, is used, where the grating constant G_{SIM} is the spatial period of the modulation. $\mathbf{k}_G$ points in the direction of the modulation, that is, it is perpendicular to the stripes of the pattern and points along the object plane. The lateral position of the illumination pattern is determined by the phase φ.

Substituting Eq. (9.8) in Eq. (9.7) yields

$$A'(\mathbf{r}) = [\rho(\mathbf{r}) \times I_{\text{ill}}(\mathbf{r})] \otimes \text{PSF}(\mathbf{r}) \tag{9.11}$$

Now, the image information $\tilde{A}'$ in the frequency space is described by

$$\tilde{A}' = \text{FT}[A'] = \text{FT}[\rho \times I_{\text{ill}}] \times \text{OTF} \tag{9.12}$$

The multiplication of ρ and I_{ill} transforms into a convolution because of the convolution theorem:

$$\tilde{A}' = \tilde{\rho} \otimes \tilde{I}_{ill} \times \text{OTF} \tag{9.13}$$

This convolution of the FTs $\tilde{\rho}$ with $\tilde{I}_{ill}$ leads to a mixing of spatial frequencies and ultimately to a potentially enhanced optical resolution, as will be shown for the case of two-beam interference in the following. The FT of the illumination pattern (Eq. (9.10)) is

$$\tilde{I}_{ill}(\mathbf{k}) = \frac{1}{2\pi} \iint I_0 [1 + m \cos(\mathbf{k}_G \mathbf{r} + \phi)] e^{-i(\mathbf{k}\mathbf{r})} \, d\mathbf{r} \tag{9.14}$$

Using the identity

$$\cos x = \frac{1}{2}(e^{ix} + e^{-ix}) \tag{9.15}$$

$\text{FT}(I_{ill})$ becomes

$$\tilde{I}_{ill}(\mathbf{k}) = \frac{1}{2\pi} \iint I_0 \left[1 + \frac{m}{2} e^{i(\mathbf{k}_G \mathbf{r} + \phi)} + \frac{m}{2} e^{-i(\mathbf{k}_G \mathbf{r} + \phi)}\right] e^{-i(\mathbf{k}\mathbf{r})} \, d\mathbf{r} \tag{9.16}$$

Expanding this equation yields

$$\tilde{I}_{ill}(\mathbf{k}) = \frac{1}{2\pi} \iint I_0 \left(e^{-i\mathbf{k}\mathbf{r}} + \frac{m}{2} e^{-i(\mathbf{k}\mathbf{r} - \mathbf{k}_G \mathbf{r})} e^{i\phi} + \frac{m}{2} e^{-i(\mathbf{k}\mathbf{r} + \mathbf{k}_G \mathbf{r})} e^{-i\phi}\right) d\mathbf{r} \tag{9.17}$$

With the Dirac delta distribution in exponential form

$$\delta(\mathbf{k} - \mathbf{a}) = \frac{1}{2\pi} \iint e^{i(\mathbf{x}\mathbf{k} - \mathbf{x}\mathbf{a})} \, d\mathbf{x} \tag{9.18}$$

the FT of the illumination pattern becomes a sum of delta functions with complex prefactors depending on ϕ:

$$\tilde{I}_{ill}(\mathbf{k}) = 2\pi \times I_0 \times \left[\delta(\mathbf{k}) + \frac{m}{2} e^{i\phi} \delta(\mathbf{k} - \mathbf{k}_G) + \frac{m}{2} e^{-i\phi} \delta(\mathbf{k} + \mathbf{k}_G)\right] \tag{9.19}$$

One delta function is located at the origin (corresponding in principle to the conventional fluorescence imaging case) and the other two are at the reciprocal period of the illumination pattern.

In the Fourier-transformed image $\tilde{A}'(\mathbf{k})$, the convolution of these delta functions with the Fourier-transformed object distribution $\tilde{\rho}(\mathbf{k})$ leads to a sum of copies of the object information shifted by the positions of the delta functions. Each copy is multiplied by its corresponding complex coefficient:

$$\tilde{A}'(\mathbf{k}) = \text{OTF}(\mathbf{k}) \times I_0 \frac{1}{\sqrt{2\pi}} \times \left[\tilde{\rho}(\mathbf{k}) + \frac{m}{2} e^{i\phi} \tilde{\rho}(\mathbf{k} - \mathbf{k}_G) + \frac{m}{2} e^{-i\phi} \tilde{\rho}(\mathbf{k} + \mathbf{k}_G)\right] \tag{9.20}$$

How far the single additional copies are shifted in the frequency space is defined by the angle of incidence α and the wave vector $\mathbf{k}$ of the incident light waves.

Note the low-pass filter property of the OTF (during image formation, the OTF is multiplied by $\text{FT}[\rho \times I_{ill}]$). As the frequencies of the additional copies of the image are shifted, a frequency region of previously irresolvable frequencies beyond the cut-off limit $k_{\text{cut-off}}$ of these copies is shifted into the passband of the OTF. This frequency information is superposed with the information of the other copies and located in the frequency region imaged by the microscope. Therefore, it has to be separated and shifted back to its appropriate position after image acquisition. The separation and reconstruction process will be discussed in Section 9.2.2.

In case it is possible to separate the superimposed, that is, added frequency information copies in the acquired image, the additional copies can be shifted back to their original high spatial frequency position. It is apparent that now the frequency region of the image spans further in the direction of the modulation of the illumination pattern compared to the original image. Therefore, the resolution has been enhanced by this process, as higher frequency information is transmitted into the image. The factor of resolution improvement is given by the factor by which the size of the frequency region of the image is expanded.

The size of the resulting frequency region depends on the period of the illumination pattern. To calculate the maximum possible resolution improvement by SIM, the minimum possible grating distance should be considered. For illumination through the objective lens, the minimum grating distance is reached when the two incident coherent excitation beams with the vacuum wavelength $\lambda_{0\text{ex}}$ span the maximum angle, that is, illumination along the maximum opening angle of the acceptance cone of the objective lens. Therefore, the minimum grating distance is given by

$$G_{\text{SIMmin}} = \frac{\lambda_{0\text{ex}}}{2n\sin(\alpha_{\max})} = \frac{\lambda_{0\text{ex}}}{2\text{NA}} = d_{\text{Abbe}} \tag{9.21}$$

which is the Abbe definition for the minimum grating distance of a yet-resolvable grating in the object plane for transmitted light microscopy.

In fluorescence microscopy, the wavelength of the emission light is close to that of the excitation light ($\lambda_{\text{em}} \approx \lambda_{\text{ex}}$). Hence, the reciprocal grating period of the smallest possible illumination pattern approximates the cut-off frequency $k_{\text{cut-off}}$. The delta functions of the Fourier-transformed illumination pattern are located at the positions $-2\pi/G_{\text{SIM}}$ and $+2\pi/G_{\text{SIM}}$. Therefore, the origins of the copies of Fourier-transformed information are shifted to the cut-off limit (Figure 9.3b). When these copies are separated and shifted back, the region of accessible information in the frequency space is twice as large as in the standard microscopy case. Thus, the resolution improvement in linear SIM can reach a factor of 2.

The term *linear* is used because in this case the dependence of the fluorescence emission intensity is considered to be linearly proportional to the excitation intensity.

Under usual microscopy conditions, saturation and photobleaching effects can be often be neglected, and the assumption of linear dependence of the fluorescence response on the excitation intensity holds true. Saturation and photobleaching effects would result in a decreasing gradient of the emission response with increasing excitation intensity. It has been shown [9, 10] that this nonlinear effect can be used to increase the resolution of SIM beyond the factor of 2 when the magnitude of the nonlinearity is maximized by the application of appropriate setup conditions such as, for example, special fluorochromes and

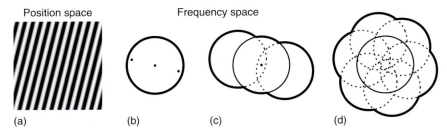

Figure 9.3 Accessible frequency region by SIM. The illumination pattern (a) contains three delta peaks in the frequency space as illustrated in the Fourier-transformed raw image (b). The circular area corresponds to the passband of the objective lens, that is, the bold outline shows the frequency limit. The raw data (b) consists of superposed original image information positioned at three different origins. After separation, the information can be shifted back to its respective position, resulting in an expanded accessible frequency area (c). To expand the resolution in the object plane not only along one direction but also isotropically, the illumination pattern is subsequently rotated to carry out the image acquisition with several illumination pattern orientations (d).

matched excitation intensities. In this case, the effective fluorescence emission pattern that is obtained from the fluorochrome distribution would contain the periodic illumination intensity pattern (Figure 9.3) and additionally higher harmonics of the fundamental pattern (not shown). In the frequency space, the higher harmonics shift the image information by twofold, threefold, or more of the shift by the fundamental pattern.

The twofold resolution improvement in linear SIM yields a lateral resolution of ~100 nm [11], whereas widefield and confocal microscopy can achieve ~230 and ~180 nm lateral resolution, respectively [12].

However, when an illumination pattern with modulation in only one direction in the optical plane (i.e., the plane of focus) is used, the resolution improvement is anisotropic and only maximum in the direction of the modulation (Figure 9.3c). To attain a more isotropic resolution, the direction of the modulation has to be applied in several directions in the optical plane (Figure 9.3d).

An alternative depictive approach to understand the resolution enhancing effect provided by SIM is to consider the Moiré effect, as shown in Figure 9.4. When two fine gratings are superposed, a third coarse grating with a larger spatial period (or grating distance) G_3 occurs. If one of the fine original patterns with a spatial period G_1 and also the resulting pattern is known, while the second fine grating with G_2 is unknown, the unknown pattern can be reconstructed mathematically.

The known and the unknown fine gratings represent the SIM excitation pattern and the high-frequency object information, respectively, whereas the coarse grating represents the detected image.

9.2.2 Extracting the High-Resolution Information

When structured illumination is applied, the total spatial frequency information transmitted through the objective lens increases. However, the additional information is overlaid with the original image information, as described in Section 9.2.1.

Figure 9.4 Moiré effect. If two fine patterns are superposed, a third, raw pattern, can occur.

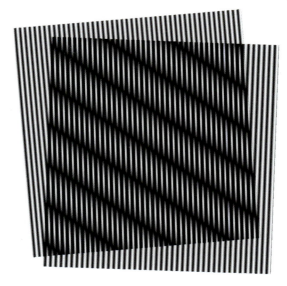

The additional information, therefore, has to be separated and shifted to the correct position in the frequency space. To separate the components, the acquisition of several images with different positions (phase ϕ, see Eq. (9.10)) of the illumination pattern is required. By this method, the complex coefficients (1, $(m/2)e^{i\phi}$, and $(m/2)e^{-i\phi}$) of the image copies depending on the grating position ϕ (see Eq. (9.20)) can be calculated for the different grating positions consecutively. The image information belonging to the single copies can be separated afterward by the respective coefficient. When using two-beam interference, three images with different grating positions are sufficient to solve the equations; for three-beam interference, five images are necessary.

After separation, the copies of the image information can be shifted to their correct position by the reciprocal vectors of the illumination grating. When the correctly processed information of the different copies is added, the resulting image has a higher resolution as compared to the original image.

As described previously, the resolution improvement in the focal plane is anisotropic if the modulation of the illumination pattern is one dimensional in the lateral direction.

In order to achieve an almost isotropic resolution improvement, nonetheless, in this case, a series of images are taken at different angles of the periodic illumination pattern, typically at 0°, 60°, and 120° (Figure 9.3d) or at 0°, 45°, 90°, and 135°.

It is also mathematically possible to use a two-dimensional (2D) grating producing a hexagonal pattern to generate the diffraction orders of the excitation light.

9.2.3 Optical Sectioning by SIM

Not only the lateral resolution, as shown in the 2D example in the previous paragraph, is improved in SIM compared to standard widefield microscopy but also the confocality can be improved, resulting in an optical section with less signal from the out-of-focus region.

This effect becomes clear when the widefield OTF (Figure 9.1) is considered. For the spatial frequencies k_x, $k_y = 0$, and $k_z \neq 0$, the OTF is always zero. A cone-shaped frequency region around the optical axis frequencies k_z in the OTF is missing. The corresponding information is not transmitted by the microscope. This "missing cone" problem results in the fact that in conventional microscopy the z-resolution is not well defined.

As shown in Figure 9.5, the additional object information copies provided by SIM are capable of covering the missing cone of the original OTF, and therefore yield an improved optical sectioning power of SIM compared to conventional fluorescence microscopy. However, in the case of two-beam illumination (Figures 9.2a,c and 9.5a,c), confocality is improved only if the illumination pattern frequency is (slightly) smaller than the cut-off frequency. At the cut-off, the missing cones overlap and therefore remain uncovered.

As is the case with lateral resolution, it is difficult to quantify the axial resolution of a microscope and it is important to find a suitable model to calculate or measure comparable quantitative values for different microscopy methods. For example, the full width at half-maximum (FWHM) of the 3D PSF in the axial dimension could be used. However, this z-direction FWHM depends on the x and y positions (Figures 2.12 and 9.12).

In the work of Karadaglić and Wilson [13], the optical sectioning powers of two different SIM setups and of a confocal microscope were compared using a model of a thin, fluorescent sheet located in the focal plane. The ability to determine the z position of the sheet by the microscopes was compared.

It was shown that both two-beam interference (fringe projection) and three-beam interference (grid projection) SIM exhibit optical sectioning strengths comparable to that of a confocal microscope. However, the three-beam interference provided a better optical sectioning capability compared to two-beam interference. Nevertheless, two-beam interference requires the acquisition of single images at three different phase positions as opposed to five different phase positions for three-beam interference, and because fewer object information copies are used, it is less susceptible to noise in the raw data.

The answer to the question whether two- or three-beam illumination is ultimately more favorable depends on the specimen that is analyzed. Two-beam

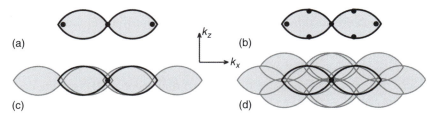

Figure 9.5 Resulting OTF in (a, c) two- and (b, d) three-beam illumination. The black spots in the microscope's widefield OTFs (a: two-beam and b: three-beam) show the origins of the shifted object information copies. When these copies are shifted back (c, d) to their correct positions, the effective OTF increases in size. The black outline shows the original widefield OTF cut-off border. The borders of the shifted OTFs are shown in gray. The corresponding illumination patterns are shown in Figure 9.2a,b.

interference is often used in cases where sectioning is achieved by other means, for example, via total internal reflection fluorescence illumination, or when the specimens are thin.

9.2.4 How the Illumination Pattern is Generated?

For structured illumination, a multitude of coherent light beams is necessary to generate the illumination conditions shown in Figure 9.2.

Basically, two techniques have been developed to arrange this: the application of a diffraction grating (Figure 9.6a), or the use of an interferometer (Figure 9.6b). In Section 9.2.6, examples for both approaches are shown.

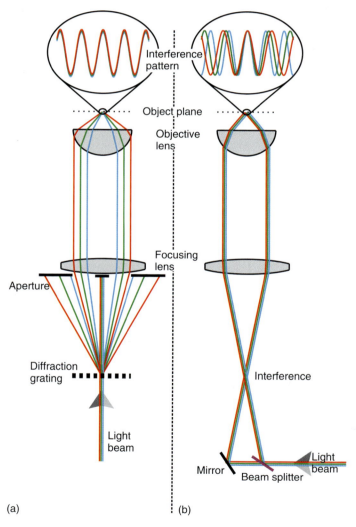

Figure 9.6 Beam path and interference pattern for (a) grating and (b) interferometer setup. In the grating setup shown here, the zero order of diffraction is blocked. Higher orders [±2, ±3, etc.] also do not enter the objective lens.

The two techniques differ in the dependence between the period of the illumination pattern and the wavelength of the excitation light (Figure 9.6).

In setups applying a diffraction grating in a conjugate image plane, an image of the grating is projected into the object plane. Therefore, the spatial period of the illumination pattern is independent of the wavelength of the illumination light. Hence, the illumination light does not need to be coherent. An incoherent light source (e.g., gas discharge lamp or light-emitting diode (LED)) can be used instead of a laser.

For an interferometer setup, on the other hand, the illumination pattern has a period proportional to the wavelength of light. The larger the wavelength, the coarser the pattern. Here, the light source has to be coherent in order to achieve a periodic intensity pattern (Figure 9.6b). For interferometer setups, the angle of the excitation light beams depends only on the angle of the interferometer. If this angle is set to a suitable value according to the objective lens's numerical aperture (NA) once, a change in interference angle is not necessary for different excitation wavelengths. For diffraction grating setups, a change in grating might be necessary if the excitation wavelength is changed. As shown in Figure 9.6a, the red beams hit the objective lens at the utmost border, whereas the blue beams are closer toward the center of the objective lens.

9.2.5 Mathematical Derivation of the Interference Pattern

In the equations for the illumination pattern (Eq. (9.9)) and the derived statements, a variable modulation strength m was used.

In order to explore the magnitude of the modulation strength, the interference pattern for a two-beam interference SIM setup is calculated in the following. The two initial beams are considered to be equally strong and have identical linear polarization $\mathbf{p}_0$. The right beam has a phase shift φ compared to the left beam.

In the following, the complex exponential form for the superposing waves is used instead of the trigonometric form. Both beams propagate identically along the z-direction before passing through the objective lens. While doing this, the left beam is tilted by the angle α and the right beam by $-\alpha$ (Figure 9.7) around the y-axis, thereby affecting the polarization of the beams.

Before passing through the lens, the electric fields of the left and the right beam are given by

$$\mathbf{E}_{l0} = \mathbf{p}_0 E_0 e^{i(\mathbf{k}_0 \mathbf{r} - \omega t)} \tag{9.22}$$

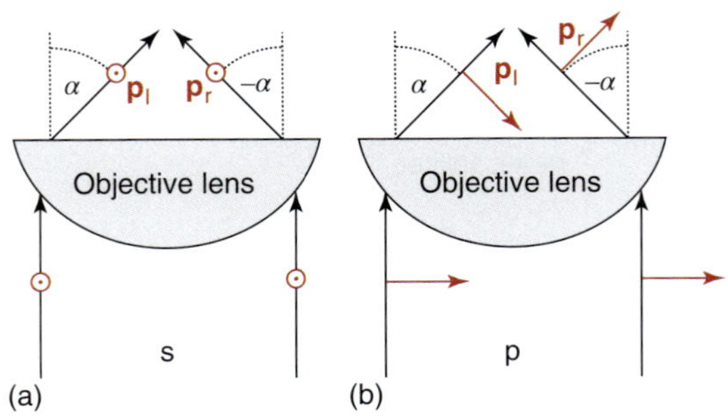

Figure 9.7 Polarization of the beams before and after passing through the objective lens for polarization (a) perpendicular (s) and (b) parallel (p) to the interferometer plane.

9.2 Structured Illumination Microscopy (SIM)

and

$$E_{r0} = p_0 E_0 e^{i(k_0 r - \omega t - \phi)} \tag{9.23}$$

The polarization vector is a unity vector ($|p_0| = 1$) pointing in the direction in which the electric field oscillates.

The original wave vector k_0 points along the z-direction:

$$k_0 = \begin{pmatrix} 0 \\ 0 \\ 1 \end{pmatrix} \frac{2\pi}{\lambda} \tag{9.24}$$

Behind the lens, the rotated polarizations of the beams are given by

$$p_l = R_y(\alpha) p_0 = \begin{pmatrix} \cos \alpha & 0 & \sin \alpha \\ 0 & 1 & 0 \\ -\sin \alpha & 0 & \cos \alpha \end{pmatrix} p_0 \tag{9.25}$$

and

$$p_r = R_y(-\alpha) p_0 = \begin{pmatrix} \cos \alpha & 0 & -\sin \alpha \\ 0 & 1 & 0 \\ \sin \alpha & 0 & \cos \alpha \end{pmatrix} p_0 \tag{9.26}$$

through application of the rotation matrix R_y around the y-axis.

The same rotation accounts for the wave vectors, which yields

$$k_l = \begin{pmatrix} \sin \alpha \\ 0 \\ \cos \alpha \end{pmatrix} \frac{2\pi}{\lambda} \tag{9.27}$$

$$k_r = \begin{pmatrix} -\sin \alpha \\ 0 \\ \cos \alpha \end{pmatrix} \frac{2\pi}{\lambda} \tag{9.28}$$

$$E_l = p_l E_0 e^{i(k_l r - \omega t)} \tag{9.29}$$

$$E_r = p_r E_0 e^{i(k_r r - \omega t - \phi)} \tag{9.30}$$

The superposition $E_s = E_l + E_r$ in the object plane yields

$$E_s = E_0 (p_l e^{i(k_l r - \omega t)} + p_r e^{i(k_r r - \omega t - \phi)}) \tag{9.31}$$

The intensity is proportional to the absolute square of the electric field:

$$I \propto |E_s|^2 \Rightarrow I = a|E_s|^2 = a E_s E_s^* \tag{9.32}$$

with the proportionality constant a. By substituting Eq. (9.31) in Eq. (9.32), the intensity becomes

$$I = 2a E_0^2 (1 + p_l p_r \cos[(k_l - k_r) r + \phi]) \tag{9.33}$$

after some rearrangements. Substituting the wave vectors k_l and k_r (Eqs (9.27) and (9.28)) yields

$$I = 2a E_0^2 \left(1 + p_l p_r \cos \left[\frac{4\pi \sin \alpha}{\lambda} x + \phi \right] \right) \tag{9.34}$$

Obviously, with $\mathbf{p}_l\mathbf{p}_r = m$ and $2aE_0^2 = I_0$, the term is identical to the illumination pattern of a two-beam interference SIM microscope (Eq. (9.9)).

When both interfering beams of a two-beam interference SIM setup are equally strong, the modulation strength depends on the primary polarization of the beams and on the angle α by which the beams are deflected by the objective. To examine these effects, we compare two extreme cases of linear polarization: polarization perpendicular (s)

$$\mathbf{p}_{0s} = \begin{pmatrix} 0 \\ 1 \\ 0 \end{pmatrix}$$

to the interferometer plane and parallel (p)

$$\mathbf{p}_{0p} = \begin{pmatrix} 1 \\ 0 \\ 0 \end{pmatrix}$$

to the interferometer plane (Figure 9.7). The interferometer plane is the plane that is spanned by the two beams.

If the polarization of the incoming beams is perpendicular (s), the polarization of the single beams ($\mathbf{p}_l, \mathbf{p}_r$) will not be changed by passing through the objective lens (application of $R_y(\alpha)$ and $R_y(-\alpha)$) and the modulation m is unity (Figure 9.8a). For beams that are polarized parallel (p) to the interferometer plane, owing to the change in beam angle, the polarization of the left beam is rotated by α and of the right by $-\alpha$. For p polarization and $\alpha = 45°$, the polarizations and therefore the electric fields of both beams are perpendicular. The modulation strength $m = \mathbf{p}_l\mathbf{p}_r$ becomes zero and the resulting intensity of the object plane becomes constant (Figure 9.8b). Usually, the interferometer plane is rotated to different angles (usually 0°, 60°, and 120°) around the optical axis in the SIM image acquisition process. To achieve high modulation strength for all angles, the polarization of the excitation light has to be rotated with the pattern to allow s polarization in every case. This can be done, for example, with a rotatable half-wave plate or an electro-optic modulator.

9.2.6 Examples for SIM Setups

In SIM, the excitation intensity in the focal plane is a periodic pattern. To achieve this illumination distribution, different setups have been established. Commonly, the excitation light is projected through the same objective lens that is used for the detection of the fluorescence in order to generate a standing wave field, even though other illumination techniques with the use of additional optical devices or even near-field excitation are possible.

Most commonly, SIM microscopes use either a physical grating [6, 7] or a synthetic grating generated by a spatial light modulator (SLM) [14, 15] in an intermediate image plane to create the modulated pattern in the object plane (Figure 9.9). Grating-based SIM is usually set up in such a way that only up to the ± first-order diffracted beams enter the objective lens. The use of higher orders (five and more beam interference) is possible but would lead to a more complicated separation of the copies of the Fourier-transformed object information

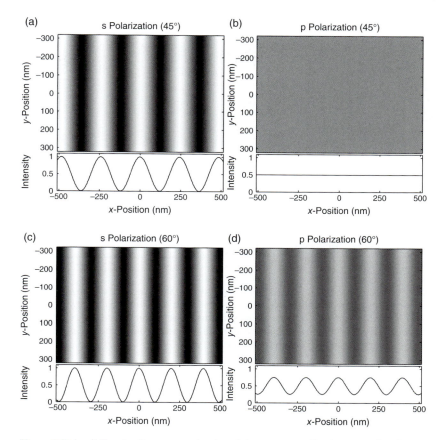

Figure 9.8 (a–d) Illumination pattern (top) and the corresponding intensity distribution (bottom) in the object plane for perpendicular (s) and parallel (p) polarization at different angles.

and a lower signal-to-noise-ratio as the number of copies would be higher. The zero-order diffraction (undiffracted beam) can optionally be blocked in order to apply two-beam interference; the use of zero, plus, and minus first order leads to the case of three-beam interference.

The phase (i.e., position) of the interference pattern in the object plane can be shifted by moving the diffraction grating in the intermediate image plane. To change the orientation of the pattern's modulation, the grating can be rotated around the optical axis.

Alternatively, a 2D grating can be applied to generate a pattern with modulation along multiple axes in the object plane, as noted in the previous section.

When an SLM is used, the pattern can be shifted and rotated by displaying different gratings on the SLM. Scientific SLMs usually consist of an array of liquid crystals to modify the polarization and/or phase of light individually for the separate pixels (a conventional liquid crystal display (LCD) is an SLM, too).

A fast SLM exhibits speed advantages over a solid grating and is also more versatile because the grating's lattice spacing can be easily adjusted to the used

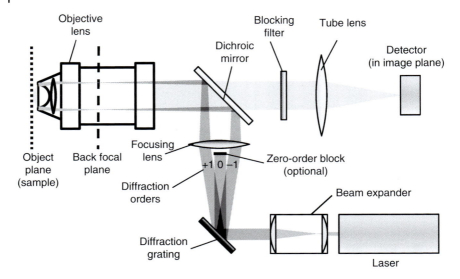

Figure 9.9 Schematic of a grating-based SIM setup. The excitation beam (dark gray) is diffracted by a grating into multiple orders. The imaging path is shown in light gray. Usually, only the zero- and first-order diffraction beams enter the objective lens. Here, the zero order can be optionally blocked by a low-frequency aperture located in front of the focusing lens to switch between three- and two-beam interference.

wavelength. Nonetheless, a solid physical grating has certain advantages over the use of an SLM. Less intensity is lost and the phase of the excitation light is less affected by spatially varying phase errors when passing through the grating, which leads to a stronger modulation of the intensity pattern in the object plane.

A different approach is to use an interferometric setup to generate the coherent beams and channel them toward the objective. A particular setup applying an interferometer [16] (Figure 9.10) is based on a Twyman–Green interferometer, which is a special case of a Michelson interferometer applying a collimated, widened beam. The used interferometer consists of a beam-splitting cube and two opposing mirrors.

This setup is based on an inversely applied, specially designed microscope. The excitation laser beam is directed to a 50% beam-splitting cube (Figure 9.10, cube A), positioned at the focal point of the focusing lens. Half of the beam is reflected by 90°, and the other half passes through the cube without any change in direction. The resulting two beams are then reflected with mirrors by 180° back into the cube. After the light passes through the beam splitter, again, two beams, each at one-fourth intensity, are generated, two of which leave the cube channeled toward the focusing lens. If the beam-splitting cube is rotated around the axis perpendicular to the table by an angle θ, one beam is deflected by 2θ and the other beam is deflected by -2θ. Thus, the interference pattern can be adjusted by rotating the beam splitter. The beams pass through the focusing lens and are then deflected by a dichromatic beam splitter toward the objective. After passing through the focusing lens and objective, the interference of the two beams in the object plane leads to a sinusoidal pattern with modulation parallel to the object

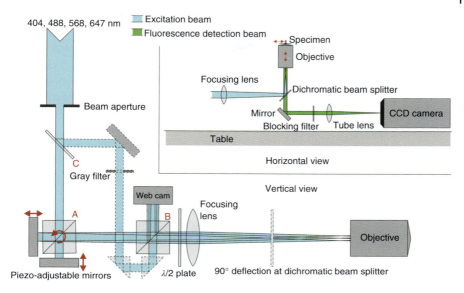

Figure 9.10 Interferometer setup for SIM. The schematic is simplified. The excitation beams are deflected by 90° onto the vertically applied objective by a dichromatic beam splitter. The deflection is not shown to improve visibility. The detection path is shown only in the horizontal view. The elements shown in dashed lines are used for the optional three-beam interference mode. The objects labeled with A, B, and C are beam splitters. The arrows indicate movable elements. (Reprinted from Best et al. [16], ©2011, with permission from Elsevier.)

plane. The orientation of the modulation can be changed by rotating the beam splitter around an axis parallel to the ground and perpendicular to the previous rotation axis. It is therefore possible to generate sinusoidal interference with arbitrary period and direction, which makes it possible to adjust the period of the pattern according to the particular task and wavelength. A webcam can be used to monitor the interference pattern directly. The microscope also offers the option to use a third beam, not deflected and positioned centrally to the two outer deflected beams, in order to generate a three-beam interference pattern. The beam splitter extracting the central beam (C) is located in the beam line before the other splitters. The phase of the pattern can be altered by shifting the relative phases of the separate beams. To accomplish this accurately and quickly, piezo-actuators are attached to the mirrors facing the rotatable beam-splitting cube (A).

9.3 Spatially Modulated Illumination (SMI) Microscopy

9.3.1 Overview

SMI is a microscopy method that can be used for precise size measurements of small fluorescent objects in the direction of the optical axis.

In SIM, the excitation light comes from one objective lens, and the interference pattern is spanned along the focal plane (e.g., x–y plane), whereas in SMI microscopy two counter-propagating waves from two opposing objective lenses are used (Figure 9.11). The specimen is located between the two objectives, where the two beams form a standing wave pattern. Here, the interference pattern occurs only along the optical axis (z-direction).

Because the two beams are counter-propagating, the period of the interference pattern fringe distance d is very small (i.e., it has a high frequency). Analogous to SIM, the finer the pattern, the higher the resolution. When the resulting OTF is considered (Figure 9.12b), as in Figures 9.3 and 9.5 for SIM, in the SMI case, copies of the OTF appear shifted far in the z-direction of spatial frequencies k_z. The OTF copies have no overlap with the widefield OTF in the center, which means that the supported frequency region of the resulting OTF is not continuously interconnected. Even though information of very high-resolution frequencies in the axial direction (dashed structures in Figure 9.12b) is transmitted by the microscope, information from large frequency regions in between is not accessible.

As a consequence, the generation of high-resolution images free of artifacts, as done by SIM, is not possible by this method because large fractions of the object's moderately high-resolution information get lost.

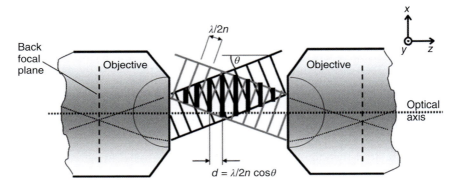

Figure 9.11 SMI setup. Two coherent light beams propagate through two opposing objectives and interfere at the object plane with a fringe distance d.

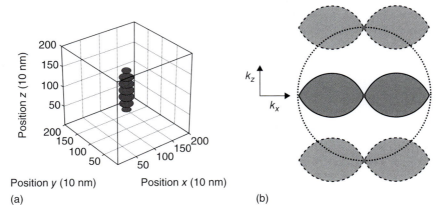

Figure 9.12 (a) Simulated SMI-PSF and (b) the corresponding OTF. While in structured illumination microscopy (SIM) the intensity is usually modulated along the optical plane, in SMI the intensity is typically modulated along the axial direction.

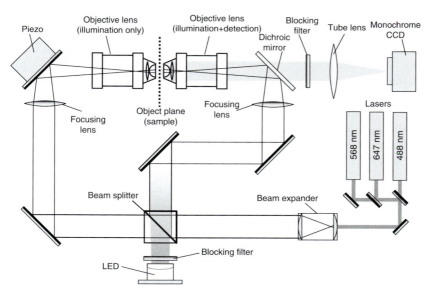

Figure 9.13 Horizontal SMI setup with three different laser sources, one LED for common transmitted light illumination, and one monochrome charge-coupled device (CCD) camera.

Therefore, *a priori* knowledge concerning the analyzed specimen is necessary in order to use SMI microscopy [17]. However, measurements of size and position along the optical axis can be done for small objects with a precision far beyond the diffraction limit – typically 40 nm – with this method.

9.3.2 SMI Setup

A typical SMI setup is shown in Figure 9.13. In order to generate two coherent and collimated beams, a collimated laser beam is divided using a beam splitter. Each beam passes through an additional lens for each objective, where the focal plane of the lens is located in the back focal plane of the objective. This converts the objective and the focusing lens into a collimator. With this optical setup, an interference pattern of sinusoidal shape along the optical axis can be generated between the objective lenses. The sample is placed between the objective lenses and can be moved along the optical axis through the interference pattern. The detection, which is similar to conventional widefield fluorescence detection, is realized by only one objective lens, which is used in conjunction with a dichroic mirror to separate the excitation light from the fluorescence signal. During image acquisition, the object is moved in precise axial steps (e.g., each 20 or 30 nm) through the standing wave field. At each step, a fluorescence image is registered [18].

9.3.3 Excitation Light Distribution

The fringe period of the standing wave located in the object plane depends on the excitation wavelength λ, the refractive index n of the sample and the slide, as well as a possible tilting angle θ of the beams relative to the optical axis (Figure 9.11).

In the following, the mathematical expression for the intensity distribution is derived.

Two counter-propagating coherent electromagnetic waves E_l and E_r interfere in the focal region. It is assumed that both beams have the same amplitude A and that both beams might be rotated around the y-axis by an angle θ. The beam passing through the right objective (Figure 9.13) has a phase delay ϕ relative to the left beam. The result is a periodic standing wave field E_s with an intensity distribution I_s.

The calculation of the intensity distribution of the SMI interference pattern can be derived from the equation for the pattern of a two-beam interference SIM (Eq. (9.33)):

$$I = I_0(1 + \mathbf{p}_l \mathbf{p}_r \cos[(\mathbf{k}_l - \mathbf{k}_r)\mathbf{r} + \phi]) \tag{9.35}$$

In the SMI case (Figure 9.11), the wave vectors of the two beams are given by

$$\mathbf{k}_l = \begin{pmatrix} \sin\theta \\ 0 \\ \cos\theta \end{pmatrix} k \tag{9.36}$$

$$\mathbf{k}_r = \begin{pmatrix} \sin\theta \\ 0 \\ -\cos\theta \end{pmatrix} k \tag{9.37}$$

where k is the absolute value of the wave vector $\mathbf{k}$.

$$k = |\mathbf{k}| = \frac{2\pi}{\lambda} \tag{9.38}$$

The polarization is considered to be parallel to the y-axis (s polarization):

$$\mathbf{p}_l = \mathbf{p}_r = \begin{pmatrix} 0 \\ 1 \\ 0 \end{pmatrix} \tag{9.39}$$

The intensity becomes

$$I = I_0(1 + \cos[2\cos\theta kz + \phi]) \tag{9.40}$$

The illumination pattern can be phase-shifted by changing the optical path length difference ϕ of the interferometer. k, the norm of the wave vector $\mathbf{k}$, is expanded by the refractive index factor n of the medium, as the wavelength of light is inversely proportional to n:

$$\lambda = \frac{\lambda_0}{n} \tag{9.41}$$

$$\Rightarrow k = k_0 n; \quad k_0 = \frac{2\pi}{\lambda_0} \tag{9.42}$$

The grating distance G_{SMI} of the pattern in the z-direction can be derived from Eq. (9.35) by

$$(\mathbf{k}_l - \mathbf{k}_r) = \begin{pmatrix} 0 \\ 0 \\ G_{SMI} \end{pmatrix} = 2k G_{SMI} \cos\theta = 2\pi \tag{9.43}$$

$$G_{SMI} = \frac{\lambda_0}{2n \cos\theta} \tag{9.44}$$

This means that for samples with a thickness larger than G_{SMI}, multiple interference fringes are located inside this volume and, as a result, no unique phase position might be distinguishable. Several maxima of the illumination pattern might be located in the depth of the sample simultaneously, irrespective of the actual phase of the pattern.

For a well-aligned system with a tilting angle below 10° ($\cos\theta \approx 1$), the cosine term in Eq. (9.44) can be neglected. A fringe period is obtained that is proportional to the vacuum wavelength λ_0 divided by 2 times the refractive index. Using an immersion medium with a high refractive index, n will be roughly 1.5. In this case, the fringe distance is λ divided by 3. Therefore, an excitation wavelength of 488 nm results in a fringe distance of ~163 nm. This means that the intensity profile of the excitation pattern has a maximum every 163 nm. When this value is compared with the axial depth of focus (~600 nm for an NA 1.4 objective lens), it can be seen that there is more than one intensity maximum within the focal depth (Figure 9.12a). Hence, it is possible to achieve information from the high-resolution region beyond the classical resolution limit. The approach is described in the following sections.

9.3.4 Object Size Estimation with SMI Microscopy

There are two major effects that occur when moving the sample through the focus of the detection objective lens. On one hand, the detected intensity is modulated by the interference of the excitation pattern while the modulation contrast depends on the object size. On the other hand, there is a variation of intensity between objects in the focus and outside the focus.

The analysis of SMI data is primarily based on these two effects. The axial PSF of a widefield microscope along the optical axis is given by a sinc^2 function (Chapter 2). As a result of two-beam interference, the excitation intensity distribution is given by a $\cos^2$ function (Equation 9.10).

When the illumination pattern is kept fixed to the object plane and a point-like object is moved along the z-direction, two effects superpose. The illumination intensity will vary sinusoidally because of the illumination pattern, and the image of the object will move through the focus. As a result, the illumination pattern and the widefield PSF are multiplied to form the resulting spatially modulated illumination-point spread function (SMI-PSF) (Figure 9.14). Figure 9.12 shows the 3D SMI-PSF and the corresponding OTF.

The axial SMI-PSF shows as the envelope the sinc^2 function, which is modulated by a $\cos^2$ interference pattern. In this example of an infinitely small point-like object, the modulation contrast $R = 1$. The modulation contrast R is defined as follows:

$$R = \frac{I_{max} - I_{min}}{I_{max}} \tag{9.45}$$

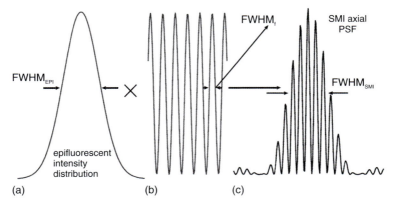

Figure 9.14 Axial detection PSF (a), excitation pattern (b), and resulting axial SMI point spread function (c).

where I_{max} and I_{min} represent the maximum and minimum intensity values of the inner and outer SMI-PSF envelopes, respectively. In other words, the detected axial intensity distribution (AID) is an overlap of the small fringes from the excitation light modulation and the envelope function from the axial detection of the objective lens [19] (see also Figure 9.14):

$$\text{AID} = \left(\frac{\sin(k_1 z)}{k_1 z}\right)^2 \times (a_1 \times \cos^2(k_2 z) + a_2) \tag{9.46}$$

In this equation, k_1 represents the FWHM of the envelope function, k_2 is the wave vector of the standing wave field, a_1 represents the modulation strength, and a_2 defines the modulation contrast. For simplification, additional phase-offset values and nonmodulated parameters are neglected in this formula.

When, instead of a tiny point object, a larger object is imaged, the modulation contrast is smaller than 1 and varies with the object's size and geometry [20–22]. When analyzing biological nanostructures, it is often convenient to assume a spherical geometry. Figure 9.15 shows the modulation contrast R as a function of the diameter of a spherical object. Obviously, the modulation contrast graph is not bijective. For a given modulation contrast, there might be several solutions for the size of the object. The modulation contrast for spherical objects is 0 for certain object sizes (240 and 415 nm for 488 nm excitation wavelength). This means

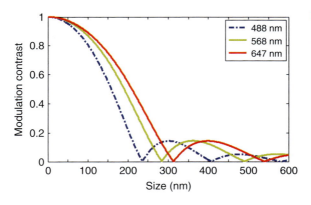

Figure 9.15 Modulation contrast function for different wavelengths.

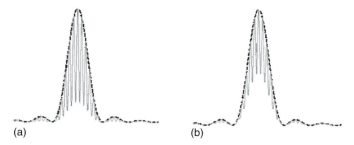

Figure 9.16 Simulation of the expected intensity distribution for a small object ((a) ø 50 nm at 488 nm excitation wavelength) and a larger object ((b) ø 150 nm at 488 nm excitation wavelength) under object scanning condition with an SMI microscope.

that the emitted intensity does not vary when the illumination pattern is shifted through the specimen for these object sizes.

If the application of SMI is constrained to the case of a modulation contrast beyond 0.18, which corresponds to an object size of roughly 200 nm, the relation is bijective and therefore can be used for size determinations. Above this size limit, conventional microscopy can be used to determine the object's size. The smallest diameter is limited by the shape of the modulation contrast function and the amount of detected photons. Typically, even sizes of around 40 nm can be measured by SMI with high precision.

The graphs in Figure 9.15 were calculated for the case of a spherical object with a homogeneous fluorophore density distribution [23]. If the object is small compared to the interference pattern, the detected fluorescence signal is similar to that in Figure 9.16a.

When the diameter of the object increases (Figure 9.16b), the modulation depth decreases, but the fringe distance stays the same and the envelope function also roughly stays the same. In the extreme case, the modulation contrast R will disappear when the object diameter becomes much larger than the fringe distance. By the use of the simulated modulation contrast graph (Figure 9.15), it is possible to extract a value for the diameter of a diffraction-limited object from the measured modulation contrast. It is important to keep in mind that this value is a calculated value under some assumptions – especially the geometry of the object has to be defined *a priori* – and that this value may differ from the "true" size if the assumptions are incorrect.

9.4 Application of Patterned Techniques

Patterned techniques have become an important tool in microscopic analysis of biomedical subjects. Today, almost every major microscope manufacturer provides a SIM system.

In this paragraph, some brief examples for applications of patterned techniques are discussed. One of the many applications of SIM is the analysis of autofluorescent structures in tissue of the eye. Degradation of the retinal pigment epithelium (RPE) is responsible for the age-related macular degeneration

(AMD), the main reason for blindness in the developed world in the elderly population. This disease, in which vision in the macula (the region of sharpest sight) is progressively lost, is linked to excessive aggregations of autofluorescent compounds in the RPE cells. RPE is a cell monolayer between the retina and the choroid. It has essential functions in sustaining the vision process. The compound called *lipofuscin* accumulating in the RPE cells is autofluorescent, which makes it easily detectable by fluorescence imaging without the need for additional labeling.

The images presented here have been generated with an interferometric SIM setup [16]. Paraffin sections of human retinal tissue have been cut, deparaffined, and prepared on object slides [24]. The specimens were analyzed using three different excitation wavelengths (488, 568, and 647 nm).The use of different excitation wavelengths showed a spatially varying composition of the fluorescent granules.

When comparing the acquired SIM images with the standard widefield data, the considerable improvement of the former is apparent (Figures 9.17 and 9.18). Not only is the contrast greatly improved but also is the optical resolution

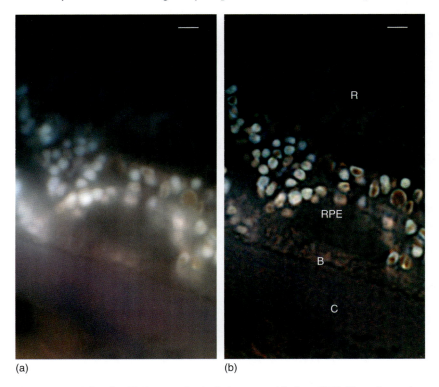

Figure 9.17 Multicolor SIM image of retinal pigment epithelium (RPE). The colors red, green, and blue represent the signal for 647, 568, and 488 nm excitation, respectively. The SIM image (b) shows more details and less out-of-focus light than the widefield image (a). The background of each channel is subtracted, and each channel is stretched to full dynamic range. The Bruch membrane (B) is located between the choroid (C) and the RPE. On the top of the image, endings of retinal rod cells (R) can be seen. The scale bar is 2 µm [16]. (Reprinted from Best et al. [16], with permission from Elsevier.)

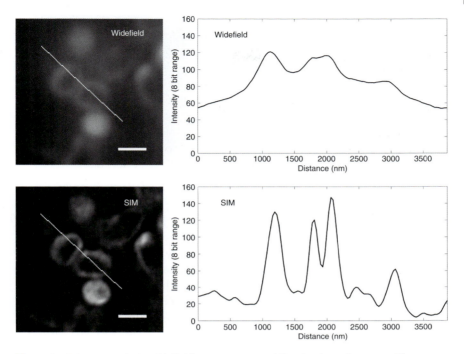

Figure 9.18 Image analysis: widefield versus structured illumination microscopy. The specimen was excited with 488 nm light. The scale bar is 1 µm. (Reprinted from [16], with permission from Elsevier.)

enhanced by a factor of 1.6–1.7 depending on the wavelength. Thus, previously nonresolvable details can be discovered. Figure 9.18 shows the intensity along a line through structures in RPE cells below 488 nm excitation each in a widefield image and the corresponding structured illumination image. The gray value (intensity) of both images was normalized to full 8-bit range, and the background intensity was subtracted.

The improvement of contrast in the SIM image is apparent by the efficient suppression of out-of-focus light, resulting in higher dynamic range in the examined region.

The lateral resolution enhancement allows additional details to be seen (e.g., the two circular structures in the center of the image can clearly be separated in the SIM image). The suppression of out-of-focus light allows the generation of 3D images. SIM images at several axial focal positions are acquired, and the resulting high-resolution data is combined into a 3D dataset. The acquisition speed of SIM is sufficient to allow super-resolution imaging of living cells, as shown in Figure 9.19.

The other microscopy technique treated in this chapter, namely the more specialized method SMI, is suitable for analysis of small nanostructures in biological specimens.

Baddeley *et al.* [25] showed that SMI can be used to measure the size of DNA replication foci in the nucleus of mammalian cells. For this study, both SMI and SIM have been used.

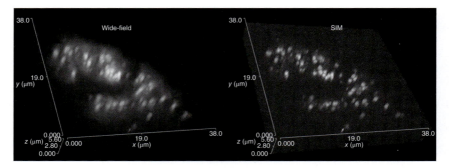

Figure 9.19 3D SIM image of autofluorescent lipofuscin granules inside of a living RPE cell.

Both methods showed an average size of 125 nm irrespective of the labeling method. Remarkably, the super-resolution methods were able to resolve three- to fivefold more distinct replication foci than previously reported. For the SMI data analysis, the data stack was first filtered for pointlike objects. For the identified objects, an axial profile was extracted, which was then analyzed with the SMI methods described in Section 9.3.4 to determine their accurate size. A spherical geometry was assumed for the single replication foci (Figure 9.20).

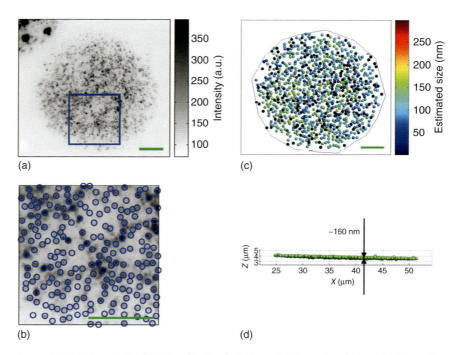

Figure 9.20 SMI analysis of DNA replication foci. Panel (b) shows thresholded foci from the raw data (a). Panel (c) shows the size of the foci (the black ones had to be ignored because of insufficient signal level or a size outside of the effective range). Panel (d) shows the axial position of the analyzed foci [25]. (Baddeley et al. 2010 [25]. Reproduced with permission of Oxford University Press.)

9.5 Conclusion

Interference techniques deliver increased optical resolution albeit at the cost of increased acquisition time and higher irradiance. The main benefit of these techniques compared to other super-resolution techniques is that no special requirements for the applied fluorochromes are necessary. No photoswitching or nonlinear effects are necessary to increase the resolution. The approach is based solely on the change in the intensity distribution of the excitation light compared to standard fluorescence microscopy. However, super-resolution methods based on nonlinear effects have the intrinsic advantage of even higher resolution than SIM (theoretically unlimited).

Often, the ease of use makes SIM the high-resolution method of choice for many biomedical applications. In contrast to most other high-resolution techniques, SIM allows living cell analysis because of relatively short acquisition times and low excitation intensities. Compared to conventional confocal microscopy, SIM delivers a superior image quality and optical resolution at a comparable applicability. Even though linear patterned techniques cannot compete directly with localization methods for single molecules (Chapter 8) in terms of positioning accuracy, they can deliver complementary information. When both methods are combined, high structural resolution, on one hand, and high localization accuracy, on the other, can be realized [26].

9.6 Summary

Fluorescence microscopy techniques using patterned illumination light offer the opportunity to extract high-resolution object information beyond the conventional resolution limit. Especially the SIM technique, where laterally modulated illumination through one objective lens is used, has evolved to become an important tool for high-resolution imaging in biomedical research. Because of the interplay of the illumination pattern with the object's spatial frequencies, a resolution improvement by a factor of 2 is possible in SIM. In this chapter, we provided a fundamental mathematical description of this method and presented different variations of the SIM setups. The secondary focus of this chapter was on the less common method of SMI, where two opposing objective lenses are used to generate a high-frequency interference pattern along the optical axis. The SMI method is used to measure the size of nanostructures with great precision. At the end of the chapter, we provided a brief description of exemplary biological applications for the two patterned illumination methods.

Acknowledgments

We thank Stefan Dithmar and Thomas Ach from the University Hospital Heidelberg for their kind cooperation and funding. We also thank Rainer Heintzmann for his support and for providing his reconstruction software for

SIM. We gratefully appreciate the support of the members of the group of Christoph Cremer, especially the help of Margund Bach.

References

1 Rayleigh, L. (1896) On the theory of optical images, with special reference to the microscope. *Philos. Mag. J. Sci.*, **42** (255), 167–195.
2 Cremer, C. and Cremer, T. (1978) Considerations on a laser-scanning-microscope with high resolution and depth of field. *Microsc. Acta*, **81** (1), 31–44.
3 Hell, S. and Stelzer, E.H. (1992) Fundamental improvement of resolution with a 4Pi-confocal fluorescence microscope using two-photon excitation. *Opt. Commun.*, **93** (5), 277–282.
4 Hell, S.W., Lindek, S., Cremer, C., and Stelzer, E.H. (1994) Confocal microscopy with an increased detection aperture: type-B 4Pi confocal microscopy. *Opt. Lett.*, **19** (3), 222–224.
5 Hänninen, P.E., Hell, S.W., Salo, J., Soini, E., and Cremer, C. (1995) Two-photon excitation 4Pi confocal microscope: enhanced axial resolution microscope for biological research. *Appl. Phys. Lett.*, **66**, 1698–1700.
6 Gustafsson, M.G. (2000) Surpassing the lateral resolution limit by a factor of two using structured illumination microscopy. *J. Microsc.*, **198** (2), 82–87.
7 Heintzmann, R. and Cremer, C.G. (1999) Laterally modulated excitation microscopy: improvement of resolution by using a diffraction grating. SPIE BiOS Europe'98, pp. 185–196, http://proceedings.spiedigitallibrary.org/proceeding.aspx?articleid=972650 (accessed 10 May 2013).
8 Abbe, E. (1873) Beiträge zur Theorie des Mikroskops und der mikroskopischen Wahrnehmung. *Arch. Mikrosk. Anat.*, **9** (1), 413–418.
9 Heintzmann, R., Jovin, T.M., and Cremer, C. (2002) Saturated patterned excitation microscopy – a concept for optical resolution improvement. *J. Opt. Soc. Am. A Opt. Image Sci. Vision*, **19** (8), 1599–1609.
10 Gustafsson, M.G. (2005) Nonlinear structured-illumination microscopy: wide-field fluorescence imaging with theoretically unlimited resolution. *Proc. Natl. Acad. Sci. U.S.A.*, **102** (37), 13081–13086.
11 Heintzmann, R. and Ficz, G. (2006) Breaking the resolution limit in light microscopy. *Briefings Funct. Genomics Proteomics*, **5** (4), 289–301.
12 Pawley, J.B. (ed.) (2006) *Handbook of Biological Confocal Microscopy*, Springer, Boston, MA, http://link.springer.com/10.1007/978-0-387-45524-2 (accessed 22 February 2016).
13 Karadaglić, D. and Wilson, T. (2008) Image formation in structured illumination wide-field fluorescence microscopy. *Micron*, **39** (7), 808–818.
14 Hirvonen, L., Mandula, O., Wicker, K., and Heintzmann, R. (2008) Structured illumination microscopy using photoswitchable fluorescent proteins. *Proc. SPIE*, **6861**, 68610L. doi: 10.1117/12.763021
15 Kner, P., Chhun, B.B., Griffis, E.R., Winoto, L., and Gustafsson, M.G.L. (2009) Super-resolution video microscopy of live cells by structured illumination. *Nat. Methods*, **6** (5), 339–342.

16 Best, G., Amberger, R., Baddeley, D., Ach, T., Dithmar, S., Heintzmann, R. et al. (2011) Structured illumination microscopy of autofluorescent aggregations in human tissue. *Micron*, **42** (4), 330–335.

17 Spöri, U., Failla, A.V., and Cremer, C. (2004) Superresolution size determination in fluorescence microscopy: a comparison between spatially modulated illumination and confocal laser scanning microscopy. *J. Appl. Phys.*, **95** (12), 8436–8443.

18 Reymann, J., Baddeley, D., Gunkel, M., Lemmer, P., Stadter, W., Jegou, T. et al. (2008) High-precision structural analysis of subnuclear complexes in fixed and live cells via spatially modulated illumination (SMI) microscopy. *Chromosome Res.*, **16** (3), 367–382.

19 Schneider, B., Upmann, I., Kirsten, I., Bradl, J., Hausmann, M., and Cremer, C. (1999) A dual-laser, spatially modulated illumination fluorescence microscope. *Microsc. Anal.*, **57** (1), 5–7.

20 Failla, A.V., Spoeri, U., Albrecht, B., Kroll, A., and Cremer, C. (2002) Nanosizing of fluorescent objects by spatially modulated illumination microscopy. *Appl. Opt.*, **41** (34), 7275–7283.

21 Albrecht, B., Failla, A.V., Heintzmann, R., and Cremer, C. (2001) Spatially modulated illumination microscopy: online visualization of intensity distribution and prediction of nanometer precision of axial distance measurements by computer simulations. *J. Biomed. Opt.*, **6** (3), 292–299.

22 Wagner, C., Spöri, U., and Cremer, C. (2005) High-precision SMI microscopy size measurements by simultaneous frequency domain reconstruction of the axial point spread function. *Optik*, **116** (1), 15–21.

23 Wagner, C., Hildenbrand, G., Spöri, U., and Cremer, C. (2006) Beyond nanosizing: an approach to shape analysis of fluorescent nanostructures by SMI-microscopy. *Optik*, **117** (1), 26–32.

24 Ach, T., Best, G., Ruppenstein, M., Amberger, R., Cremer, C., and Dithmar, S. (2010) Hochauflösende Fluoreszenzmikroskopie des retinalen Pigmentepithels mittels strukturierter Beleuchtung. *Ophthalmologe*, **107** (11), 1037–1042.

25 Baddeley, D., Chagin, V.O., Schermelleh, L., Martin, S., Pombo, A., Carlton, P.M. et al. (2010) Measurement of replication structures at the nanometer scale using super-resolution light microscopy. *Nucleic Acids Res.*, **38** (2), e8.

26 Rossberger, S., Best, G., Baddeley, D., Heintzmann, R., Birk, U., Dithmar, S. et al. (2013) Combination of structured illumination and single molecule localization microscopy in one setup. *J. Opt.*, **15** (9), 094003.

10

STED Microscopy

Travis J. Gould[1], Lena K. Schroeder[2], Patrina A. Pellett[2], and Joerg Bewersdorf[2]

[1] Bates College, Department of Physics & Astronomy, 44 Campus Ave, Lewiston, ME 04240, USA
[2] Yale University, Department of Cell Biology, 333 Cedar Street, New Haven, CT 06520-8002, USA

10.1 Introduction

The consensus for over a century was that resolution in far-field microscopy is fundamentally limited by diffraction. Ernst Abbe's famous formula

$$\Delta r \approx \frac{\lambda}{2n \sin \alpha} \quad (10.1)$$

was considered to be the resolution limit, with λ being the wavelength, n the refractive index of the medium, and α the half aperture angle of the objective lens. Equation (10.1) represents the full width at half-maximum (FWHM) of the point spread function (PSF) and is commonly used as a quantitative definition of resolution in far-field light microscopy. Identical fluorescent emitters separated by less than Δr, which is typically ~200–250 nm, appear merged into one image and therefore cannot be resolved.

Fortunately, this long-standing barrier has not prevented researchers from pursuing new techniques that significantly improve resolution in the far field. Over the last two decades, the field of fluorescence imaging has seen a number of developments that have pushed resolution beyond the limits imposed by diffraction. The concept of stimulated emission depletion (STED) microscopy, first introduced by Stefan Hell in 1994, demonstrated that the diffraction barrier could, in fact, be surpassed [1]. In STED microscopy, the diffraction limit is overcome by exploiting the inherent photophysical properties of the fluorescent molecules to reduce the size of the effective focal volume of a laser scanning microscope (LSM).

In concert with other super-resolution microscopy methods, STED microscopy is currently revolutionizing the field of biomedical imaging, offering resolutions that previously had been limited to the realm of electron microscopy. The momentous impact of STED microscopy on improving the resolving power of fluorescence imaging was celebrated (along with single-molecule super-resolution microscopy) with the Nobel Prize in Chemistry in 2014 [2].

Fluorescence Microscopy: From Principles to Biological Applications, Second Edition.
Edited by Ulrich Kubitscheck.
© 2017 Wiley-VCH Verlag GmbH & Co. KGaA. Published 2017 by Wiley-VCH Verlag GmbH & Co. KGaA.

10.2 The Concepts behind STED Microscopy

10.2.1 Fundamental Concepts

10.2.1.1 Switching between Optical States

Generally, image generation in a light microscope involves three consecutive steps:

- illumination of the sample
- interaction of the illuminating light with the sample
- detection of the light emitted by the sample.

In the far field of fluorescence microscopes, illumination and detection are both ruled by diffraction, as described by Eq. (10.1). The key to achieving diffraction-unlimited resolution lies in the light–sample interaction, more specifically the interaction between illumination photons and probe molecules. If populations of probe molecules can be switched on and off independently, molecules previously indistinguishable in detection can become distinguishable. This concept is exploited in different super-resolution microscopes developed over the last two decades, with STED microscopy having been the first to be realized. The difference between STED microscopy and the group of approaches using single-molecule localization lies in the spatial organization of switching: STED microscopy switches off molecules in the outer area of the excitation volume in a targeted manner, allowing only molecules close to the focus center to fluoresce. In contrast, localization-based techniques exploit stochastic switching processes and rely on imaging individual molecules to determine their positions in a sequential manner (Chapter 8).

The concept of a targeted optical switching process, which can – when saturated – create arbitrarily small fluorescent spots as realized in STED, has also been generalized to include other reversible saturable optical fluorescence transitions (RESOLFT) [3], for example, reversibly switching fluorescent proteins (FPs). By scanning the sample with many foci in parallel, image recording can be speeded up by several orders of magnitude [4], which holds great promise for live-cell super-resolution imaging.

10.2.1.2 Stimulated Emission Depletion

Stimulated emission (SE), first described by Einstein in 1916 and represented by the "SE" in LASER (light amplification by *stimulated emission* of radiation), describes the quantum mechanical phenomenon of a fluorophore in an electronic excited state releasing a photon when resonating with illuminating light of a certain wavelength. The photon energy of the illuminating light matches the energy gap between the excited state and the electronic ground state of the fluorophore (Figure 10.1a). The emitted light carries the same energy and phase as the illuminating light and can therefore be easily separated from the much broader spectrum of spontaneously emitted fluorescence light (usually, by using a bandpass filter which blocks the SE wavelength).

STED microscopy uses SE to locally force fluorescent molecules out of the excited state before they can spontaneously fluoresce. This active

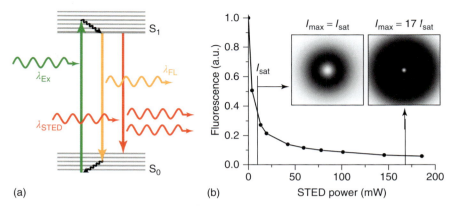

Figure 10.1 (a) Simplified Jablonski diagram. Molecules are excited from the electronic ground state (S_0) to the excited state (S_1) by absorbing light (λ_{Ex}). They can then spontaneously return to S_0 by emitting a fluorescence photon (λ_{Fl}) unless forced down to the ground state by stimulated emission (λ_{STED}). (b) Depletion curve measured for the organic dye ATTO647N in aqueous solution. The normalized fluorescence signal is plotted as a function of the STED laser power ($\lambda_{STED} = 770$ nm; measured at the objective back aperture). The insets show simulated depletion efficiency profiles for different values of maximum STED intensity I_{max}.

off-switching – the *depletion* – of the excited state is accomplished using a laser of appropriate wavelength known as the *STED laser*.

If the ground and first excited states are denoted as S_0 and S_1, respectively, then the probabilities, P_i, of finding a fluorescent molecule in either state are governed by the following differential equation:

$$\frac{dP_1}{dt} = -k_{10}P_1 + k_{01}P_0 = -\frac{dP_0}{dt} \tag{10.2}$$

where k_{10} and k_{01} are the transition rates for $S_1 \to S_0$ and $S_0 \to S_1$, respectively. Relaxation is the sum of spontaneous and stimulated processes and can be summarized as

$$k_{10} = k_{10,\,\text{spont}} + \sigma I_{STED} \tag{10.3}$$

where $k_{10,\,\text{spont}}$ is the rate of spontaneous emission (including fluorescence), σ is the transition cross section at the STED laser wavelength, and I_{STED} is the photon flux per unit area of the STED laser. To efficiently switch off ("deplete") an excited fluorophore, SE therefore has to outcompete spontaneous relaxation of the excited state (fluorescence as well as nonradiative decay).

If a fluorophore initially resides in S_0, the probability of finding it in S_1 at a later time t is then

$$P_1(t) = \frac{k_{01}}{k_{10} + k_{01}} \left(1 - e^{-(k_{10}+k_{01})t}\right) \tag{10.4}$$

For $t \gg (k_{10} + k_{01})^{-1}$, an equilibrium population is reached:

$$P_1 = \frac{k_{01}}{k_{10} + k_{01}} = \frac{1}{1 + (k_{10}/k_{01})} \tag{10.5}$$

This equation shows that the population P_1 depends on the ratio k_{10}/k_{01}, which in turn can be controlled by the intensity of the STED laser (Eq. (10.3)). Increasing the STED laser intensity therefore allows effective depopulation of the excited state (Figure 10.1b).

10.2.1.3 Stimulated Emission Depletion Microscopy

The goal in STED microscopy is to produce effective focal volumes of sub-diffraction size. This is achieved by confining the region in which molecules can emit fluorescence. Fluorescence confinement is achieved using a doughnut-shaped STED laser focus, with an intensity profile I_{STED} featuring a local zero centered on the regular excitation focus (see, e.g., Figure 10.2a). More specifically, $I_{STED}(r) = I_{max} f(r)$, where I_{max} is the maximum STED beam intensity and $f(r)$ is the radial profile of the focused STED beam, which features $f(r=0) = 0$. As $f(r)$ is governed by diffraction, it is a continuous and differentiable function that varies only gradually and on size scales of conventional microscope resolution. Fluorescence depletion by SE is therefore less and less efficient the closer the position is to $r = 0$, and fluorescence will be primarily emitted from the vicinity of the center. By increasing the depletion intensity I_{max}, even relatively

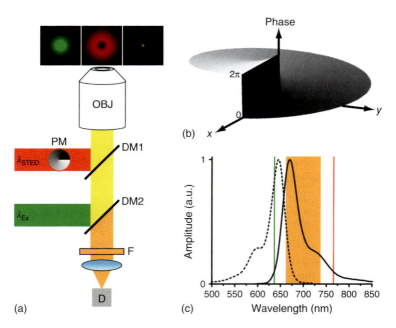

Figure 10.2 (a) Simplified schematic of a STED microscope. The excitation (λ_{Ex}) and STED (λ_{STED}) beams are merged by dichroic mirrors DM1 and DM2 and focused by the objective lens (OBJ) into a common focus. A helical phase mask (PM) in the STED beam path creates a doughnut-shaped STED focus in the sample. Fluorescence is collected by the objective, separated from laser light by DM1 and DM2, bandpass-filtered (F), and focused onto a detector (D). (b) Detailed view of the helical phase ramp used to produce a doughnut-shaped STED focus. (c) Hypothetical excitation (dotted line) and emission spectra of a fluorophore showing wavelengths used for excitation (green line), depletion (red line), and the spectral window for fluorescence detection (orange box).

low amplitudes of the depletion profile are sufficient to cause significant SE and decrease the area from which fluorescence can be emitted (Figure 10.1b). The FWHM of the remaining effective fluorescent spot (and hence the resolution of the system) is well approximated by an expanded form of Eq. (10.1):

$$\Delta r_{STED} \approx \frac{\lambda}{2n \sin \alpha \sqrt{1 + (I_{max}/I_{sat})}} \quad (10.6)$$

where $I_{sat} \equiv 1/(\tau_{fl}\sigma)$ is defined to be the characteristic saturation intensity for a particular fluorophore at which the emitted fluorescence is reduced by a factor of $1/e$, τ_{fl} is the fluorescence lifetime of the excited state, and n, λ, α, and σ are as defined in Eqs. (10.1) and (10.3) [3]. For typical fluorophores, $I_{sat} \sim 10\,\mathrm{MW\,cm^{-2}}$, which requires $I_{max} \geq 100\,\mathrm{MW\,cm^{-2}}$ to push the resolution significantly beyond the diffraction limit. Figure 10.3 shows a comparison between images obtained by confocal and STED microscopy of small fluorescent beads, in which the resolution of the STED images is on the order of 20 nm.

Equation (10.6) shows that by increasing I_{max}, in principle, an arbitrarily small Δr_{STED} can be achieved and therefore the diffraction barrier broken [1]. Consequently, the resolution of a STED microscope is not fundamentally limited by diffraction any longer but rather by how well the theoretical conditions

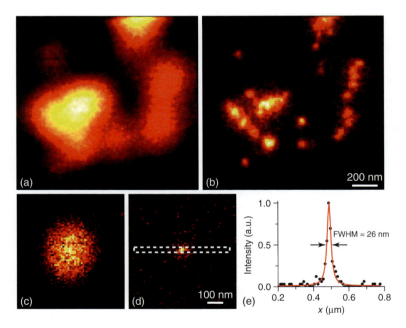

Figure 10.3 (a) Confocal and (b) STED images of 20 nm diameter crimson beads (Invitrogen) demonstrating the superior resolving power of STED microscopy. (c) Confocal image of a single bead and (d) the corresponding STED image. (e) Line profile (black circles) of the boxed area in (d) generated by summing over the short axis of the box. Fitting to a Lorentzian model function (red line) yields an FWHM of 26 nm, which represents an upper bound for the resolution of the STED microscope. Panels (a) and (b) were smoothed by a 2 × 2 pixel FWHM Gaussian filter.

underlying Eq. (10.6) can be realized in practice and by the largest value of I_{max} that may be tolerated by the sample.

10.2.2 Key Parameters in STED Microscopy

What resolution can be achieved in STED microscopy in practice? To answer this question, the influence of a number of key parameters has to be understood. As can be seen from Eq. (10.6), STED resolution depends on a number of classical optical parameters, wavelength, refractive index, and aperture angle, and also on the ratio of STED laser intensity I_{max} and saturation intensity I_{sat}, which reflects the properties of the fluorescent probe. The first set of parameters represents the classical diffraction limit. The resolution of STED microscopy consequently scales with the size of the conventional PSF of the microscope. The dependence on I_{max} and I_{sat} reflects the phenomenon that the improvement of the resolution beyond the diffraction limit is achieved by switching off the fluorescent molecules, which is a photophysical process. In practice, the resolution of a STED microscope is also influenced by imperfect imaging conditions, which can make experimental performance differ significantly from the ideal world of Eq. (10.6).

10.2.2.1 Pulsed Lasers and Fluorophore Kinetics
For $I_{max} \gg I_{sat}$, Eq. (10.6) can be approximated as

$$\Delta r_{STED} \approx \frac{\lambda}{2n \sin \alpha \sqrt{I_{max}/I_{sat}}} \tag{10.7}$$

which indicates that the STED resolution scales with the inverse square root of the ratio I_{max}/I_{sat}. Quadrupling the STED laser intensity (or a fourfold reduction of I_{sat}) therefore improves (under ideal conditions) the STED resolution by a factor of 2. Increasing I_{max} is straightforward and mainly limited by the available laser power and the intensity level tolerated by the imaged sample. On the other hand, the time-averaged power P_{sat} required to reach I_{sat} depends on many photophysical factors.

As indicated by Eq. (10.3), SE competes with spontaneous emission of fluorescence. The latter does not contribute to resolution improvement and in fact counteracts it. SE is therefore most effective (resulting in lower P_{sat}) when the STED laser light is concentrated to a time interval that immediately follows excitation and is shorter than the fluorescence lifetime. This can be realized with laser beams where the fluorophores are first exposed to an excitation laser pulse and immediately afterward by a STED laser pulse. In addition to the fluorescence lifetime τ_{fl} (typically ~2 ns), vibrational relaxation lifetimes τ_{vib} (typically <1 ps) and nonlinear absorption occurring at high peak intensities have to be considered in this context. The STED pulse length, τ_{STED}, should be significantly shorter than τ_{fl} to dominate over the fluorescence process, but much longer than τ_{vib}. The reason for the latter is that a fluorescent molecule is susceptible to re-excitation by the STED light (which would counteract the depletion effect) directly after SE because it is still in an excited vibrational state and the energy gap back to S_1 matches the STED photon energy. Only after vibrational relaxation occurs is the molecule protected from re-excitation by the STED laser because of the larger energy gap to S_1 (see Figure 10.1a). Stretching the STED pulses to $\tau_{STED} \gg \tau_{vib}$ reduces the likelihood of another STED laser photon interacting with the molecule right after SE. Additionally, long τ_{STED} also avoids unwanted multiphoton processes, such as two-photon excitation, which require peak intensities on the order of gigawatts per square centimeter. In practice, τ_{STED} values of several hundred picoseconds to 1 ns have been shown to work well in STED microscopy. As the excitation pulse has far lower intensity than the STED pulse, the excitation pulse length τ_{exc} does not play a major role as long as it is significantly shorter than τ_{fl}.

The use of pulsed laser sources is, however, not an absolute requirement for successful STED microscopy. Using pulsed lasers allows concentrating the available power into increased peak powers for time intervals immediately following excitation where the depletion process is most efficient, but using higher average powers with continuous wave (CW) lasers achieves similar depletion effects.

Another factor to consider is the length of the time period between laser pulses: it is typically ~12.5 ns because of the availability of lasers with ~80 MHz repetition rates. This, however, means that fluorophores with $\tau_{fl} = 2$ ns, for example, are put to work for only approximately one-sixth of the available time. The repetition rate therefore influences how much fluorescence can be recorded per unit time, which in turn affects the imaging speed. The ideal STED laser system from this perspective would therefore feature high-intensity laser pulses at a repetition rate of $\sim 1/\tau_{fl}$.

The relevant photophysics of the dyes are, however, even more complicated, as alternative transitions are possible. Of particular concern is intersystem crossing into the triplet state. While intersystem crossing rates are low, the relatively long life time of the triplet state can lead to significant build-up in this state, which results in decreased fluorophore populations in the singlet state system and correspondingly lower fluorescence signal. Furthermore, photobleaching is accelerated for molecules in the triplet state. Using lasers with lower pulse repetition rates (1 MHz range) results therefore not only in much lower P_{sat} values but also in significantly reduced photobleaching, although at the expense of increased image recording times.

An elegant, however technically challenging, alternative to avoid populating the triplet state without compromising imaging speed is to sweep the (CW or high repetition rate) laser beams across the sample at a high enough speed that only a few subsequent pulses can hit the same molecule before the beam has moved on [5]. To accumulate enough signal for each pixel, the field of view can be scanned multiple times and the data from several recorded images summed up.

Following this discussion of instrument parameters, it should be pointed out that selecting suitable fluorescent probes or improving their characteristics shows promise for improving the quality of STED imaging. The identification of color centers (nitrogen vacancies) in diamond as highly suitable STED probes, for example, has led to a demonstrated STED resolution below 10 nm [6].

10.2.2.2 Wavelength Effects

The effect that the STED laser beam has on a given fluorophore depends not only on the intensity but also on the wavelength. SE depends strongly on the illuminating wavelength since the corresponding energy has to match the energy gap of the desired state transition (i.e., from the excited state to the ground state). However, the STED wavelength can also induce other transitions. Of these, direct excitation from the electronic ground state by the STED laser has to be especially emphasized. Even though the STED beam does not overlap with the peak of the excitation spectrum, its high intensity can compensate for a low excitation cross section at this wavelength, which can result in background in the image and jeopardize the resolution improvement. This effect becomes especially visible when the excitation intensity is low. The STED wavelength is therefore generally shifted as far to the red as possible to find a compromise between minimizing excitation and maximizing SE (Figure 10.2c). On the other hand, recent work has shown that STED wavelengths closer to the peak of the emission spectrum can be used if the anti-Stokes background signal, that is, unwanted fluorescence resulting from excitation by the STED laser, is subtracted from the STED image to recover the desired signal generated only by the excitation laser [7].

10.2.2.3 PSF Shape and Quality

As described previously, STED resolution improvement is usually achieved by depleting excited fluorophores in a doughnut-shaped profile that allows molecules in the doughnut center to still emit fluorescence (Figure 10.2a). Saturating this depletion profile sharpens the nondepleted area and pushes the resolution beyond the diffraction limit (Figure 10.1b).

Several STED PSF shapes have been reported and more are easily imaginable. The central minimum is produced by destructive interference of the focused wave, which is modified by phase filters (Figure 10.2b) in the beam path in front of the objective. STED PSFs can be toroidal (doughnut shaped) as shown in Figures 10.2a and 10.4a, or they can feature additional lobes above and below to quench fluorescence axially to improve resolution in all three dimensions (Figure 10.4b). In either case, the width of the STED PSF minimum – not considering the saturation effect – is at best diffraction-limited.

Resolution improvement is strongly influenced by imperfections of the STED PSF, that is, an imperfect "zero" in the STED PSF minimum. Even an intensity value in the PSF minimum that is only a few percent of the peak intensity can increase to a significant level when $I_{max} \gg I_{sat}$ and result in an undesired suppression of fluorescence. Strong care is therefore required in experimentally producing a STED PSF. In particular, polarization effects have to be considered. For example, two parallel y-polarized beamlets that enter the back aperture close to its edge in the positive and negative x-direction with a π phase shift will cancel out in the focus center (Figure 10.5a). Even a small additional x-polarization component with the same phase shift, however, will result in z-polarization components, which will interfere constructively (Figure 10.5b), thereby increasing the intensity at the intended zero at the center of the PSF. Similarly, distortions such as a rotationally nonsymmetric intensity distribution of the

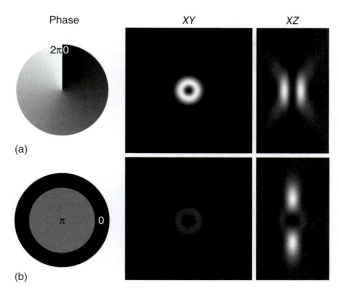

Figure 10.4 Phase masks commonly used in STED microscopy. (a) Helical phase ramp and resulting toroidal (doughnut-shaped) PSF. Note that this phase distribution does not confine fluorescence emission in the axial direction. (b) Central π-step phase mask and the resulting PSF. Note that circular polarization of the laser beam is required to produce the shown PSFs.

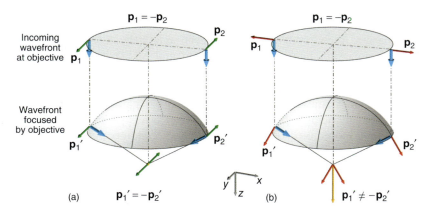

Figure 10.5 Correct polarization of the STED beam is essential for destructive interference in the STED focus center. Two linearly polarized beamlets of a collimated beam ("1" and "2") entering the objective at opposing sides, one phase-shifted by 180°, can interfere destructively when their polarization vectors (p_1 and p_2) are azimuthally oriented (a). The same scenario with radial polarization orientation leads to a z-polarization component (see p'_1 and p'_2) caused by focusing the incoming beam into a spherical wavefront. This z-component interferes constructively (b). Using circularly polarized light with a suitable phase mask allows eliminating this z-polarization component again through destructive interference with other beamlets (not shown).

incoming STED beam can cause imperfect destructive interference at the focus center.

Additionally, beam aberrations caused by the instrument itself or by samples with an inhomogeneous refractive index can create different phase delays for different parts of the incoming beam, which distorts the wavefront and results in imperfect STED foci. It has been shown that adaptive optics allows correcting for these wavefront distortions [8].

10.3 Experimental Setup

A STED microscope is generally based on an LSM to which a third beam path for the STED beam is added to the excitation and detection beam paths (Figure 10.2a). In a typical STED setup, a phase mask shaped like a vortex, varying the phase delay linearly with the rotation angle from 0 to 2π (Figures 10.2b and 10.4a), is placed in the path of the STED laser and imaged into the back aperture of the objective lens. The modified STED laser beam thereby generates a doughnut-shaped focus in the sample, with an intensity zero at its center as described previously. The intensity zero is co-aligned with the intensity maximum of the excitation laser.

10.3.1 Light Sources and Synchronization

Optimal resolution enhancement in STED microscopy has typically been achieved using pulsed laser systems. In comparison to STED microscopes using CW lasers, these systems are, however, relatively complex, as they require synchronization between the excitation and STED laser pulses and often additional components to stretch the typically subpicosecond pulses provided by the laser source to >100 ps (Section 10.2.2.1). Pulsed laser diodes or mode-locked titanium:sapphire lasers have been most commonly used, but they are increasingly being replaced by pulsed fiber lasers which directly emit pulses of several hundred picosecond length and do not require pulse stretching. Lasers with short (i.e., femtosecond) laser pulses, such as titanium:sapphire lasers, require pulse stretching, which typically is accomplished using diffraction gratings or more simply by using dispersion in long single-mode optical fibers. Pulse synchronization requires triggering between the excitation and STED laser sources (e.g., by electronically triggering pulsed laser diodes to the pulses of a mode-locked titanium:sapphire laser) or the use of inherently synchronized sources (e.g., by creating excitation light from the white light spectrum of a photonic crystal pumped by a mode-locked titanium:sapphire laser or using a commercially available super-continuum light source). Synchronized pulses also require adjustment of the timing in which each pulse arrives at the sample to ensure that the depletion pulse immediately follows the excitation pulse. This adjustment is achieved using either an electronic delay between electronically triggered laser sources or an optical delay stage to change the relative length of the beam paths.

STED microscopy with CW lasers greatly simplifies the experimental setup by eliminating the need for pulse stretching and synchronization. However, to produce similar resolution as with pulsed lasers, three- to fivefold higher depletion intensities are required with CW lasers.

In addition to synchronizing excitation and STED lasers to each other, the detection windows can be synchronized to the laser pulses. This restricts detection to fluorescence that has been emitted after the STED laser has had sufficient time to act on the excited fluorescent molecules and suppresses background light stemming from molecules outside the doughnut zero that have emitted fluorescence before they could be depleted. Time-gated detection both enhances image contrast and reduces collection of background signal, and consequently it is increasingly used in both CW and pulsed STED configurations (including commercial STED systems) [9].

When choosing the ideal laser source, the available wavelengths also have to be taken into account. Owing to photophysical wavelength restrictions (Section 10.2.2.2), the STED wavelength, in particular, has to be matched to the used fluorophore. In practice, however, it is often the case that the fluorophores to be used are chosen to be compatible with the available STED wavelength.

10.3.2 Scanning and Speed

STED microscopy is typically implemented in a point scanning geometry, and therefore image pixels are generated by scanning either the laser focus (using galvanometer-driven mirrors) or the sample (using a piezo-stage). While a setup that implements sample scanning certainly results in a simpler optical arrangement, this mode of scanning is usually too slow for live-cell imaging. Much faster imaging is possible using scanning mirrors to translate the laser foci through the sample.

To satisfy the Shannon/Nyquist sampling frequency, the pixel size in confocal microscopy must be less than one-half of the resolution of the system. It is important to note that the effective PSF used in STED microscopy inherently requires smaller pixels, which, at a given scanning speed, results in a reduced dwell time for each pixel. Hence, for a STED image with an m-fold improved resolution over the diffraction limit, an m^2 slower frame rate is required to maintain the same pixel dwell time as in a diffraction-limited recording. To compensate this loss in imaging speed, the intensity of the excitation laser can be increased to a certain degree; however, saturation of the fluorophore emission ultimately limits the number of photons that can be detected in a given time interval. In any case, the size of the scanned field of view can be reduced to increase imaging speeds.

STED images can also appear dimmer than conventional microscope images owing to the fact that only a fraction of an object might fluoresce at any given time as a result of the sharper STED PSF. It is important to note, however, that this loss in signal does not correspond to a loss of structural information (Figure 10.6). In fact, small objects below the STED resolution will appear only marginally dimmer when imaged by STED microscopy compared to the regular confocal image but appear much sharper (Figure 10.6e,i). Large clusters of these objects, on the other hand, which might appear much brighter in the

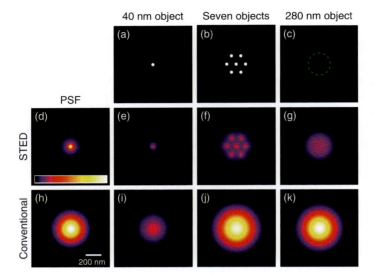

Figure 10.6 The relative brightness of STED and conventional microscope images can vary strongly depending on the shape of the imaged object as simulated here for three different structures (a–c) and two PSFs (STED: Lorentzian shaped with 50 nm FWHM; conventional: Gaussian shaped with 235 nm FWHM) (d, h). The peak brightness of an individual object of sub-STED resolution size is comparable in both microscopes (e, i). For larger objects, however, the conventional images are much brighter because the larger PSF integrates over neighboring areas (j, k). Objects with the same total amplitude but different subdiffraction structures (b, c) are practically not distinguishable. The higher STED resolution, on the other hand, allows resolving the different structure but leads to significantly dimmer images because the integration effect is avoided (f, g). The structures shown in (b) and (c) have the same total signal. The dashed circle in (c) was added to make the dimension of the 280 nm circle more discernible.

conventional image owing to an overlap of the fluorescence images of each of the cluster components, can be resolved in the STED microscope even though the signal drops significantly (from the level of summed up cluster signal to that of an individual object) (Figure 10.6f,j). In both cases, the structural information content of the STED images has been much improved over that of the conventional image, even though the image brightness has been significantly decreased in the latter case.

10.3.3 Multicolor STED Imaging

As with any modern fluorescence microscopy technique, there is a strong need for multicolor imaging in order to investigate the spatial relationship of different sample components. Unfortunately, extending STED microscopy to dual-color imaging is hampered by the fact that two laser wavelengths are required for each staining. For example, a conventional dual-color imaging scheme using two dyes required complex laser systems and the precise alignment of four laser beams [10]. Furthermore, this approach required sequential imaging of each channel owing to the efficient excitation of the red-shifted dye by the high-powered STED laser used for depletion of the blue-shifted dye. Consequently, the red-shifted dye was strongly bleached when imaging the blue-shifted channel. Imaging the red channel first allows recording a single image for this channel but no repeated imaging after recording the other channel. Dual-color imaging is also possible using just one STED laser to deplete both dyes. This method simplifies the experimental setup, but the fact that the emission spectra of the two dyes overlap at least partially has to be taken into account. Cross-talk between the two detection channels can be reduced to practically negligible levels with optimized excitation wavelengths and spectral detection ranges along with sequential imaging of the two channels where the microscope quickly switches between the two excitation wavelengths [11]. Figure 10.7 shows an example of a dual-color STED image taken using this approach. The advent of lasers offering a super-continuum spectrum provides flexibility for dual-color STED, which previously had not been available. Another option is to distinguish fluorophores by their different fluorescence lifetimes [12]. In this case, technically only one laser is required for excitation and one for stimulated depletion, which simplifies the setup significantly.

10.3.4 Improving Axial Resolution in STED Microscopy

Increasing lateral resolution using a doughnut-shaped depletion beam as discussed above does not confine the focal volume in the axial direction. As typically

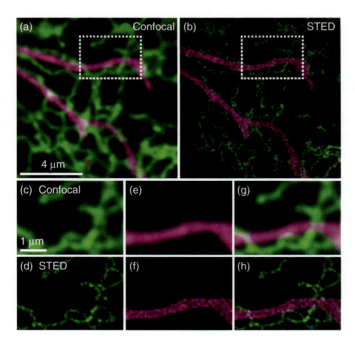

Figure 10.7 Example of a dual-color STED image of the endoplasmic reticulum (ER) and mitochondria. (a) Overlay of confocal images of ER labeled with Atto594 (green) and mitochondria labeled with ATTO647N (magenta). (b) Corresponding STED image. (c–h) Zoom-in of boxed areas in (a) and (b). Images were obtained using a custom-built STED instrument and smoothed by Gaussian filters.

implemented in a confocal geometry, STED inherently provides optical sectioning. However, without additional effort to axially confine the focal volume, the extent of this sectioning is limited to a conventional confocal resolution of typically >500 nm.

Increasing the axial resolution in STED microscopy requires confining the axial extent of the focal volume by appropriately modulating the phase of the depletion laser. Introducing a π phase shift into the central region of the depletion laser, for example, results in a focus that features a central intensity zero with high intensity above and below the focal plane, which reduces the axial extent of the effective PSF to sub-diffraction size [13] (Figure 10.4b). This depletion pattern can be combined with an additional depletion beam with the typical doughnut shape to efficiently reduce the size of the effective PSF below the diffraction limit in all directions [14].

To increase axial resolution further, STED microscopy can be implemented in a 4Pi geometry (i.e., using two opposing objective lenses). This approach has been demonstrated to generate spherical focal volumes of diameter <45 nm in the so-called isoSTED configuration [15].

As an alternative approach to increase the optical sectioning power in STED microscopy, the lateral resolution improvement provided by a single STED beam can be combined with evanescent field excitation to limit the axial extent of the focal volume [16]. While this method increases the axial resolution without the use of multiple depletion beams or objective lenses, it is limited to imaging structures at the sample–substrate interface.

10.4 Applications

The usefulness of any new microscopy method is measured by its ability to address scientific questions. STED microscopy has been one of the most successful super-resolution methods in terms of biological applications, but several factors must be considered when applying this technique. In particular, the choice of fluorophores (Section 10.4.1) and the labeling strategy (Section 10.4.2) are essential when planning a STED microscopy experiment.

10.4.1 Choice of Fluorophore

Several dye properties must be considered when choosing a fluorophore for STED microscopy. First, the emission spectrum of the fluorophore must be compatible with the wavelength of the STED laser being used. Generally, the STED laser should be in the range of the red-shifted tail of the emission spectrum of the dye (Figure 10.2c). If the emission spectrum and STED wavelength are matched poorly, the increase in resolution will be compromised, as SE will not be as efficient or excitation by the STED wavelength will cause background fluorescence (Section 10.2.2.2). Second, fluorophores with relatively long fluorescent lifetimes are desired because they allow more time to efficiently stimulate emission. The ideal fluorophore for STED microscopy has an emission spectrum that matches the STED wavelength of the instrument being used, long fluorescence lifetimes, high quantum yield, and a strong resistance to photobleaching.

Because the resolution of STED microscopy relies on the characteristic saturation intensity of the fluorophore (Eq. 10.6) and its ability to switch between different optical states, the photobleaching behavior and photostability of the fluorophores used are critical for optimal STED imaging (Section 10.2.2.1). Fluorophores with small intersystem crossing rates from the singlet into the triplet state are especially well suited for STED microscopy, as this property helps to avoid photobleaching of the probe, which often is linked to the triplet state population. Reducing and oxidizing systems (ROXSs) can also be used to help depopulate the photobleaching-prone triplet states, resulting in increased fluorophore photostability and allowing the use of higher intensity STED beams [17].

The two major classes of fluorophores available for STED microscopy are FPs and small-molecule organic dyes. While FPs are typically not as bright or photostable as organic dyes, green fluorescent protein (GFP) and yellow fluorescent protein (YFP) variants have been used successfully for STED microscopy experiments [18]. Small-molecule organic dyes (especially ATTO, DY, and Abberior STAR dyes) have been more widely used for STED microscopy, as they are available with emission spectra that are compatible with commonly used STED wavelengths, are extremely bright, are photostable, and have long fluorescence lifetimes. For example, ATTO647N is unlikely to decay into a triplet state which results in photobleaching, making it one of the most popular fluorophores for STED microscopy. While most organic fluorophores are not compatible with live-cell STED imaging of intracellular structures, a small number of dyes have been identified that are cell-permeable and have minimal nonspecific binding. For example, ATTO590 and SiR dyes can label structures deep inside cells and have been used for live-cell STED imaging utilizing a 775 nm STED laser [11].

Using a high-intensity STED beam in the far-red region for imaging results in less photodamage to the sample; therefore, imaging in this region is generally preferable for live-cell applications. A list of fluorophores successfully used for STED microscopy and the corresponding references can be found at the following website from the Hell lab at the Max Planck Institute of Biophysical Chemistry: http://nanobiophotonics.mpibpc.mpg.de/old/dyes/.

10.4.2 Labeling Strategies

The implementation of STED microscopy in biological imaging applications requires that the STED probes are specifically targeted to the protein of interest. Three main strategies are used for targeting STED probes: FPs, protein-mediated labeling, and immunofluorescence. FPs and protein-mediated labeling tags are fused to a protein of interest since they are genetically encodable. While these tags are live-cell-compatible, their large size can potentially perturb the localization and function of the protein of interest. Immunofluorescence, on the other hand, is generally not compatible with live-cell imaging except when labeling extracellular targets. FPs are susceptible to photobleaching since they are usually not nearly as photostable as the small-molecule organic dyes used for protein-mediated labeling and immunofluorescence. Protein-mediated labeling

Table 10.1 Comparison of labeling strategies for STED microscopy.

Labeling strategy	Advantages	Disadvantages
Immunofluorescence	• Compatible with organic dyes • Some great antibodies available • Does not rely on overexpression	• Fixed cells only, except for surface targets • Good antibodies not always available for protein of interest • Can add 10–20 nm on structure
Fluorescent proteins	• Genetically encodable – specific targeting • Live-cell compatible • Low cytotoxicity • Many multicolor options available	• Brightness/quantum yield not great – difficult to record many frames because of photobleaching • Few options for far-red region • Large and bulky – some proteins of interest will not tolerate tag • Tendency to aggregate
Protein-mediated (SNAP-tag, HaloTag, etc.)	• Genetically encodable – specific targeting • Compatible with organic dyes • Live-cell compatible	• Many STED-compatible organic dyes are not cell-permeable • Large and bulky – some proteins of interest will not tolerate tag • Additional staining and washing steps required

strategies such as SNAP, CLIP, and Halo tags combine the robustness of organic dyes with the live-cell compatibility of genetically encodable tags. These labeling approaches have been used with cell-permeable dyes for live-cell STED imaging of intracellular targets [11]. All of the above labeling techniques are discussed in detail in Chapter 4. Their advantages and disadvantages for STED microscopy experiments are summarized in Table 10.1.

10.5 Summary

The diffraction barrier no longer imposes a limit on the resolution of far-field fluorescence microscopes. The key to surpass the diffraction limit resides in the interaction between light and fluorophores. Through SE, the targeted switching of fluorophores effectively allows for the production of focal volumes no longer limited by diffraction. The resolving power of STED microscopy is limited only by how well its theoretical foundations can be put into practice. Under optimal conditions, resolution down to the nanometer scale has already been reported [6].

In recent years, STED microscopy has been extended to three-dimensional, multicolor, and live-cell imaging. These innovations are already proving to be invaluable tools for cell biology. RESOLFT microscopy using reversibly switching FPs or other fluorophores with on/off transitions that require low light intensities is a compelling alternative to STED microscopy. Given recent advances in imaging speed for this super-resolution modality, the ability to image subcellular dynamics with reduced phototoxicity should be of broad appeal. Future developments in the availability of lasers and fluorophores optimized for targeted-switching are sure to push these techniques even further in addressing biological questions.

References

1 Hell, S.W. and Wichmann, J. (1994) Breaking the diffraction resolution limit by stimulated emission: stimulated-emission-depletion fluorescence microscopy. *Opt. Lett.*, **19**, 780–782.
2 Hell, S.W. (2015) Nanoscopy with focused light (Nobel lecture). *Angew. Chem. Int. Ed.*, **54**, 8054–8066.
3 Hell, S.W. (2009) Microscopy and its focal switch. *Nat. Methods*, **6**, 24–32.
4 Chmyrov, A., Keller, J., Grotjohann, T., Ratz, M., d'Este, E., Jakobs, S., Eggeling, C., and Hell, S.W. (2013) Nanoscopy with more than 100,000 'doughnuts'. *Nat. Methods*, **10**, 737–740.
5 Schneider, J., Zahn, J., Maglione, M., Sigrist, S.J., Marquard, J., Chojnacki, J., Krausslich, H.G., Sahl, S.J., Engelhardt, J., and Hell, S.W. (2015) Ultrafast, temporally stochastic STED nanoscopy of millisecond dynamics. *Nat. Methods*, **12**, 827–830.
6 Rittweger, E., Han, K.Y., Irvine, S.E., Eggeling, C., and Hell, S.W. (2009) STED microscopy reveals crystal colour centres with nanometric resolution. *Nat. Photonics*, **3**, 144–147.
7 Vicidomini, G., Moneron, G., Eggeling, C., Rittweger, E., and Hell, S.W. (2012) STED with wavelengths closer to the emission maximum. *Opt. Express*, **20**, 5225–5236.
8 Gould, T.J., Burke, D., Bewersdorf, J., and Booth, M.J. (2012) Adaptive optics enables 3D STED microscopy in aberrating specimens. *Opt. Express*, **20**, 20998–21009.
9 Eggeling, C., Willig, K.I., Sahl, S.J., and Hell, S.W. (2015) Lens-based fluorescence nanoscopy. *Q. Rev. Biophys.*, **48**, 178–243.
10 Donnert, G., Keller, J., Wurm, C.A., Rizzoli, S.O., Westphal, V., Schonle, A., Jahn, R., Jakobs, S., Eggeling, C., and Hell, S.W. (2007) Two-color far-field fluorescence nanoscopy. *Biophys. J.*, **92**, L67–L69.
11 Bottanelli, F., Kromann, E.B., Allgeyer, E.S., Erdmann, R.S., Wood Baguley, S., Sirinakis, G., Schepartz, A., Baddeley, D., Toomre, D.K., Rothman, J.E., and Bewersdorf, J. (2016) Two-colour live-cell nanoscale imaging of intracellular targets. *Nat. Commun.*, **7**, 10778.
12 Buckers, J., Wildanger, D., Vicidomini, G., Kastrup, L., and Hell, S.W. (2011) Simultaneous multi-lifetime multi-color STED imaging for colocalization analyses. *Opt. Express*, **19**, 3130–3143.
13 Klar, T.A., Jakobs, S., Dyba, M., Egner, A., and Hell, S.W. (2000) Fluorescence microscopy with diffraction resolution barrier broken by stimulated emission. *Proc. Natl. Acad. Sci. U.S.A.*, **97**, 8206–8210.
14 Harke, B., Ullal, C.K., Keller, J., and Hell, S.W. (2008) Three-dimensional nanoscopy of colloidal crystals. *Nano Lett.*, **8**, 1309–1313.

15 Schmidt, R., Wurm, C.A., Jakobs, S., Engelhardt, J., Egner, A., and Hell, S.W. (2008) Spherical nanosized focal spot unravels the interior of cells. *Nat. Methods*, **5**, 539–544.
16 Gould, T.J., Myers, J.R., and Bewersdorf, J. (2011) Total internal reflection STED microscopy. *Opt. Express*, **19**, 13351–13357.
17 Kasper, R., Harke, B., Forthmann, C., Tinnefeld, P., Hell, S.W., and Sauer, M. (2010) Single-molecule STED microscopy with photostable organic fluorophores. *Small*, **6**, 1379–1384.
18 Willig, K.I., Kellner, R.R., Medda, R., Hein, B., Jakobs, S., and Hell, S.W. (2006) Nanoscale resolution in GFP-based microscopy. *Nat. Methods*, **3**, 721–723.

11

Fluorescence Photobleaching Techniques
Reiner Peters

The Rockefeller University, Laboratory of Mass Spectrometry and Gaseous Ion Chemistry, 1230 York Avenue, New York 10065 NY, USA

11.1 Introduction

In the 1960s, biological membranes were generally believed to consist of lipid bilayers covered over both surfaces by unfolded protein monolayers (Davson–Danielli–Robertson model). This view was superseded in the early 1970s by the fluid mosaic model [1]. On the basis of thermodynamic considerations, the protein–lipid ratio of membranes, electron microscopic studies, and cell fusion experiments [2], that model assumed that fluid lipid bilayers form the basis of cellular membranes while membrane proteins occur in their natively folded conformations and are either attached to or integrated into the bilayer. The fluid mosaic model furthermore implied that membranes are fluids rather than solids and that membrane lipids and proteins are mobile in the membrane plane. The far-reaching implications of lateral mobility in membranes triggered off an intensive search for methods by which this hypothesis could be tested. Cone combined flash photolysis with microscopic absorption measurements to determine the rotational and lateral mobility [3] of rhodopsin in photoreceptor membranes. Simultaneously, Peters *et al.* [4] established the method now known as *fluorescence recovery after photobleaching* (FRAP) or *fluorescence photobleaching recovery* (FPR). In the original FRAP method, an intense light beam was used to rapidly bleach the fluorescence of a circumscribed membrane region, for example, a hemisphere of a resealed red cell membrane or a circular spot of the plasma membrane of cultured cells. The dissipation of the fluorescence inhomogeneity in the membrane was recorded by fluorescence measurements and used to derive diffusion coefficients. In the following decades, FRAP experienced a profound development and grew into a large family of methods widely applied in biomedicine and physical sciences. Tellingly, some recent family members rely neither on photobleaching nor on fluorescence recovery any longer.

Fluorescence Microscopy: From Principles to Biological Applications, Second Edition.
Edited by Ulrich Kubitscheck.
© 2017 Wiley-VCH Verlag GmbH & Co. KGaA. Published 2017 by Wiley-VCH Verlag GmbH & Co. KGaA.

11.2 Basic Concepts and Procedures

In this section, the fundamentals shared by the diverse photobleaching techniques are described. These include the initial concept for turning photobleaching into a useful analytical tool, basic instrumentation, and the extension of the initial concept to binding processes and transmembrane transport.

11.2.1 One Principle, Several Modes

Photobleaching techniques depend on the intimate amalgamation of microscopy and fluorescence. We will first consider the relevance of fluorescence, and subsequently that of microscopy in this liaison.

Fluorescence provides a high degree of sensitivity and specificity and is also the handle for photochemical manipulations such as photobleaching. Also referred to as *fading*, photobleaching is a long-known property of fluorophores and was, before FRAP, conceived as purely negative. When fluorophores are illuminated, absorbing and re-emitting photons, and cycling between ground and higher singlet states, there is always a small probability that they enter a long-lived state, undergo a chemical reaction such as oxidation, and become nonfluorescent. The *quantum efficiency of photobleaching* , defined as the probability of photobleaching per absorbed photon, varies among different fluorophores between wide limits. For organic dyes optimized for photostability, it is in the order of 10^{-6} but for fluorescent proteins it is only 10^{-5}. Thus, the average maximum number of photons emitted by a given fluorophore before photobleaching, that is, the quotient of fluorescence quantum yield and quantum efficiency of photobleaching, varies roughly between 50 000 and 500 000.

An example of a lateral diffusion measurement by FRAP is given in Figure 11.1. Later, we will discuss (Sections 11.3 and 11.5) how such experiments are quantitatively evaluated for kinetic parameters such as lateral diffusion coefficients, binding constants, and membrane transport rates.

The experiment shown in Figure 11.1 was performed with a confocal laser scanning microscope (CLSM). However, at the time FRAP was conceived, CLSMs were not yet available. Instead, fluorescence microscopes were employed, in which a single stationary light beam was used for illumination and photobleaching and a photomultiplier tube for measuring the integral fluorescence of the illuminated spot. A FRAP measurement employing such an instrument is shown schematically in Figure 11.2a,c. In the hypothetical experiment, the sample is a peripheral part of a cultured cell, and the scheme at the top of the figure indicates its tremendous complexity. Thus the thin cytoplasmic layer contains many copies of a freely diffusing fluorescent molecule (green dots) but also filaments carrying binding sites for the fluorescent molecules (red dots) as well as a number of membrane-bounded vesicles permeable for the fluorescent molecules (Figure 11.2a,b). The fluorescent molecules are distributed throughout the system according to their diffusion, binding, and permeability properties. A small region of the sample, referred to in the following as the *region of photolysis* (ROP), is illuminated by a focused light beam, and the integral fluorescence originating from the ROP is measured by a photomultiplier.

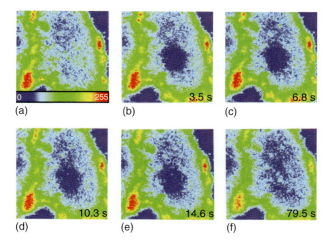

Figure 11.1 An example of a FRAP measurement of lateral diffusion in membranes. Cultured cells were incubated with a fluorescent lipid analog that spontaneously inserts into plasma membranes, and one of the cells was imaged by confocal laser scanning microscopy. At zero time, a circular membrane area of ~7 μm diameter was photobleached. Subsequently, the equilibration of the fluorescence inhomogeneity was recorded at the times indicated in (b)–(f). The fluorescence intensity was color-coded according to the scale shown in (a). (Wedekind et al. 1994 [5]. Reproduced with permission of John Wiley & Sons.)

In FRAP experiments (Figure 11.2a,c), the intensity of illumination I is varied over time, as indicated at the bottom of Figure 11.2c. Initially, the illumination intensity is very small, yielding the signal $F(-)$. Then, the intensity is suddenly increased by three to four orders of magnitude for a short time, causing "instantaneous" photobleaching and depletion of the ROP from fluorescence. Starting immediately after bleaching ($t = 0$), the dissipation of the local fluorescence inhomogeneity is monitored by recording the integral fluorescence of ROP at the initial small light intensity. This gives rise to a fluorescence recovery curve, which starts at level $F(0)$ and approaches the level $F(\infty)$ at long times. If some fluorescent molecules are immobile on the time scale of the experiment, $F(\infty)$ is smaller than $F(-)$, and the immobile fraction is

$$f_{\text{immobile}} = \frac{F(-) - F(\infty)}{F(-) - F(0)} \tag{11.1}$$

Figure 11.2 also indicates that photobleaching experiments can be performed in a different mode, referred to as *continuous fluorescence photobleaching* or *continuous fluorescence microphotolysis* (CFM) (Figure 11.2b,d). In that case, the ROP is illuminated at a constant intensity while the emitted fluorescence is recorded simultaneously. The intensity of illumination is adjusted to an intermediate level, inducing a substantial rate of photobleaching, so that the resultant fluorescence gradient between the ROP and surroundings is counteracted by diffusion of fresh fluorescent molecules from the surroundings into the ROP. Together, the two competing processes yield a fluorescence decay, which is steep in the beginning and becomes increasingly shallow with time. Immobile molecules are bleached

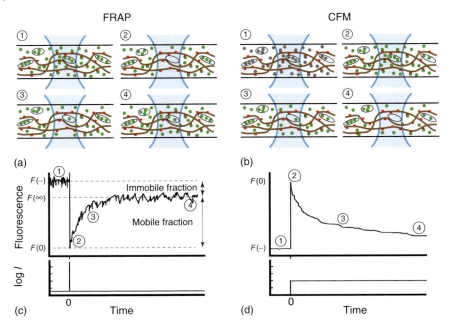

Figure 11.2 Schematic representation of FRAP and CFM measurements. A thin cytoplasmic layer containing many copies of a fluorescent protein (green dots), filaments (brown lines) with binding sites for the fluorescent protein (red dots), and vesicles (black shapes) is shown in cross section together with a focused laser beam. In FRAP experiments (a, c), the laser beam power I is modulated between a very low level and a very high level, as indicated at the bottom. This results in a tripartite fluorescence intensity curve comprising the prebleach fluorescence $F(-)$, the fluorescence immediately after photobleaching $F(0)$, and the fluorescence recovery toward the equilibrium $F(\infty)$. In CFM experiments (b, d), the object is continuously illuminated at a power intermediate between the levels employed in FRAP. The resultant fluorescent decay reflects depletion of the illuminated area from fluorophore by photobleaching and their simultaneous partial replenishment by diffusion from adjacent areas into the bleached area. The numbers indicate different phases of the experiments.

at the onset of illumination so that the immobile fraction can be determined from the initial slope.

On comparing FRAP with CFM, it is obvious that CFM is the general mode of the photobleaching methodology, while FRAP is a special case characterized by the conditions of "instantaneous" photobleaching during the photolysis step and "negligible" photobleaching during fluorescence measurements. In fact, FRAP is just one of several special photobleaching modes, as will be shown later on (cf. Section 11.4.3). The comparison of FRAP and CFM also suggests that CFM instrumentation is simpler because there is no need for creating high-intensity laser flashes. Finally, using the same sample, the measuring signal will be much stronger in the CFM mode because a much higher excitation intensity is permitted.

To understand why photobleaching techniques, by necessity, are microscopic methods, the time required for a molecule to cross the ROP by diffusion may be considered. This time essentially determines how fast the fluorescence recovery in FRAP experiments and the fluorescence decay in CFM experiments can be. The mean time for crossing a distance l by diffusion is

$$\tau_{\text{diff}} = \frac{l^2}{nD} \qquad (11.2)$$

where D is the diffusion coefficient and n equals 2, 4, or 6 depending on whether one, two, or three spatial dimensions are involved. The effective diffusion coefficient of lipids in membranes is typically $\sim 1\ \mu m^2\ s^{-1}$, while that of proteins in aqueous phases of the cell varies between zero in the case of tight binding to immobile structures to 20 $\mu m^2\ s^{-1}$ in the case of free diffusion in aqueous phases of the cell. Employing a microscopic ROP of 2 µm, τ_{diff} is ~ 1 s for lipid diffusion and ~ 30 ms for free protein diffusion. However, if the ROP had macroscopic dimensions (e.g., 1 mm radius), τ_{diff} would be impractically large, namely $\sim 2.5 \times 10^5$ s for lipids and $\sim 10^4$ s for fast proteins.

11.2.2 Setting up an Instrument

A basic FRAP/CFM instrument can be easily set up from commercial components. It is based (Figure 11.3) on a regular fluorescence microscope with phase-contrast or differential-interference-contrast optics and components for the visual observation of fluorescence in various spectral regions. These comprise an intense light source with a broad spectral output and a vertical illuminator with interchangeable sets of dichroic mirrors and emission filters. Visual observation by phase or interference contrast and fluorescence is essential to bring the sample into focus, to identify its features, and to select the site at which measurements are done.

The fluorescence microscope is equipped with lasers providing emission lines at different wavelengths. Each laser line should have a power of up to 50–100 mW, which is necessary and sufficient to make photobleaching sufficiently fast for FRAP experiments. The laser beams are coaxially aligned and passed through an acousto-optical modulator (AOM). With the help of the AOM, the intensity of

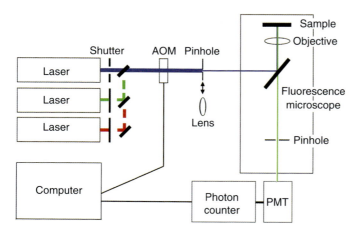

Figure 11.3 Scheme of a basic instrument for FRAP and CFM experiments. For details, see text.

the lasers beam can be changed within microseconds by a factor of up to ~2000, and thus the photobleaching flash typical for FRAP experiments can be created. In the scheme shown in Figure 11.3, the laser beam leaving the AOM is used to illuminate a field diaphragm, which is imaged by the objective into the object, thus defining the ROP. The ROP is frequently circular. However, other geometries such as strips are widely used. Arrays of parallel strips and other periodic patterns have also been employed and are advantageous when studying extended objects such as whole cells or large artificial membranes.

An alternative way of creating the ROP does not involve a field diaphragm. Instead, the laser beam is focused before entering the microscope, and the laser beam waist is imaged by the objective into the object. ROPs created by the field diaphragm or the prefocus methods differ in an important aspect. The field diaphragm method yields, in principle, a uniform profile, while the prefocus methods yields a Gaussian profile. Consequences of this difference for data evaluation are considered in Section 11.3.

Another essential component of a FRAP instrument is a sensitive photodetector. This may be a photomultiplier tube operated in the single-photon counting mode or an avalanche photodiode. For mounting the photodetector, the microscope has to have a suitable port, which includes a second field diaphragm (Figure 11.3). The field diaphragm on the detector side should match the one on the illumination side such that their images in the object coincide in position and size. This provides a confocal effect, partly rejecting background fluorescence.

Finally, a computer is required to coordinate the laser shutters, to time the AOM, to down-gate the photodetector during the bleaching step, to store signals generated by the photodetector, to plot measuring curves, and to fit them by appropriate model functions.

11.2.3 Approaching Complexity from Bottom up

As indicated in Figure 11.2, biological systems are extremely heterogeneous and complex. The cytoplasm, for instance, consists of an aqueous phase, which is crowded by molecules of all sizes and shapes, and in addition contains an abundance of large structures such as macromolecular complexes, cytoskeletal filaments, membrane-bounded vesicles, and cell organelles. The diffusion of fluorescent probes in such media is restricted by steric parameters and influenced by reactions with other molecules and transport across membranes. All these parameters shape the fluorescence recovery and decay curves obtained by FRAP or CFM and, in principle, can be recovered from the experimental data by fitting them with appropriate theoretical models. When approaching this complexity, it is useful to take a pathway that was also the historical one: analyzing limiting cases first, and then proceeding from limiting to more general cases (Figure 11.4).

In the case of live-cell measurements, at least the following parameters are involved: diffusion coefficients, on- and off-rates, and membrane transport coefficients. These parameters can be analyzed separately when diffusion, binding, or membrane transport is rate-limiting. Limiting cases can be sometimes arranged for because transport parameters depend differently either on basic system parameters such as temperature or on experimental conditions such as

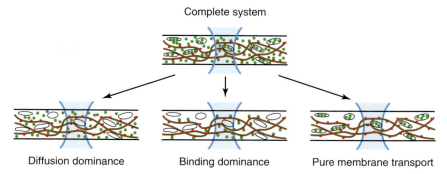

Figure 11.4 Parameters determining the shape and kinetics of FRAP and CFM curves in live cell measurements. A thin cytoplasmic layer with several copies of a fluorescent protein (green dots), filaments with binding sites for the fluorescent protein (red dots), and vesicles is shown. The mobility of probe molecules in biological systems is a convolution of free diffusion, chemical reactions, and membrane transport. The parameters of these processes all shape the measuring curves obtained by FRAP and CFM. Conceptually, the complexity can be deconstructed into limiting cases such as diffusion dominance, binding dominance, and pure membrane transport. Knowledge the characteristics of the limiting cases helps in analyzing the complete system.

the size of the ROP. For instance, diffusion times depend on the square of size of the ROP, while the kinetics of pure binding is independent of the ROP size. The membrane transport kinetics, on the other hand, depends on the volume of the membrane-bound vesicles. Nevertheless, a complete analysis of lateral transport in biological systems remains a tremendous challenge because of a complex submicroscopic topography and a multitude of different binding and membrane transport processes.

11.3 Fluorescence Recovery after Photobleaching (FRAP)

In this section, FRAP is considered in more detail. First, an often-used theoretical model for diffusion measurements [6] is recapitulated, and then the extension of FRAP to reversible binding and to membrane transport processes is summarized.

11.3.1 Evaluation of Diffusion Measurements

The theoretical description of FRAP measurements, in which a small circular spot of a fluorescently doped or labeled membrane is rapidly photobleached and the recovery of the fluorescence in the bleached spot is followed by repetitive measurements, has four parts: the relative concentration of fluorophores after photobleaching is computed, the relationship between fluorophore concentration and measured fluorescence is established, the diffusion equation is solved for appropriate initial and boundary conditions, and the diffusion coefficient is extracted by fitting the theoretical model to experimental data.

Experimental studies show that the photobleaching reaction can be approximated by an irreversible first-order reaction. The time dependence of the fluorophore concentration $C(r, t)$ is then

$$\frac{\partial C(r,t)}{\partial t} = -\alpha I(r) C(r,t) \tag{11.3}$$

where α is a constant and $I(r)$ is the bleaching intensity. If the bleach interval T is much smaller than the characteristic time of diffusion, the fluorophore concentration at the end of the bleach interval ($t = 0$) is

$$C(r, 0) = C_0 e^{-\alpha T I(r)} \tag{11.4}$$

where C_0 is the initial uniform fluorophore concentration. The normalized concentration at the center of the ROP at the end of bleaching is referred to as the *amount of bleaching* given by

$$K = \alpha T I(0) \tag{11.5}$$

As mentioned in Section 11.2.2, the photobleaching light beams can be created in two ways. Focusing a TEM_{00} mode laser beam directly into the sample yields a spot with a Gaussian intensity profile. In contrast, if an illuminated circular aperture is imaged into the sample, the intensity profile of the image is ideally uniform, that is, box shaped. It is therefore useful to consider both cases. For a Gaussian intensity profile, $I(r)$ is given by

$$I(r) = \frac{2P_0}{\pi w^2} e^{-\frac{2r^2}{w^2}} \tag{11.6}$$

where w is the half width at e^{-2} height and P_0 is the total laser power. For a uniform circular beam, the intensity distribution is

$$I(r) = \begin{cases} \frac{P_0}{\pi w^2}, & r \leq w \\ 0, & r > w \end{cases} \tag{11.7}$$

Here w corresponds to the circle radius. Together, Eqs (11.4)–(11.7) permit computing initial concentration profile $C(r,0)$. Examples for the intensity profile at different K-values are shown in Figure 11.5a,b. It can be recognized that for Gaussian beams the diameter of the ROP increases with K, while the concentration decreases relatively slowly. In contrast, for uniform beams, the size of the ROP remains the same with increasing K, while the concentration decreases much more rapidly.

The time-dependent fluorescence $F(t)$ observed after bleaching is given by

$$F(t) = \frac{q}{A} \int I(r) C(r,t) d^2 r \tag{11.8}$$

where q is the product of all quantum efficiencies of light absorption, emission, and detection; A is the attenuation factor during measurement; and $C(r, t)$ is fluorophore concentration.

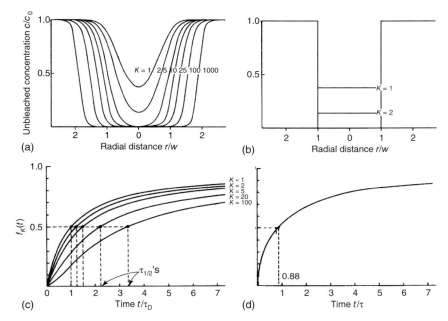

Figure 11.5 Theoretical analysis of 2D diffusion measurements by FRAP according to Axelrod et al. [6]. The light beam used to induce fluorescence and photobleaching in a small circular spot may either have a Gaussian (a, c) or a circular uniform (b, d) intensity profile. (a) In the case of a Gaussian beam, the intensity profile of the bleached spot immediately after bleaching is also approximately Gaussian at small values of K, a factor indicating the amount of bleaching. With increasing K, the bleach profile becomes broader and approaches a rectangular shape. (b) In the case of a uniform beam profile, the bleach profile is rectangular with constant diameter for all K values. (c) The fluorescence recovery curves also depend on K, and the halftime of recovery $\tau_{1/2}$ is a function of K. (d) At this condition, the fluorescence recovery kinetics are independent of K and characterized by a constant halftime ($\tau_{1/2} = 0.88$). (Adapted from Axelrod et al. 1976 [6]. Reproduced with permission of Elsevier.)

For an axially symmetric, two-dimensional (2D) system, the diffusion equation may be written as

$$\frac{\partial C(r,t)}{\partial T} = D\nabla^2 C(r,t) \tag{11.9}$$

where ∇^2 is the Laplace operator and D is the diffusion coefficient. Solving Eq. (11.9) with the initial conditions for a Gaussian beam profile yields

$$F(t) = F(-)\sum_{\infty}^{n=0}\left[\frac{(-K)^n}{n!}\right]\left[1+n\left(1+\frac{2t}{\tau_D}\right)\right]^{-1} \tag{11.10}$$

where $\tau_D = w^2/4D$, the characteristic diffusion time, and the bleach parameter K can be determined from the equation

$$F_K(0) = F(-)K^{-1}(1-e^{-K}) \tag{11.11}$$

For the uniform intensity profile, solving Eq. (11.9) yields

$$f(t) = 1 - \left(\frac{\tau_D}{t}\right) e^{\left(-\frac{2\tau_D}{t}\right)} I_0\left(\frac{2\tau_D}{t}\right) + I_2\left(\frac{2\tau_D}{t}\right) + 2 \sum_\infty^{k=0} \frac{(-1)^k (2k+2)!(k+1)!\left(\frac{\tau_D}{t}\right)^{k+2}}{(k!)^2 [(k+2)!]^2}$$

(11.12)

However, Eq. (11.12) does not yield stable results at small times. Therefore, an alternative solution was derived [7], which is stable over the complete time range:

$$f_t = \exp\left(-\frac{2\tau_D}{t}\right) I_0\left(\frac{2\tau_D}{t}\right) + I_1\left(\frac{2\tau_D}{t}\right)$$

(11.13)

In this formulation, $f(t)$ is the fractional recovery defined as $(F(t) - F(0))/(F(\infty) - F(0))$, and $I_0(t)$ and $I_1(t)$ are modified Bessel functions.

Plots of Eqs (11.10) and (11.12) are shown in Figure 11.5c,d. They emphasize once more the difference between recovery curves obtained by Gaussian and uniform bleaching profiles. The uniform profile has the advantage that the recovery kinetics is independent of the bleaching parameter.

To evaluate experimental data in terms of diffusion coefficients, Eqs (11.10), (11.12), or (11.13) may be fitted to the experimental data. However, frequently a simplified analysis is employed, in which $\tau_{1/2}$, the halftime of fluorescence recovery (t at $f = 1/2$), is determined from the experimental data, and the diffusion coefficient is obtained from

$$D = \left(\frac{w^2}{4\tau_{1/2}}\right) \gamma_D$$

(11.14)

The factor γ_D has the value 0.88 for a uniform beam. For Gaussian beams, γ_D depends on K; for a table of the dependence of γ_D on K-values, see Axelrod et al. [6].

When some of the fluorophores are immobile, their fraction is given by Eq. (11.1).

11.3.2 Binding

The analysis is now extended to association reactions in the presence of diffusion. Here, the simplest case and starting point is a reversible first-order association–dissociation reaction between a free fluorescent ligand S and its binding sites L according to

$$L + S \underset{k_{on}}{\overset{k_{off}}{\rightleftarrows}} = H$$

(11.15)

H is the L–S complex, and k_{on} and k_{off} are the on- and off-rates. To account for this condition, the diffusion equation has to be replaced by a set of coupled diffusion–reaction equations [8]:

$$\frac{\partial l}{\partial t} = D_l \nabla^2 l - k_{on} ls + k_{off} h \tag{11.16a}$$

$$\frac{\partial s}{\partial t} = D_s \nabla^2 s - k_{on} ls + k_{off} h \tag{11.16b}$$

$$\frac{\partial h}{\partial t} = D_h \nabla^2 h - k_{on} ls + k_{off} h \tag{11.16c}$$

where l, s, and h are the concentrations of L, S, and H, respectively. Equation (11.16) has been solved for the case in which the binding sites L and thus the L–S complexes are immobile. However, that solution involved simulation by numerical methods, which are commented on in connection with the evaluation of CFM experiments (Section 11.4.1). Thus, theoretical curves have to be derived for particular initial and boundary conditions and compared with experimental data.

However, Sprague et al. [8] also provided analytical approximations for limiting cases with important implications. Thus, in case diffusion is rate-limiting (diffusion dominance), FRAP curves are formally described by the solution for pure diffusion (Eq. (11.13)). The diffusion coefficient determined in this way is, however, an effective quantity, and related to the coefficient D of free diffusion by

$$D_{eff} = \frac{D}{1 + \dfrac{k_{on}}{k_{off}}} \tag{11.17}$$

When the "true" diffusion coefficient D can be measured or estimated independently, the k_{on}/k_{off} ratio can be determined from Eq. (11.26). In contrast, when the ligand is tightly bound so that the dissociation time $(1/k_{off})$ is much larger than the characteristic diffusion time τ_D (binding dominance), fluorescence recovery is independent of the size of the bleached area and given by

$$f(t) = 1 - C_{eq} e^{-k_{off} t}, \quad \text{with } C_{eq} = \frac{k_{on}}{k_{on} + k_{off}} \tag{11.18}$$

A closed-form analytical solution of Eq. (11.15), (11.16b), and (11.16c) is also available [9]. That solution allows for a mobile complex H and can be applied to experiments with a Gaussian or uniform beam.

11.3.3 Membrane Transport

An example for membrane transport measurements by FRAP is shown in Figure 11.6, while experimental configurations used in this and other transport measurements are illustrated in Figure 11.7. These include isolated membranes spanning artificial micro- or nanowells, isolated cell ghosts, cellular and artificial vesicles attached to a solid substrate, single living cells in culture, and dense monolayers of cultured cells or arrays of epithelial cells *in vivo*.

The configurations shown in Figure 11.7A–D can all be described as two-compartment systems consisting, as indicated in Figure 11.7E, of two volumes, V_1 and V_2, which share a membrane of surface area S. A fluorescent solute is present in compartments 1 and 2 at concentrations c_1 and c_2, respectively. In the simplest case, transport between V_1 and V_2 across the membrane is passive,

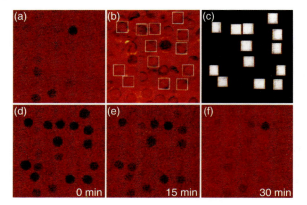

Figure 11.6 Example of a flux measurement by FRAP. Resealed red cell membranes (ghosts) were titrated with small amounts of the pore-forming bacterial toxin perforin and incubated with the small hydrophilic fluorophore Lucifer Yellow, which can penetrate the perforin pores. (a) A fluorescence scan of a ghost layer acquired by a confocal laser scanning microscope. A few ghosts have not incorporated perforin pores and remained void of Lucifer Yellow. (b) Using a scan obtained in transmitted light, several cells are chosen for photobleaching (white boxes). (c) During the bleach scan, the photobleached areas display a very bright fluorescence. (d–f) The depletion of the ghosts from fluorescence and the subsequent influx of fresh fluorophores were recorded by repetitive scans obtained at the indicated times. (Tschodrich-Rotter et al. 1996 [10]. Reproduced with permission of Elsevier.)

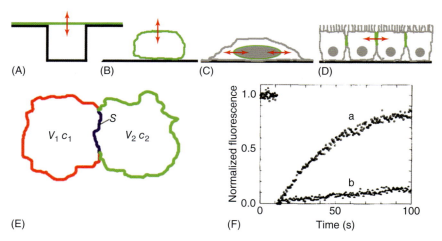

Figure 11.7 Schematic representation of flux measurements by FRAP and CFM. Different experimental configurations for flux measurements include (A) isolated cell membranes covering micro- or nanowells, (B) isolated resealed plasma membranes, (C) intact cells, and (D) cell monolayers. (E) The theoretical evaluation of flux measurements is based on two-compartment systems, in which compartments 1 and 2 with volume V_1, V_2 and probe concentration c_1, c_2 share a common membrane of area S. (F) An example of flux measurements by FRAP. Red cell ghost were incubated with the fluorescent anion NBD-taurine, which is transported by the red blood cell anion transporter. Experiments (a) and (b) were carried out in the absence and presence of the anion transport inhibitor 4,4′-diisothiocyanatostilbene-2,2′-disulfonic acid (DIDS). (Peters and Passow 1984 [11]. Reproduced with permission of Elsevier.)

that is, proportional to the concentration difference. A further simplification can be made by assuming that diffusion is much faster than membrane transport so that $c_1(t)$ and $c_2(t)$ are only dependent on time but not on their respective positions within V_1 or V_2. Then, the *flux across the membrane*, defined as the net number of molecules crossing from one compartment to the other per time and membrane surface, is

$$\Phi = \frac{dn}{S\,dt} = P\Delta c \tag{11.19}$$

P is the permeability coefficient, which depends on the specific permeability of the transporter for the substrate and the area density of transporters. Concentration changes are described by

$$\frac{dc_1}{dt} = \frac{PS}{V_1}(c_2 - c_1) \tag{11.20a}$$

$$\frac{dc_2}{dt} = \frac{PS}{V_2}(c_1 - c_2) \tag{11.20b}$$

Accounting for a FRAP experiment, we assume that compartment 1 is partially depleted of fluorescent substrate at $t=0$ so that the initial concentrations are $c_1(0) \equiv c_{10}$ and $c_2(0) \equiv c_{20}$. Then, the solutions to Eqs (11.20a) and (11.20b) are

$$c_1(t) = \frac{c_{10}V_1}{V_1 + V_2} + \frac{c_{20}V_2}{V_1 + V_2} + \frac{(c_{10} - c_{20})V_2}{V_1 + V_2}e^{-k_1 t} \tag{11.21a}$$

$$c_2(t) = \frac{c_{10}V_1}{V_1 + V_2} + \frac{c_{20}V_2}{V_1 + V_2} - \frac{(c_{10} - c_{20})V_1}{V_1 + V_2}e^{-k_2 t} \tag{11.21b}$$

where k_1 and k_2 are the transport rate constants, given by

$$k_1 = k_2 = PS\left(\frac{1}{V_1} + \frac{1}{V_2}\right) \tag{11.22}$$

Assuming that the recorded fluorescence signal is directly proportional to concentration, the fluorescence intensity of compartment 1, F_1, is given by

$$F_1(t) = \frac{F_{10}V_1}{V_1 + V_2} + \frac{F_{20}V_2}{V_1 + V_2} + \frac{(F_{10} - F_{20})V_2}{V_1 + V_2}e^{-k_1 t} \tag{11.23}$$

An analogous relation holds for $F_2(t)$.

Frequently, one of the compartments is much larger than the other, for example, $V_2 \gg V_1$. Then F_2 is constant, whereas F_1 follows

$$F_1(t) = F_{20} + (F_{10} - F_{20})e^{-kt} \tag{11.24}$$

In order to derive transport rate constants, the experimental data are fitted by Eq. (11.23) or (11.24), an easy task with mono-exponential functions. An example is shown in Figure 11.7F.

So far it has been assumed that transport is passive, that is, directly proportional to the concentration difference. However, many biological membrane transport processes display saturation kinetics according to the Michaelis–Menten equation. The kinetics of Michaelis–Menten-type transport processes in two-compartment systems have extensively been analyzed, and solutions and even simulation programs that can be directly applied to FRAP and CFM experiments are available (http://www.physiome.org/).

Another assumption made previously was that diffusion is much faster than membrane transport. That condition is valid for

$$\tau_T = 1/k \gg \tau_{D_{\text{eff}}} = \frac{d^2}{6D_{\text{eff}}} \qquad (11.25)$$

where τ_T is the effective transport time and d is the smallest linear dimension of compartment 1. This condition holds frequently for intracellular measurements [12]. However, in some cases, diffusion times cannot be neglected. Then, it is still possible to deconvolve membrane transport and diffusion and recover the membrane transport rates [13].

11.4 Continuous Fluorescence Microphotolysis (CFM)

In this section, CFM is described in more detail. First, the evaluation of CFM data in terms of diffusion coefficients, binding constants, and membrane transport rates is presented. Then, the combination of CFM with complementary techniques such as fluorescence correlation spectroscopy (FCS), 4Pi microscopy, and total internal reflection fluorescence (TIRF) microscopy is discussed. Finally, variants of CFM using refined illumination schemes are considered.

11.4.1 Theoretical Background and Data Evaluation

In CFM experiments [14], the ROP is illuminated at a constant power while simultaneously recording ROP fluorescence (cf. Section 11.2.1, Figure 11.2b,d). The observed fluorescence decay is, in general, the result of photobleaching, lateral diffusion, binding, and membrane transport. As in the case of FRAP experiments, it is instructive to first consider pure diffusion and then proceed to binding in the presence of diffusion and to membrane transport.

In the case of pure diffusion, the CFM curve has three time regimes (Figure 11.8a). At the onset of irradiation, photobleaching is the only relevant process, although only for a very short time. If some of the fluorophores are immobile, they will be photobleached during that period. In the second time regime, the fluorescence gradient between ROP and its surroundings induces diffusion of fresh fluorophores from the surroundings into the ROP, thus slowing down the fluorescence decay. The regime is characterized by the condition that the time scales of bleaching and diffusion are approximately equal. In the third time regime, the fluorescence gradient has spread from the ROP far into the surroundings (Figure 11.8b), ultimately approaching the boundary of the diffusion space. These qualitative considerations suggest that the first time regime is best suited for determining the order of the photobleaching reaction and the immobile fraction. The second time regime is best for determining the diffusion coefficient, while the third time regime may be used to obtain information about the size of the diffusion space.

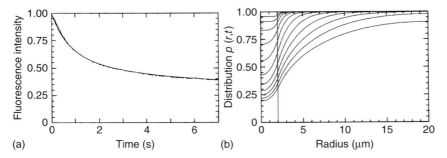

Figure 11.8 Example of diffusion measurements by CFM. The diffusion of a fluorescent lipid analog in the membranes of large artificial bilayers was measured. (a) The experimental curve is shown together with the best-fitting theoretical curve. (b) The computed fluorescence intensity profile is shown at $t = 0.0625, 0.125, 0.25, 0.50, 1.0, 2.0, 4.0, 8.0, 16.0, 32.0$, and 64.0 s. The illuminated area had a radius of 2 μm, indicated in (b) by the vertical line. (Peters and Schulen 1981 [14].)

On a quantitative level, much of what has been said in Section 11.3.1 about the theoretical framework of FRAP measurements holds for CFM. However, to account for the removal of fluorophores by photobleaching, the differential equation for diffusion has to be complemented. In the simplest case, the photobleaching reaction can be described by

$$C \underset{}{\overset{h\nu}{\rightleftarrows}} C^* \rightarrow P \tag{11.26}$$

where C^* is the excited singlet state of the fluorophore, which can either return to the ground state by emitting a photon or undergo an irreversible reaction, yielding the nonfluorescent product P. Then, the diffusion–reaction equation assumes the form

$$\frac{\partial c(r,t)}{\partial T} = D\nabla^2 C(r,t) - kc \tag{11.27}$$

An analytical solution of Eq. (11.27) is not available. Therefore, the development of CFM [14] depended on the use of numerical methods. The numerical approach has subsequently been extended [15–17] and used to evaluate CFM as well as FRAP experiments involving diffusion, photobleaching, binding, and/or membrane transport. Although numerical simulation of diffusion–reaction systems was previously somewhat demanding, the progress in computational power and the availability of software packages such as MATLAB have made it the method of choice.

The essence of numerical simulation of reaction–diffusion systems is spatiotemporal discretization: The diffusion–reaction space is subdivided into a finite number of spatial elements. Likewise, time is subdivided into intervals. The spatial elements are filled with probe molecules according to the initial conditions. A frequently used initial condition is a random distribution of probe molecules. However, frequently the initial distribution is inhomogeneous. In these cases, it may be possible to estimate the distribution from prebleach confocal scans and to use that information as initial condition [17]. Then, the molecules are endowed with relevant properties, for example, probabilities per interval for moving into a neighboring cell, becoming nonfluorescent by photobleaching, associating temporarily with binding sites, being translocated through a membrane, and so forth. The time dependence of the process is then simulated by computing the spatial position of each probe molecule for each time interval. The result depends, of course, on the degree of discretization. If spatial elements and/or intervals are chosen too large, the process may be distorted. If too small, the computation time may become too large. Therefore, choosing an appropriate degree of discretization is an important consideration.

An example for the simulation of CFM curves and their fitting to experimental curves is shown in Figure 11.8a.

In the case of diffusion in the presence of binding, the corresponding differential equations have to be complemented by the bleaching term in an analogous manner:

$$\frac{\partial l}{\partial t} = D_l \nabla^2 l - k_{on} ls + k_{off} h \tag{11.28a}$$

$$\frac{\partial s}{\partial t} = D_s \nabla^2 s - k_{on} ls + k_{off} h - ks \tag{11.28b}$$

$$\frac{\partial h}{\partial t} = D_h \nabla^2 h - k_{on} ls + k_{off} h - kh \tag{11.28c}$$

To solve these equations, the numerical simulation of this system of coupled differential equations is again the method of choice. However, several analytical approximations are also available. Thus, for both effective diffusion and binding dominance, approximations involving simple exponential functions were derived [15]. For the long-time regime of CFM measurements on spherical vesicles, Delon et al. [18] obtained

$$\tau_1 = \frac{0.4 + \frac{\pi D}{\sigma_B} R^2}{D} \tag{11.29}$$

where τ_1 is the decay time as obtained by a mono-exponential fit of the data, σ_B is the photobleaching cross section, P is the power of the irradiation light beam, and R is the vesicle radius.

In the case of membrane transport, the flux equation is also complemented by the photobleaching term [19]:

$$\frac{dc_1}{dt} = \frac{PS}{V_1}(c_2 - c_1) - k_p c_1 \tag{11.30}$$

Solving Eq. (11.30) is straightforward and yields

$$F_1(t) = F_{20} + (F_{10} - F_{20})e^{-kt} \tag{11.31}$$

with rate constant k given by

$$k = \frac{PS}{V_1} + k_p \tag{11.32}$$

An example for a CFM flux measurement is shown in Figure 11.9.

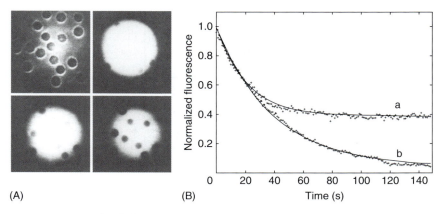

Figure 11.9 Example of a flux measurement by CFM. Resealed erythrocyte membranes were incubated with NBD-taurine. (A) A cell layer is illuminated while recoding fluorescence with a charge-coupled device (CCD) camera. The fluorescence inside the ghosts is progressively bleached with time, while the external medium shows no bleaching effect owing to the rapid diffusion of the fluorescent probe. (B) CFM experiments were done (a) in the absence and (b) in the presence of the anion transport inhibitor DIDS. (Scholz et al. 1985 [19]. Reproduced with permission of Elsevier.)

11.4.2 Combination of CFM with Other Techniques

Because of its technical simplicity, CFM can be easily combined with other techniques. For instance, the microscopic setups for CFM and FCS are virtually identical, so that both types of experiment can be performed with the same instrument and even on the same region of a given sample [15, 20, 21]. Furthermore, CFM and FCS complement each other ideally. Diffusion coefficients in the range of 0–10 µm² s⁻¹ are most easily determined by CFM, while FCS is better suited for diffusion coefficients ranging from 0.1 to 100 µm² s⁻¹; immobile fractions are easily quantified by CFM while not detectable by FCS.

CFM can also be combined with 4Pi microscopy [21]. In a 4Pi microscope, two opposing focused laser beams are brought to interference in a common focal plane. This yields a point spread function (PSF) with a complex three-dimensional (3D) shape comprising a main peak and two side lobes. In combination with two-photon excitation, the main peak has a width of ~100 nm in direction of the z-axis. Combining 4Pi microscopy (and other super-resolution methods such as stimulated emission depletion (STED) microscopy) with CFM provides new possibilities for extending diffusion and binding measurements into the nanoscopic domain.

TIRF is a method to selectively illuminate thin layers at glass/fluid interfaces (Box 12.2). Combining TIRF with CFM [22] permits the measurement of molecular mobility and binding with an axial resolution of ~100 nm. This was used, for instance, to measure protein mobility in a single bacterium.

11.4.3 CFM Variants

So far, the basic version of CFM technique was described. However, the concept can be refined to serve special purposes. A selection of illumination schemes and

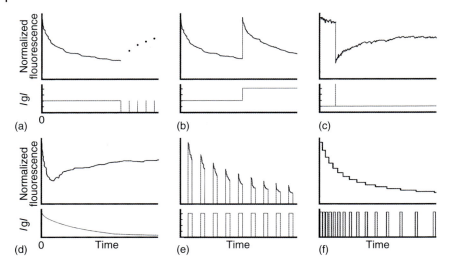

Figure 11.10 Variants of the CFM technique. In each of the panels (a–f), the illumination scheme is shown at the bottom and the resulting measuring curve at the top. According to [14] – (a), [23] – (b), [24] – (c), [25] – (d), [20] – (e), and [26] – (f). (Panels (b), (c), and (f): With kind permission from Springer Science and Business Media.)

the resulting fluorescence signals are presented in Figure 11.10. These include adding a number of short spikes to a period of continuous illumination, using two intensity levels of continuous illumination, applying a high-intensity spike during continuous illumination, reducing the intensity during continuous irradiation according to a hyperbolic function, and using pulsed irradiation with elaborate time schemes. Benefits of these procedures have been demonstrated, for instance, in the case of membrane transport measurements, in which a stepped intensity eliminates background effects [19], and in the case of diffusion measurements in small cells, in which pulsed irradiation permitted the characterization of diffusive barriers [20].

11.5 CLSM-Assisted Photobleaching Methods

In this section, the use of CLSMs for photobleaching experiments is considered. In particular, the implementation of the photobleaching methodology on CLSMs, the modalities of conducting CLSM-assisted photobleaching experiments, and a range of new opportunities for analyzing molecular transport in cellular systems are discussed. Two common artifacts and their correction are also mentioned.

11.5.1 Implementation

As fully discussed in Chapter 5 of this volume, a CLSM employs a focused laser beam to rapidly scan a sample in an orderly manner, that is, point by point and line by line. The fluorescence emitted during scanning is collected by the objective, de-scanned, passed through a confocal aperture, and recorded by a sensitive photodetector. Images are then reconstructed from the large number of sequential point measurements. The spatial resolution of the reconstructed images depends on both the illuminating PSF and the imaging PSF and amounts at optimal conditions to ~220 nm in the focal plane (x- and y-axis) and 600 nm in direction of the optical axis (z-axis). The time resolution is determined by the scanning speed, which is usually ~2 μs per pixel. This amounts to ~1 ms for a line of 512 pixels and 0.5 s for a frame of 512 × 512 pixels. The scanning time can be increased by slowing down the scanning speed or by averaging the lines or frames. It can be also decreased in certain limits by decreasing the number of pixels per frame or increasing the scanning speed.

The application of CLSMs in FRAP and CFM experiments was worked out to a great deal by complementing a commercial CLSM with powerful lasers and a device referred to as *scamper* [5]. The scamper is essentially an algorithm that monitors the scanning process and modulates the laser beam power during scanning according to a freely programmable image mask. Today, FRAP modules based on this principle are incorporated in virtually all commercial CLSMs.

CLSM-assisted FRAP experiments involve the following steps: acquisition of an image; selection of one or several ROPs; initiation of one or several bleaching scans, in which the laser beam is set to a high power level but only switched on when inside an ROP; and acquisition of a series of imaging scans at a low, non-bleaching beam power. For a thorough discussion of the theoretical basis of CLSM-assisted FRAP experiments and step-by-step protocols for data acquisition in protein diffusion and trafficking measurements, see [4].

CLSM-assisted CFM experiments may be performed in different ways. In the simplest case, an image is acquired, an image point selected, and the fluorescence of that point monitored over time without scanning. This is equivalent to a conventional CFM measurement, making use, however, of the high resolution and the optical sectioning capabilities of the CLSM. In a different scheme, an image is acquired and ROPs are selected. Then the sample is continuously scanned. However, the laser beam is switched on only when inside ROPs. This enables one to spatially resolve the fluorescence decay inside the ROPs. In yet another mode, the sample is continuously scanned, setting the laser beam power to a photobleaching level when inside a ROP and to a low, non-bleaching level when outside ROPs. This enables one to monitor the loss of fluorescence from areas adjacent to the ROPs.

11.5.2 New Opportunities

The use of CLSM in photobleaching experiments provides many new opportunities for analyzing molecular transport in biological systems. Among these, the creation of multiple and/or arbitrarily shaped ROPs, the resolution of the time- and space-dependent fluorescence intensity inside ROPs, the generation of a high time resolution, and the progression from two to three spatial dimensions stand out and will be discussed in the following.

11.5.2.1 Multiple ROPs

CLSMs permit the creation of not only a single ROP at a time but also a large number of ROPs simultaneously. An example concerning lateral diffusion measurements with two circular disk-shaped ROPs is illustrated in Figure 11.11a. Such experiments may be used to map the plasma membrane or other subcellular

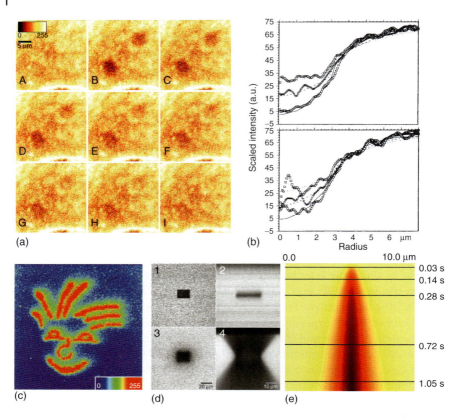

Figure 11.11 New possibilities provided by confocal FRAP and CFM. (a, b) Using spatial information to improve data evaluation. A diffusion measurement by confocal FRAP is shown in which two circular areas of the plasma membrane of a single cell were photobleached simultaneously. The time-dependent intensity profiles of the bleached area, as shown in (b) for each of the bleaches areas, to derive diffusion coefficients from these profiles. (Kubitscheck et al. 1994 [27]. Reproduced with permission of Elsevier.) (c) Arbitrary bleach pattern. Scanning FRAP permits the creation of almost arbitrary shapes in immobile fluorescent objects. The pattern has a sharpness of ∼0.5 µm. (Wedekind et al. 1994 [5]. Reproduced with permission of John Wiley & Sons.) (d) 3D pattern by two-photon excitation. Comparison of bleach pattern in a 3D object (gel slice) obtained (1, 2) by two-photon excitation, and (Dc, Dd) by single photoexcitation. (1) and (3) are vertical (xy) scans. (2) and (4) are horizontal (xz) scans. It can be recognized that photobleaching by two-photon excitation results in a pattern that is well defined in all three dimensions. In contrast, single-photon excitation yields a pattern that is less well defined in the horizontal section, and degenerated into a large cone in the vertical section. (Kubitscheck et al. 1996 [28]. Reproduced with permission of John Wiley & Sons.) (e) 3D diffusion measurements at high time resolutions. A 3D diffusion measurement by confocal CFM in the line scanning mode is shown. The line scanning mode yields a "picture" in which the x-axis represents the coordinate on the scanned segment while the y-axis represents time. The fluorescence intensity is color coded similar to Figure 11.1. In this example, a very small segment (0.5 µm) of a 10 µm long line was continuously bleached. The expansion of the bleach profile with time is seen. The time per single scanned line was 2 ms. (Kubitscheck et al. 1996 [28]. Reproduced with permission of John Wiley & Sons.)

components for diffusional anisotropy and discontinuity, which may arise, for instance, by lipid phase separation (lipid rafts) or by interactions of cytoskeletal filaments with membranes.

In measurements of transport through the membranes of artificial lipid vesicles, cell organelles, cell ghosts, or intact cells, the possibility of creating many ROPs simultaneously enables one to perform many measurements in parallel (cf. Figure 11.6 in Section 11.3.3), thus greatly speeding up data collection.

11.5.2.2 Arbitrarily Shaped ROPs

In CLSM-assisted photobleaching, the ROP does not need to be circular. The creation of rectangular ROPs is particularly easy with CLSMs, which usually rely on rectangular scanning schemes. Rectangular ROPs are of great advantage in lateral mobility measurements because the fluorescence recovery kinetics can be described by closed-form equations, thus simplifying and accelerating data evaluation [29]. However, rectangular shape is only one possibility and more intricate ROP shapes can be easily created, as illustrated Figure 11.11c.

11.5.2.3 Spatially Resolved Bleaching and Recovery

In conventional FRAP and CFM experiments, the integral ROP fluorescence is monitored and used to derive transport and binding coefficients. However, in CLSM-assisted photobleaching experiments, the time-dependent fluorescence intensity of the ROP and its surroundings can be resolved at confocal resolution. In the case of FRAP experiments as illustrated in Figure 11.11a, the intensity profile of the ROP (Figure 11.11b) can be used to derive the diffusion coefficient. Here, the particular advantage is that knowledge of the initial conditions and the precise timing of bleaching and measurements is not required. In addition, diffusion during the bleaching interval is permissible and does not distort the outcome of the analysis [27]. In the case of diffusion measurement by CFM, even a single intensity profile is sufficient to derive the diffusion [30]. The application of rectangular ROPs is particularly well suited for discriminating between free and anomalous diffusion.

11.5.2.4 Millisecond Time Resolution

In CLSMs, complete image frames are generated at a frequency of a few hertz only. However, most CLSMs may be operated in a mode in which single lines are scanned at a frequency of up to 2 kHz. The data of such line scans are displayed as "images" in which the abscissa represents the spatial coordinate (usually the x-axis) and the ordinate the time. In CLSM-assisted FRAP, the line scanning mode can be applied to bleach one or several segments of the scanned line at high laser beam power and to monitor the dissipation of the bleach pattern at a time resolution of up to 1 ms [31]. Similarly, line scanning can be employed in CLSM-assisted CFM, as illustrated in Figure 11.11e [32].

11.5.2.5 Three-Dimensional Photobleaching

The 3D intensity distribution around the focal point of a focused laser beam is rather complicated. However, to a first approximation it can be described as a circular double cone with the axis parallel to the optical axis and an apex in the focal plane. The total beam power in planes parallel to the focal plane is constant, implying that the intensity has a maximum in the apex and falls off rapidly along the optical axis. When operating a CLSM in the "normal" mode, in which fluorescence is excited by single-photon absorption, the photobleached volume will closely reflect the intensity distribution of the exciting beam: The bleached volume has a cone-like shape with boundaries that are sharp only in the focal plane. This is demonstrated in Figure 11.11d3,4. The blurring of the bleach pattern above and below the focal plane prohibits 3D studies in both FRAP and CFM experiments.

The situation is different when, instead of single-photon-absorption, two- or multiphoton absorption is employed. Then, fluorescence intensity is not directly proportional to excitation intensity but to its second or higher power, and photobleached volumes have sharp boundaries in all three dimensions. This is demonstrated in Figure 11.11d1,2. FRAP and CFM studies have confirmed that 3D diffusion coefficients can be faithfully measured in thick gel slabs when employing two-photon excitation [28]. The evaluation of such experiments relies in general on numerical simulation. For 3D diffusion measurements by line scanning, an analytical solution is also available [33].

11.5.3 Two Common Artifacts and Their Correction

When employing commercial CLSMs in FRAP experiments, laser power is frequently a limiting parameter. Obtaining a sufficient degree of photobleaching may require several bleach scans and take several seconds. As a consequence, diffusion occurs during bleaching. This violates the instantaneous-bleaching condition of common data analysis schemes. Weiss [34] showed by numerical simulation that this leads to a gross underestimation of diffusion coefficients. An approximate analytical solution of the diffusion equation has been derived [35], which takes diffusion during bleaching in FRAP experiments into account. The theory holds for a circular ROP and incorporates an effective radius in addition to the nominal radius of the ROP. An analytical solution for the evaluation of FRAP experiments, which formally resembles the solution of Axelrod *et al.* [6], was provided by [36].

Typically, the concentration of fluorophores in biological samples is very low. In addition, laser scanning microscopes (LSMs) have a complex optical pathway, reducing their detection efficiency. Therefore, it may be necessary to raise the laser beam power during imaging to a degree at which photobleaching cannot be neglected. Fortunately, the correction for bleaching during imaging is simple. According to [37], the time dependence of the fluorophore concentration $C(r,t)$ is simply the product of a bleaching term and a diffusion term. The bleaching term can be easily measured and used to correct the FRAP data [38].

11.6 Summary and Outlook

The chapter presented the following material:
- An introduction to the basic concepts and procedures of photobleaching techniques, including theoretical methods and instrumentation;

- A detailed description of FRAP and its extension from diffusion to binding and membrane transport measurement;
- A detailed description of CFM, including methodological variants, and the combination of CFM with other techniques such as FCS;
- An account of the use of CLSMs in photobleaching experiments, comprising new possibilities and potential artifacts.

The material presented in this chapter revealed future tasks and directions:

- The optimization of confocal microscopes for photobleaching experiments, including more efficient bleaching, smaller lag times between bleaching and imaging, and less photobleaching during imaging;
- The integration of CFM and FCS into CLSMs in a way that permits measurements by both techniques at the same location;
- The extension of photobleaching techniques to the nanoscopic domain.

References

1 Singer, S.J. and Nicolson, G.L. (1972) The fluid mosaic model of the structure of cell membranes. *Science*, **175** (23), 720–731.
2 Frye, L.D. and Edidin, M. (1970) The rapid intermixing of cell surface antigens after formation of mouse-human heterokaryons. *J. Cell Sci.*, **7**, 319–335.
3 Poo, M. and Cone, R.A. (1973) Lateral diffusion of rhodopsin in the visual receptor membrane. *J. Supramol. Struct.*, **1** (4), 354.
4 Peters, R., Peters, J., and Tews, K.H. (1974) Microfluorimetric studies on translational diffusion of proteins in erythrocyte-membrane. *Pflugers Arch. Eur. J. Physiol.*, **347**, R36-R.
5 Wedekind, P., Kubitscheck, U., and Peters, R. (1994) Scanning microphotolysis: a new photobleaching technique based on fast intensity modulation of a scanned laser beam and confocal imaging. *J. Microsc.*, **176**(Pt. 1), 23–33.
6 Axelrod, D., Koppel, D.E., Schlessinger, J., Elson, E., and Webb, W.W. (1976) Mobility measurement by analysis of fluorescence photobleaching recovery kinetics. *Biophys. J.*, **16** (9), 1055–1069.
7 Soumpasis, D.M. (1983) Theoretical analysis of fluorescence photobleaching recovery experiments. *Biophys. J.*, **41**, 95–97.
8 Sprague, B.L., Pego, R.L., Stavreva, D.A., and McNally, J.G. (2004) Analysis of binding reactions by fluorescence recovery after photobleaching. *Biophys. J.*, **86** (6), 3473–3495.
9 Kang, M. and Kenworthy, A.K. (2008) A closed-form analytic expression for FRAP formula for the binding diffusion model. *Biophys. J.*, **95** (2), L13–L15.
10 Tschodrich-Rotter, M., Kubitscheck, U., Ugochukwu, G., Buckley, J.T., and Peters, R. (1996) Optical single-channel analysis of the aerolysin pore in erythrocyte membranes. *Biophys. J.*, **70** (2), 723–732.
11 Peters, R. and Passow, H. (1984) Anion transport in single erythrocyte-ghosts measured by fluorescence microphotolysis. *Biochim. Biophys. Acta*, **777**, 15.

12 Peters, R. (1984) Flux measurement in single cells by fluorescence microphotolysis. *Eur. Biophys. J.*, **11** (1), 43–50.
13 Kiskin, N.I., Siebrasse, J.P., and Peters, R. (2003) Optical microwell assay of membrane transport kinetics. *Biophys. J.*, **85**, 12.
14 Peters, R., Brünger, A., and Schulten, K. (1981) Continuous fluorescence microphotolysis: a sensitive method for study of diffusion-processes in single cells. *Proc. Natl. Acad. Sci. U.S.A.*, **78** (2), 962–966.
15 Wachsmuth, M., Weidemann, T., Muller, G., Hoffmann-Rohrer, U.W., Knoch, T.A., Waldeck, W. *et al.* (2003) Analyzing intracellular binding and diffusion with continuous fluorescence photobleaching. *Biophys. J.*, **84** (5), 3353–3363.
16 Sprague, B.L. and McNally, J.G. (2005) FRAP analysis of binding: proper and fitting. *Trends Cell Biol.*, **15** (2), 84–91.
17 Beaudouin, J., Mora-Bermudez, F., Klee, T., Daigle, N., and Ellenberg, J. (2006) Dissecting the contribution of diffusion and interactions to the mobility of nuclear proteins. *Biophys. J.*, **90** (6), 1878–1894.
18 Delon, A., Usson, Y., Derouard, J., Biben, T., and Souchier, C. (2006) Continuous photobleaching in vesicles and living cells: a measure of diffusion and compartmentation. *Biophys. J.*, **90** (7), 2548–2562.
19 Scholz, M., Schulten, K., and Peters, R. (1985) Single-cell flux measurement by continuous fluorescence microphotolysis. *Eur. Biophys. J.*, **13** (1), 37–44.
20 van den Bogaart, G., Hermans, N., Krasnikov, V., and Poolman, B. (2007) Protein mobility and diffusive barriers in *Escherichia coli*: consequences of osmotic stress. *Mol. Microbiol.*, **64** (3), 858–871.
21 Arkhipov, A., Hüve, J., Kahms, M., Peters, R., and Schulten, K. (2007) Continuous fluorescence microphotolysis and correlation spectroscopy using 4Pi microscopy. *Biophys. J.*, **93**, 12.
22 Slade, K.M., Steele, B.L., Pielak, G.J., and Thompson, N.L. (2009) Quantifying green fluorescent protein diffusion in Escherichia coli by using continuous photobleaching with evanescent illumination. *J. Phys. Chem. B*, **113** (14), 4837–4845.
23 Scholz, M., Schulten, K., and Peters, R. (1985) Single-cell flux measurement by continuous fluorescence microphotolysis. *Eur. Biophys. J. Biophys. Lett.*, **13** (1), 37–44.
24 Hagen, G.M., Roess, D.A., de Leon, G.C., and Barisas, B.G. (2005) High probe intensity photobleaching measurement of lateral diffusion in cell membranes. *J. Fluoresc.*, **15** (6), 873–882.
25 Glazachev, Y.I. and Khramtsov, V.V. (2006) Fluorescence recovery under decaying photobleaching irradiation: concept and experiment. *J. Fluoresc.*, **16** (6), 773–781.
26 Glazachev, Y.I. (2009) Fluorescence photobleaching recovery method with pulse-position modulation of bleaching/probing irradiation. *J. Fluoresc.*, **19** (5), 875–880.
27 Kubitscheck, U., Wedekind, P., and Peters, R. (1994) Lateral diffusion measurements at high spatial resolution by scanning microphotolysis in a confocal microscope. *Biophys. J.*, **67**, 9.

28 Kubitscheck, U., Tschodrich-Rotter, M., Wedekind, P., and Peters, R. (1996) Two-photon scanning microphotolysis for three-dimensional data storage and biological transport measurements. *J. Microsc.*, **182**, 225–233.

29 Deschout, H., Hagman, J., Fransson, S., Jonasson, J., Rudemo, M., Loren, N. et al. (2010) Straightforward FRAP for quantitative diffusion measurements with a laser scanning microscope. *Opt. Express*, **18** (22), 22886–22905.

30 Cutts, L.S., Roberts, P.A., Adler, J., Davies, M.C., and Melia, C.D. (1995) Determination of localized diffusion-coefficients in gels using confocal scanning laser microscopy. *J. Microsc.*, **180**, 131–139.

31 Wedekind, P., Kubitscheck, U., Heinrich, O., and Peters, R. (1996) Line-scanning microphotolysis for diffraction-limited, measurements of lateral diffusion. *Biophys. J.*, **71**, 1621–1632.

32 Kubitscheck, U., Heinrich, O., and Peters, R. (1996) Continuous scanning microphotolysis: a simple laser scanning microscopic method for lateral transport measurements employing single- or two-photon excitation. *Bioimaging*, **4**, 10.

33 Braeckmans, K., Remaut, K., Vandenbroucke, R.E., Lucas, B., De Smedt, S.C., and Demeester, J. (2007) Line FRAP with the confocal laser scanning microscope for diffusion measurements in small regions of 3-D samples. *Biophys. J.*, **92**, 2172–2183.

34 Weiss, M. (2004) Challenges and artifacts in quantitative photobleaching experiments. *Traffic*, **5** (9), 662–671.

35 Braga, J., Desterro, J.M., and Carmo-Fonseca, M. (2004) Intracellular macromolecular mobility measured by fluorescence recovery after photobleaching with confocal laser scanning microscopes. *Mol. Biol. Cell*, **15** (10), 4749–4760.

36 Kang, M., Day, C.A., Drake, K., Kenworthy, A.K., and DiBenedetto, E. (2009) A generalization of theory for two-dimensional fluorescence recovery after photobleaching applicable to confocal laser scanning microscopes. *Biophys. J.*, **97** (5), 1501–1511.

37 Endress, E., Weigelt, S., Reents, G., and Bayerl, T.M. (2005) Derivation of a closed form analytical expression for fluorescence recovery after photo bleaching in the case of continuous bleaching during read out. *Eur. Phys. J. E*, **16** (1), 81–87.

38 Bancaud, A., Huet, S., Rabut, G., and Ellenberg, J. (2010) in *Live Cell Imaging: A Laboratory Manual*, 2nd edn (eds R.D. Goldman, J.R. Swedlow, and D.L. Spector), Cold Spring Harbor Laboratory Press, pp. 67–93.

12

Single-Molecule Microscopy in the Life Sciences

Markus Axmann[1], Josef Madl[2], and Gerhard J. Schütz[3]

[1] Institute of Medical Chemistry and Pathobiochemistry, Center for Pathobiochemistry and Genetics, Medical University of Vienna, Währinger Straße 10, 1090 Vienna, Austria
[2] Albert-Ludwigs University Freiburg, Faculty of Biology and BIOSS, Schänzlestraße 18, 79104 Freiburg, Germany
[2] Institute of Applied Physics, TU Wien, Wiedner Hauptstraße 8-10, 1040 Wien, Austria

12.1 Encircling the Problem

The first microscopes were used to explore biological samples, and the key findings in early biology – the discovery of cells and organelles, for example – were linked to advances in microscopy. With increasing magnification, it soon became clear that there is a fundamental limit to the resolving power. In the words of Abbe [1]: "A microscope does not allow for resolving objects (or features of an object), if they were so close that the zeroth and first diffraction orders cannot enter the objective simultaneously." He set a milestone by linking the empirical term *resolution* to the quantifiable term *diffraction*.

However, there is enormous biological complexity below the optical resolution limit. Approaching and exploring this nanocosm – the collection of phenomena on length scales below the diffraction limit of visible light – has become the main research focus of many life scientists nowadays; indeed, the efforts have been rewarding. Meanwhile, we know that biomaterials are structured within modules at all length scales down to the level of multimolecular complexes and even single molecules.

We have some information on the complexity to be expected: about 20 000–25 000 genes and the corresponding mRNA molecules are present in human cells, which, due to alternative splicing and post-translational modifications, may be converted to a much larger number of different types of proteins (about 10^6 can be expected). Cellular membranes may host an estimated 10^5 distinct species of glycerophospholipids and glycosphingolipids. A light microscopist may thus be concerned whether the sample would provide sufficient contrast between these compounds even if a hypothetical technique allowed improving the resolution dramatically beyond the diffraction limit. More precisely, molecular biology demands chemical contrast, that is, the identification and localization of different molecular species within the sample.

Fluorescence Microscopy: From Principles to Biological Applications, Second Edition.
Edited by Ulrich Kubitscheck.
© 2017 Wiley-VCH Verlag GmbH & Co. KGaA. Published 2017 by Wiley-VCH Verlag GmbH & Co. KGaA.

However, it is currently impossible for optical microscopy to discriminate between two arbitrary biomolecules based on their natural photophysical contrast *per se*. It is feasible, though, to artificially introduce contrast into the system by specifically labeling a small subfraction of the sample using fluorescent dyes. Fluorescence labeling thus paved the way for modern cellular imaging, as it allows singling out individual components within complex biological samples. For this, the microscopist needs the fluorophores to be sufficiently bright and the background to be sufficiently dim. The term *sufficient* should indicate that different experimental approaches may impose different constraints. It turned out that most biological samples are rather tolerant toward fluorescent labeling: in general, even bulky attachments such as the green fluorescent protein (GFP) (a β-barrel with $\sim$3 nm diameter and $\sim$4 nm height, $\sim$27 kDa) hardly affect the functionality of the molecules of interest. Therefore, fluorescence microscopy is the method of choice when it comes to studying the dynamics of biological systems *in vivo*.

Even more so, fluorescent markers allow the localization of the molecules of interest with high precision. In classical light microscopy on ensembles of molecules, there was no direct route to access the positions of the emitter molecules: the obtained image was given by the convolution of the individual emitter distributions and the point spread function of the imaging system. Essentially, the photons from the various emitters get scrambled and cannot be attributed to the original source any longer, resulting in Abbe's diffraction limit. However, if the emitters are sparsely distributed over the sample – so that the distance between them is much larger than the diffraction limit – and if the imaging system provides the required sensitivity, we can directly measure their positions.

Indeed, fluorophores turned out to be quite strong light sources, so that single-molecule strategies could be reached. The history of single-molecule detection goes back to the 1960s, when Boris Rotman measured the activity of single enzymes by the generation of a fluorescent substrate [2]. Imaging of single proteins in a fluorescence microscope was achieved by Hirschfeld in 1976 [3]; in this experiment, each protein was labeled with about 100 fluorescein molecules. Improvements in the detection schemes and the sensitivity of the instrumentation enabled scientists to reach the ultimate limit of directly observing and imaging single dye molecules. Milestones on this way were the achievement of single-molecule detection at low [4] and room temperature [5], measurements in fluid membranes [6], and studies on living cells [7, 8].

Next, photons can be even further exploited; for example, they also show color and polarization and may be used to obtain information on the lifetime of the excited state of the emitter. Each labeled molecule *i* thus features time-dependent three-dimensional (3D) locations $r_i(t)$ and additional parameters $\{a_i(t), b_i(t), …\}$ reporting on the local environment, probe conformation, and so on. In consequence, the resolution problem of optical microscopy can be rephrased: *To which precision (in space and time) can we determine the parameters $r_i(t)$ or $\{a_i(t), b_i(t), …\}$?*

In this chapter, we provide some of the essential considerations when building a single-molecule microscope. Next, the key parameters accessible from a single-molecule measurement are discussed: position, orientation, color, and brightness. Finally, we put these parameters in the context of biological experiments. We begin the chapter with an overview of the specific benefits a single-molecule approach provides.

12.2 What is the Unique Information?

All of us have probably experienced the following situation: we enter a large ballroom full of people talking with each other and generating a wall of sound; no individual voices can be identified, and the content of the different communications cannot be extracted. Some individuals appear heavily involved in their arguments, but it is barely possible to detect a single word. Being conversant with opening mixers or conference dinners, scientists know how to mingle, making such gatherings a useful source of information. A more timid colleague, however, who follows the activities from the gallery, will hardly catch details from the conversations and will probably leave the evening only with the experience of the extreme loudness.

We can regard this scenario as a metaphor for our biological sample. In addition, a plethora of unsynchronized players (molecules) are moving, interacting, grouping, and switching between specific states. Let us follow now what a single-molecule mingling experiment provides us as additional information compared to the ensemble-type overview from the gallery.

12.2.1 Kinetics Can Be Directly Resolved

Resolving the transitions between different molecular states has become a key question in current molecular biology. Experimentally, it can be approached by synchronizing the sample and measuring the progression of the signal in time. Fluorescent reporters can be synchronized, for example, by photobleaching or photoactivation, which allows capturing the dynamics of transport processes. However, it is more difficult or even impossible to prepare a protein sample in a particular state; we may think of the open versus closed state of an ion channel, the active versus inactive conformation of an enzyme, the various conformations within a power stroke of molecular motors, or the ligated versus unligated state of a diffusing receptor. In such cases, conventional ensemble-based methods typically fail to provide insights into the transition kinetics. In contrast, single-molecule measurements allow the direct resolution of the sequence of transitions in time individually and the quantification of the characteristic transition rate constants. Figure 12.1a shows, as an example, the transitions between two conformational states of a protein detected via the concomitant change in the Förster resonance energy transfer (FRET) efficiency E_{FRET}, from which the lifetimes of the states can be directly monitored.

12.2.2 Full Probability Distributions Can Be Measured

Many ensemble experiments are limited to the determination of the mean value of a molecular parameter. In contrast, single-molecule experiments provide

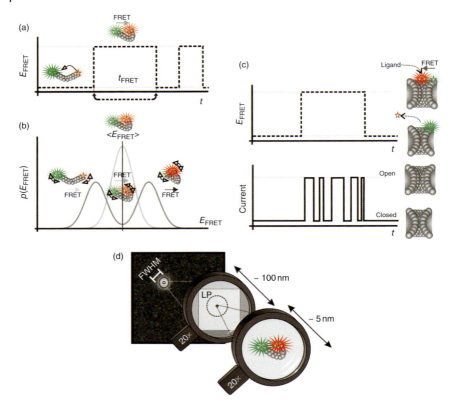

Figure 12.1 Four key characteristics of single-molecule detection. (a) A FRET signal increases after a conformational change of the observed protein, so that the donor dye (green) gets closer to the acceptor dye (red). Observation of the individual transitions between the protein conformations allows direct monitoring of the lifetimes of the states (indicated for the packed conformation by t_{FRET}). (b) Ensemble measurement of the FRET efficiency yields its mean value (black vertical line), which may originate from a homogeneous population (light gray line) or a heterogeneous population (dark gray line). Single-molecule measurements reveal the full probability distribution $p(E_{FRET})$ and thus allows discriminating the two cases. Wobbling of the molecular conformation may further broaden the observed peaks. (c) The binding of a ligand to its specific ligand-operated ion channel can be observed using single-molecule FRET. Simultaneously performed electrophysiological measurements allow the correlation of the structural signal with the functional consequence, for example, by measuring the single-channel currents. (d) The Gaussian intensity distribution arising from single diffraction-limited signals of fluorophores.

direct access to the full statistical distribution functions and the full sequence of transitions. Let us consider here, as an example, the distance between two sites of a protein measured via FRET (Figure 12.1b). A single-molecule experiment would include many observations of E_{FRET}, which can be analyzed, for example, by plotting the probability distribution $p(E_{FRET})$; the maximum of the curve gives the most likely conformation. The corresponding ensemble experiment, in contrast, yields only the average value ($\langle E_{FRET} \rangle$), which *a priori* has no further structural meaning.

For demonstration, Figure 12.1b compares two extreme cases, both yielding the same average FRET efficiency. The light gray curve shows the E_{FRET} distribution for a protein that adopts a single conformation only (around which it can wobble a bit). In this case, the mean value would be equal to the maximum; therefore, the outcome of an ensemble experiment would be the E_{FRET} of the most likely conformation. The dark gray curve shows the E_{FRET} distribution for a protein with two equally populated conformations. With the single-molecule experiment, the two conformations can be discriminated. The average E_{FRET}, however, lies in between the two peaks of the curve; therefore, the ensemble experiment would characterize a conformation that is not adopted by the protein. In addition to the risk of a misleading result, the ensemble experiment does not allow discriminating between the two cases: the same E_{FRET} value may originate from a homogeneous (one conformation) or a heterogeneous distribution (multiple conformations).

In addition to the FRET efficiency, other parameters have also been frequently used for statistical analysis by single-molecule methods: mobility or anisotropy analysis allows the identification of transient binding, and brightness analysis yields the stoichiometry of molecular complexes. In summary, recording distribution functions with single-molecule tools has multiple advantages:

- Minute differences of subpopulations – for example, single base-pair differences on DNA – can be identified in heterogeneous samples.
- Minority populations can be resolved. In many experiments on purified proteins only a minority remains functional due to denaturation partly during purification and partly during surface immobilization. However, even in a live-cell context, active proteins are frequently outbalanced by inactive ones. Such minority populations would be masked by the signal of the majority in an ensemble experiment, but can be detected rather easily at the single-molecule level.
- Ergodic and nonergodic behaviors can be discriminated. Ergodicity-breaking is frequently observed in biology, indicating that the probes do not necessarily explore the entire phase space. For example, a particular protein may remain in the same conformation from its production until its degradation, while another molecule may experience multiple conformational and functional transitions. Heterogeneities in the molecular behavior can be detected in single-molecule experiments but not in ensemble studies.

12.2.3 Structures Can Be Related to Functional States

State-of-the-art molecular biology, biochemistry, and biophysics paved the way for a mechanistic understanding of biomolecular function. Many proteins (or multimolecular complexes) show well-defined functional states, with characteristic transitions between the states. The term *function* denotes here any distinguishable and quantifiable behavior of interest: an active versus inactive state of an enzyme, an open versus closed state of an ion channel, a state of high versus low binding affinity of a receptor. The task is to correlate structural features with functional states. At first glance, this may appear nontrivial, given the large phase space of protein conformations. For simplification, it is frequently assumed that proteins adopt only a limited set of stable conformations with a direct map to their functional states. Researchers primarily aim at preparing proteins in certain functional states for X-ray crystallography to obtain a high-resolution structure for each state. Frequently, however, studies suggest a more complex picture with multiple structural intermediates accompanying functional transitions. In general, it will be difficult to prepare each state and to ensure that no state is missed.

A single-molecule approach allows the simultaneous recording of the sequence of structural and functional transitions, thereby enabling the direct correlation of the two datasets. For example, the conformation of an ion channel could be interrogated by single-molecule FRET and its functional state by simultaneous electrophysiological measurements (Figure 12.1c).

12.2.4 Structures Can Be Imaged at Super-Resolution

Light microscopy is limited by diffraction: a plane wave with wavelength λ is focused by an objective with a certain numerical aperture (NA) to a spot with an extension full width at half-maximum (FWHM) $\geq 0.51 \lambda/\text{NA}$. If the sample interacts with the excitation light according to a linear mathematical relation, FWHM sets an actual limit to the resolution of the microscope. If the relation is nonlinear (e.g., due to saturation effects of the fluorophores), higher resolution also becomes possible.

Single-molecule tools opened up alternative imaging concepts. These concepts are based on the utilization of prior knowledge: the position of a fluorophore is encoded by the centroid of the photon distribution and can be determined to an accuracy much below the diffraction limit by fitting with the point spread function [9].

We assume in the following a Gaussian photon distribution for a single molecule; for simplicity, the molecules are assumed to be located on a planar surface identical with the focal plane of the imaging system. Then, the image is given by

$$\text{Im}(r) = \sum_i F_i \frac{1}{2\pi\sigma^2} \exp\left[-\frac{(r-r_i)^2}{2\sigma^2}\right] \quad (12.1)$$

where F_i and r_i denote the fluorescence brightness and position of the ith fluorophore, and σ the standard deviation of the distribution (FWHM $= 2.35\sigma$). If there is no noise and the signals are high, unambiguous deconvolution is possible and the precise coordinates of all fluorophores can be directly calculated. However, both background noise and photon emission fluctuations render such direct deconvolution impossible. Still, there are ways to obtain information on the spatial organization of the sample using single-molecule approaches. Eric Betzig summarized these approaches long before they were implemented: "We can resolve multiple discrete features within a focal region of m spatial dimensions by first isolating each on the basis of $n \geq 1$ unique optical characteristics and then measuring their relative spatial coordinates" [10]. Such characteristics could be the color of the fluorophore, but also its emissive state. In Box 12.1, we summarize the concepts for using single-molecule microscopy for super-resolution imaging (see Chapter 8 for a more extensive description).

Box 12.1 Strategies for Super-resolution Imaging Based on Single-Molecule Microscopy

1. *Multicolor imaging:* Different color channels can be used to measure the distances between spectrally distinguishable molecules (indicated in Figure 12.2 by r_{GB}, r_{BG}, and r_{GR}) below the diffraction limit.

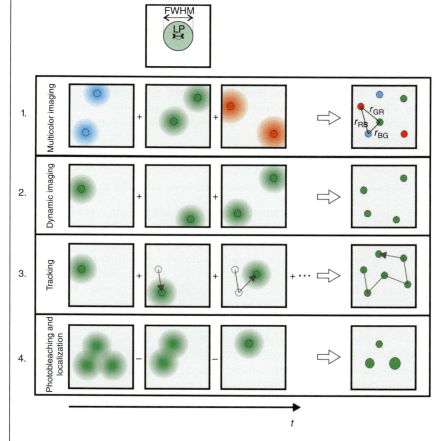

Figure 12.2 Strategies for super-resolution imaging based on single-molecule microscopy. The size of the spots indicates the point spread function; dashed circles represent the localization precision.

2. *Dynamic imaging:* The biomolecule or structure of interest is labeled transiently by a weakly binding fluorescent probe, so that single binding events can be identified and localized. An overlay of all determined positions provides a super-resolution map of the labeled structure.
Novel photoactivation or photoswitching localization microscopy approaches are conceptually similar (Chapter 8). Activatable or switchable fluorophores are stochastically turned on, imaged, and photobleached; the whole process is repeated until most of the fluorophores of the sample have been addressed.

(Continued)

> **Box 12.1 (Continued)**
>
> The obtained positions are mapped, yielding an image with a resolution limited only by the localization precision (LP).
> 3. *Single-molecule/-particle tracking:* The thermal motion of particles can be exploited to report on matrix properties. Ideally, the density of labeled probes is low enough so that single-molecule trajectories can be reconstructed unambiguously. Particularities of the movements can be detected below the diffraction limit. This strategy has been used to identify, for example, step sizes of motor proteins, and confinements in artificial membranes or in the cellular plasma membrane.
> 4. *Photobleaching localization microscopy:* If only a few chromophores are located within the diffraction-limited resolution volume, successive photobleaching may allow the reconstruction of their positions: each photobleaching step is accompanied by a shift in the signal distribution, which depends on the location of the bleached dye.

12.2.5 Bioanalysis Can Be Extended Down to the Single-Molecule Level

The detection and identification of the smallest sample amounts – ultimately single molecules – has become of central interest to state-of-the-art bioanalysis. Global analysis of mRNA or protein expression profiles is typically performed using microarrays. Its current applicability, however, is limited by the large amounts of sample required. In particular, the minute number of cells in subpopulations that can be isolated from small clinical samples still poses a significant problem when it comes to processing for global analysis; ultimately, only a few molecules will be bound to distinct loci on a biochip. Indeed, bioanalysis in the regime of the lowest sample amounts can be robustly performed by single-molecule detection assays, for example, for mRNA expression profiling or fluorescence-linked immunosorbent assays. One may further obtain the whole sequence of single DNA molecules. Single-molecule detection brings also cytometry to a new level of sensitivity so that even extremely low expressers can still be analyzed.

12.3 Building a Single-Molecule Microscope

Devices for the detection and continuous observation of signals emitted from single dye molecules have been developed in the last 20 years in various laboratories. Different designs of such devices reflect the differences in the scientific focus of those groups, ranging from purely spectroscopic investigations at low and room temperature, over measurements in fluid systems, to studies of single molecules in living cells. For biological applications, nature itself defines the requirements for implementation of single-molecule microscopy. Active transport in living cells occurs at the speeds of several micrometers per second; Brownian motion of cell components ranges from diffusion constants of $\sim 100\,\mu m^2\,s^{-1}$ for proteins in solution, over $\sim 1\,\mu m^2\,s^{-1}$ for lipids in a membrane, and down to values smaller than 10^{-4} µm² s^{-1} for immobilized proteins. In all these cases, millisecond time resolution is required for resolving submicrometer displacements. In addition, this is the very time regime of many molecular transitions, such as the gating of ion channels or low-affinity protein–protein interactions.

12.3.1 Microscopes/Objectives

For the first pioneering single-molecule spectroscopy studies, researchers built their setups from scratch, primarily to enable operation at low temperature. With the advent of room-temperature single-molecule microscopy, commercially available microscope configurations became increasingly attractive. Commonly, inverted epi-fluorescence microscopes are used.

12.3.1.1 Dual View

In its standard configuration, a microscope allows recording the time sequences of images. Between two consecutive images, the excitation or emission path may be altered (e.g., by changing the polarization and inserting filters), so that different parameters can be interrogated sequentially. To enable such interrogation simultaneously, one may split the image in the emission path according to specific properties (e.g., polarization by inserting a polarizing beam splitter or color by using a dichroic mirror) and record the resulting images on different detectors. A more efficient version is now commercially available in various designs (Figure 12.3a): in such a dual-view system, a dichroic mirror in the parallel beam path is used to split the image according to color; or mirrors are used to overlay the two images with a slight adjustable shift, so that they can be positioned next to each other on the same camera. This configuration provides full access to both beam paths, so that additional optical elements (e.g., filters and lenses) can be inserted separately.

Alternatively, one may insert a Wollaston prism or a dichroic wedge into the infinity space in order to generate two images of orthogonal polarization or different color, respectively.

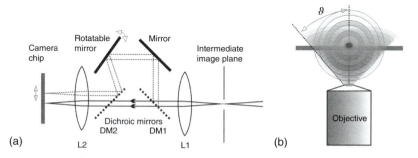

Figure 12.3 (a) Sketch of a dual-view system. A relay lens L1 focused at the intermediate image plane produces a parallel beam path, in which two spectral channels are split by a dichroic mirror (DM1). Two additional rotatable mirrors and a second dichroic mirror (DM2) are used for overlaying the two beams with a slight shift. A second lens L2 with same focal length as L1 is used for image formation on the camera chip. (b) The collection efficiency of an objective, η_{obj}, can be estimated by the amount of light entering a cone of angle ϑ, where ϑ is defined by the objective's numerical aperture $NA = n \times \sin \vartheta$.

12.3.1.2 Objective

The objective is the main imaging tool and thus one of the key elements in microscopy. The following figures of merit should be considered when selecting the objective:

- *Magnification/resolution:* The size of the point spread function and thus the resolving power of an objective are determined by its NA; the FWHM in the image plane is given by FWHM = $M(0.51\lambda/\text{NA})$, where M denotes the magnification. To optimize single-molecule LP, this number has to be matched with the pixel size of the detector (see Section 12.3.3.1 for a discussion on the optimal pixel size).
- Many researchers use objectives with NA > n_{sample}, as they allow objective-type total internal reflection fluorescence (TIRF) microscopy [11] (Box 12.2 and Figure 12.4a).

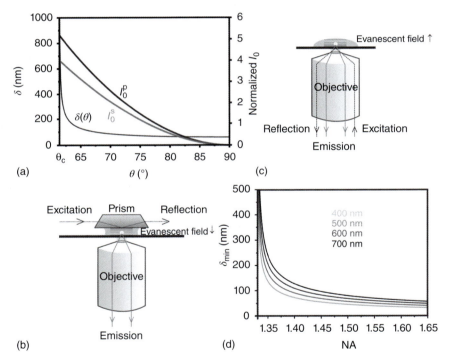

Figure 12.4 TIRF microscopy. (a) The dependence of the TIR illumination intensity at the interface, I_0, on the angle of incidence for p- and s-polarized light. The intensity values are normalized to the incident intensity at $\theta_{\text{in}} = 0$. The panel also includes the θ_{in}-dependence of the penetration depth δ. Plots were calculated for $n_{\text{glass}} = 1.515$ and $n_{\text{water}} = 1.33$, and $\lambda = 550$ nm. (b) In prism-type TIR excitation, a glass prism is coupled to the coverslip to generate the evanescent field at the coverslip surface. A separate objective is used for imaging from the opposite direction. (c) In objective-type TIR, the same objective is used for excitation and imaging. TIR excitation is achieved by offsetting the excitation beam in the back focus of the objective, yielding a tilt of the emanating beam. (d) Minimum achievable penetration depth δ_{min} in objective-type TIR configuration, plotted for different objective NAs and excitation wavelengths.

- *Collection efficiency:* We define here collection efficiency as the fraction of emitted photons that pass the objective. It is thus a critical number for the brightness of the obtained signals. In general, the objective collects all light rays emitted within a cone of angle ϑ, which is specified by the relation $NA = n \times \sin \vartheta$. If the emission is isotropic and photons are not scattered in the sample ("ballistic light"), we can estimate the collection efficiency η_{obj} by the volume fraction of the acceptance cone and the transmittance of the objective η_{trans} according to (Figure 12.3b)

$$\eta_{obj} = \frac{1}{2}(1 - \cos \vartheta) \times \eta_{trans} = \frac{1}{2}\left[1 - \sqrt{1 - \left(\frac{NA}{n}\right)^2}\right] \times \eta_{trans}$$

For large NAs ($NA/n \to 1$), the collection efficiency approximates $\eta_{obj} \approx 0.5 \eta_{trans}$, as a full hemisphere can be collected. For small NAs ($NA/n \to 0$), we obtain $\eta_{obj} \approx (NA/2n)^2 \times \eta_{trans}$. High-quality objectives show transmittances $\eta_{trans} \approx 0.9$. If emitters are located within a distance of a few nanometers from an interface, they emit preferentially into the medium of high refractive index (typically the glass coverslip) at the critical angle of total reflection. Using objectives designed for TIRF microscopy allows collecting such supercritical angle fluorescence; by this, collection efficiencies up to ~80% can be achieved.

- If samples are observed within an aqueous solution micrometers away from the coverslip surface, the use of oil-immersion objectives yields spherical aberrations of the point spread function and focal shifts, which have to be considered for 3D imaging.

Box 12.2 Total Internal Reflection Fluorescence (TIRF) Microscopy

Reducing the out-of-focus background signal has been a central issue for microscopy in general, and for single-molecule microscopy in particular. Total internal reflection fluorescence (TIRF) microscopy takes advantage of the fact that most biological samples are placed on glass coverslips, which allows exploiting the characteristic optical properties of interfaces. In a nutshell, if the coverslip surface is illuminated at a very shallow angle, the incident beam is completely reflected, so that only fluorophores close to the interface get excited. In the following, we give a brief introduction into the optics of total reflection, and discuss some practical problems.

In general, Snell's law of refraction defines the propagation of light at the interface between two media of different refractive indices n_1 and n_2: $n_1 \sin \theta_{in} = n_2 \sin \theta_{refr}$, with θ_{in} and θ_{refr} denoting the angle of incidence and the angle of the refracted beam. For collimated light propagating from glass (refractive index n_{glass}) to the aqueous environment of our biological sample ($n_{sample} < n_{glass}$), the beam is refracted toward the interface so that the angle of incidence becomes larger than the angle of refraction.

In conventional epi-configuration, a well-collimated laser beam aligned with the optical axis is used for illumination of the sample; in that case, $\theta_{in} \sim 0°$, and

(Continued)

Box 12.2 (Continued)

there will be hardly any effects due to refraction. In contrast, if the incident beam is tilted, the refracted beam will progressively incline toward the surface, until it finally propagates parallel to the surface. The corresponding angle of incidence is called the *critical angle* and is specified by $\sin\theta_c = n_{sample}/n_{glass}$. Further tilting the incoming beam leads to its reflection back into the glass coverslip. This regime, $\theta_{in} > \theta_c$, is called *total internal reflection* (TIR) and features some interesting optical phenomena [11]:

i) The excitation intensity decreases exponentially with increasing distance from the surface z, according to $I = I_0 \cdot \exp(-z/\delta)$, where $\delta = (\lambda/4\pi)(n_{glass}^2 \sin^2\theta_{in} - n_{sample}^2)^{-1/2}$ specifies the decay length, and I_0 is the intensity at the surface. In other words, only objects very close to the surface get excited. Depending on the indices of refraction and the angle of incidence, δ varies between 0.1λ and λ. The penetration depth thus scales linearly with wavelength. Figure 12.4a shows a comparison of penetration depths for various angles of incidence.

ii) The excitation intensity at the surface, I_0, depends on the angle of incidence (Figure 12.4a). At the critical angle it is maximum, and exceeds the incident intensity by a factor of 5 or 4 for p- or s-polarized excitation. I_0 decreases to zero for parallel incidence ($\theta_{in} \to 0$).

iii) Linear polarization is preserved for s-polarized incident light. p-Polarized incidence leads to an elliptically polarized evanescent field, where the ellipse is parallel to the plane of incidence. Circular polarized incident light yields an elliptically polarized evanescent field, with the ellipse being tilted against the plane of incidence.

The three phenomena give rise to considerable advantages when it comes to applications to single-molecule microscopy: (i) Because of the exponential decay of the excitation profile, out-of-focus background is strongly reduced. It is thus possible to detect the weak signal of single dye molecules at the glass surface even under unfavorable conditions such as high dye concentrations in the buffer solution. (ii) The enhanced excitation intensity substantially increases the brightness of the observed fluorophores. (iii) Using p-polarized incidence allows the excitation of dipoles perpendicular to the surface that are difficult to excite by standard means.

Practically, there are two common methods for setting up a TIR microscope. In *prism-type TIR*, the coverslip is connected to a glass prism to couple in the excitation beam. In this configuration, the sample chamber is sandwiched between two glass coverslips, with the objective being placed opposite to the illuminated surface (Figure 12.4b). In *objective-type TIR*, the imaging objective is also used for generating the evanescent wave (Figure 12.4c). In that case, an objective with $NA > n_{sample}$ has to be used. Then, excitation rays can emanate from the objective at an angle $> \theta_c$, so that total reflection can occur at the interface between the glass slide and the sample. In practice, the excitation beam is focused slightly off-centrally into the objective's back focal plane, which leads to the beam tilt at the front focus. This can be achieved either by tilting it at a conjugate plane of the sample plane (e.g., at the position of the field stop), or by offsetting the excitation laser parallel to the optical axis in a conjugate plane to the back focal plane of the objective.

For choosing the appropriate TIR objective for a given application, however, more aspects should be considered:

1. The refractive index of the aqueous solution may deviate from $n_{water} = 1.33$. For example, the high protein content of the cellular cytoplasm yields a rather high refractive index $n_{cytoplasm} \approx 1.38$, so that objectives with NA > 1.4 are recommended.
2. If the excitation light is not perfectly collimated, a fraction of photons will irradiate the interface below the critical angle and thus penetrate into the sample. To minimize such propagating waves, most researchers opt for objectives with NA well above n_{sample}.
3. Objectives with a given NA allow reducing the penetration depth to a minimum value $\delta_{min} = (\lambda/4\pi)(NA^2 - n_{sample}^2)^{-1/2}$. For the same angle θ, one can therefore achieve substantially reduced penetration depth by increasing n_{glass} (and consequentially NA). Olympus offers an objective with NA = 1.65, which requires the use of immersion liquid and glass slides with $n = 1.78$. Figure 12.4d shows a comparison of the penetration depths for various values of NA.

12.3.2 Light Source

Ideally, for fluorescence microscopy the region of interest is uniformly illuminated with well-defined polarization, wavelength, and excitation intensity I for a given amount of time, t_{ill}. The choice of these parameters critically affects the outcome of the experiment and needs to be considered when selecting the light source.

12.3.2.1 Uniformity

Köhler illumination can be regarded as a cornerstone of modern light microscopy: by imaging the light source into the back focal plane of the condenser, the field of view can be uniformly illuminated and is hardly perturbed by structures of the light source itself. The best performance is achieved using low coherence light sources such as arc lamps or incandescent lamps. However, Köhler illumination includes neither the entire surface nor the full angular distribution of the light source, rendering it rather inefficient. An alternative way of excitation is critical illumination, where the light source is directly imaged onto the specimen plane. This mode of operation is highly efficient, yet all nonuniformities of the light source become directly visible.

Lasers operating in TEM_{00} mode allow illumination of the sample with a Gaussian profile, which can be focused to arbitrary size and is limited only by the NA of the condenser (in epi-configuration the condenser is the same as the objective). In practice, however, highly coherent light sources such as lasers frequently generate speckles originating from interferences from rays reflected from the various optical elements. A practical way for reducing speckles is the use of light scramblers (e.g., fiber-optic light guides, rapidly rotating diffusers, or spatial filters).

12.3.2.2 Intensity

The excitation intensity limits the number of photons that can be emitted by the fluorophores. To estimate the excitation intensity required for single-molecule imaging, let us first introduce some basic spectroscopic considerations. To a good approximation, fluorescent dyes can be described as three-level systems, with a singlet ground state (S_0), a singlet excited state (S_1), and a triplet excited state (T_1; see Figure 12.5a). Excitation via absorption of a single photon occurs from S_0 to S_1, and spontaneous emission of a red-shifted photon from S_1 to S_0 takes place at a characteristic rate τ^{-1}. The intersystem crossing to the triplet state is spin-forbidden and, therefore, highly unlikely, with rate constants k_{isc} for the population and k_T for the depopulation of T_1. In time equilibrium, the number of photons F detected from a single dye molecule on excitation with monochromatic light of intensity I is given by

$$F = \frac{f_\infty}{1 + \frac{I_s}{I}} t_{ill}$$

with

$$f_\infty = \frac{\tau^{-1}}{1 + \frac{k_{isc}}{k_T}} \times \eta \times \Phi$$

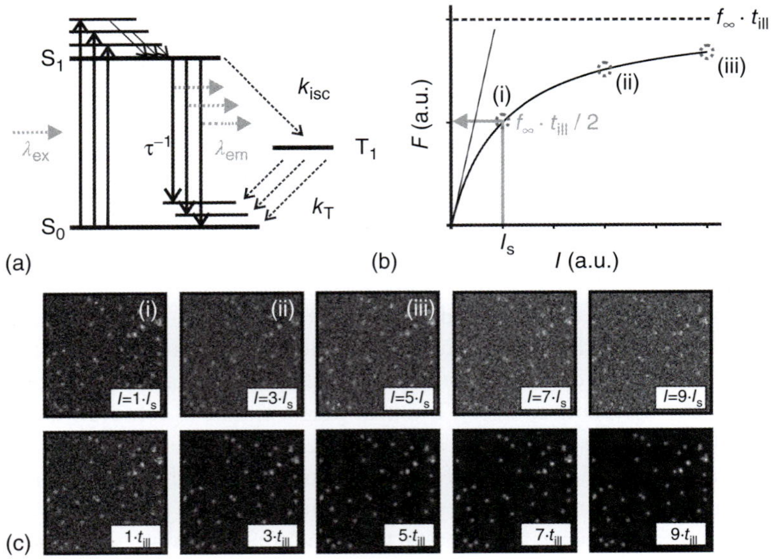

Figure 12.5 Effect of saturation on the image quality. (a) Jablonski diagram illustrating the transition between different states S_0, S_1, and T_1. Excitation and emission wavelengths λ_{ex} and λ_{em} are indicated. τ denotes the fluorescence lifetime of the excited singlet state, and k_{isc} and k_T represent the intersystem crossing to and from the triplet state, respectively. (b) Saturation curve of the fluorescence signal F as a function of the excitation intensity I. (c) Simulation of single-molecule signals at increasing illumination intensity (first row) or illumination time (second row). In both cases, nonsaturable background signal and readout noise were added, and the images were subjected to shot noise. In the top row, image quality deteriorates as a result of saturation of the specific signal; in the bottom row, image quality improves because of negligible readout noise at long illumination times.

and

$$I_S = \frac{\tau^{-1} + k_{isc}}{1 + \frac{k_{isc}}{k_T}} \frac{hc}{\lambda_{ill}\sigma}$$

where t_{ill} defines the illumination time, Φ denotes the quantum yield, and η is the detection efficiency of the apparatus. The excitation cross section σ is related to the extinction coefficient ε via $\sigma = \varepsilon \times \ln(10)/N_A$, with N_A the Avogadro constant. At low illumination intensity I, we can expect the dye molecule to reside in its ground state at the time the excitation photon hits the molecule's absorption cross section σ. In this regime, the single-molecule brightness is mainly determined by the absorption properties; it scales linearly with I according to

$$F \approx \frac{\lambda_{ill}\sigma}{hc} I \times \Phi \times \eta \times t_{ill}$$

with

$$\frac{\lambda_{ill}\sigma}{hc} I$$

specifying the excitation rate. For high illumination intensity I, there is a high probability that the dye is already excited when it is hit by the photon; in this case, the signal saturates and approaches a maximum value $f_\infty \times t_{ill}$ (Figure 12.5b). As examples, rhodamine shows a saturation intensity of $I_s \sim 7\,\text{kW cm}^{-2}$ and Alexa647 a saturation intensity of $I_s \sim 19\,\text{kW cm}^{-2}$. Good organic fluorophores have an excitation rate of $\sim 10^6\,\text{s}^{-1}$ at $I = 1\,\text{kW cm}^{-2}$.

Increasing the excitation intensity, therefore, yields increased single-molecule brightness, but the effect levels off beyond I_s. Since brightness is a key determinant of single-molecule visibility against the readout noise of the detector, high excitation intensity may be recommended. Most biological samples, however, contain not only the specific signal but also nonsaturatable unspecific background arising from cellular autofluorescence. On increasing the excitation intensity beyond I_s, this background signal becomes more and more dominant, finally reducing the single-molecule visibility (Figure 12.5c); see also Box 12.3 for a discussion on signal-to-noise ratio.

Furthermore, some fluorophores show photoinduced blinking or bleaching. In the case of GFP, the probability for detecting the molecule in the on-state decreases with increasing excitation intensity: equal on- and off-times were found for intensities $\sim 1.5\,\text{kW cm}^{-2}$. The dark-state lifetime is independent of the excitation intensity and rather long, with typical off-times of $\sim 1.6\,\text{s}$. Photoinduced blinking has two consequences: first, it affects tracking, as it is difficult to re-identify mobile molecules after a dark period; second, integration over multiple on and off periods yields a reduced brightness of the average signal. Both effects are clearly visible in Figure 12.6: GFP fused to the Orai1 channel was imaged in Chinese hamster ovary (CHO) cells at 10 and at $1\,\text{kW cm}^{-2}$ with 10-fold increased illumination time; thus, the overall excitation energy was the same in both cases. Measurements at low excitation power yielded much longer and slightly brighter trajectories.

In summary, choosing the appropriate excitation intensity is crucial for single-molecule imaging. For example, by focusing a 10 mW laser to a width

$\sigma = 12.5\,\mu m$, a peak excitation intensity of $\sim 1\,kW\,cm^{-2}$ can be obtained. Recall, however, that the intensity of a TEM_{00} laser has a Gaussian profile and thus decreases to $\sim 66\%$ within the σ range, rendering the excitation profile rather inhomogeneous. One may thus increase the illumination area and power to obtain a uniformly illuminated field of view and confine it by a field stop to the central part of the Gaussian profile: with $\sigma = 25\,\mu m$ and a power of 40 mW, the same excitation intensity can be achieved, but now the signal decreases only by 10% within the central circle of a radius of $12.5\,\mu m$. Increased excitation intensity can also be obtained by illuminating the sample with an evanescent wave: directly at the glass/water interface, the illumination intensity is increased up to fivefold (see Box 12.2) [11].

12.3.2.3 Illumination Time

Increasing the illumination time (t_{ill}) is an alternative way for enhancing the single-molecule brightness. One has to ensure though that t_{ill} is shorter than both the molecular kinetics to be measured and the photobleaching time (integration over dark periods would only add the background signal). In particular, long illumination times allow averaging out dark intermediates of blinking fluorophores, thereby improving the traceability (Figure 12.6).

12.3.2.4 Polarization

While the radiation of arc lamps or incandescent lamps is unpolarized (more precisely, it contains a variety of different polarization angles perpendicular to the wave vector), lasers emit polarized light. Fluorophores interact with electromagnetic radiation just as dipole absorbers. Accordingly, the excitation rate needs to be modified to $(\lambda_{ill}\sigma/hc)I\cos^2\varphi$, which scales with the angle φ between the

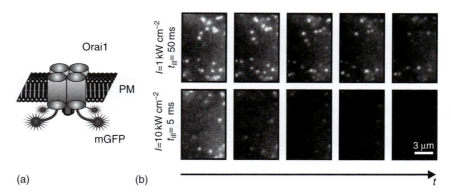

Figure 12.6 Single-molecule imaging of GFP works better at low excitation intensity. (a) Sketch of the tetrameric ion channel Orai1 fused to mGFP used for this experiment. (b) Tracking of single Orai1–mGFP channels in Chinese hamster ovary cells. Upper panel: Images from a sequence recorded at low excitation power (1 kW cm^{-2}) and long illumination time (50 ms). Lower panel: Images from a sequence recorded at 10-fold increased excitation power and 10-fold reduced illumination time. Despite the same excitation energy per illumination, the channels recorded at low power are visible for more observations and appear brighter at the first image.

molecular dipole moment and the electromagnetic polarization. Optimum excitation is obtained at angle $\varphi=0°$. Freely rotating fluorophores will be excited at a reduced rate, determined by the average of $\cos^2 \varphi$ ($\cos^2 \varphi = 1/3$ for free rotation in three dimensions and $\cos^2 \varphi = 1/2$ for free rotation in a 2D plane perpendicular to the optical axis). In this case, any orientation of the polarization will work for excitation.

However, if there are constraints for the orientation of the fluorophore, one needs to consider the polarization of the excitation light in more detail. For example, the amphipathic carbocyanine dye DiI integrates in lipid membranes, so that the chromophore dipole lies parallel to the membrane surface; it may thus rotate in the membrane plane, but out-of-plane motion is impeded. Consequentially, DiI is not excitable by linearly polarized light with polarization perpendicular to the membrane surface. In order to capture more fluorophore orientations, researchers frequently revert to circular polarized light by inserting a quarter-wave plate in the excitation path. In turn, such effects can be used to obtain insights into the membrane topology.

In general, Köhler illumination does not provide an efficient way to excite fluorophores with a dipole moment parallel to the optical z-axis. However, there are illumination configurations with significant components of the electrical field in z-direction: in TIR configuration, the evanescent field contains a z-component if p-polarized light is used for excitation (see Box 12.2) [11]; a focused laser beam (as used in confocal microscopy) contains longitudinal field components, which can be further enhanced, for example, by annular illumination. Light emerging from the tip of a tapered fiber (as used in near-field scanning optical microscopy (NSOM)) contains components along the fiber axis.

12.3.2.5 Wavelength

Laser light is monochromatic and therefore provides the power in the spectral region where it is needed, whereas lamps are polychromatic so that only a small fraction of light is emitted in the required spectral region. Currently, lasers are available over the whole visible range of the spectrum at sufficient power (hundreds of milliwatts) and decent prize. The used wavelength is thus determined mainly by the scientific question.

In cell biological applications, for example, the molecule of interest has to be discriminated from endogenous autofluorescence. Cellular autofluorescence has been characterized in terms of spectral properties, lifetime, and spatial distribution. In the visible regime, flavins and lipofuscin are regarded as the major source. Flavins are mainly located in mitochondria, while lipofuscins predominantly reside in lysosomes. In fluorescence images, both organelles appear as spots randomly distributed in the cytoplasm of the cell. The high variability of the fluorescence intensity of such spots, even within one cell, makes unambiguous distinction between fluorophores and autofluorescence a challenging task.

In general, it turns out that the brightness of autofluorescent structures decreases with increasing wavelength, rendering red excitation beyond 600 nm the pragmatic choice. However, frequently the biological problem defines the excitation wavelength to use. In this case, the most promising strategy discriminates the used label from autofluorescence via its characteristic color.

Alternatively, photoswitchable fluorophores may be used as labels, which allow lock-in techniques to specifically identify the probe.

12.3.2.6 Collimation

The highly collimated beam is another major plus point of lasers compared to lamps. On one hand, this allows focusing the excitation light to small regions down to the diffraction limit. The illumination can thus be confined to regions of interest without the inevitable losses when using a lamp in combination with a field stop. On the other hand, Köhler illumination is possible with extremely high quality of the planar wave; in other words, the range of wave vectors is extremely narrow. This is an important requirement for setting up high-quality TIRF microscopy (see Box 12.2).

12.3.3 Detector

A good camera is crucial for optimum performance of the imaging system. Pixel size and noise are the key parameters that define the quality and thus the information content of single-molecule images. We will discuss here the specific demands of single-molecule experiments on the pixel size and give a brief description of the noise characteristics of typical camera types.

12.3.3.1 Pixel Size

Cameras divide an image into small boxes called *pixels*, each being assigned certain intensity numbers. Manufacturers offer cameras with a wide range of pixel sizes between ~5 and ~30 μm; in the following, we define the pixel size s in the object plane; therefore, we divide the physical pixel size by the magnification M. In standard microscopy applications, the pixel size is typically chosen in accordance with the Nyquist theorem, which states that faithful reconstruction of a periodic sine wave of frequency f can be achieved at a sampling frequency $\geq 2f$. Applied to microscopy, the criterion defines a maximum pixel size $s \approx \text{FWHM}/2$: using larger pixels ("undersampling") would constrain the obtained resolution. In turn, using smaller pixels ("oversampling") does not yield additional structural information. Whenever the number of photons is limiting, oversampling would be even detrimental: light emanating from a diffraction-limited spot would be distributed over many pixels, so that shot noise would increase and the spot visibility decrease.

Single-molecule applications add additional constraints to the appropriate choice of the detector pixel size. The most distinguished variants make use of the localization of single fluorophores with a precision below the diffraction limit (Section 12.2.4 and Chapter 8). Thompson *et al.* provided a formula for the pixel size at which optimum LP of a point emitter can be achieved: $(s/\text{FWHM})^4 = 3000\pi(N^2/F)$, with N denoting background noise and F the number of detected photons (note that we specified here the spot size by its FWHM, thereby yielding a different prefactor than that obtained by Thompson *et al.*) [9]. Figure 12.7 shows that the minimum is rather broad, with steep inclines toward smaller and larger pixels: too small pixels yield an increased photon noise, whereas too large pixels generate essentially a single bright pixel that does not provide further insights into the molecule's subpixel position.

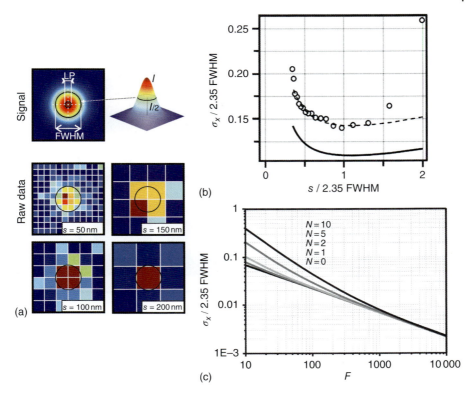

Figure 12.7 (a) The Gaussian intensity distribution (we assumed here an FWHM = 200 nm) arising from a single diffraction-limited fluorophore is imaged with different pixel sizes s. LP denotes the localization precision. (b) Dependence of localization precision σ_x on the pixel size s. Both axes were normalized by the spot size, which was defined as the standard deviation of the point spread function (FWHM/2.35). The curve has a pronounced minimum: for pixel sizes smaller than the spot size, σ_x increases with decreasing pixel size because of increased background noise; for pixel sizes larger than the spot size, σ_x increases with increasing pixel size because of pixelation noise. The theoretical prediction is shown as solid line. Results from computer-generated images (circles) yield a 30% bias but reproduce the overall shape of the theoretical curve well (the theoretical prediction plus 30% is shown as dashed line). Both analytical theory and the computer-generated images used $F = 100$ photons and $N = 0.7$. (c) Dependence of localization precision σ_x on the number of counted photons, F, for various noise levels N (for the calculation we set $s = \text{FWHM} = 1$). Over the whole range of the plot, subpixel localization precision is possible at all noise levels. (Reprinted from [9] with permission from Elsevier.)

Single fluorophores are electric dipole emitters, which – when placed near the interface of two media with different refractive indices – interact with their back-reflected electromagnetic field. As a result, the obtained single-molecule images may deviate massively from Gaussian profiles, depending on the inclination of the dipole moment toward the sample surface and on the defocusing. In particular, inclination angles of ~45° generate asymmetric patterns, with an offset centroid position [12] (Figure 12.8). Only oversampling allows capturing such effects.

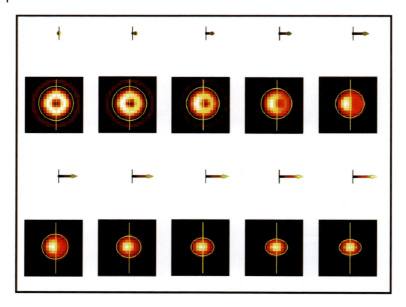

Figure 12.8 Ideal intensity distributions of molecules on the CCD for different dipole inclinations toward the sample surface, as indicated by the arrows depicted above each image. The image was calculated for an air/glass interface. (Reprinted from [12] with permission from the Optical Society.)

12.3.3.2 CCD Cameras

Most researchers in the single-molecule field use state-of-the-art charge-coupled device (CCD) cameras. High quantum yield and low readout noise are prerequisites for the best image quality. High quantum yield can be achieved by thin etching of the CCD chip, so that the light passes through the back layers (back-illuminated); quantum yields up to 90% are specified by manufacturers. The image quality depends critically on the noise levels, which are essentially determined by the shot noise of specific signal (N_{signal}) and unspecific background ($N_{background}$), by the dark signal (N_{dark}), and by the readout noise ($N_{readout}$): $N_{total} = \sqrt{N_{signal}^2 + N_{background}^2 + N_{dark}^2 + N_{readout}^2}$. Dark current is minimized by deep cooling of the detector and can usually be neglected. Readout noise is largely determined by the digitization rate of the detector, that is, the number of pixels that can be read out within 1 s. Low readout noise is difficult to achieve and possible only at low digitization rates (typically 50–100 kHz), rendering such devices rather slow. Systems are available with a readout noise of about two electrons per pixel at 50 kHz readout rate. For such detectors, the readout noise becomes negligible and the shot noise of the signal and of the background will dominate.

With low digitization speed (below 100 kHz), full-frame readout requires several seconds. Even readout of a subregion of 100 × 100 pixels, sufficient for imaging a single small cell, would take at least 200 ms. Although this time resolution is high enough to observe, for example, intracellular transport, reorganization processes on a molecular level typically happen on a much faster timescale. In order to resolve such rapid kinetics, some manufacturers offer a special operation mode based on multiple frame transfer. In this mode, only a subregion of $n_x \times n_y$ pixels is used for observation. On illumination, this subregion is shifted into a masked area of the chip rapidly, where it is temporarily stored; photoelectrons can be shifted to the camera-chip within a few microseconds per line. The readout process is performed subsequently after several illumination/shifting cycles and is independent of this timing protocol.

12.3.3.3 Electron-Multiplying CCD Cameras

Continuous and rapid imaging has been realized by the use of image-intensified cameras since the 1990s; such devices, however, had the disadvantages of lower quantum efficiency, poor resolution, and additional spontaneous intensifier noise, rendering the quantification of such images difficult. The recently introduced electron-multiplying charge-coupled device (EM-CCD) cameras make use of on-chip electron amplification before fast readout: an additional serial multiplication register after the readout register provides a low noise gain before the charge-to-voltage conversion. With EM-CCDs, fast and continuous readout can be achieved at high quantum yield ($\leq 90\%$) and optimum resolution, at an effective readout noise $<1\,e^-$ per pixel. Because of the amplification, however, the effective noise of any detected signal will also be increased.

The multiplication register consists of n stages, each amplifying photoelectrons with a probability $\alpha \approx 0.01$, yielding a total gain $G = (1+\alpha)^n$; for the following calculations, we used $n = 536$. Fluorescence signal F and noise N of EM-CCDs are typically referenced to the image plane to allow comparison with non-amplified detectors ($F \rightarrow F/G$, $N \rightarrow N/G$); in the following, we will use these referenced values. Statistical analysis and experimental data revealed an increased noise due to amplification by a factor of 2 ($N^2_{\text{EM-CCD}} = 2 N^2_{\text{CCD}}$). EM-CCDs are thus characterized by $N_{\text{total}} = \sqrt{2(N^2_{\text{signal}} + N^2_{\text{background}} + N^2_{\text{dark}}) + (N_{\text{readout}}/G)^2}$. There are two main differences from non-amplified CCDs: readout noise can be reduced deliberately by increasing the gain (the last term), but specific and unspecific signal noise is larger (the first two terms); again, dark noise is less important, as it can be essentially eliminated by cooling the chip.

The advantage compared to non-amplified CCDs is the option to read out the chip at extremely high rates and low readout noise: a 5-MHz chip may be run at $N_{\text{readout}} = 50$ electrons per pixel; thus, for typically applied gains up to $G \sim 1000$, the last noise term will vanish (effective readout noise $\ll 1$ electron per pixel), so the device can be operated in the shot noise limit. Note, however, that compared to a slow-scan CCD, the image quality will be deteriorated due to the additional amplification noise. This may be acceptable if the detection of a molecule is the primary goal and no additional background affects the image. When it comes to discriminating specific signals from unspecific homogeneous background, however, the decreased visibility becomes relevant (Figure 12.9).

In Figure 12.9b, we compare the performance of a CCD and an EM-CCD by a simulation of the imaging process. Single emitters of different brightness were arranged on a background gradient. For CCD imaging, we added readout noise of

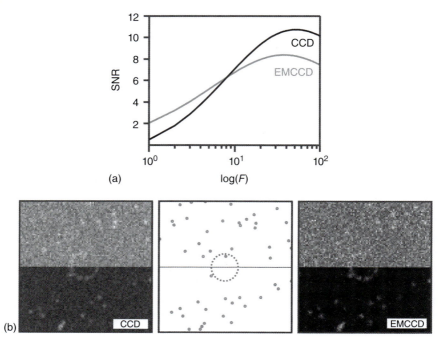

Figure 12.9 Simulation of CCD and electron-multiplying charge-coupled device (EMCCD) imaging. (a) The signal-to-noise ratio was calculated for an EMCCD and a conventional CCD as a function of the number of incident photons, F. We assumed a rather high readout noise of 15 electrons and low background noise. The EMCCD outperforms the conventional CCD at low photon numbers because of the virtually zero readout noise, whereas at high photon numbers this gain vanishes and the improved photon-counting noise of a conventional CCD becomes significant. At high background noise, the CCD would yield better signal-to-noise ratio even over the whole range of signal intensity (not shown). Included in this plot is also signal saturation at high excitation intensity, which leads to a decline of the signal-to-noise ratio. (b) Point emitters of different total brightness were distributed randomly on a surface (middle image) and a ring of molecules was placed in the center. Background signal was added only to the top half of the image, and the whole images were subjected to shot noise and readout noise. While the low effective readout noise of EMCCDs yields a clearer image of the bottom part, a conventional CCD outperforms the EMCCD when background is present.

$N_{readout} = 2$ electrons, and for EM-CCD imaging we simulated the multiplication with gain $G = 1000$ and added $N_{readout} = 50$ electrons.

Note that analytical expressions of the localization and brightness errors (discussed below) derived for CCD cameras cannot be directly applied for EM-CCDs; however, replacing the detected signal by $F/\sqrt{2}G$ provides a reasonable estimate. Note further that comparison of data obtained with different devices may be hampered by unknown absolute gain settings. Camera manufacturers try to fix this problem by providing gain calibration options.

12.3.3.4 CMOS Detectors

While the massively parallel readout possibilities of complementary metal–oxide–semiconductor (CMOS)-based imagers enable extremely high frame rates, the rather poor noise characteristics rendered single-molecule applications so far not feasible. Novel scientific CMOSs operating at extremely low readout noise of ~2–3 electrons appear highly promising.

> **Box 12.3 Signal and Noise**
>
> Identification and quantification of single-molecule signals mainly depend on the brightness of the specific signal F, that is, the sum over all detected photons, and the noise N, that is, the fluctuations of the signal over unspecific background; a figure of merit is the signal-to-noise ratio (F/N).
>
> As discussed in Section 12.3.2, fluorescence signals increase linearly with excitation power at low laser intensities and saturate when increasing the laser intensity beyond I_s. Noise is more difficult to estimate, as there are multiple contributions; the total noise is the root of the sum of all squared contributions: $N = \sqrt{\sum N_i^2}$:
>
> - *Detector noise:* This includes readout noise ($N_{readout}$) and dark noise (N_{dark}). $N_{readout}$ is a detector-specific constant, and N_{dark} increases with the illumination time (Section 12.3.3).
> - *Signal fluctuations* (N_{signal}): Because of the low photon count rates, the signals fluctuate as a result of shot noise according to $N_{signal} = \sqrt{F}$ (or $N_{signal} = \sqrt{F/G}$ for-EM-CCDs).
> - *Unspecific background noise from the environment:* There are not only multiple sources for background noise in biological samples, including Rayleigh and Raman scattered light, but also autofluorescence originating from weakly fluorescing molecules at high concentrations. These contributions are, in general, not saturatable and thus scale linearly with the excitation intensity and illumination time. In addition, background noise scales with the square root of the brightness ($N_{background} = \sqrt{F_{background}}$ or $N_{background} = \sqrt{F_{background}/G}$ for EM-CCDs).
>
> Increasing the laser power, therefore, increases the signal-to-noise ratio at low excitation intensities, where unspecific background signal is negligible. At high laser power, background increases linearly while specific signal saturates, so that the signal-to-noise ratio declines (Figure 12.9a).

12.4 Analyzing Single-Molecule Signals: Position, Orientation, Color, and Brightness

There are multiple parameters that may be used to characterize a dye molecule i: for example, the time-dependent position $[x_i(t), y_i(t), z_i(t)]$, the orientation of its excitation or emission dipole $[\vartheta_i(t), \varphi_i(t)]$, its absorption and emission spectra $[Ab_i(t), Em_i(t)]$, or its brightness $F_i(t)$. Considering dye molecules as reporters, one may use these parameters for obtaining insights into the local environment, for example, the viscosity $\eta(x, y, z, t)$, the type of motion (directed transport, random diffusion, obstructed diffusion, etc.), the spatial arrangement, the local proximity to other molecules, the degree of association, or conformational changes.

12.4.1 Localizing in Two Dimensions

It is possible to locate a single molecule to an accuracy that is substantially less than the actual signal width; for a high signal-to-noise ratio, even subnanometer precision has been reported. When it comes to estimating the position of a molecule, two questions become relevant:

- What are the errors in the lateral coordinates when fitting a known function to the image?
- What is the fit function?

There are various approaches available in the literature, many of them originating from single-particle tracking studies. Indeed, the signal distributions of freely rotating dyes are well approximated by Gaussian functions (Eq. (12.1)) [13]; therefore, those algorithms and the according characterizations are also valid for single-molecule tracking. In one of the first approaches, Bobroff based his treatment on chi-squared fitting and calculated the LP for a Gaussian point spread function; he identified the signal-to-noise ratio as the major determinant for errors [14]. Further methods for localizing single-molecule signals have been introduced and compared: nonlinear least-squares fits to a Gaussian function yielded less bias and higher precision than centroid estimations. Thompson et al. tested a Gaussian mask algorithm, which – interestingly – performed only marginally worse than a nonlinear least-squares fit [9].

Furthermore, Thompson et al. provided a very detailed treatment of the LP of a Gaussian signal on a detector with pixel size s, yielding a total localization precision in each dimension of

$$\sigma_x^2 = \left(\frac{\text{FWHM}^2}{5.52} + \frac{s^2}{12}\right)/F + \frac{0.82 \text{FWHM}^4 N^2}{s^2 F^2}$$

(see also Figure 12.7 and discussion on pixel size above) [9]; F and N denote the total fluorescence photon counts and the background noise, respectively, and FWHM is the full width at half-maximum of the Gaussian. The first term essentially accounts for photon-counting noise and pixelation noise: even in the absence of a noisy background ($N = 0$), there is a limit to the localization of a single fluorophore, which originates from the Poissonian fluctuations of the signal itself. This particular aspect was further addressed by Ober, who calculated a fundamental limit for the LP of a Gaussian-shaped signal distribution:

$$\sigma_x = \frac{\lambda}{2\pi \text{NA} \sqrt{F}} = \frac{\text{FWHM}}{3.27 \sqrt{F}}$$

[15]. The second term in the equation accounts for background noise; at high noise levels N, the localization errors scale with the inverse of the signal-to-noise ratio.

When the dipole emitters are fixed or somehow constrained in their orientational degrees of freedom, however, deviations from a Gaussian function can become significant. For example, the molecules of interest are frequently measured next to a water/glass interface, and an objective with high NA is used for TIR excitation. As described in Section 12.3.3, the emission of surface-proximal dipoles is highly anisotropic, with a pronounced maximum at the critical angle of total reflection. Because of their high NA, TIR objectives capture even supercritical angle electromagnetic radiation. Emission patterns were calculated for different scenarios including slightly aberrated or defocused images. In particular, asymmetries of the central maximum are observable for emission dipoles tilted against the optical axis. In such cases, fitting with a symmetric Gaussian function yields a significant bias in the position, which can be as high as 10 nm [12] (Figure 12.8).

The most comprehensive theory for single-molecule localization was presented by Mortensen et al., who treated individual emitters as dipoles [13]. They provided frameworks for the analysis of immobilized or freely rotating dipoles excited in TIR configuration. Estimates for the x- and y-positions (and the azimuthal and polar angles in case of immobile emitters) were derived based on maximum likelihood estimators. Different fitting methods were compared, yielding ~70% higher uncertainties for a least-squares estimator compared to the maximum likelihood estimator. Finally, formulas were provided to estimate the localization errors for maximum likelihood and least-squares fitting. Smith et al. showed how such maximum likelihood estimators can be implemented on graphics processing units from a video card, yielding ~80-fold faster computation [16].

Many researchers optimized the mathematical fitting process to enhance the LP. Ultimately, however, one is not so much interested in the absolute coordinates but in relative distances, which could be the displacements of a moving particle or the distances between different nearby fluorophores that are resolvable as individual spots (e.g., blinking molecules or molecules in different color channels). After correcting for drifts, it turned out that the photoresponse of different pixels on the camera-chip is not uniform, which sets the ultimate limit to localizing single molecules at the nanometer length scale.

12.4.2 Localizing along the Optical Axis

Localizing point emitters in two dimensions is conceptually straightforward, but the determination of the position along the optical axis is more demanding, as the centroid of the emission pattern is not affected by the z-coordinate. In the following, we describe two approaches.

12.4.2.1 Analysis of the Shape of the Point Spread Function

A defocused point emitter generates a characteristic image which is described by the point spread function of the microscope and encodes the dye position along the optical axis (a detailed treatment would have to account for the dipole orientation of the dye as well, but it can be neglected for rapidly rotating molecules). In general, the lateral width of the defocused signal scales with the position z_0 according to

$$\text{FWHM}(\delta z) = \text{FWHM}_0 \sqrt{1 + \left(\frac{\delta z - z_0}{\text{DOF}}\right)^2}$$

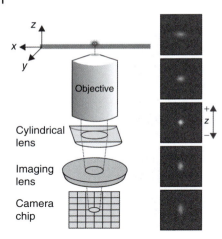

Figure 12.10 On-axis localization of single dye molecules using astigmatic distortion. Astigmatism is generated using a cylindrical lens in the emission beam path. The images of a single dye molecule at various z-positions are shown, yielding elliptical distortions of the intensity profile.

where δz denotes the defocusing distance, $FWHM_0$ is the width of the point spread function of the focused image, and $DOF = \pi \sigma_0^2/\lambda$ is the depth of focus; it yields a parabolic behavior for small defocusing below the depth of field (DOF) and a linear regime above DOF.

One may obtain the z-coordinate from multiple images recorded of different focal planes. This can be implemented by using a z-piezo that rapidly shifts the objective. Alternatively, one may record multiple planes simultaneously by splitting the emission into multiple paths with slightly shifted focal planes.

The z-dependence of the point spread function can be further accentuated by inserting a cylindrical lens with long focal length (~1 m) in the emission path of the microscope, thereby generating artificial astigmatism [17] (Figure 12.10). Let us assume the cylindrical lens to be aligned with the lateral coordinate frame; thereby two foci are generated, one for the components along the axis of the cylindrical lens (e.g., the x-axis) and the other for the perpendicular components (y-axis). The two foci have a distance of 2γ, where γ is termed the astigmatism and is a function of the focal length of the cylinder lens. In this configuration, the intensity profiles are elliptically distorted $\propto \exp\{-[(x-x_0)^2/2\sigma_x^2] - [(y-y_0)^2/2\sigma_y^2]\}$ (recall $FWHM_{x,y} = 2.35\sigma_{x,y}$), with the distortion swapping from the x- to the y-axis during focusing. Thus, the distortion specifies whether the molecule is located above or below the current plane of focus and the distance to this plane. The z-position can be calculated from the recording of a single focal plane via the elliptical distortion $z_0 = (DOF/\sigma_0)\sqrt{FWHM_x^2 - FWHM_0^2} - \gamma$ for $\sigma_x > \sigma_y$ and $z_0 = (DOF/\sigma_0)\sqrt{FWHM_y^2 - FWHM_0^2} - \gamma$ for $FWHM_x < FWHM_y$. This approach allowed 3D super-resolution imaging in fixed cells. Note that additional spherical aberrations may arise from the use of oil-immersion objectives, which have to be corrected before reconstructing the 3D image.

A much stronger modification of the point spread function was recently introduced by the group of W. E. Moerner (Figure 12.11): they used a spatial light modulator to generate a double-helix point spread function [18]. In this configuration, a point emitter is imaged as two adjacent spots, with the angle between

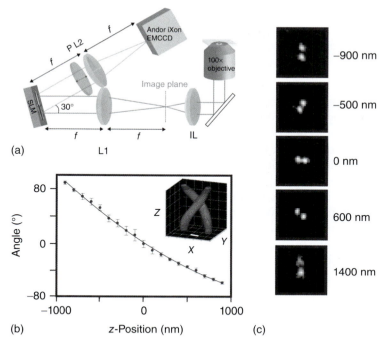

Figure 12.11 Determination of the z-position of a single molecule by imaging via a double-helix point spread function. (a) Collection path of the setup. IL is the imaging (tube) lens of the microscope, L1 and L2 are focal-length-matched achromatic lenses, and SLM is a liquid crystal spatial light modulator. (b) Calibration curve of the angle between two lobes versus the emitter's position along the optical axis measured with a piezo-controlled objective. (Inset) 3D plot of the double-helix point spread function intensity profile. (Scale bar: 400 nm.) (c) Images of a fluorescent bead used for the calibration curve at different axial positions, with 0 being in focus. (Reprinted from [18] with permission from *Proceedings of the National Academy of Sciences of the United States of America*.)

the spots encoding the z-position. Single-molecule localization errors as low as 20 nm could be specified along the z-dimension (in the x- and y-dimension errors of ∼12 nm were obtained). The double-helix approach proved particularly superior for localizing objects that are not exactly in the focal plane.

In general, single-molecule localization along the optical axis yields a lower precision than along radial dimensions. Isotropic resolution demands recording a side view of the object, which can indeed be achieved by using a tilted mirror near the emitter; in this case, both front and side views can be obtained simultaneously on the same detector.

12.4.2.2 Intensity Patterns along the Optical Axis

Alternatively, one may use variations in the excitation intensity or detection efficiency to encode the z-coordinate by brightness. If the specimen is located near the interface with a glass slide, the exponential decay of the evanescent field renders dyes increasingly bright with decreasing distance from the surface and – for known decay length – provides a relative distance measure. Note that in objective-type TIR configuration not only the excitation intensity but also the detection efficiency varies along the z-coordinate, so that an independent calibration of the z-dependence of the single-molecule brightness is favorable.

12.4.3 Brightness

The number of photons F detected from a single point light source is an important parameter in single-molecule microscopy, as it represents the basis for the study of a variety of molecular properties such as molecular association, dye orientation, or color (see below for more detailed discussions). Operationally, it is the integral of the point spread function over background in a given focal plane (Eq. (12.1)). The brightness *per se*, however, is not well defined, as it depends on experimental properties such as the detection efficiency of the optics and the excitation intensity. Yet, for a given setting, it may be a valuable property in a comparative experiment. Essentially, the number of photons in a single spot can be calculated by the same mathematical approaches as its lateral position.

When it comes to brightness analysis, researchers will encounter two main questions:

1. What is the number of photons captured from a single-point emitter? Thompson *et al.* used least-square fitting to a Gaussian function [9]. They provided an error analysis for the determination of F as a function of the total detected number of photons, which for high signal-to-noise ratios approximates the squared counting error $N_F^2 \approx F$. Gaussian fitting works well for a background that fluctuates around zero signal and does not show spatial variations. In live-cell experiments, however, data often contain substantial contributions of autofluorescence. In such cases, some scientists prefer calculating the sum of the photon counts over all pixels minus the local background.
2. Are there variations in the calculated brightness of *a priori* identical molecules, which add to the errors? For example, lateral or temporal variations in the excitation intensity or in the orientation of the transition dipole will lead to increased brightness variations. Variations in the distance to quenchers may further increase brightness fluctuations. In consequence, the single-molecule brightness of mobile molecules in cells shows much stronger variations than expected from shot noise.

12.4.4 Orientation

Organic dyes are dipoles with a characteristic orientation and may thus be used to obtain information on the orientation of the moieties they are linked to. Conformational changes or rotations can be addressed. There are two standard approaches for obtaining information on the dipole orientation of single molecules.

12.4.4.1 Polarization Microscopy

Light absorption of a fluorophore depends on the angle between the polarization of the excitation light and the orientation of the absorption dipole of the dye. Using linearly polarized light for excitation allows the determination of the orientation of a single fluorescent dye by measuring the modulation of the fluorescence signal, which is equivalent to measuring the linear dichroism of a single dye molecule. In addition, the polarization of the fluorescence signal emitted from the fluorophore can be determined by introducing a polarizing beam splitter into the emission path of the microscope. The combination of polarized excitation and the determination of the emission polarization allows steady-state anisotropy measurements on a single molecule, which opens up the possibility to access nanosecond timescales for probing single-molecule reorientation.

The basic limitation of standard polarization microscopy is its restriction to the study of the in-plane component of the orientation of a dye molecule. Yet, illumination in TIR configuration contains a strong component along the z-axis, which can be used to obtain information on both azimuth and elevation angles.

12.4.4.2 Defocused Imaging

As delineated earlier, defocused imaging (typically ~500 nm) of single molecules on surfaces via high NA objectives yields characteristic side lobes that contain information on the 3D orientation of the emission dipole.

12.4.5 Color

Multicolor microscopy has become routine in live-cell biology and accordingly in single-molecule microscopy and spectroscopy. The idea is to discriminate different fluorophores (and thereby the biomolecules they are attached to) by their characteristic excitation or emission spectra. Ideally, one would like to record simultaneously the well-separated signals of multiple fluorophores. Yet, typical excitation and emission spectra are rather broad, yielding substantial overlap between different dyes. If we assume noninteracting emitters, the total detected signal on each pixel, F_{tot}, is given by a linear combination of the spectral contributions of the individual fluorophores, F_i: $F_{total}(\lambda_{ex}, \lambda_{em}) = \sum_i \alpha_i F_i(\lambda_{ex}, \lambda_{em})$, where α_i denotes the contributions from the ith fluorophore. The signals depend on the chosen excitation wavelength λ_{ex} and the settings of the emission filters λ_{ex}.

Linear unmixing is the standard procedure to determine the coefficients α_i. Spectral information can be accessed similarly at the single-molecule level. However, the single-molecule approach offers additional advantages. Let us consider a sample labeled in multiple colors. First, if the excitation or emission spectra of the used dyes are well separated, so that the different dye species can be addressed individually without cross-talk, super-resolution microscopy becomes possible [10] (Box 12.1 and Figure 12.2). The conceptually simplest example would be the distance measurement between two differently colored emitters. At low temperature, van Oijen *et al.* made use of the extremely sharp homogeneous line width to spectrally select and localize many individual pentacene molecules [19]. Second, the emission signature of single fluorophores can be determined from the count ratio measured in two detection channels, for example, by using a dual-view system. Thereby, even highly similar emitters are still distinguishable based on slight differences in their emission spectra [20] (Figure 12.12).

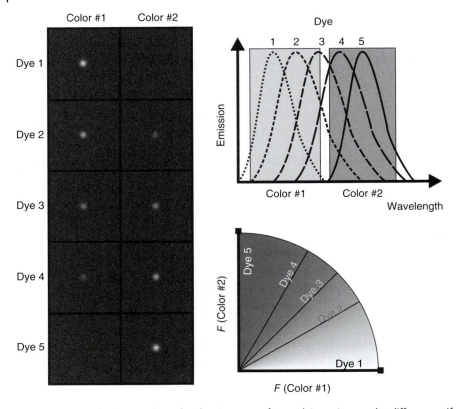

Figure 12.12 Multicolor single-molecule microscopy for resolving minute color differences. If a single dye molecule is imaged in two spectral channels, the intensity ratio encodes its color. We simulated here five different dyes with slightly shifted emission spectra; the according brightness difference in the two color channels is clearly visible. When plotting the single-molecule brightness values of the two color channels against each other, the dyes can be nicely discriminated by their polar angle.

12.5 Learning from Single-Molecule Signals

The recorded single-molecule parameters can be further processed to obtain insights into the biological sample. Here, we discuss, based on real-life scenarios, the strengths and potential pitfalls of single-molecule approaches.

12.5.1 Determination of Molecular Associations

Prima facie, it appears attractive to use single-molecule assays for interaction studies. The rational is simple: if one can detect a single molecule, it should be possible to quantify colocalizations. Indeed, assigning to each observed fluorescent spot a statement on its actual composition has been a key goal since the advent of ultrasensitive bioanalytics. Depending on the underlying question, one may choose from different approaches:

1. *Does the observed signal originate from a single-molecule emitter?* The most prominent and unique signature of single-molecule emitters is the irreversible disappearance in a single step. The presence of any quantal transitions between well-defined bright and dark states further indicates the observation of a single emitter molecule. In order to discriminate single emitters from ensembles, one may also use the effect of photon antibunching – the anticorrelated emission of photons within the fluorescence lifetime. Using real-time feedback for 3D tracking of quantum dots, antibunching even of freely diffusing particles can be measured. Finally, the presence of an absorption dipole of a dye molecule can be measured, for example, by rotating the excitation polarization. Yet, as multiple images have to be recorded, this method is applicable only to slowly rotating molecules.
2. *How many emitters are colocalized in an isolated, diffraction-limited spot?* Well-defined brightness levels can often be distinguished in the photobleaching curve of an oligomeric structure [21] (Figure 12.13a). The number of levels gives directly the cluster size. This method is in particular applicable for characterizing immobile molecules, where signal and background variations are manageable.

For the characterization of mobile clusters in highly fluorescent environments – for example, proteins diffusing in the cell membrane – brightness levels may be difficult to assign unambiguously. The reason is that mobile molecules encounter different environments, which may influence the brightness levels F: unspecific background may vary between different regions of a cell, thereby rendering a precise estimation of the specific signal difficult and prone to errors; if TIR excitation is used, movements of the target molecule along the optical axis will lead to variations in the excitation intensity; additional nonhomogeneities of the excitation intensity may occur because of laser speckles.

In such cases, one may revert to the analysis of brightness histograms to estimate the presence and the fractions of monomers, dimers, and so on. Let us assume that the probability distribution of the brightness levels recorded from the individual spots is characterized by $\rho(F)$, and the brightness of single dye molecules is given by $\rho_1(F)$. If the dyes emit independently, the distribution of n colocalized emitters can be calculated numerically by the convolution $\rho_n(F) = \int \rho_1(F')\rho_{n-1}(F - F')dF'$. $\rho(F)$ can then be fitted with a linear combination of the n-mer signal distributions $\rho(F) = \sum_{n=1}^{n_{\max}} \alpha_n \cdot \rho_n(F)$ [22] (Figure 12.13b).

Finally, when two different colors can be used, the presence of dimers can be detected to high accuracy and sensitivity via colocalization measurements.

12.5.2 Determination of Molecular Conformations via FRET

Generally, large biomolecules such as proteins or nucleic acids do not adopt a single conformation determined by a well-defined energetic minimum; in contrast, most macromolecules show highly dynamic transitions between distinct conformations, some of which can be directly related to their function. It is of central interest to characterize conformational states and transitions and to link them to biomolecular function.

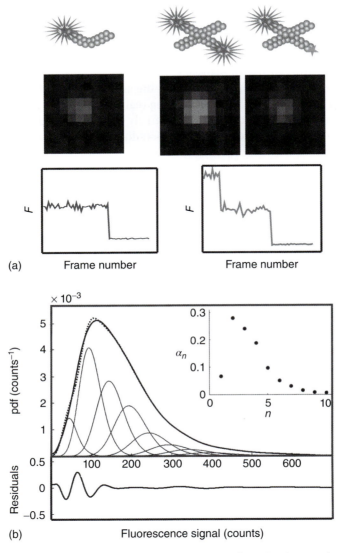

Figure 12.13 Determining the association state of a molecular complex. (a) Counting the photobleaching steps in a fluorescence time trace allows direct inference of the number of colocalized fluorophores. (b) From a pool of single-molecule observations, one can construct a brightness histogram, which can be fitted by a linear combination of the contributions of monomers, dimers, and so on (dashed line). The data are shown for single IgG molecules labeled randomly with fluorescein. The statistical weights α_n are given in the inset. The dominant species was found to be dimers; however, a long tail toward high N shifts the average load to $N = 3.5$. (Reprinted with permission from [22]. ©2005, The American Institute of Physics.)

FRET has been one of the first methods to obtain structural insights into a biomolecule in a solvent (Chapter 13). It was soon realized that FRET is perfectly suited as readout parameter in single-molecule experiments: it allows capturing static or dynamic intermediates in a heterogeneous sample [23]. Ideally, one would like to record conformations (i) precisely; (ii) quantitatively; (iii) in time; (iv) directly in the natural environment; and (v) noninvasively. Experimental constraints – both biochemical and optical – limit the compatibility of these issues.

Box 12.4 Förster Resonance Energy Transfer (FRET)

FRET describes the transfer of energy from an excited donor to a nearby acceptor dye molecule. The strong distance dependence of the transfer efficiency – which scales as $1/[1 + (r/r_0)^6]$ – makes it a practical nanoscale ruler with a sensitive range below ~10 nm; r and r_0 denote the donor–acceptor distance and the Förster radius, respectively. The Förster radius is given by $r_0 = [\text{const} \times (J\kappa^2 \Phi_D/n^4)]^{1/6}$, which depends on the following parameters:

- Φ_D, the donor quantum efficiency without FRET;
- n, the refractive index of the medium between donor and acceptor;
- J, the overlap integral as function of the wavelength λ, $J = \int_0^\infty F_D(\lambda)\varepsilon_A(\lambda)\lambda^4 \, d\lambda$, with ε_A and F_D the acceptor extinction and normalized donor emission, respectively;
- κ^2, an orientation factor that can range between 0 and 1; if all orientations are equally likely and averaged during signal integration, the mean value is $\kappa^2 = 2/3$;
- The constant term, which is equal to $9 \ln(10)/128\pi^5 N_A$.

---

1. *Precise measurements*: Most researchers calculate the apparent FRET efficiencies from the sensitized acceptor emission:

$$E_{FRET} = \frac{F_{D_{ex}}^{A_{em}}}{F_{D_{ex}}^{A_{em}} + \gamma F_{D_{ex}}^{D_{em}}}$$

where $F_{D_{ex}}^{A_{em}}$ and $F_{D_{ex}}^{D_{em}}$ denote the detected acceptor and donor fluorescence upon excitation of the donor, respectively. Data can be recorded, for example, with a dual-view system as described in Section 12.3.1, which allows the simultaneous measurements of $F_{D_{ex}}^{D_{em}}$ and $F_{D_{ex}}^{A_{em}}$. $\gamma = \Phi_A \eta_A / \Phi_D \eta_D$ is a detection-correction factor, which depends on the donor and acceptor quantum yields (Φ_D, Φ_A) and the detection efficiencies (η_D, η_A) [24]; typically, $0.5 < \gamma < 2$. In many cases, such analysis provides sufficient information to identify different molecular conformations and the respective transitions in time (Figure 12.14a). Yet, there are biochemical and photophysical constraints that render this direct ratiometric approach prone to errors: each target molecule has to be labeled in a 1 : 1 stoichiometry with a donor and an acceptor; each dye has to be active, that is, not photobleached and not in a

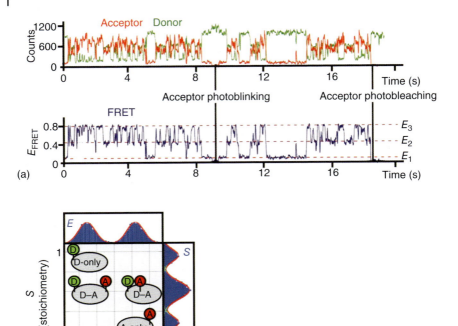

Figure 12.14 Single-molecule FRET. (a) Example of two-color single-molecule fluorescence resonance energy transfer (smFRET) data. A mutant hairpin ribozyme that carries the donor and acceptor on different arms of the same molecule undergoes transitions between three FRET states (E_1, E_2, and E_3). The recorded acceptor and donor traces are shown in the top panel, from which the apparent FRET efficiencies were calculated (bottom panel). The anticorrelated nature of the donor and acceptor signals indicates that these intensity changes are due to energy transfer. Dye molecules also show transitions to dark states: in this particular trace, the acceptor intensity transiently dropped to zero at ~9 s and completely photobleached at ~18.5 s. (b) E–S histogram for donor-only, acceptor-only, and donor–acceptor species with different interdye distance $R_{donor-acceptor}$. E sorts species according to FRET and $R_{donor-acceptor}$, reporting on structure; S sorts species according to donor–acceptor stoichiometry, reporting on interactions. (Reprinted from [24] with permission from *Proceedings of the National Academy of Sciences of the United States of America*.)

transient dark state; the quantum yields have to be stable, so that γ and the Förster radius r_0 do not vary.

It has, therefore, become common practice in single-molecule FRET experiments to subsequently excite also the acceptor with a separate laser [24]. Such datasets provide information on the number of active fluorophores per target molecule and allow the calculation of γ. In 2D histograms, E_{FRET} can be plotted versus $S = (F_{D_{ex}}/F_{D_{ex}}) + F_{A_{ex}}$, with $F_{D_{ex}}$ and $F_{A_{ex}}$ being the sums of donor and acceptor excitation-based emissions, respectively (Figure 12.14b). Such plots allow, for example, the discrimination of conformational subpopulations; moreover, data originating from molecules that do not contain active donor and acceptor fluorophores can be identified and rejected.

2. *Quantitative measurements*: Most available protein conformations are currently based on high-resolution X-ray crystallography data. Those data, however, are obtained in static experiments (flexible domains do not produce analyzable signals) and in non-native environments. It appears thus attractive to complement crystallographic data with single-molecule FRET studies. In principle, from the recorded FRET efficiencies, donor–acceptor distances can be directly calculated. One may thus envision an experiment in which two positions on a protein are labeled site-specifically with a donor and an acceptor dye and the inter-dye distance is calculated; from experiments on many sites, information on the protein conformation can be obtained. The idea was nicely demonstrated by Muschielok et al., who localized a particular domain of a macromolecular complex (labeled with the FRET donor) by measuring the distances to acceptors placed on well-defined positions on the complex [25]; triangulation was used to compute the probability distribution for the donor location.

   A few further considerations are required before applying this approach to other target molecules: First, one has to ensure that the labeling does not affect the protein structure. Second, the FRET efficiency is not a direct measure of the donor–acceptor distance but depends on multiple other parameters (Box 12.4 and Chapter 13). In particular, it is difficult to account for the orientation of the donor with respect to the acceptor; therefore, most researchers attempt to link the dyes via flexible spacers to the protein (so that $\kappa^2 = 2/3$ can be assumed). This has the detrimental effect that the position of each dye is not well defined, yielding errors in the distance determination.

3. *Kinetic measurements*: FRET allows direct observation of the conformational changes of a macromolecule that influence the distance between donor and acceptor. Camera-based systems were used in combination with TIRF microscopy to capture single-molecule trajectories containing a series of transitions between the individual states, so that kinetic rate constants could be determined [23]. The time resolution has been boosted by the introduction of fast cameras, so that nowadays FRET changes as low as $\Delta E_{app} = 0.08$ can be detected at 20 ms temporal resolution.

   Moreover, the picosecond time resolution of single-photon avalanche diodes (SPADs) can be exploited in a confocal configuration so that the acquisition time of the detector is not limiting for resolving conformational transitions. Confocal detection schemes indeed allow capturing FRET changes even for molecules in solution; but they have to occur within the dwell time of the molecule in the laser spot.

4. *In vivo/in vitro measurements*: By far most applications of single-molecule FRET microscopy were performed *in vitro*, that is, in well-defined buffer environments with low background and additives to prevent photobleaching. Besides the more problematic environment, it is also not straightforward to label a biomolecule in a live-cell context. Standard molecular biology approaches using fusion proteins are hardly applicable at the single-molecule level due to inappropriate photophysics in particular of the available blue chromophores. One approach is to purify the protein of interest, label it *in vitro* with an organic dye, and microinject it into the cells.

5. *Noninvasive measurements*: It is important to ensure that the molecule of interest preserves its conformation during the experiment; in particular, interactions with surfaces may transiently stabilize substrates or affect the dye photophysics, thereby modulating the transition kinetics. An alternative way for immobilizing single biomolecules is their encapsulation in surface-anchored vesicles, which precludes interactions with the glass coverslip. As a minor drawback, this approach makes changes of the biomolecule's environment difficult. Introduction of pores (either proteins or intrinsic lipid pores) helps to facilitate the exchange of small molecules.

12.5.3 Single-Molecule Tracking

After determining the coordinates of all molecules in the frame recorded at time t, one would like to correlate the images and track the individual molecules over time. Nearest-neighbor algorithms are frequently used for automatic reconstruction of single-molecule trajectories. We have recently described the following algorithm, which works robustly even at high surface densities [26]:

- *Tracking*: The fitted single-molecule positions in consecutive images are compared. For this, we test whether a single molecule in image $i+1$ can be found at a distance $r < R_{max}$ from any molecule identified in image i. In other words, all molecules in image i are circumscribed by a circle with radius R_{max}; if exactly one molecule in image $i+1$ can be found within a circle, it is connected to the respective molecule in image i (Figure 12.15a). If a circle contains more than one molecule, the trajectory is terminated at the particular image i. The procedure is continued through the whole sequence, yielding the single-molecule trajectories. Note that at high surface densities, some trajectories may be cut in two or more fragments or be deleted.
- *Test of the tracking algorithm*: Trajectories are analyzed by calculating the displacement r for different time lags t_{lag}. From the increase of the mean-square displacement (msd) with t_{lag}, the diffusion constant can be estimated by $msd = 4D_{est} t_{lag}$. D_{est} is a function of the radius of the circle, R_{max}, according to

$$D_{est} = D \left[1 - \left(1 + \frac{R_{max}^2}{4Dt_{lag}} \right) \exp\left(-\frac{R_{max}^2}{4Dt_{lag}} \right) \right] \times \left[1 - \exp\left(-\frac{R_{max}^2}{4Dt_{lag}} \right) \right]^{-1}$$

Apparently, D_{est} approaches zero for small values of R_{max}, as long steps are in this case excluded from the analysis. For large R_{max}, D_{est} shows a plateau at the correct value of the diffusion constant (Figure 12.15b,c).

Various additional nearest-neighbor-based tracking algorithms have been introduced, which explicitly consider blinking and high fluorophore densities. A fuzzy-logic-based approach was introduced for vesicle tracking, which quantifies image similarities. For an objective comparison of the different algorithms, see [27]. Finally, it should be noted that the determination of single-molecule mobility – even in the case of different mobile fractions – does not require the reconstruction of the trajectories, as correlation of the molecular positions provides already the required information [28].

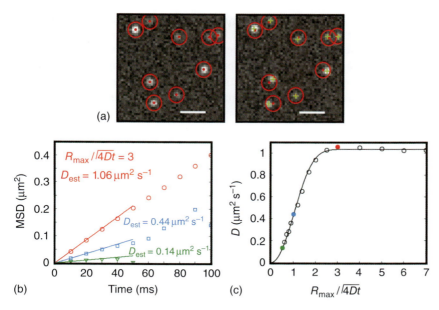

Figure 12.15 Single-molecule tracking. (a) The two images show simulations of single-molecule signals. The pixel size was set to 160 nm (scale bar = 2 μm). The simulated proteins were allowed to diffuse freely with a diffusion constant of 1 μm² s⁻¹; the second image shows the same detail after 10 ms. In the first image, the positions of the molecules are indicated as red points. For the tracking algorithm, a circle is drawn around each molecule, here with a radius of $R_{max} = 3\sqrt{4Dt_{lag}}$. In the second image, some of the molecules have considerably moved, but no molecule has left the circle. The new positions are indicated as green crosses, and the circles and red dots are overlaid from the left image for comparison. (b) The mean-square displacement is shown as a function of the time lag for different sizes of the circle radius: $R_{max} = 0.5\sqrt{4Dt_{lag}}$ (∇), $R_{max} = 1\sqrt{4Dt_{lag}}$ (□), and $R_{max} = 3\sqrt{4Dt_{lag}}$ (○). The first two data points were fitted by a straight line, yielding the estimated diffusion constants D_{est}; the dependence on R_{max} is shown in (c). For small values of R_{max}, D_{est} significantly underestimates the correct value of D due to exclusion of the long step sizes. At $R_{max} \approx 3\sqrt{4Dt_{lag}}$, a plateau region is reached, where the diffusion coefficient is correctly estimated. The examples shown in (b) have been included with the respective color code. Data points show the results of the simulation; the solid line is plotted after the analytical model described in the main text. (Reprinted from [26] with permission from Elsevier.)

12.5.4 Detecting Transitions

Single-molecule trajectories may be further analyzed by identifying transitions between different states. Such transitions could be the changes in the mobility of the tracer, in the brightness, color, or any other recorded parameter. A first attempt for spatial analysis – that is, for identifying transient confinements within a single trajectory – was reported by Saxton [29], who calculated the probability for a random walker with mobility D to be accidentally confined for a time t within a circle of radius R: $\log \psi = 0.2048 - 2.5117(Dt/R^2)$. In turn, this formula can be used to determine probability profiles for transient confinements, characterized by the parameter $L = -\log(\psi) - 1$. Segments within the trajectory above a certain threshold ($L > 3.16$) are typically classified as confined and the rest of the trajectory as freely diffusing. The algorithm is reasonably robust against the detection of false positive events; however, confinements are easily missed. More recently, Montiel et al. introduced a test-based approach for identifying transition points between two different diffusion constants [30]. Finally, segmenting the trajectory allows local msd analysis on each segment to identify, for example, regions of confined diffusion.

It is also interesting to detect steps in the brightness of an immobile spot, for example, for analyzing conformational changes due to transitions in the FRET efficiency, for counting photobleaching steps in association analysis, or for determining the segments of n active fluorophores in photobleaching localization microscopy. Besides a vast amount of *ad hoc* solutions developed by different research groups, there are rather general approaches available based on maximum likelihood estimations or Bayesian statistics.

Acknowledgments

JM and GJS acknowledge the support by the Austrian Science Fund (Ph.D. Program W1201 "Molecular Bioanalytics" and project Y250-B3) and the GEN-AU project of the Austrian Federal Ministry for Science and Research. MA was supported by a Schrödinger fellowship of the Austrian Science Fund (J3086-B11), and thanks the Max-Planck Society for financial and administrative supports.

References

1 Abbe, E. (1873) Beiträge zur Theorie des Mikroskops und der mikroskopischen Wahrnehmung. *Arch. Mikrosk. Anat.*, **9** (1), 413–468.
2 Rotman, B. (1961) Measurement of activity of single molecules of beta-D-galactosidase. *Proc. Natl. Acad. Sci. U.S.A.*, **47**, 1981–1991.
3 Hirschfeld, T. (1976) Optical microscopic observation of single small molecules. *Appl. Opt.*, **15** (12), 2965–2966.
4 Moerner, W.E. and Kador, L. (1989) Optical-detection and spectroscopy of single molecules in a solid. *Phys. Rev. Lett.*, **62** (21), 2535–2538.
5 Shera, E.B., Seitzinger, N.K., Davis, L.M., Keller, R.A., and Soper, S.A. (1990) Detection of single fluorescent molecules. *Chem. Phys. Lett.*, **174** (6), 553–557.
6 Schmidt, T., Schütz, G.J., Baumgartner, W., Gruber, H.J., and Schindler, H. (1996) Imaging of single molecule diffusion. *Proc. Natl. Acad. Sci. U.S.A.*, **93** (7), 2926–2929.
7 Schütz, G.J., Kada, G., Pastushenko, V.P., and Schindler, H. (2000) Properties of lipid microdomains in a muscle cell membrane visualized by single molecule microscopy. *EMBO J.*, **19** (5), 892–901.
8 Sako, Y., Minoghchi, S., and Yanagida, T. (2000) Single-molecule imaging of EGFR signalling on the surface of living cells. *Nat. Cell Biol.*, **2** (3), 168–172.

9 Thompson, R.E., Larson, D.R., and Webb, W.W. (2002) Precise nanometer localization analysis for individual fluorescent probes. *Biophys. J.*, **82** (5), 2775–2783.
10 Betzig, E. (1995) Proposed method for molecular optical imaging. *Opt. Lett.*, **20** (3), 237–239.
11 Axelrod, D., Burghardt, T.P., and Thompson, N.L. (1984) Total internal reflection fluorescence. *Annu. Rev. Biophys. Bioeng.*, **13**, 247–268.
12 Enderlein, J., Toprak, E., and Selvin, P.R. (2006) Polarization effect on position accuracy of fluorophore localization. *Opt. Express*, **14** (18), 8111–8120.
13 Mortensen, K.I., Churchman, L.S., Spudich, J.A., and Flyvbjerg, H. (2010) Optimized localization analysis for single-molecule tracking and super-resolution microscopy. *Nat. Methods*, **7** (5), 377–381.
14 Bobroff, N. (1986) Position measurement with a resolution and noise-limited instrument. *Rev. Sci. Instrum.*, **57** (6), 1152–1157.
15 Ober, R.J., Ram, S., and Ward, E.S. (2004) Localization accuracy in single-molecule microscopy. *Biophys. J.*, **86** (2), 1185–1200.
16 Smith, C.S., Joseph, N., Rieger, B., and Lidke, K.A. (2010) Fast, single-molecule localization that achieves theoretically minimum uncertainty. *Nat. Methods*, **7** (5), 373–375.
17 Kao, H.P. and Verkman, A.S. (1994) Tracking of single fluorescent particles in three dimensions: use of cylindrical optics to encode particle position. *Biophys. J.*, **67** (3), 1291–1300.
18 Pavani, S.R., Thompson, M.A., Biteen, J.S., Lord, S.J., Liu, N., Twieg, R.J. et al. (2009) Three-dimensional, single-molecule fluorescence imaging beyond the diffraction limit by using a double-helix point spread function. *Proc. Natl. Acad. Sci. U.S.A.*, **106** (9), 2995–2999.
19 van Oijen, A.M., Kohler, J., Schmidt, J., Muller, M., and Brakenhoff, G.J. (1999) Far-field fluorescence microscopy beyond the diffraction limit. *J. Opt. Soc. Am. A*, **16** (4), 909–915.
20 Bossi, M., Folling, J., Belov, V.N., Boyarskiy, V.P., Medda, R., Egner, A. et al. (2008) Multicolor far-field fluorescence nanoscopy through isolated detection of distinct molecular species. *Nano Lett.*, **8** (8), 2463–2468.
21 Ulbrich, M.H. and Isacoff, E.Y. (2007) Subunit counting in membrane-bound proteins. *Nat. Methods*, **4** (4), 319–321.
22 Moertelmaier, M., Brameshuber, M., Linimeier, M., Schütz, G.J., and Stockinger, H. (2005) Thinning out clusters while conserving stoichiometry of labeling. *Appl. Phys. Lett.*, **87**, 263903.
23 Roy, R., Hohng, S., and Ha, T. (2008) A practical guide to single-molecule FRET. *Nat. Methods*, **5** (6), 507–516.
24 Kapanidis, A.N., Lee, N.K., Laurence, T.A., Doose, S., Margeat, E., and Weiss, S. (2004) Fluorescence-aided molecule sorting: analysis of structure and interactions by alternating-laser excitation of single molecules. *Proc. Natl. Acad. Sci. U.S.A.*, **101** (24), 8936–8941.
25 Muschielok, A., Andrecka, J., Jawhari, A., Bruckner, F., Cramer, P., and Michaelis, J. (2008) A nano-positioning system for macromolecular structural analysis. *Nat. Methods*, **5** (11), 965–971.

26 Wieser, S. and Schütz, G.J. (2008) Tracking single molecules in the live cell plasma membrane – do's and don't's. *Methods*, **46** (2), 131–140.

27 Chenouard, N., Smal, I., de Chaumont, F., Maska, M., Sbalzarini, I.F., Gong, Y. et al. (2014) Objective comparison of particle tracking methods. *Nat. Methods*, **11** (3), 281–289.

28 Semrau, S. and Schmidt, T. (2007) Particle image correlation spectroscopy (PICS) retrieving nanometer-scale correlations from high-density single-molecule position data. *Biophys. J.*, **92** (2), 613–621.

29 Saxton, M.J. (1993) Lateral diffusion in an archipelago single-particle diffusion. *Biophys J.*, **64** (6), 1766–1780.

30 Montiel, D., Cang, H., and Yang, H. (2006) Quantitative characterization of changes in dynamical behavior for single-particle tracking studies. *J. Phys. Chem. B*, **110** (40), 19763–19770.

13

Förster Resonance Energy Transfer and Fluorescence Lifetime Imaging

Fred S. Wouters

University Medical Center Göttingen, Laboratory for Molecular and Cellular Systems, Department of Neuropathology, Waldweg 33, 37073 Göttingen, Germany

13.1 General Introduction

Molecular physiology, that is, the understanding of the working of molecular machines that mediate complex cellular responses, requires high optical resolution to identify the location of events and reactions, high temporal resolution to delineate their dynamic behavior, and, most importantly, a means for the identification of the event or reaction itself. Modern imaging equipment and recent site-directed staining techniques with genetic and synthetic dyes meet the spatiotemporal demands. The identity of the components under study is encoded by the choice of the fluorescent labels they carry. However, the way in which these components engage in a specific reaction, that is, what they do, cannot be easily concluded from the identity of the probe alone. In order to obtain this information, label fluorescence has to change its properties conditional on the reaction that acts on the probe.

All fluorescence properties can be used to encode a sensing function, and scores of "functional" dyes have been used that are based on changes in the yield, wavelength, or polarization of the fluorescence emission. One other metric of fluorescence that can be used to sense changes in the direct molecular environment of the dye, that is, to sense a reaction occurring, is the fluorescence lifetime. The fluorescence lifetime describes the duration of the excited state, the delay between an excitation and emission of a photon. Its advantages and properties will be described in the following. The imaging technique that measures and maps the fluorescence lifetime is called *fluorescence lifetime imaging microscopy* (FLIM). One of the most useful properties of the fluorescence lifetime is that it offers a view on the decay kinetics of fluorescent probes. This can be used to detect and quantify a highly useful and popular photophysical phenomenon among the life scientists: Förster resonance energy transfer (FRET), the relocation of energy from the excited state of one fluorophore to a second suitable acceptor fluorophore. This transfer of energy can be observed from the change in transition rates that act on the excited state of the fluorophores in the coupled system, and also from the redistribution of fluorescence yields between the coupled fluorophores. This chapter will outline the origins, consequences, measurements, and implications of FRET in the life sciences.

13.2 Förster Resonance Energy Transfer

13.2.1 Physical Basis of FRET

Upon absorption of a photon of sufficient energy content, electrons of the absorbing chemical compound can be excited to higher electronic states. When, upon relaxation to the energetic ground state, energy is released by the emission of a photon, the chemical compound is called a *fluorophore*.

The excited fluorophore can be understood as a transmitting (Hertzian dipole) antenna. The absorbed energy manifests itself as an oscillating electromagnetic field around the fluorophore (in a distance regime known as the *near field*), which leads to the generation of electromagnetic radiation, here light, which escapes from the emitter into the *far field*. Most uses of fluorescence – and of transmitting antennas in general – lie in the properties of the escaping radiation, as it allows the transmission of information over long distances. In fluorescence microscopy, information with diffraction-limited resolution (about 200 nm) is transmitted over distances approximately a million times longer (the optical path length in a microscope).

In the fluorophore's near field, generally considered to extend to $1/2\pi$ times the wavelength of the emitted light, energy is maintained that is not radiated as light. FRET acts on this energy in the near field. The fact that this interaction can be read out in the far field, that is, from the changes in the radiation that can be imaged using a microscope, makes the method so powerful. The near field is capable of electromagnetic induction much like the coupled circuits in a transformer. Absorption of energy by resonance coupling with a molecule in the near field drains the excited state energy of the fluorophore. An immediate consequence of this coupling is therefore a reduction in the fluorescence yield of the donating fluorophore, an effect that is used in many of the FRET microscopy methods. This reactive component in the interaction between the emitter and absorber is the basis of FRET. In FRET, the emitter is called the *donor* and the absorber is any compound that can absorb energy in the field around the donor (and thus, the wavelength of emission). The absorber is known as the *acceptor*. When the acceptor is a fluorophore, the energy that is absorbed by FRET is again emitted as light. Some FRET methods use this acceptor emission was excited by FRET coupling. Sections 13.3 and 13.4 deal with the different methods for detecting FRET.

13.2.2 Historical Development of FRET

Although a number of people were involved in the last steps of deriving a valid theory for FRET (see for excellent reviews [1–3]), for simplicity's sake only the contributions of Jean and Francis Perrin (father and son) and Theodor Förster are considered. Förster managed to bring all the puzzle pieces together, and therefore deserves to be recognized for his seminal work on FRET. It is instructive to follow the historical course.

In the 1920s and 1930s, it became apparent that optically excited molecules and atoms could transfer excited state energy not only via the transfer of electrons during physical collisions but also without "binding to matter of any kind" (quotes are from the seminal paper by Theodor Förster who provided the theoretical and mathematical framework for the understanding of the process of FRET) [4]. This research was motivated by the observation that approximately 1000 molecules of chlorophyll were involved in the assimilation of one molecule of CO_2 into glucose during photosynthesis. The energy absorbed by the individual molecules thus had to find their way to a single "reduction site" where the chemistry of assimilation takes place.

Solutions of fluorescent molecules at high viscosity provided a convenient experimental condition for the observation of the transfer of energy between optically excited molecules by the measurement of the (de)polarization of fluorescence upon excitation with polarized light. Fluorescence of a dilute fluorophore solution maintains the polarization direction of the excitation source because little rotation occurs in the viscous solvent. The polarized excitation light photoselects a pool or favorably oriented "primary" fluorophores that will emit equally polarized light. However, the polarization was observed to reduce strongly with increasing concentration of the fluorophore. Since the relaxation time of the orientation of the molecules is not expected to depend on concentration in these ranges, and there is no reason for the duration of the excited state to increase, it can only be concluded that molecules other than the primary excited ones take over the fluorescence radiation. A transfer of the excitation energy through the "indifferent solvent" could be observed between molecules that were approximately 50 Å apart. One trivial possibility would be the re-absorption of the primary fluorescence radiation, followed by secondary fluorescence. This could be excluded, as re-absorption of only a very small portion of the primary fluorescence could take place and give rise to depolarized secondary fluorescence. This is because of the large Stokes' shift between fluorescence emission and excitation of the same fluorophore.

The Perrins were the first to suggest that, besides emission and re-absorption of photons, energy could be transferred without collision over distances greater than molecular diameters by a direct electrodynamic dipole–dipole interaction between the primary excited molecule and its neighbor (transfer d'activation). Francis Perrin derived an equation that described the critical molecular separation distance R_0. At this distance, the rate of FRET equals the rate of radiation: that is, the efficiency of FRET is 50%. He assumed that the two identical fluorophores behaved as two identical electron oscillators, oscillating at a fixed frequency, but taking into account that the frequency would be broadened by solvent interactions [5].

$$R_0 = \frac{\lambda}{2\pi} \sqrt[6]{\frac{\bar{t}}{\tau}} \tag{13.1}$$

where λ is the wavelength of the radiation of the oscillators, $\bar{t}$ is the average time between collisions of the solvent molecules and the fluorophores, which gives rise to a broadening of the emission frequency (assumed to be between 10^{-14} and 10^{-13} s), and τ is the duration of the excited state (the fluorescence lifetime). The sixth-order dependence stems from the fact that the excitation probability of the acceptor is proportional to the square of the local electric field amplitude, which falls off proportionally to the third power of the distance from the dipole.

For a solution of fluorescein ($\lambda = 500$ nm), the critical molecular distance would be 15–25 nm – corresponding to a concentration of 2×10^{-5} to 1×10^{-4} mol l^{-1}. However, experiments show that these concentrations are a factor of three below the concentration at which depolarization actually occurs. Consequently, the Perrin equation for FRET underestimated the dependence of energy transfer on the separation distance.

Perrin had treated the system as interacting, pure oscillators in which identical molecules have the same fundamental oscillation frequency. His equation assumes only a (broadened) emission line spectrum and did not consider the difference in absorption and fluorescence frequencies of the molecules due to the Stokes' shift.

As the spectra of absorption and fluorescence are broadly dispersed, not only by solvent interactions, and the available energy is thus distributed over broad spectra, the probability that the frequencies of two molecules will be identical at the same time is small. Förster took into account the true dispersion of the spectra and the interaction probability in his famous "overlap integral."

The extreme distance dependence of FRET is therefore explained not only by the fact that it acts in the near field but also by the fact that even *within* this near field the probability of resonance is not very high. The near field of a typical green emitting fluorophore like green fluorescent protein (GFP) or fluorescein would still be about 80 nm, a resolution that can easily be reached by the new super-resolution microscopy techniques that have been developed over the last years and have earned the 2014 Nobel Prize in Chemistry to Moerner, Betzig, and Hell, as it is only about half the diffraction limit!

Förster realized that the relation describing the dipole coupling strength as a function of oscillation frequency should be multiplied by the resonance probability. Perfect resonance between the dyes implies that for every frequency the number of donor molecules oscillating with that frequency exactly matches the number of acceptor molecules capable of oscillating at this frequency. For Förster's fluorescein experiment, the Stokes' shift prohibits such a perfect match.

Although later manuscripts are more detailed, Theodor Förster's first publication of energy transfer [6], written in the nonspecialist journal "Naturwissenschaften" (natural sciences, Springer Verlag), clearly illustrates the essential role of the overlap integral for the resonance condition in energy transfer as described by the critical distance (in little more than 1 page of the 10-page manuscript!). Förster himself calls his early estimation an "Überschlagsrechnung" which can best be translated by the colloquial term "back-of-the-envelope calculation." A translation of the relevant section of this paper is included as Box 13.3. Even though this paper appears primarily qualitative, its derivation strategy – which is different from his later, more rigorous work [7] – can be followed to obtain the modern FRET equation (see Box 13.1, the Förster equations). In this manuscript, Förster considered the electron system of interacting fluorophores as classical mechanical oscillators in the sense of the original theory of J. Perrin. A later quantum mechanical treatment yielded the same results.

In this paper, he formulates his first corrected estimation of R_0 (which he calls d_0):

$$d_0 \approx \frac{\lambda}{2\pi} \sqrt[6]{\frac{\Omega'}{\tau_0 \Omega^2}} \tag{13.2}$$

where Ω'/Ω^2 represents the overlap integral with Ω being the width of the absorption and emission spectra and Ω' the overlap between both spectra.

The relative area overlap expressed by Ω'/Ω^2 represents the probability that a donor and an acceptor molecule oscillate at the same frequency, and, hence can transfer energy. This approach was also followed in [4]. In the absence of knowledge on the precise form of the emission spectrum, Förster had to approximate the emission spectrum by mirroring the absorption spectrum around the arithmetic average of the maximal absorption and emission frequencies (v_0):

$$\text{Absorption integral}: \int_0^\infty \epsilon(v)\,dv$$

$$\text{Emission integral}: \int_0^\infty \epsilon(2v_0 - v)\,dv$$

$$\text{Overlap integral}: \frac{\int \epsilon(2v_0 - v)\epsilon(v)\,dv}{\left(\int \epsilon(v)\,dv\right)^2} \tag{13.3}$$

The fractional overlap is thus given by the ratio of the true overlap (Abs·Em, Ω') and Perrin's assumption of complete overlap (Abs = Em; Abs·Abs, Ω^2), and represents the probability distribution for pairs with similar energy content. Förster has therefore scaled Perrin's estimation of the critical molecular distance by this fraction. The dependence of the energy transfer rate k_T on the spectral overlap could be demonstrated using a FRET pair linked by a rigid steroid spacer. The extent of the spectral overlap was modulated by changing the solvent, which altered the spectral properties of the dyes [8].

The major contribution of Theodor Förster was to provide a quantitative explanation for radiationless energy transfer that relates the efficiency of energy transfer by electronic Coulombic coupling on the molecular scale to simple physical terms and constants. Taking into account the Stokes' shift in the emission of fluorescein, Förster calculated a critical molecule distance (which he denotes R_0) of 5 nm, a value that he could experimentally verify [4].

Förster later derives his now-famous formula for the transfer rate $n_{A,B}$:

$$n_{A,B} = \frac{9 \ln 10}{128\pi^5} \frac{\kappa^2 c^4}{n^4 N' \tau_e R^6} \int_0^\infty F_Q^{(A)}(v)\epsilon^{(B)}(v) \frac{dv}{v^4} \tag{13.4}$$

in original notation from his 1951 book "Fluoreszenz organischer Verbindungen" [7] (a typographical error shows π^6), where N' is Avogadro's number (6.02×10^{20} molecules mmol^{-1}), τ_e is the average radiative lifetime, R is the separation distance (r), and $F_Q^{(A)}$ is the "Quantenspektrum," describing the proportion of the donor, here denoted as "molecule A," fluorescence yield that falls in the unit interval dλ. This is the donor emission spectrum normalized to unity. $\epsilon^{(B)}$ is the absorption spectrum of the acceptor, "molecule B." Spectra are used with base frequency (ν), which can be easily converted to wavelength by $\lambda = c/\nu$ with c being the speed of light. The radiative lifetime τ_e denotes the duration that the excited state would possess in the absence of nonradiative losses. Because there always are losses, this is a theoretical constant. The inverse gives the radiative rate Γ. The radiative lifetime τ_e can be expressed in terms of donor quantum yield Q_D and the lifetime of the donor τ_D. The donor lifetime τ_D is the duration of the donor excited state including nonradiative processes, that is, the real measured lifetime:

$$\frac{1}{\tau_e} = \frac{Q_D}{\tau_D} \tag{13.5}$$

Using this conversion, we obtain the familiar modern notation for the FRET transfer rate k_T:

$$k_T(r) = \frac{9 \ln 10 \kappa^2 Q_D}{128 \pi^5 n^4 N_A r^6 \tau_D} \int_0^\infty F_{qD}(\lambda) \epsilon_A(\lambda) \lambda^4 \, d\lambda \tag{13.6}$$

where $F_{qD}(\lambda)$, the donor quantal spectrum, describes the proportion of the fluorescence yield that falls in the unit interval dλ, that is, the spectrum is normalized to unity. N_A is Avogadro's number in 6.02×10^{23} molecules mol^{-1}.

van der Meer points out in his excellent chapter [3] that Förster's use of the N' notation for Avogadro's number (6.02×10^{20} mmol^{-1}) lies at the basis of a common mistake in quoting the Förster equation. Although they are in fact identical because the number of molecules is scaled, the difference in the expression per millimole or mole has led many to "correct" the formula by multiplication with a factor of 1000. A common, but wrong notation therefore shows 9000 ln 10 rather than 9 ln 10.

This formula can be simplified by grouping the constant factors in the critical distance R_0 (d_0 in Eq. (13.2)). When $r = R_0$, then $k_T = \Sigma(k_r + k_n)$, that is, transfer is equal to all other radiative (k_r) and nonradiative (k_n) depopulation rates. Note that this condition is slightly different from the original treatment in which nonradiative decay rates were ignored and the critical distance was defined at $k_T = k_r$ (see Box 13.3).

$$k_T(r) = \frac{1}{\tau_D} \left(\frac{R_0}{r}\right)^6 = \sum (k_f + k_n) \left(\frac{R_0}{r}\right)^6 \tag{13.7}$$

in which

$$R_0 = \sqrt[6]{\frac{9 \ln 10 \kappa^2 Q_D}{128 \pi^5 n^4 N_A} \int_0^\infty F_{qD}(\lambda) \epsilon_A(\lambda) \lambda^4 \, d\lambda} \tag{13.8}$$

The integral term is often abbreviated as $J(\lambda)$, the overlap integral. It should be noted, however, that $J(\lambda)$, which has the units $M^{-1}cm^{-1}nm^4$ is not identical to the true overlap integral, that is, the dimensionless oscillation probability, as first introduced in [6] and as used in Eqs (13.2) and (13.3) (see Box 13.1).

Jares-Erijman and Jovin have stated that the expansion of $1/\tau_e$ to the ratio of Q_D and τ_D (Eq. (13.5)) in the common modern notation was not strictly necessary [9]. The expansion was performed in order to obtain an equation (Eq. (13.6)) containing experimentally verifiable parameters rather than the unknown τ_e. Without this expansion, the original definition of the critical distance would have been maintained as the distance at which the rate of transfer k_t equals the rate of radiation, $1/\tau_e$.

$$k_T(r) = \frac{1}{\tau_e}\left(\frac{F_0}{r}\right)^6 = k_f\left(\frac{F_0}{r}\right)^6$$

$$k_T(r) = \frac{1}{t_e}\left(\frac{F_0}{r}\right)^6 = k_f\left(\frac{F_0}{r}\right)^6 \tag{13.9}$$

$$\text{with } F_0 = \sqrt[6]{\frac{9\ln 10\kappa^2}{128\pi^5 n^4 N_A}\int_0^\infty F_{qD}(\lambda)\epsilon_A(\lambda)\lambda^4\,d\lambda} \tag{13.10}$$

in which the quantum yield Q_D is absent – which has a certain appeal as there is no compelling reason to include the quantum yield in the definition of a new critical distance F_0.

In honor of his contributions to the quantitative understanding of FRET, the critical distance R_0 is called the *Förster distance*. In a convenient notation that combines all constants, R_0 is expressed as

$$R_0^6 = 8.79 \times 10^{-5}\frac{\kappa^2 Q_D}{n^4}J(\lambda)$$

$$J(\lambda) = \int_0^\infty F_{qD}(\lambda)\epsilon_A(\lambda)\lambda^4\,d\lambda \tag{13.11}$$

If the wavelength is expressed in nanometers, the equation returns R_0 in angstroms.

Förster's pioneering work on FRET took place on the special experimental condition of identical fluorophores, which we now call homo-FRET. Today, most FRET measurements are performed between chemically nonidentical dyes, that is, between different donor and acceptor fluorophores (hetero-FRET). Here, the acceptor is also red-shifted to allow the separate measurement of donor and acceptor emission.

The development of FRET theory by Jean Perrin helps us to understand the derivation of Förster's corrected equation. Förster's elegant first introduction of the probabilistic overlap integral Ω'/Ω^2 [6] uses a simplified approach that appears to produce a different result from the later more rigorous derivations [4, 7]. One obvious difference is the dependence on wavelength. The wavelength dependence seems to be different between the older (Perrin-like) and newer descriptions of R_0; Eq. (13.2) describes a λ^6 relationship and Eq. (13.8) includes a λ^4 term in $J(\lambda)$. Box 13.1 shows how they are the same.

Box 13.1 The Förster Equations

Jean Perrin's Equation

Jean Perrin first recognized that radiationless transfer of energy between fluorophores could be explained by the Coulombic coupling of emission and absorption dipoles. He modeled the interacting fluorophores as Hertzian oscillating dipoles and derived the first description of the transfer rate. It should be noted that all of this work was performed on homo-FRET between fluorophores in viscous solutions, observed from the concentration-dependent loss of polarization in the emission. Following the strategy in Theodor Förster's first paper on FRET ([6], Box 13.3), two physical laws are needed to derive his relationship:

1. Coulomb's law applied to the interaction between two static transition electric dipoles

$$U = \frac{M^2 \kappa}{4\pi\epsilon_0 n^2 d^3} \tag{13.12}$$

describing the interaction energy U in terms of the transition dipole moment M, the refractive index n, and the dipole–dipole distance d. κ describes the orientation between the dipoles, and ϵ_0 is the vacuum permittivity constant.

2. Fermi's golden rule

$$\Gamma(\omega) = \frac{M^2 \omega^3 n}{3\pi\epsilon_0 \hbar c^3} \tag{13.13}$$

which describes the radiative rate Γ of an oscillating charge, here the emission transition dipole of the fluorophore, as a function of the (angular) frequency ω, the dipole moment M, the refractive index n, the vacuum permittivity constant ϵ_0, the reduced Planck constant $\hbar$, and the vacuum light speed c.

The emission transition dipole moment is therefore

$$M^2 = \frac{\Gamma 3\pi\epsilon_0 \hbar c^3}{\omega^3 n} \tag{13.14}$$

The energy of a photon is given by $\hbar\omega$. The time t_0 needed to transfer coupling energy U from the donor to the extent of the energy content of a photon is

$$t_0 = \frac{\hbar\omega}{U\omega} = \frac{\hbar}{U} \tag{13.15}$$

in which $U\omega$ is the amount of energy transferred per unit time, as U is transmitted per time period $1/\omega$. This is assuming perfect resonance.

The rate of transfer, k_T, is the inverse of the time t_0. Using Eqs (13.12) and (13.13), and expressing the frequency as wavelength using $\lambda = 2\pi c/\omega$ and the radiative rate as lifetime $\Gamma = 1/\tau_e$, we obtain Jean Perrin's equation:

$$k_T = \frac{3\kappa\lambda^3}{32\pi^3 n^3 d^3 \tau_e} \tag{13.16}$$

At the critical distance $d = d_0$, the transfer rate k_T equals the radiative rate Γ, that is, $k_T = 1/\tau_e$, so that d_0 becomes

$$d_0 = \sqrt[3]{\frac{3\kappa\lambda^3}{32\pi^3 n^3}} \qquad (13.17)$$

This is approximately equal to 0.2λ, that is, the critical distance would be about 100 nm for 500 nm emission!

Overlap Integral and Francis Perrin's Equation

However, in fluorescence, dipoles do not resonate perfectly. Their frequencies are distributed in spectra. This reduces the chance for resonance because not all interacting molecules will have matching frequencies, even when both spectra overlap perfectly.

Förster defines the probability of overlapping frequencies W as

$$W = \frac{U}{\hbar} \frac{\Omega'}{\Omega^2} \qquad (13.18)$$

in which Ω is the width of the spectra, assuming they are equally broad, and Ω' is the width of the overlap between these spectra. This probability contains two components: $U/\hbar\Omega$ is the chance that coupling takes place in the time needed to transfer the energy content of a photon, and Ω'/Ω is the fractional overlap between the frequency spectra. Note that Ω'/Ω^2 represents the fractional *area* overlap between the spectra equivalent to the definitions in Eq. (13.3) and Eqs (13.25) and (13.26). Inserting the probability W into Eq. (13.15), with $k_T = 1/t_0$, the transfer rate becomes

$$k_T = \frac{U}{\hbar} W \text{ or } k_T = \left(\frac{U}{\hbar}\right)^2 \frac{\Omega'}{\Omega^2} = \left(\frac{U}{\hbar}\right)^2 J(\omega) \qquad (13.19)$$

which is the rate of transfer for perfectly coupled oscillation, adjusted for the probability of coupling.

$J(\omega)$ is the overlap integral. k_T becomes dependent on $(U/\hbar)^2$ as the two incoherently oscillating molecules both have to interact in the same time $\hbar/U$.

Francis Perrin followed the same argument. He included the effect of line broadening, that is, the distribution of frequencies in spectra. He did, however, not take into account that fluorophores also undergo a Stoke's shift in their emission. Without Stoke's shift, both spectra have the same and overlapping widths, making $\Omega' = \Omega$ so that $W = U/(\hbar\Omega)$. As the reason for the width Ω of the spectra, he invoked molecular collisions with solvent molecules occurring with an average time $\bar{t}$ between collisions of $10^{-13} - 10^{-14}$ s. In this case, as $\omega_A = \omega_B$ and assuming a quantum yield of 1, the critical distance becomes (using Eqs (13.12), (13.13), and (13.19))

$$d_0 = \sqrt[6]{\frac{9\pi\kappa^2}{8n^6}} \frac{\lambda}{2\pi} \sqrt[6]{\frac{\bar{t}}{\tau}} \approx 0.81 \frac{\lambda}{2\pi} \sqrt[6]{\frac{\bar{t}}{\tau}} \qquad (13.20)$$

which is close to Perrin's equation (Eq. (13.1)). d_0 is dependent on λ^6, and the critical distance contains the lifetime of the unperturbed donor molecule.

(Continued)

> **Box 13.1 (Continued)**
>
> Still, this amounts to a critical distance that is a factor of 2 too large.
>
> ### Nonequal Dipole Moments
>
> In all the considerations above, the dipole moments M^2 of the interacting fluorophores in the interaction energy U (Eq. (13.12)) were taken to be equal (see Eq. (13.14)). These were, in fact, emission dipole moments. In FRET coupling, we need to consider emission and absorption dipole moments, and these are NOT the same.
>
> The absorption dipole moment equals
>
> $$M^2_{01} = \frac{3e^2\hbar}{4\pi m\nu} \cdot f_a(\nu) \qquad (13.21)$$
>
> The subscripts 0 and 1 indicate that we are considering the transition from the ground state (0) to the excited state (1). e is the elementary charge and $f_a(\nu)$ is the *oscillator strength function* for absorption. It can be related to the molar extinction coefficient using
>
> $$f_a(\nu) = \frac{2n(\ln 10)\epsilon_0 mc}{N'\pi e^2} \epsilon(\nu) \qquad (13.22)$$
>
> The emission dipole moment M^2_{10} was given by Fermi's golden rule that was introduced earlier (Eq. (13.13)) and is shown in Eq. (13.14).
>
> The dipole moments are therefore
>
> $$M^2_{01} = \frac{3\hbar n(\ln 10)\epsilon_0 c}{2\pi^2 \nu N'} \epsilon(\nu)$$
>
> $$M^2_{10} = \frac{\Gamma 3\epsilon_0 \hbar c^3}{8\pi^2 \nu^3 n} \qquad (13.23)$$
>
> As the emission and absorption dipole moments are different, the transfer rate k_T becomes dependent on the product of the interaction energies for emission U_{10} and for absorption U_{01}. In the same general form of Eq. (13.19)
>
> $$k_T = \left(\frac{U_{01}}{\hbar}\right)\left(\frac{U_{10}}{\hbar}\right) J(\omega) \qquad (13.24)$$
>
> $J(\omega)$ is the dimensionless probability of two molecules sharing the same angular frequency. It is the product of the emission and absorption probability spectra, that is, the spectra in quanta, normalized to unity and equivalent to Ω'/Ω^2. These spectra show the proportion of fluorescence yield or absorption that falls in the unit interval $d\omega$.
>
> They can be constructed by dividing the measurable spectra $F_D(\lambda)$ and $\epsilon(\nu)$ by their frequency integrals, as was done for the "Quantenspektrum" $F_q(\nu)$. Similar to Eq. (13.3), the overlap integral $J'(\nu)$ is constructed as
>
> $$J'(\nu) = \frac{\int_0^\infty \epsilon(\nu) F_D(\nu) d\nu}{\int_0^\infty \epsilon(\nu) d\nu \int_0^\infty F_D(\nu) d\nu} \qquad (13.25)$$

from the definition of the donor "Quantenspektrum" $F_{qD}(v) = F_D(v)/\int_0^\infty F_D(v)\,dv$:
so that

$$J'(v) = \frac{\int_0^\infty \epsilon(v) F_{qD}(v)\,dv}{\int_0^\infty \epsilon(v)\,dv} \tag{13.26}$$

Using Eq. (13.12) for the interaction energies, Eq. (13.24) becomes

$$k_T = \frac{M_{01}^2 \kappa}{4\pi\epsilon_0 n^2 d^3 \hbar} \frac{M_{10}^2 \kappa}{4\pi\epsilon_0 n^2 d^3 \hbar} 2\pi \frac{\int_0^\infty \epsilon(v) F_{qD}(v)\,dv}{\int_0^\infty \epsilon(v)\,dv} \tag{13.27}$$

The factor of 2π originates from the fraction $2\pi/t$ of molecules with matching frequencies even in the case of completely overlapping spectra, that is, moving from angular frequency ω to frequency v.

Equation (13.27) can be rearranged for clarity as

$$k_T = \left(\frac{\kappa^2}{8\pi\epsilon_0^2 n^4 d^6 \hbar^2}\right) M_{01}^2 \, M_{10}^2 \frac{\int_0^\infty \epsilon(v) F_{qD}(v)\,dv}{\int_0^\infty \epsilon(v)\,dv} \tag{13.28}$$

and with the definitions of the emission dipole M_{10}^2 (Eq. (13.14)) and absorption dipole M_{01}^2 (Eqs (13.21) and (13.22)), integrated over frequency, as

$$k_T = \frac{9(\ln 10)\Gamma \kappa^2}{128\pi^5 N_A n^5 d^6} \int_0^\infty \frac{c^4}{v^4}\,dv \int_0^\infty \epsilon(v) F_{qD}(v)\,dv \tag{13.29}$$

Converting between frequency and wavelength and expressing the radiative rate Γ as Q_D/τ_D (the donor quantum yield/the lifetime of the donor in the absence of FRET) leaves us with the well-known Förster equation:

$$k_T = \frac{9(\ln 10)\kappa^2 Q_D}{128\pi^5 N_A n^5 d^6 \tau_d} \int_0^\infty \epsilon(\lambda) F_{qD}(\lambda) \lambda^4\,d\lambda \tag{13.30}$$

Note that k_T now depends on λ^4 (which is incorporated into $J(\lambda)$).

For a recent detailed and excellent treatment of the classical and quantum mechanical derivation of the Förster equation, see [3].

Two major breakthroughs had been achieved by Förster: the concept of the overlap integral as a probability in the transfer rate, and its relation to real-life measurable quantities. With his equation, Förster could, for the first time, predict and test the distance dependence of FRET.

The impact of Förster's work on modern FRET applications is remarkable. Even more so, considering how much of his early work was written in German. One has to assume that most researchers have never been able to read these publications. As the derivations in this section heavily rely on the developments laid out in his very first publication [6]: the first probabilistic definition of the overlap integral

(Continued)

Box 13.1 (Continued)

and the derivation of the transfer rate using Coulomb's law and dipole moments, I provide a translation of the relevant section in Box 13.3.

It should be noted that the modern definition of the overlap integral $J(\lambda)$ and the fractional overlap of the frequency spectra Ω/Ω' are different. The latter represents the true oscillation probability, but is only valid for spectra in appropriately normalized quanta. The expected λ^6 relationship in the FRET rate stems from the use of the square emission dipole moment, which depends on v^3. Equation (13.23) shows that a factor λ^2 is "lost" due to the fact that the absorption dipole depends on v rather than v^3. This λ^4 is now part of the $J(\lambda)$ "overlap integral." As a consequence, the modern overlap integral $J(\lambda)$ no longer represents the oscillation probability. Confusingly, many reviews still maintain that $J(\lambda)$ represents a probability.

13.2.3 Spectral and Distance Dependence of FRET

Förster also described the rate of the FRET process at a given separation distance $k_T(r)$ as a function of the separation distance r and the decay rate of the donor (given by $1/\tau_D$) in the absence of acceptor and FRET, as in Eq. (13.7):

$$k_T(r) = \frac{1}{\tau_D} \left(\frac{R_0}{r} \right)^6$$

Thus, the first requirement for high FRET efficiencies is the close vicinity of both molecules. For a given distance r, higher FRET efficiencies are achieved with increasing values of R_0. From this equation, it is clear that R_0 describes the distance ($R_0 = r$) at which the rate of transfer is equal to the rate of decay without FRET, and the efficiency of FRET is therefore 50%.

The other requirements for efficient FRET are given in the definition of R_0. As we have seen in the previous section, one of the major determinants of R_0 is the overlap integral $J(\lambda)$. If two dipole oscillators are to resonate, they should contain similar energy contents, as the probability of resonance depends on the occurrence of matching frequencies during the time in which coupling can take place, that is, during the duration of the donor excited state τ_D. One should therefore strive for a high spectral overlap between the emission spectrum of the donor and the excitation spectrum of the acceptor. Note that the overlap integral encompasses the extinction coefficient of the acceptor ϵ_A and that the donor emission spectrum is scaled by its quantum yield Q_D in the definition of R_0 (Eq. (13.8)).

As Eqs (13.6) and (13.8) show (and Box 13.1 explains), the overlap integral contains the term λ^4. The FRET efficiency is thus highly dependent on the wavelength. This is the reason why seemingly little overlap between the spectra, but over a long wavelength band, can still contribute significantly to the FRET efficiency. This is sometimes referred to as the *red tail effect*. It also dictates that red-shifted FRET pairs generally present higher FRET efficiencies than would be expected from their overlap only. A simple simulation shows how FRET scales with overlap and wavelength.

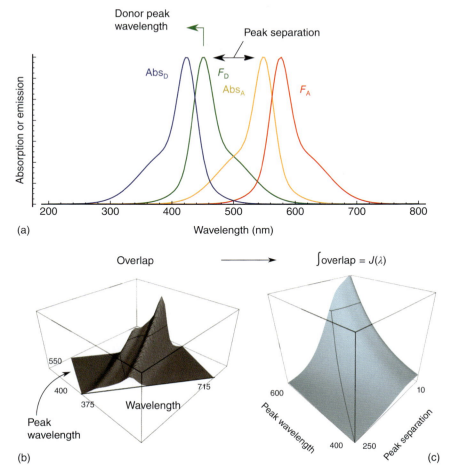

Figure 13.1 Overlap integral and FRET. The effect of wavelength ("peak" = donor peak wavelength) and peak-to-peak distance between the donor emission and acceptor absorption spectra ("P–P") were calculated for synthetic spectra modeled by two Gaussian distributions. In line with Förster's approach in [4] and Eq. (13.3), emission spectra were obtained by mirroring the absorption spectra. (a) Absorption and emission spectra of the donor (Abs_D and F_D) and acceptor (Abs_A and F_A) shown for donor peak wavelength of 450 nm and a peak-to-peak distance of 100 nm. (b) Spectral overlap in the range 375–715 nm for a donor–acceptor pair with a peak-to-peak distance of 100 nm calculated from the F_D and Abs_A spectra as in [4]. Shown are the overlaps for different donor peak emission wavelengths from 400 to 550 nm. (c) Overlap integral $J(\lambda)$ calculated over the range of donor peak wavelengths 400–595 nm and peak-to-peak distances of 10–245 nm. As the rate of FRET scales with the overlap integral, (b) and (c) also reflect the dependence of FRET on wavelength and spectral separation.

The "superman cape" function in Figure 13.1c shows the positive effect of using longer wavelength fluorophores and high spectral overlap for FRET. It should be noted that FRET measurements for smaller spectral separations between the donor and the acceptor become increasingly difficult as the spectral contaminations increase, as will be discussed in Section 13.3.

The description of the coupling strength of the transition dipole of the FRET pair in terms that are connected to fluorescence can be misunderstood to mean that FRET is based on the emission and re-absorption of a photon. Förster discounted this possibility of trivial re-absorption exactly on the basis of the Stokes' shift between emission and absorption that allowed him to correct Perrin's gross overestimation of the FRET efficiency. The process of FRET is nonradiative in nature. The use of absorption and emission spectra, and of quantum yield and extinction coefficient, applies equally well to a description of the process in terms of energy management.

The refractive index of the medium through which energy is transferred is expected to scale the efficiency of FRET. As most biological components have a very similar refractive index – water 1.33, lipid membranes 1.46, proteins 1.5 – an effect of maximum 40% is to be expected between biological extremes. However, as some imaging takes place in high-refractive-index mounting media and some chemical measurements could be performed in solvents of polymers with high refractive index, the refractive index could become a confounding factor.

The orientation factor κ^2 describes the spatial relationship between the interacting donor emission dipole and the acceptor absorption dipole and is an important factor in determining the coupling strength (see Figure 13.2). A collinear arrangement allows the strongest interaction, and its κ^2 value reaches a maximum value of 4. A parallel orientation is considerably less optimal, reaching a value of 1. As the value of κ^2 scales with R_0^6, switching between these two geometrical orientations would cause a 26% change in the Förster distance. In contrast to the effect of the refractive index, however, κ^2 can reach a value of zero when the dipoles are oriented perpendicularly (and at least one of the dipoles is perpendicular to the connecting line). In this orientation, FRET coupling does not occur irrespective of short distance, large $J(\lambda)$, or large Q_D and ϵ_A.

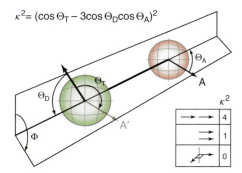

Figure 13.2 Transition dipole orientation. Three angles between the transition dipoles of the donor (D) and acceptor (A) determine the value of κ^2. These are the angle Θ_D between the donor dipole and the line connecting D and A, the angle Θ_A between the acceptor dipole and the line connecting A and D, and the angle Θ_T between D and A', which is the image of A at the base of D. Θ_T is therefore the same as the angle Φ between the plane in which the dipoles of D and A lie. The maximum case for κ^2 is a collinear arrangement with $\kappa^2 = 4$, and the minimum case is a perpendicular arrangement with $\kappa^2 = 0$. A parallel orientation of the dipoles possesses a κ^2 value of 1.

The value of R_0 for a given FRET pair is often calculated assuming a value for κ^2 of 2/3, as this is the statistical average of all possible orientations. Under the assumption that the dipoles can indeed sample all possible orientations during the lifetime of the donor, this allows the estimation of r from the FRET efficiency E (Eq. (13.32)).

The FRET efficiency can be defined in terms of rates: the transfer rate k_T, the fluorescence (or radiative) rate k_f, and the rate of nonradiative decay k_n (see Box 13.2 for more details):

$$E = \frac{k_T}{k_f + k_n + k_T} = \frac{k_T}{\tau_D^{-1} + k_T} \tag{13.31}$$

Combining this with Eq. (13.7), we obtain the relationship between FRET efficiency, R_0, and the separation distance r:

$$E = \frac{R_0^6}{r^6 + R_0^6}; \quad r = R_0 \sqrt[6]{\frac{1}{E} - 1} \tag{13.32}$$

This equation shows the extreme distance dependence of FRET, which is graphically represented in Figure 13.3. At distances r below R_0, E quickly increases to 1. Conversely, above R_0, E quickly falls to zero. At $R_0 = r$, 50% of the molecules engage in FRET rather than following the fluorescence emission path, and the transfer efficiency E is consequently 50%.

It is exactly this extreme distance dependence behavior that makes FRET such an attractive tool for cell biologists. Over a range of only a fraction of R_0, FRET switches from easily detectable to barely detectable. The distance sensitivity is highest at R_0. Assuming a typical fluorophore FRET pair with a value of R_0 equal to 5 nm and assuming the experience value of 5% as measurement resolution for the FRET efficiency, the spatial resolution of distance changes around R_0 amounts to ~1.6 Å, at a FRET separation sensitivity of 10% the spatial resolution around R_0 is still ~3.3 Å.

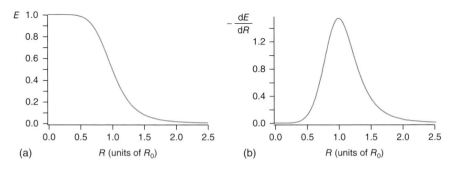

Figure 13.3 Distance and FRET efficiency. The sixth-order dependence of FRET on the separation distance between the donor and acceptor chromophore results in a very steep distance dependence (a). Distance is shown in units of the Förster critical distance R_0. FRET has maximum sensitivity toward distance changes at a distance R_0 (b). Shown is the distance sensitivity (the derivative) of E as a function of distance R.

Using oligomers of poly-L-proline, which forms a rigid helix of known dimensions, Stryer and Haugland could experimentally verify Förster's prediction of the dependence of k_T on the inverse sixth power of the distance [10].

13.2.4 FRET is of Limited Use as a Molecular Ruler

When considering all the determinants of the FRET efficiency stated previously, it is obvious why FRET has been considered to be a molecular ruler. When all factors other than the separation distance can be precisely controlled, the determination of the FRET efficiency contains spatial information on the scale of molecular size. Furthermore, the distance scale at which the spatial resolution is highest (around R_0, see Figure 13.3) for most biologically relevant fluorophores corresponds to the typical size of proteins. FRET microscopy is therefore the optical imaging technique with the highest spatial discriminating power, higher than even achieved by the recently developed optical sub-diffraction techniques [11, 12] like stimulated emission–depletion (STED) microscopy, or the family of localization imaging techniques like photoactivated localization microscopy (PALM) and stochastic optical reconstruction microscopy (STORM) [13–15], which are outlined in earlier chapters of this book. The catch is, of course, that the FRET images are diffraction-limited as they are produced with the optical resolution of the microscope that is used for this purpose. The ultrahigh spatial information is contained in an additional information layer in the image.

It is clear that molecular-scale distances represent important information for cell biology. In particular, biological applications involving a yes/no answer to the question of protein–protein interactions or conformational states can de directly addressed using FRET microscopy.

It is tempting to extend the use of FRET microscopy to map distances quantitatively, for example, in protein complexes. Given the spatial resolution of FRET, structural biological information can be obtained on a distance scale that can otherwise be addressed only by NMR or crystallography techniques.

Tempting as this may be, the uncertainty on the κ^2 orientation factor severely limits this otherwise exciting application possibility. The argument given above that the statistical average value of $2/3$ can be taken as a default value holds true only under one condition: namely the dipoles can sample all available orientations within the lifetime of the donor. This condition might be met for small fluorophores in solution, but is not a likely scenario for labeled larger cellular components such as proteins and for larger fluorophores such as the fluorescent proteins. Especially for the latter, it cannot be assumed that the chromophore, buried deep within the barrel structure of the fluorescent protein, can undergo free rotation. The same argument can be made for the entire structure fused to a protein of interest, as it is relatively large and it is very likely that the fluorescent protein interacts with its host protein in one way or another. Even if it were to enjoy a high level of rotational freedom, the rotations would be on a time scale longer than the fluorescence lifetime, and thus still be essentially immobile (GFP itself with a fluorescence lifetime of 2.5 ns exhibits a rotational diffusion correlation time of ~20 ns in cells) [16].

We therefore have to deal with a preferred narrow range of orientations in most biological measurements. Unfortunately, when the distribution of all possible orientations is considered, the probability of obtaining an unfavorable orientation is very much higher than that of obtaining a favorable orientation. This can already be seen from the fact that the statistical average lies at $2/3$, and not closer to the arithmetic average of the two extreme values of 0 and 4, that is, 2.

One should therefore be careful in applying Eq. (13.29) because R_0 can vary. In fact, most users of FRET have grown so accustomed at seeing R_0 as a constant, which describes the "strength" of an FRET pair, that its dependence on κ^2 is often forgotten. The same problem that, in most cases, reduces the effective FRET in a measurement because of a fixed κ^2 lower than $2/3$ can also work in the other way; the cyan fluorescent protein (CFP)/Venus-YFP (yellow fluorescent protein) fusion protein called Cy5.11 [17] exhibits a FRET efficiency of ~98%. In this construct, the C-terminus of CFP, shortened by five amino acids, was fused directly to the N-terminus of Venus, shortened by 11 amino acids (hence the name cyan 5 yellow 11, Cy5.11). Because of this extraordinarily high FRET efficiency and the high pH sensitivity of the cyan donor, we have used Cy5.11 as a FRET-based intracellular pH sensor [18]. Given the diameter of the barrel structure of 24 Å (this is thus the minimum separation distance) and R_0 for the pair of 54 Å, this would mean that the two fluorescent proteins have to align side by side at a distance below 1 nm; however, they are fused head to tail, and therefore possess a minimum separation distance of 4.2 nm (the length of the barrel), which would correspond to "only" 80%. However, the fusion apparently fixed the fluorescent proteins and therefore the dipoles in a highly favorable orientation. A value of R_0 closer to 8 nm is therefore expected, that is, the value for κ^2 is likely closer to 4 (from Eqs (13.8)–(13.32)).

This argument brings us to one of the most difficult practical aspects of FRET-based assays. Without an easy means of controlling κ^2 in an interaction between two labeled proteins, or between two labels on one polypeptide construct, there is no easy way to optimize the dynamic range of the measurement. This is especially troublesome in the currently popular biosensor design strategy of sandwiching a polypeptide sensing domain between donor and acceptor fluorescent protein moieties in a single polypeptide chain [19]. Such a sensor is based on a change in conformation in the sensing domain upon the occurrence of a reaction/modification, but the outcome in FRET efficiencies between the native (inactive) and changed (active) conformation is unpredictable. It is clear, also from experimental observations, that any significant changes in FRET between two states in these intramolecular single-chain biosensors have to be dominated by κ^2 rather than r. The situation is less restricted in the case for two separate interacting proteins. Here, the magnitude of the FRET efficiency might still be very well dominated by κ^2, but the difference between bound and unbound molecules is clearly a matter of distance.

This does not mean that FRET can never be used as a molecular ruler; it just adds an uncertainty in the determination of the separation distance when κ^2 is allowed to vary. The extent of variation in κ^2 can be estimated from anisotropy measurements (in the absence of FRET), which provide information on the orientational freedom of fluorescent molecules during the excited state [20].

We refer to the publication by van der Meer [21] for an extensive treatment of the statistical distribution of the values of κ^2 and the effect of rotational mobility. The influence of the relative dipole orientation on the FRET efficiency, and from this on calculations of distances, is a complicated matter which depends highly on the orientational averaging regime. If all molecules are static but randomly oriented, or when all molecules are dynamic and can randomly sample all orientations, the errors are relatively small. For static random and (partly) dynamic regimes, the most probable distance still lies close to the value obtained under the assumption $\kappa^2 = 2/3$. In a dynamic situation, κ^2 is averaged and will eventually reach the exact value of 2/3.

13.2.5 Special FRET Conditions

The distance dependence of FRET, as discussed previously, is stated for the case in which a donor molecule interacts with the same single acceptor during the lifetime of the donor. Furthermore, the two fluorophores can approach each other unhindered in all three dimensions. There are, however three conditions that affect the efficiency of FRET: diffusion during the donor lifetime, interaction of the donor with multiple acceptors, and geometries that restrict the orientational freedom of the donor–acceptor interaction.

13.2.5.1 Diffusion-Enhanced FRET

The effect of diffusion on the extent of energy transfer depends on the comparison between molecular motion and the donor decay time. At the so-called rapid diffusion limit, the diffusive motion of the excited donors allows them to sample a given space around them, thereby extending the distance range of FRET, which is also called *diffusion-enhanced energy transfer*. For fluorophores with nanosecond lifetimes this limit cannot be reached, but for dyes with long lifetimes such as lanthanides, which exhibit decay times in the range of milliseconds, the effect of diffusion can be significant. We will not consider this condition here, but refer to the literature [22, 23].

13.2.5.2 Multiple Acceptors

The presence of multiple acceptors within the FRET radius of a donor molecule will increase the probability for energy transfer. In general, FRET interactions with multiple acceptors reduce the dependence on orientational effects because of the averaging over multiple donor–acceptor pairs. Consider a simple system, say a solution of noninteracting donor and acceptor fluorophores. Increasing the concentration of the acceptor will increase the probability of FRET. FRET is now dependent on the acceptor concentration, and a critical acceptor concentration can be defined as that which is needed to statistically place an acceptor within a distance R_0 from a donor. This concentration typically lies in the range 2–50 mM, well beyond the expected dye concentrations in biological tissues. At higher concentrations, more than one acceptor molecule will occupy the FRET radius of the donor.

In three dimensions, this situation is difficult to reach. In conditions of lower dimensionality, however – where we cannot really speak of concentrations (number of molecules per unit volume) but more generally in terms of labeling density (number of molecules per unit area) – interactions of donor with multiple acceptors becomes more easily possible.

For interacting fluorescently labeled molecules, the number of acceptor fluorophores per donor fluorophore influences the distance dependence of the FRET efficiency, E_{line}. For this situation, consider a protein labeled with one donor and multiple acceptors, or a donor-labeled protein recruiting multiple copies of an acceptor-labeled protein. The operational R_0^6 is now $\sim n$ times that for the single donor–acceptor pair (for equidistant acceptors). The multiple acceptors (Figure 13.4a) that are available for FRET can be envisaged to form a virtual acceptor molecule with increased FRET coupling to the donor (Figure 13.4b).

Increasing R_0^6 is the same as reducing the influence of separation distance, given by r^6, by the number of acceptors at this distance, as they all contribute with equal weight (Eq. (13.33)). The effect on the distance dependence of the FRET efficiency is shown in Figure 13.6a.

$$E_{line} \approx \frac{nR_0^6}{r^6 + nR_0^6} = \frac{R_0^6}{\left[\left(\frac{1}{n}\right)r^6\right] + R_0^6} \tag{13.33}$$

13.2.5.3 FRET in a Plane

The biologically most interesting case of restricted geometry is where fluorophores are embedded in a membrane, that is, for labeled lipids or membrane-associated proteins. In this case, coupling of multiple acceptors with one donor is possible already at a relatively low labeling density. The system rapidly becomes very complicated, as the planar distribution of molecules might not be homogeneous and the physical space that the labeled components take up might have to be taken into account by defining an exclusion zone called the *distance of closest approach*. A detailed description of the planar FRET problem can be found in a recent elegant study that compares analytical predictions with Monte Carlo simulations and real experimental data that were obtained using a range of different FRET quantification methods [24].

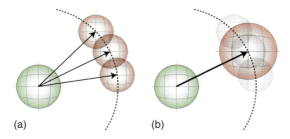

(a)   (b)

Figure 13.4 FRET with multiple acceptors. When a donor (green ball) can couple with multiple acceptors (red balls) within its FRET radius (a), the resulting increase in FRET efficiency can be envisaged by "adding up" the influences of the individual acceptors in a new higher FRET, the virtual acceptor (b). The increased size of the acceptor in (b) represents the improved coupling strength.

An important aspect of FRET in a plane is that it depends on both the separation distance and the acceptor labeling density. A donor molecule at a distance r from an acceptor molecule will undergo less FRET than a donor molecule at the same average distance from multiple acceptor molecules. Furthermore, it should be noted that the plane is also easier filled than the volume, so that nonspecific FRET by molecular crowding – something that can be safely excluded in the volume – now can no longer be excluded in cellular membranes. Caution should therefore be taken in membrane-based FRET assays: for instance, in the popular study of the oligomerization of membrane proteins, for example, raft-resident proteins and receptors. How big is this problem?

For a two-dimensional situation, where a donor approaches a planar distribution of acceptors at distance r, the FRET efficiency E_{plane} (from [25, 26]) can be expressed as a function of n_a, the dimensionless planar acceptor density in number of acceptors per unit area R_0^2 (Eq. (13.34)). The influence of the acceptor density on FRET in the planar case can be best appreciated by expressing n_a as ρR_0^2, where ρ is the acceptor density in the number of acceptors per square nanometers when R_0 is in nanometers. This substitution shows that the FRET efficiency still scales with the sixth power of R_0, but with the fourth power of r.

This relationship reduces to the state of a single interacting donor–acceptor pair when the density ρ drops below a critical value ρ_{crit} where there is only one acceptor at a distance r from a donor: this is, if we exclude the effect of multiple acceptors communicating with a donor. The deviation in the distance dependence of FRET between in-plane FRET and the 3D single donor–acceptor interaction case is depicted in Figure 13.6b.

$$E_{plane} = \frac{R_0^4}{\left(\frac{2}{\pi n_a}\right)r^4 + R_0^4} = \frac{R_0^6}{\left[\left(\frac{2}{\pi \rho}\right)r^4\right] + R_0^6} \tag{13.34}$$

Also here, a visualization of the problem is helpful. One can envisage rings of constant width at increasing diameters from the point of nearest approach of the donor to the acceptor plane. The more distant rings will contain more acceptors, increasing the FRET efficiency, but these are also at a larger distance – partially counteracting the "summation effect" shown in Figure 13.5b. The resulting

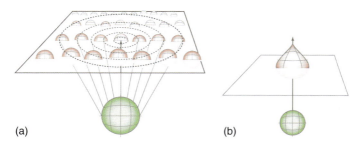

(a) (b)

Figure 13.5 FRET with a plane of acceptors. When a donor (green ball) can couple with multiple acceptors (red balls) in a plane (a), the resulting increase in FRET efficiency can be envisaged by "adding up" the influences of the individual acceptors in a new virtual drop-shaped acceptor structure (b), which represents the combined effects of increasing amounts of acceptor, but at larger distances. This structure is not shown to scale, but intended solely as a visual aid for the effect on FRET.

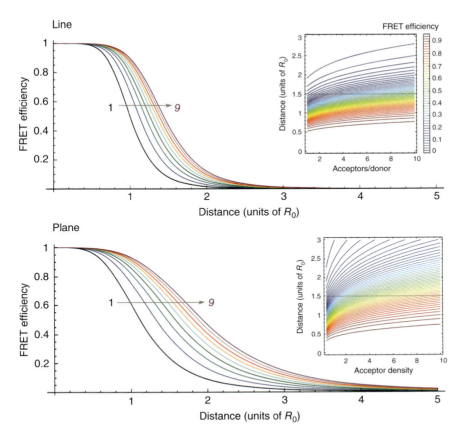

Figure 13.6 Factors that influence the FRET efficiency. (a) The effect of different acceptor-to-donor excess conditions (n) on the distance dependence of the FRET efficiency. Distances are expressed in units of R_0 and assume an equidistant distribution of the multiple acceptors to the donor. Shown are the conditions with an acceptor/donor ratio of 1 to 9 in curves from colder to warmer colors. The insert shows a contour plot of acceptor/donor and distance relations with equal FRET efficiencies. A distance threshold of $1.5 \times R_0$ (dashed line) is typically assumed for the maximum allowable separation in the 1 : 1 D/A case. The iso-FRET lines curve up toward longer separation distances at increasing acceptor/donor ratio, that is, the same FRET efficiencies are reached at longer distances for higher acceptor/donor ratios. (b) The effect of different acceptor densities in a planar distribution on the distance dependence of the FRET efficiency with an approaching donor. The acceptor density is given as the number of acceptors within the range R_0. Higher densities increase the FRET efficiency. The insert with iso-FRET curves again shows that longer separation distances are tolerated, especially for lower FRET efficiencies. The effect is larger in magnitude than for the "linear" acceptor/donor ratio increase (a).

situation can be visually represented by a virtual drop-like acceptor structure, whose influence on the distance dependence of the FRET efficiency is shown in Figure 13.6b.

An acceptor bleaching experiment can determine whether the acceptor concentration in the sample is below or above ρ_{crit}. If the FRET efficiency reduces linearly with the reduction in acceptor fluorescence (Eq. (13.39)), the density is below ρ_{crit}. Above ρ_{crit}, bleaching will have a larger effect.

The effect of multiple acceptors on the FRET efficiency in the linear (Figure 13.4) and planar (Figure 13.5) case, as shown in Figure 13.6, shows that in both cases the contribution of all acceptors within the FRET range of the donor has to be considered and the effective range of FRET increased by reducing the bracketed distance term. Note that the increase in acceptor labeling density in the linear case shifts the familiar FRET efficiency curve to longer distances. Increasing the acceptor density in planar FRET also increases the distance range for which FRET can be found, but by changing the shape of the distribution, especially for the lower FRET efficiencies.

13.3 Measuring FRET

The depolarization of fluorescence emission is one consequence of FRET. Such measurements allowed Förster to experimentally verify his description of the FRET efficiency on distance and can be used to derive information on FRET between spectrally identical donor–acceptor pairs.

Polarization (or anisotropy) measurements are best performed in spectrofluorometry, that is, in cuvettes, because the high NA objectives in microscopes can introduce a significant distortion in the polarization of the focused light.

There are also consequences in other fluorescence parameters by the changes in the coupled donor–acceptor system. They can be divided into steady-state changes, for example, in the emission yields, and changes in the decay kinetics which can be observed by time-resolved measurement techniques.

13.3.1 Spectral Changes

13.3.1.1 FRET from Donor Quenching

The transfer of energy from the excited donor to the acceptor results in a loss of donor fluorescence, which is proportional to the FRET efficiency. The loss of donor emission due to FRET ($F_D - F_{DA}$, the subscripts D and DA denoting the fluorescence of the donor D in absence and presence of the acceptor A) relative to the donor emission that would be achieved without FRET (F_D) gives the FRET efficiency:

$$E = \frac{F_D - F_{DA}}{F_D} = 1 - \frac{F_{DA}}{F_D} \tag{13.35}$$

This measurement assumes that the experimenter has a means of observing the donor-containing sample in the absence and presence of FRET. One way is that the acceptor-labeled component is added after the F_D measurement is made. The inverse way is to remove the acceptor after the F_{DA} measurement is made.

The acceptor occupies a red-shifted position in the optical spectrum and, as organic dyes typically do, the donor exhibits a sharp edge at the red side of its excitation spectrum. The acceptor can therefore be excited independently of the donor. Using high radiation intensities, the acceptor can therefore be photobleached selectively. After having bleached the acceptor, the donor fluorescence can be recorded again in the region in which the acceptor was bleached to obtain the desired F_D measurement. From the images of the donor fluorescence before and after photobleaching of the acceptor, the FRET efficiency can be calculated using Eq. (13.35) [27, 28].

The most convenient way to perform this measurement is by using a confocal microscope, as this allows the selection of a bleaching zone and the automation of the acquisition steps. In the bleach region, the donor fluorescence intensity is F_{DA} before the bleaching step and F_D after the bleaching step: that is, unquenched by removal of FRET. In the remainder of the cell, the donor fluorescence F_{DA} is unchanged by the bleaching step. Image division according to Eq. (13.35) will therefore return an image of the distribution of FRET efficiencies in the bleaching region. Outside the bleaching region, the average (should be zero) and the width (should be narrow) of the distribution of calculated FRET values are diagnostic tools for the quality of the measurement. Lateral translations in the planar field can be easily identified from apparent increases and decreases in FRET in the direction of translation in structures in the control field. Mechanical instability that leads to a z-drift can be identified but not corrected for in a single image plane [29]. The application of this technique is depicted in Figures 13.7c,d and 13.10a,b.

13.3.1.2 FRET-Induced Acceptor Emission

The transfer of energy to the acceptor leads to the population of its excited state. Decay from the acceptor excited state will therefore generate acceptor emission. This emission in the acceptor wavelength range upon excitation in the donor excitation wavelength range is called *sensitized emission*.

Figure 13.7 Quantitative detection of FRET by acceptor photobleaching and sensitized emission. (a) Fluorescence images of a cell expressing a fusion protein containing a CFP donor and a YFP acceptor. A control cell in the upper right corner expresses YFP only. An indicated region was subjected to the selective bleaching of YFP. SE' represents the fluorescence collected in the YFP window upon excitation at the CFP wavelength. This contains the true sensitized emission (SE) and unknown fractions of CFP bleed-through emission (α*CFP) and YFP emission originating from direct excitation at the CFP wavelength (β*YFP). The fractions α and β are estimated from the pure CFP emission in the YFP-bleached region and from the YFP control cell, respectively. The successive subtraction of these factors is shown in (b). Intensity counts in the regions representing donor (ROI1) and acceptor (ROI2) contaminating controls show a sharp, symmetrical distribution around zero, indicating successful removal of the spectral contaminations. Panels (c) and (d) show the calculation of FRET efficiency distributions. The FRET efficiency can be calculated from the CFP images before (CFPpre) and after (CFPpost) photobleaching of YFP in the indicated region (c), by the division of the difference of CFP fluorescence (ΔCFP) by the non-FRET, unquenched, CFP fluorescence (CFPpost). The result is shown in the left-most image of (d). The FRET efficiency can also be calculated from the spectrally purified sensitized emission image (SE) by comparison with the corresponding change in CFP emission. The correction factor γ, derived from the sensitized emission in the bleached region, and the ΔCFP signal can now be used to transform the SE information in the remainder of the cell to FRET efficiencies according to Eq. (13.26) (right-most image in (d)). Note that the FRET efficiencies are comparable between the two FRET images and that the (complementary) FRET control regions show an even color that corresponds to zero. The images were not subjected to filtering. The higher noise in the FRET image from acceptor photobleaching stems from the lower pixel resolution that was used for these images. Fluorescence images are shown in an inverted gray scale look-up table.

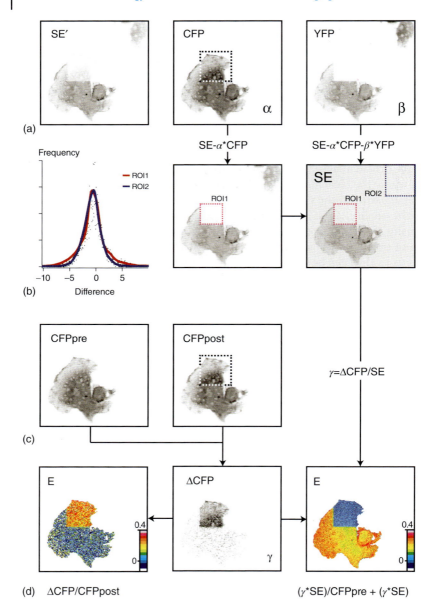

However, the shape of excitation and emission spectra that allowed the selective excitation and bleaching of the acceptor occupying a red-shifted part of the optical spectrum now works against a clean FRET measurement from sensitized emission. The blue edge of excitation spectra shows a long tail, so a portion of the acceptors is excited in the wavelength region that is optimal for excitation of the donor. Some of the acceptor photons emitted during FRET are therefore not generated by the FRET process but by direct acceptor excitation. Conversely, the red edge of the emission spectra also shows a long tail. Therefore, the red-shifted emission of the donor extends to well below the emission spectrum of the acceptor. Some of the photons that are collected in the emission bandwidth of the acceptor are actually donor photons. As the excitation and emission spectra of organic dyes, including the genetically encoded fluorescent proteins, are fairly broad and necessarily have to overlap to generate FRET, these spectral contaminations appear to be unavoidable.

As long as the donor and acceptor are physically linked, as they typically are in most FRET sensors, their stoichiometry is fixed and these spectral contaminations can be tolerated. The effect of these spectral contaminations is that they introduce an offset in the apparent sensitized emission, which will decrease the dynamic range of the spectral changes that occur as a consequence of FRET. The absolute values of the (contaminated) acceptor-sensitized emission are imprecise, but at least they will change only because of changes in FRET; they will provide FRET-based contrast.

However, for all other conditions in which the magnitude of the spectral contamination depends on the relative amount of donor (giving rise to bleed-through) and acceptor (giving rise to direct excitation), these spectral contaminations have to be corrected before the sensitized emission can be used for a FRET determination (as discussed in [19, 29–31]). These corrections require two pieces of knowledge: (i) the local concentration of donor and acceptor at each location in the image, and (ii) the extent of contamination caused by the fluorophores at the given spectral selections in the microscope. The first piece of information can be obtained from measurements at the optimal donor and acceptor excitation–emission conditions. Put simply, one needs an image with a donor excitation and emission filter and an image with the acceptor excitation and emission filters in addition to the sensitized acceptor image $F_A^{\text{exc D}}$ taken with the donor excitation filter and the acceptor emission filter (or laser lines and selected spectral emission bands for confocal microscopes). The second information has to be obtained from the same microscope, with identical settings but with two additional samples: one expressing only the donor fluorophores, and the other expressing only the acceptor fluorophores. These samples allow the reference measurements from which the correction factors for spectral contamination can be derived. The bleed-through correction factor α is the intensity observed using an acceptor emission filter upon excitation of the donor-only sample with a donor excitation filter $F_D^{\text{em A}}$ relative to the intensity under the same excitation conditions, but imaged through the proper donor emission filter $F_D^{\text{em D}}$. This factor describes the ratio of donor quantum yields at the acceptor and donor emission wavelength, $Q_D(\lambda_A)$ and $Q_D(\lambda_D)$, respectively. The direct excitation correction factor β can be determined from the acceptor-only sample:

the intensity observed upon acceptor emission using the donor excitation filter $F_A^{\text{exc D}}$ relative to the intensity with the proper excitation filter $F_A^{\text{exc A}}$. This factor describes the ratio of acceptor extinction coefficients at the donor and acceptor excitation wavelengths, $\varepsilon_A(\lambda_D)$ and $\varepsilon_A(\lambda_A)$, respectively. Arithmetic subtraction of the product of the intensities measured in the control acquisitions and their correction factor leaves the sensitized emission SE:

$$\alpha = \frac{F_D^{\text{em A}}}{F_D^{\text{em D}}} = \frac{Q_D(\lambda_A)}{Q_D(\lambda_D)}$$

$$\beta = \frac{F_A^{\text{exc D}}}{F_A^{\text{exc A}}} = \frac{\varepsilon_A(\lambda_D)}{\varepsilon_A(\lambda_A)}$$

$$SE = F_A^{\text{exc D}} - \alpha F_D^{\text{exc D}} - \beta F_A^{\text{exc A}} \tag{13.36}$$

In an experiment, it is advisable to have cells expressing the donor-only and acceptor-only situation directly side by side on the same coverslip. This allows immediate visual inspection of the spectral corrections. This is also very useful when the sensitized emission experiment is performed on a confocal microscope, where the reference situations depend on the laser power and photomultiplier (PMT) gain settings, which can vary between experiments. Such an experiment is shown in Figure 13.7a,b.

13.3.1.3 Contrast in Intensity-Based FRET Measurements

The major difference between donor quenching (DQ) and sensitized acceptor (SE) FRET determinations is that the former is immediately quantitative because the lower and higher limits are available: 0% FRET causes 100% donor fluorescence, and 100% FRET causes 0% donor fluorescence. The fractional loss of donor fluorescence, which is also often denoted with the popular term $\Delta F/F$, is therefore immediately interpretable. Sensitized emission, however, is only limited on one side: 0% FRET generates 0% SE. The amount of SE at 100% FRET cannot be likely estimated or experimentally measured because, in contrast to a situation without FRET, most systems cannot be expected to be driven to 100% FRET. Without this reference limit, any amount of SE is meaningless.

Furthermore, the SE is concentration-dependent. An apparent FRET efficiency E_{app} can be obtained by normalization to a chosen reference fluorescence intensity, making the measurement concentration-independent. DQ measurements are intrinsically normalized and provide the real FRET Efficiency. For SE measurements, such a normalization is however not straightforward, and multiple possibilities have been used in the literature (for a side-by-side comparison, see [24]). There are two obvious normalization strategies: to the donor fluorescence $F_D^{\text{exc D}}$ like for the DQ situation (E_{app}^1), or to the acceptor fluorescence $F_A^{\text{exc A}}$ (E_{app}^2). However, both ratio measurements have different meanings: $SE/F_D^{\text{exc D}}$ describes the fraction of donor molecules that successfully transferred energy to an acceptor, and $SE/F_A^{\text{exc A}}$ describes the fraction of acceptors that received energy from a donor. The latter is the preferred normalization in the literature. Both contain information on the equilibrium interaction between donor and acceptor molecules, and since all information is available from the measurement and correction of spectral contamination, it is advisable to show both. One other normalization can be made that also carries additional information: ideally, one would like to normalize the SE to the fraction of active FRET couples in the sample so that SE represents the FRET efficiency in the complexes. Since the number of FRET complexes is exactly what the SE measurement is supposed to measure, this information is not available. The next best thing, however, is to normalize to the geometrical average of the donor and acceptor concentration given by the square root of the product of $F_D^{exc\,D}$ and $F_A^{exc\,A}$, as this is related to the probability for the formation of a FRET couple. The latter normalization method (E_{app}^3) is also preferred if the relative abundance of donor and acceptor molecules is not known or cannot be experimentally driven into saturation.

$$E_{app}^1 = \frac{SE}{F_D^{exc\,D}}; \quad E_{app}^2 = \frac{SE}{F_A^{exc\,A}}; \quad E_{app}^3 = \frac{SE}{\sqrt{F_D^{exc\,D} F_A^{exc\,A}}} \quad (13.37)$$

The choice in normalization already shows that the relative abundance of donor or acceptor can influence the FRET determination. In the case of DQ measurements, the ideal spectroscopic setting is an abundance of acceptor molecules relative to the donor molecules ($A/D > 1$). With acceptor saturation, each donor can participate in FRET with an acceptor. In comparison, the most sensitive SE measurement is expected when the acceptors are saturated by donors ($D/A > 1$): each acceptor has the chance to generate SE from a FRET interaction with a donor. In that optimal case, normalization to donor or geometrical average of donor and acceptor becomes, of course, less meaningful.

13.3.1.4 Full Quantitation of Intensity-Based FRET Measurements

Even with these normalizations, the outcome is a value that is related to, but not identical with, the FRET efficiency. There have been many mathematical studies trying to derive true FRET efficiencies from SE measurements, but there does not seem to be a consensus. There is, however, a fairly simple way of obtaining true FRET efficiencies, which requires one more independent reference measurement. In this method, the SE is related to donor quenching by measuring a FRET change in an intramolecular construct containing the same identical fluorophores as the intermolecular FRET situation that is to be quantified [31, 32]. One should be able to experimentally induce a (substantial) FRET change in this reference construct; one example is the use of a calcium-sensing FRET construct of the Cameleon family. By relating the spectrally purified SE (Eq. (13.36)) increase upon a change in calcium to the medium to the co-occurring donor quenching, which can now be easily measured, a translation factor γ can be obtained that can be used to express the SE changes as DQ changes, thereby gaining full quantitation of the FRET efficiency:

$$\gamma = \frac{\Delta F_{DA}^{exc\,D}}{\Delta SE}$$

$$E = \frac{\gamma \Delta SE}{\gamma \Delta SE + F_{DA}} \quad (13.38)$$

Another method that has been suggested for obtaining the factor γ is to compare the FRET efficiencies between different intramolecular double-fluorophore constructs in (different) cells [33]. However, this method assumes that these different constructs fold identically when expressed in different amounts and in different cells. In our experience, this however can be a considerable source of variation. Perhaps the easiest method to determine γ is to photobleach the acceptor in one cell and record the corresponding increase in donor fluorescence. As spectrally purified, sensitized emission photons can be obtained from the bleached cell prior to photobleaching the acceptor, both can be related [34]. The application of this method of transformation is shown on the sensitized emission data in Figure 13.7c,d.

13.3.1.5 Occupancy Errors in FRET

Failure to saturate the donors with acceptors for DQ measurements and, vice versa, for SE measurements leads to an underestimation of FRET [35]. When the biologically relevant stoichiometry is known, one can make a judicious choice of what component should be labeled with what fluorophore. If this is not possible, one could choose the optimal situation by choosing between DQ versus SE methods.

In DQ measurements, a lower A/D ratio due to missing acceptors for interaction with donors scales the apparent FRET efficiency with the fraction f_A of donor-interacting acceptors (donor saturation corresponds to $f_A = 1$):

$$E_{app} = \left(1 - \frac{F_{DA}}{F_D}\right) f_A \tag{13.39}$$

At high FRET efficiencies, when $F_{DA}/F_D \ll 1$, small amounts of unlabeled acceptor have a large influence on the FRET efficiencies. In contrast, in SE measurements, non-FRET-coupled acceptors do not generate SE. Therefore, a reduction of the optimal D/A ratio reduces the number of acceptor molecules that produce SE but not the absolute amount of SE per acceptor molecule. Of course, in the normalizations this distinction cannot be made, and it will still appear as if the FRET efficiency has reduced. Note that the occupancy problem above also occurs when cells contain an excess of unlabeled acceptor components that compete with the labeled acceptors for binding to the donors.

13.3.2 Decay Kinetics

13.3.2.1 Photobleaching Rate

Photobleaching is an irreversible excited-state photochemical reaction with a rate constant that is very low in comparison with all the other depopulation transition rates. The rate k_{pd} reflects the photo-induced decomposition of D^*, which irreversibly forms a nonfluorescent form $D^\bullet$.

The two competing reactions here are

$$D^* \xrightarrow{k_f + k_n} D + h\nu$$

$$D^* \xrightarrow{k_{pd}} D^\bullet \tag{13.40}$$

Photobleaching is observed as the loss of fluorescence under conditions of *constant* illumination:

$$D \xrightarrow{k_e} D^* \tag{13.41}$$

where k_e is the excitation rate, which is very much faster than the other rates involved. In the absence of photobleaching, the concentration [D*] would remain constant.

As FRET reduces the excited state, it effectively protects the pool of donors against photobleaching.

Photobleaching introduces an exponential decay in the fluorescence. In accordance with its low rate, the apparent photobleaching lifetime τ_{bleach} ($1/k_{bleach}$) is on the seconds to minutes scale compared to the fluorescence decay on the nanosecond scale. As τ_{bleach} is inversely proportional to the donor fluorescence lifetime, it represents an experimentally highly convenient measure for FRET-induced changes in the fluorescence lifetime. It can be obtained by the timed acquisition of donor fluorescence images upon constant illumination. A pixel-by-pixel fit to an exponential decay model will then produce an image of the distribution of the photobleaching rates and, from this, of the FRET efficiency. The inverse relationship between τ_{bleach} and τ_D or τ_{DA} can be shown by solving the differential equations governing the time-dependent change in the concentration of the excited state D* [36]:

$$\frac{1}{\tau_{D,\,bleach}} = (k_e k_{pd})\tau_D; \quad \frac{1}{\tau_{DA,bleach}} = (k_e k_{pd})\tau_{DA} \tag{13.42}$$

As k_e and k_{pd} are constant,

$$E = 1 - \frac{\tau_{D,bleach}}{\tau_{DA,bleach}} \tag{13.43}$$

Equation (13.43) is reached by the definition of FRET from donor excited state lifetimes, as given in Eq. (13.48). For practical reasons, it is useful to perform this reference measurement on control cells in the same sample or on (parts of) cells where the acceptor has been photobleached before.

Donor photobleaching is a very sensitive measurement but requires a mechanically stable setup, as the photobleaching takes some time. The donor photobleaching technique can be easily implemented in a confocal microscope to obtain a large number of time points in a brief but complete bleaching time [37]. This permits a very precise analysis of the bleaching curves. An implementation of photobleaching kinetics measurements in a confocal microscope is given in Figure 13.8.

The donor photobleaching technique was historically the first time-resolved measurement and was the immediate cause for the development of the acceptor photobleaching technique. One of the samples that were subjected to donor photobleaching displayed an increase in donor fluorescence before the expected exponential decay was recorded [28]. As the photostability relationship in the fluorophore pair in the experiment – Cy3 as donor for Cy5 as acceptor – was in favor of the Cy3 donor, and as this sample exhibited very high FRET efficiencies, the acceptor was bleached by its excitation via FRET, whereas the already photostable

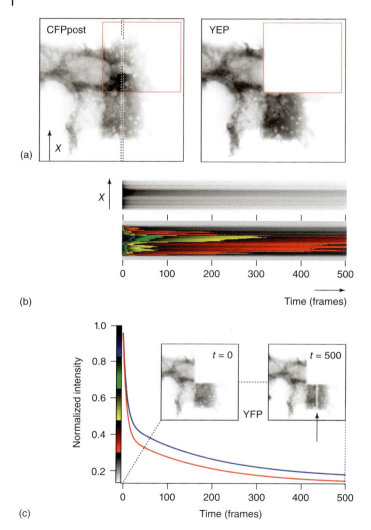

Figure 13.8 Donor photobleaching kinetics. (a) CFP and YFP images of a cell expressing a fusion protein containing a CFP donor and a YFP acceptor after bleaching of YFP in the indicated region (to generate a CFP-only zero FRET control). The indicated line in the CFP image was bleached by repeated line scanning with a confocal microscope in the x–t mode. The CFP fluorescence intensity traces are shown in grayscale and false-color representation in (b): the line is in the same orientation as in the CFP image, and the horizontal direction now represents time as all 500 lines are concatenated. (c) Graph shows the reduction in CFP fluorescence time-averaged over the part of the scan line that was in the bleached (red trace) and the unbleached (blue trace) areas. Note that the region with FRET (blue curve) bleaches more slowly than the region without FRET (red curve). Small inserts show the YFP fluorescence at the first and last bleached line in the CFP image. An arrow marks the position of the bleach line in the YFP image. Note that CFP bleaching also resulted in YFP bleaching. This is due to bleaching by FRET and by direct YFP excitation. The slow decay phases in the curves run parallel, indicating that the donor bleaching behavior in this phase is already dominated by the loss of acceptor, and that the initial bleaching phase (∼50 frames), in which the difference developed, is sensitive to the difference in FRET. This example demonstrates the general difficulty with quantitation from bleaching on the donor side. Fluorescence images are shown in an inverted gray scale look-up table.

donor was protected against photobleaching because of FRET. This observation then led to the experimental step of comparing the donor fluorescence before and after complete photobleaching of the acceptor.

13.3.2.2 Fluorescence Lifetime Changes

The FRET efficiency describes the increase in the decay rate of the excited state of the donor according to Eq. (13.16). The rate constants governing this process can be expressed in terms of lifetimes, which is the inverse of the sum of the rate constants (see also Chapter 3):

$$\tau_D = \frac{1}{k_f + k_n} > \tau_{DA} = \frac{1}{k_f + k_n + k_T} \tag{13.44}$$

where k_f is the rate of fluorescence or radiative transfer (τ_0^{-1}), k_n is the rate of nonradiative transfer, and k_T is the rate of FRET. The donor lifetime reduces proportionally with an increase in the rate of FRET.

Box 13.2 FRET and Reaction Kinetics

Lifetime measurements provide direct access to the decay rates that act on the excited donor. The process of FRET can be regarded as a competing reaction to fluorescence and be expressed in terms of rate equations:

Donor fluorescence is generated at the rate with which the excited state D* returns to the ground state D with the generation of a photon $h\nu$. The rate constants for radiative (k_f) and nonradiative (k_n) decay define the rate of this process (dD*/dt) in the absence of FRET. In the presence of FRET, the donor excited state is depopulated faster ($dD^*_{FRET}/dt > dD^*/dt$) due to the additional decay rate constant k_T, which describes FRET (Eqs (13.45) and (13.46)). [D*] indicates the concentration of donor molecules in the excited state at the start of the decay, as, for instance, introduced after an excitation pulse.

The occurrence of FRET is a parallel or competing reaction with the generation of fluorescence, as energy transfer also acts on the "reactant" D* but does not generate fluorescence.

$$D^* \xrightarrow{k_f + k_n} D + h\nu$$

$$D^* \xrightarrow{k_T} D \tag{13.45}$$

The differential rate laws governing the depopulation of D* in the absence and presence of FRET are given by

$$\frac{d[D^*]}{dt} = -(k_f + k_n)[D^*] = -\tau_D^{-1}[D^*]$$

$$\frac{d[D^*_{FRET}]}{dt} = -(k_f + k_n + k_T)[D^*] = -(\tau_D^{-1} + k_T)[D^*] \tag{13.46}$$

FRET increases the de-excitation rate of D*. We can define the FRET efficiency as the relative increase in the de-excitation of D*, yielding

$$E = 1 - \frac{\frac{d[D^*]}{dt}}{\frac{d[D^*_{FRET}]}{dt}} = \frac{k_T}{\tau_D^{-1} + k_T} \tag{13.47}$$

(Continued)

Box 13.2 (Continued)

At very low FRET, the de-excitation rate of D^*_{FRET} approaches that of D^*, and at 100% FRET, $k_T \gg k_f + k_n$ such that $dD^*_{FRET}/dt \gg dD^*/dt$) and their ratio approaches zero. This relationship and Eq. (13.28) show that the FRET efficiency depends on the unperturbed donor lifetime, that is, in the absence of FRET. Intuitively, increased FRET for longer donor lifetimes makes sense, as it provides the acceptor with more time to engage with the donor excited state. A longer τ_D signifies lower decay rates k_f and k_n, which are available for competition with k_T. From the direct relationship between τ_D and Q_D (Eqs (13.5) and (13.52)), it follows that energy transfer also increases with high donor quantum yields. Also this makes intuitive sense, as low losses translate into increased interaction possibilities.

Similarly, the FRET efficiency can be expressed in terms of changes in the donor (D) fluorescence lifetime in the absence (τ_D) and presence of FRET with an acceptor (τ_{DA}):

$$\frac{d[D^*]}{dt} = -\tau_D^{-1}[D^*]$$

$$\frac{d[D^*_{FRET}]}{dt} = -\tau_{DA}^{-1}[D^*]$$

$$E = 1 - \frac{-\tau_D^{-1}[D^*]}{-\tau_{DA}^{-1}[D^*]} = 1 - \frac{\tau_{DA}}{\tau_D} \tag{13.48}$$

Note that this is similar to the definition of the FRET efficiency based on donor fluorescence changes (Eq. (13.35)). Figure 13.9 depicts the relevant reaction kinetics of a donor fluorophore undergoing FRET.

The differential equations can be solved for the time-dependent concentrations of the reactants by the integrated first-order rate laws. Treating the fluorescence photons $h\nu$ generated in the decay reaction $D^* \rightarrow D + h\nu$ as reaction products

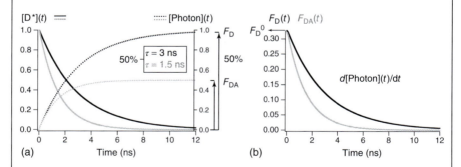

Figure 13.9 Reaction kinetics of Fluorescence decay. The reaction kinetics of the time-dependent decay of [D*], production of [Photon] (a), and generation of donor fluorescence F_D (b) is shown for a 3-ns fluorescence lifetime donor in the absence and presence of 50% FRET. FRET reduces the fluorescence lifetime to 1.5 ns. Note that the number of photons produced with FRET, F_{DA}, is also halved with respect to the number of photons produced in the absence of FRET, F_D. The plateau values of [Photon] in (a) represent the area under the curve of $F(t)$ in (b).

Photon$_D$ in the absence and Photon$_{DA}$ in the presence of the competing reaction $D^* \xrightarrow{FRET} D$, we obtain

$$[D^*(t)] = [D_0^*]e^{-\frac{t}{\tau_D}}$$

$$[Photon_D(t)] = [D_0^*]\left(1 - e^{-\frac{t}{\tau_D}}\right)$$

$$[Photon_{DA}(t)] = \frac{(k_f + k_n)}{(k_f + k_n) + k_T}[D_0^*]\left(1 - e^{-\frac{t}{\tau_{DA}}}\right) = \frac{\tau_{DA}}{\tau_D}[D_0^*]\left(1 - e^{-\frac{t}{\tau_{DA}}}\right)$$

(13.49)

in which $[D_0^*]$ is the concentration of excited state donors at the beginning of the decay, that is, at the end of an illumination pulse. The function [Photon(t)] describes the time-integrated generation of fluorescence, that is, the *number of* photons that were generated in the reaction *till* time point t, from the decaying $[D_0^*]$. These and the following relationships can best be understood by considering decay from a concentration $[D_0^*]$ of donor molecules in the excited state immediately after excitation is switched off.

What is observed as fluorescence, however, is not the build-up of the photons that are generated by the decaying concentration $[D_0^*]$ but the rate of fluorescence generation as a function of time, that is, the derivative of $[Photon_D(t)]$, referred to as $F_D(t)$.

$$\frac{d[Photon_D(t)]}{dt} = F_D(t) = \frac{[D_0]}{\tau_D}e^{-\frac{t}{\tau_D}} = F_D^0 e^{-\frac{t}{\tau_D}}$$

$$\frac{d[Photon_{DA}(t)]}{dt} = F_{DA}(t) = \frac{[D_0]}{\tau_D}e^{-\frac{t}{\tau_{DA}}} = F_D^0 e^{-\frac{t}{\tau_{DA}}}$$

(13.50)

where F_D^0 represents the fluorescence generated immediately after excitation is switched off for a donor in the absence of FRET. Note that the time-dependent generated fluorescence and F_D^0, in accordance with their nature as time-derivative signals, have the unit photons per second.

The shape of the donor *fluorescence* decay curve is directly described by the value of the exponential factor τ_D or τ_{DA}, and at $t = \tau$, the intensity has decayed to a fraction 1/e of the initial intensity. The FLIM technique of time-correlated single photon counting (explained in the next section) aims at reconstructing $d[Photon_D(t)]/dt$ (Figure 13.9b).

The equivalence of the definition of the FRET efficiency from fluorescence intensities (Eq. (13.35)) and from lifetimes (Eq. (13.48)) is easily demonstrated from the integrated rate laws describing the generation of photons, that is, fluorescence intensity, in the absence and presence of FRET.

This is equivalent to observing the fluorescence decay after a short excitation pulse. The faster decay of $F_{DA}(t)$ compared to $F_D(t)$ will cause a smaller (total) number of photons ($F_{DA} < F_D$) being produced when the system decays from the same initial fluorescence F_D^0. This is expressed as the area under the curve of the derivative functions $F(t)$ or from the values reached at long time points of the integrative rate laws for the time-dependence of $[Photon_D]$ or $[Photon_{DA}]$, that is, the plateau

(Continued)

> **Box 13.2 (Continued)**
>
> value reached. At long times, Eq. (13.50) becomes
>
> $$F_D = \lim_{t\to\infty}[F_D(t)] = [D_0^*]$$
>
> $$F_{DA} = \lim_{t\to\infty}[F_{DA}(t)] = [D_0^*]\frac{\tau_{DA}}{\tau_D} \text{ such that}$$
>
> $$E = 1 - \frac{F_{DA}}{F_D} = 1 - \frac{\tau_{DA}}{\tau_D} \qquad (13.51)$$
>
> An intuitive alternative demonstration of the equivalence is that lifetime measurements are effectively determinations of the donor quantum yield, as the lifetime is directly related to the quantum yield by
>
> $$Q_D = \frac{k_f}{\sum_i k_i} = \frac{\tau_D}{\tau_f} \text{ where } \tau_f = \frac{1}{k_f}; \quad \tau_D = \frac{1}{\sum_i k_i} \qquad (13.52)$$
>
> The lifetime term τ_f is a constant factor called the radiative lifetime, that is, the lifetime that a fluorophore would exhibit in the absence of nonradiative decay transitions.
>
> As the molar extinction coefficient of the donor does not change with FRET, the fluorescence brightness ($\varepsilon_D Q_D$) is directly proportional to the donor quantum yield (or lifetime) changes. Therefore, one can replace the lifetimes and rates in Eq. (13.31) by the corresponding fluorescence intensities, yielding Eq. (13.35).

The equivalence between intensity changes and lifetime changes (Box 13.2 and Figure 13.9) shows that donor lifetimes reflect the (time-averaged) brightness of donor molecules undergoing FRET. Lifetime measurements of FRET determine FRET-induced quenching, but they are conveniently independent of absolute intensities, concentrations, and light paths, obviating the need for calibration measurements.

The shortening of the donor fluorescence lifetime with increasing FRET (see also Figure 13.10) also means that the donor emission probability per unit time increases with FRET in the coupled donor–acceptor system. The occurrence of FRET thus accelerates the fluorescence cycle by forcing the depopulation of molecules from the excited state (after an average delay time corresponding to τ_{DA}) before they would have done so spontaneously (which would have taken an average delay time corresponding to τ_D).

The acceptor fluorescence lifetime does not change with FRET, but the acceptor "sensitized emission" lifetime does. The acceptor excited state has no knowledge of how it became populated and will depopulate with its characteristic lifetime, which is equal to the lifetime that one would measure when directly exciting the acceptor, and measuring the decay time. However, the sensitized emission is generated by a FRET event some time after the donor became excited. This additional delay time increases the observed sensitized emission lifetime and causes a characteristic increase in the lifetime when measured in the spectral acceptor window.

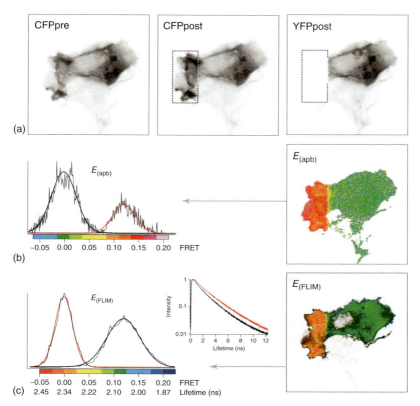

Figure 13.10 Quantitative detection of FRET by FLIM. (a) Fluorescence images of a fixed cell expressing a fusion protein containing a CFP donor and a YFP acceptor. The indicated region was subjected to the selective bleaching of YFP to provide an internal control and for comparison with the FRET efficiency determination from acceptor photobleaching as in Figure 13.4 (right-most image of (b), $E_{(apb)}$). The FRET efficiency in this cell was determined by time-correlated single-photon counting (TCSPC) FLIM. Normalized CFP fluorescence decay curves for the bleached region (red curve, zero FRET control for FLIM) and the remainder (black curve) of the cell are shown (c). The pixel-by-pixel decay analysis returns the lifetime map shown in false color (intensity-coded lifetime distribution) in the right-most image of the lower row ($E_{(FLIM)}$). The distributions of FRET efficiencies in the two cellular regions are shown. The distribution in (b) is obtained by acceptor photobleaching, the distribution in (c) by FLIM (red curve: bleached region, black curve: unbleached region). Fluorescence images are shown in an inverted gray scale look-up table.

As lifetime increases cannot be caused by other trivial influences that could falsify the detection of FRET, it can be taken as a highly confident diagnostic sign for FRET [38].

13.4 FLIM

FLIM measures the average duration of the excited state, that is, the fluorescence lifetime τ. As most biologically relevant fluorophores exhibit fluorescence lifetimes in the nanosecond range, their accurate measurement requires specialized equipment and approaches. There are mainly two ways of detecting these changes, both of which rely on the same principle and are functionally equivalent [39]. See also [23, 40] for detailed descriptions of the techniques. Conceptually, both methods can be regarded as describing the temporal distortion in the emission signal when the probe is excited with a defined periodic intensity pattern. If the frequency of the modulated intensity in the excitation signal is sufficiently high, the (typically nanosecond) delay between excitation and emission, that is, the fluorescence lifetime, will cause a distinctive distortion which can be measured at a time scale that is much larger than the fluorescence lifetime itself. These defined excitation patterns can be a pulse, or a periodic signal like a sine wave, or a block wave.

Two types of periodic signals are used: trains of short pulses, or periodic high-frequency modulated signals. In order to detect the effect of the (very short) lifetime in the resulting emission signal, typically femtosecond pulses of mode-locked lasers or pulsed lasers are used as the excitation source. The fluorescence lifetime will "smear out" the excitation pulse in time in the exponential decay curve. One of the methods measures this decay curve in time and is called *time-domain FLIM*. A FRET experiment using time-domain FLIM imaging is shown in Figure 13.10.

Lasers can also be used to generate the periodic high-frequency modulated excitation pattern, which can be used alternatively. The output of continuous wave lasers can be modulated into a sine wave by feeding it through an acousto-optical modulator or a Kerr cell, but modern diode lasers can also be modulated directly at a sufficiently high frequency (MHz) to be used in FLIM. The effect of the fluorescence lifetime is twofold. First, the emission is delayed compared to the excitation by a phase shift ϕ_ω, but will still exhibit the same fundamental modulation frequency as the excitation signal. A sine wave will therefore result in an emission sine wave with the same frequency. Periodic signals with higher frequency components, for example, a square wave, will lose their higher frequency components and will, with a sufficiently long lifetime, result in a sine wave at the fundamental modulation frequency. Second, the amplitude of the emission signal will be reduced with increasing lifetime. These are two parameters from which the fluorescence lifetime τ can be estimated:

$$\tan \phi_\omega = \omega \tau$$
$$m_\omega = \frac{1}{\sqrt{1 + \omega^2 \tau^2}} \qquad (13.53)$$

where ω is the modulation frequency of the excitation, ϕ_ω is the phase, and m is the modulation as given by the ratio of the relative excursion from the average intensity for the emission over that of the excitation signal. These effects are caused by the addition of a periodic component to the exponential decay and can be easily converted into the simulated response on a light pulse. In an everyday analog (provided in Figure 13.11), these effects and their relation can be intuitively understood: consider the tire tracks that a bicycle makes in the snow when the handle bar is rhythmically moved to make the bicycle follow a sinusoidal path. Inspection of the tire tracks will reveal that the track of the hind wheel is shifted in phase and is also demodulated. Both are caused by the fact that the excitation signal, that is, the movement of the front wheel, undergoes a delay – corresponding

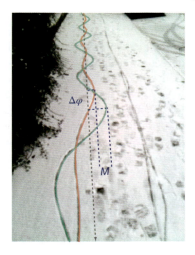

Figure 13.11 Everyday analog for the phase-shift and demodulation in harmonic excitation in frequency domain FLIM. The image shows the imprints of bicycle tire-tracks in the snow, as an analogy for the changes in fluorescence response in frequency-domain FLIM as discussed in the text. The direction of movement is indicated in the "average fluorescence level" line (dashed gray). The green trace represents the excitation (front wheel), and the red trace represents the emission response (hind wheel) of a system exhibiting a finite lifetime (i.e., bike length divided by speed).

to the distance between the wheels – before the emission signal is generated, here the track left by the hind wheel.

From this analog, it is also apparent that the phase shift and demodulation are dependent on the relation between the lifetime and the modulation frequency. When the modulation frequency is too low, the fluorescence will have decayed well before a full excitation cycle is completed (the distance between the tires is small relative to the long periodic movement that the slow modulation of the handle bars create) and the emission will follow the excitation closely. When the modulation frequency is high, the fluorophores that were excited at the maximum will still be emitting during the completion of the excitation cycle (the handle bar is moved too fast for the hind wheel to respond to), "smearing out" the signal over time and causing a loss in modulation to reach the constant average value. Every fluorescence lifetime therefore has an optimal modulation frequency where the averaging is still low and the phase shift and demodulation are optimally sensitive to changes in the lifetime (and modulation frequency). This frequency dependence is used in phase fluorimetry to obtain information on the presence of multiple lifetime emitters in a sample; by performing phase and modulation measurements at multiple frequencies, the different behaviors of the different species will allow their identification and quantification.

13.4.1 Frequency-Domain FLIM

13.4.1.1 Operation Principle and Technical Aspects

The phase modulation technique that is exploited in frequency-domain FLIM is historically the earliest method that was used for the determination of the fluorescence lifetime. Already in the 1920s, the Argentinean astrophysicist Enrique Gaviola accurately determined the lifetime of a number of fluorophores [41]. The difficulty with transferring these measurements to imaging lies in the demodulation of the signal over the entire image, as the detection efficiency of the camera has to be modulated with the same high frequency as the excitation signal. During the 1980s, technology became available that allowed just this. With the use of a multichannel plate (MCP) image intensifier, Joseph Lakowicz was the first to show that calcium indicator dyes, traditionally used on the basis of their fluorescence yield response, could provide highly sensitive measurements by the detection of their lifetime changes [42].

In recent years, all-solid-state solutions have been proposed that make the MCP obsolete. In a first solution, the charge-transfer read-out mechanism of CCD chips was used to achieve detection modulation in commercial CCD chips [43]. However, the modulation frequency stayed modest (several tens of kilohertz) and was not high enough to achieve sub-nanosecond time resolution. Faster modulation required a new design of the detection chip, and this was first demonstrated in the use of a time-of-flight camera for nanosecond lifetime measurements [44, 45]. The pixels of this camera directly act as demodulation devices as the photoelectrons generated in the photosensitive substrate of the pixel are continuously sorted between two collection bins that are charged at opposite phases and whose charge is modulated at the excitation frequency. From these two bins, two images are created that suffice for the determination of the average fluorescence lifetime. In contrast to this "single shot" mode, multiple images can be taken at different phase delays, permitting more detailed fluorescence lifetime measurements.

A detailed discussion on frequency-domain FLIM measurements is given elsewhere [39, 46].

13.4.2 Time-Domain FLIM

In time-domain FLIM, the time-encoded excitation pattern is an (or a train of) ultrashort pulse(s). The exponential "smearing out" of these pulses provides a direct and perhaps more intuitive connection to the fluorescence lifetime (Figure 13.9b). At the time of the excitation pulse, the fluorescent molecules are raised to the excited state, and will return from there by emission.

Care has to be taken that the fluorescence lifetime is matched by the repetition rate of the laser such that the fluorophore can deplete its excited state entirely before the next excitation pulse arrives. As most modern light sources, for example, mode-locked and two-photon lasers operate at a repetition rate ~80 MHz, which means that the fluorescence lifetime should not exceed a few nanoseconds. For practical purposes, this restraint is perfectly suited to the distribution of lifetimes among common fluorophores whose lifetimes are below 10 ns. In fact, it is difficult to find suitable fluorophores with higher lifetimes without switching to the 100 times slower microsecond decay time process of phosphorescence. Fluorophores that decay very slowly would lead to a considerable build-up of photons that would spill over into the next counting period, with no connection to the originating pulse.

13.4.2.1 Time-Correlated Single-Photon Counting

There are two major ways to time the arrival times of the incoming fluorescence photons; they are either individually registered, or they are counted in defined time windows. In the first method, called *time-correlated single-photon counting*, the arrival times of the individual fluorescence photons are determined [47]. At suitable count rates, in reality, there are fewer photons that are detected than there are cycles. However, rather than timing the arrival of the first photon to arrive after the last pulse, the time to the next pulse after the arrival of a photon is recorded. The reason for this "reverse start/stop" arrangement is that not all excitation pulses produce photons. By starting the counting upon the collection of a photon, and measuring the time to the next pulse, only those cycles are counted in which a photon was detected.

The reason for the "empty cycles" is that the count rate with TCSPC is limiting. Photons have to arrive with a temporal separation that allows them to be counted as individual signals. Furthermore, the counting electronics is slower than the pulse rate and would not be able to correctly count photons if they were produced with every cycle. At high count rates, pulses tend to pile up, and this deteriorates the measurement. In practice, this means that the excitation power and/or labeling density have to be low enough to allow an accurate measurement. The limiting count rate is one of the reasons why time-domain measurements are relatively slow.

As with frequency-domain FLIM, the development of solid-state imagers promises to bring down the cost and acquisition time of time-domain FLIM.

13.4.2.2 Time Gating

Rather than timing single emitted photons, photons can be counted in consecutive time bins, or gates, covering the time between pulses [48–50]. As with TCSPC, the intensity decay is repeatedly sampled. This, effectively, results in a "bar graph" representation of the intensity decay, with the number of bars corresponding to the number of time bins used. It is therefore obvious that the method is intrinsically coarser than TCSPC. However, when the objective is to image an average lifetime to provide contrast, rather than decomposing the decay into its multiple lifetime species components, a relatively small number of bins suffice. The advantage of the method is that it can accept multiple photons after a single pulse, allowing higher count rates and making the method generally faster. With the minimum number of bins (i.e., 2), the analysis also becomes mathematically simple and affords significantly shorter calculation time, as it does not require least-square fitting. This method, called the *rapid lifetime determination* (RLD), is surprisingly robust and effective for biological imaging and can provide real-time analysis of the average lifetime of a sample [46, 51].

$$\text{for } F(t) = F_0 e^{-\frac{t}{\tau}}$$

$$\tau = \frac{-\Delta t}{\ln(D_1/D_0)} \qquad (13.54)$$

where $F(t)$ describes the exponential time-dependent decay in the detection of fluorescence (see Eq. (13.50)) from the fluorescence observed immediately after switching off the excitation F_0. D_1 and D_0 are two contiguous areas of width Δt that divide the decay curve.

The operation of an MCP device as a fast shutter is the basis of most modern widefield time-gating FLIM systems [50]. The number of gates (≥ 2) can be chosen arbitrarily, and the measurements are made in series during the repetitive pulsed excitation. In a scanning microscope, gated counters are connected to the detector and perform the same task.

13.5 Analysis and Pitfalls

13.5.1 Average Lifetime, Multiple Lifetime Fitting

A mixture of emitters with similar emission spectrum but different lifetimes will produce a complicated fluorescence decay with an average lifetime that is weighted by the relative contribution (concentration, quantum yield) of the single emitters. The main drawback of the RLD is that it only delivers an average lifetime without the possibility to extract the underlying lifetime differences. Also with multiple time gates, time-gated time-domain measurements generally lack the resolution to decompose the complex signal. The same is true for frequency-domain measurements when performed at a single modulation frequency, although here the user is informed about underlying heterogeneity by the disparity between the lifetime as determined from the phase shift τ_ϕ and that from the demodulation τ_m, which increases with heterogeneity. Furthermore, even when the excitation is driven at a single harmonic frequency, the detector introduces higher frequencies, which allows the resolution of two independent lifetimes in a mixture. These methods were developed by several groups and are collectively known as *global fitting approaches*. The only method that grants direct access to multiple lifetime species is TCSPC; however, the inclusion of multiple lifetime species comes at the cost of a sharp increase in the photon demand for each additional species (one lifetime requires ~1000 photons, two lifetimes require 10 times more). Especially at lower photon numbers, the fitting routine might not return the correct results even though standard fitting quality estimators would suggest a successful determination. For an overview of aspects that need to be taken into account in order to optimize FLIM measurements, see [52].

FRET is a condition that produces an intrinsically heterogenous sample, as only a certain fraction of the donor molecules is expected to undergo FRET in a realistic setting. For most cases, the FRET efficiency estimated from an average lifetime, however, will provide sufficient and useful contrast for the investigation of the process that it reports on. The user will typically observe a homogeneous and known lifetime for cells that express only the donor fluorophore. This value then serves to estimate the apparent FRET efficiency from the shorter lifetime that is observed when both donor and acceptor are expressed, indicating, for example, the association of donor- and acceptor-labeled proteins or changes in the conformation of a single-molecule FRET biosensor that incorporates both fluorophores.

13.5.2 From FRET/Lifetime to Species

Many biological applications go beyond contrast; they aim at obtaining knowledge on the fraction of the engaging molecules. The objective is therefore to extract the FRET efficiency of the complex and the relative fraction of the complex. Furthermore, the absolute FRET efficiency is hardly ever informative, as it depends on the way the FRET assay was designed more than on the configuration of the interacting proteins. The difficulty that the – difficult to predict, determine, or manipulate – dipole orientation affects the FRET efficiency adds to this.

One difficulty with global analysis in the frequency domain [53] or lifetime fitting in the time domain is that it relies on fitting routines that might not produce the correct answer. Another is that these fitting routines are performed under a number of assumptions that might not hold.

In general, a further difficulty arises due to cellular autofluorescence contamination, which by itself is already heterogeneous. The system quickly becomes very complicated, and the number of photons required from the biological system to unequivocally solve it might become unattainably high. For this reason, a method for FLIM analysis was recently developed that avoids fitting (i.e., immediately interpretable) and provides feedback coupling with the localization information contained in the fluorescence image. A graphical representation method, called the *Phasor plot*, which is a polar representation of sine and cosine transforms of lifetime data, was developed, which allows the quantitative interpretation of lifetime heterogeneity – and from this the FRET efficiencies and fractional contributions [54]. We refer to a short review on the background and history of the Phasor approach [55]. Perhaps the most useful aspect of the method, which offers great utility to the biomedical user, is the possibility to connect the information obtained from the polar representation to the spatial location(s) in the fluorescence image. The user can select a data cloud in the polar plot and be shown where these coherently behaving pixels are located in the image, and vice versa. This kind of intuitive image-driven analysis holds great promise for the wider adoption of fluorescence lifetime investigations in modern cell biology and other biomedical applications.

> **Box 13.3 Appendix – Theodor Förster on Energy Transfer**
>
> The following is a translation of the relevant page of Theodor Förster's first paper on FRET in which he introduces the concept of the overlap integral.
>
> The swinging electrical charge couples with the radiation field to give rise to *radiation* of the energy to the surrounding. It can, however, also couple to an oscillator of a nearby molecule by the electrostatic forces that act between them. At equal frequencies, this will lead to energy *transfer* as occurs between two mechanically coupled pendulums with equal oscillation times (resonance pendulums), in which the energy from the pushed pendulum will be transferred to the second. At large separation distances, only radiation will occur, and at very short distances, only transfer. They merge at a critical distance d_0 at which radiation and transfer are equally probable (Figure 13.12). 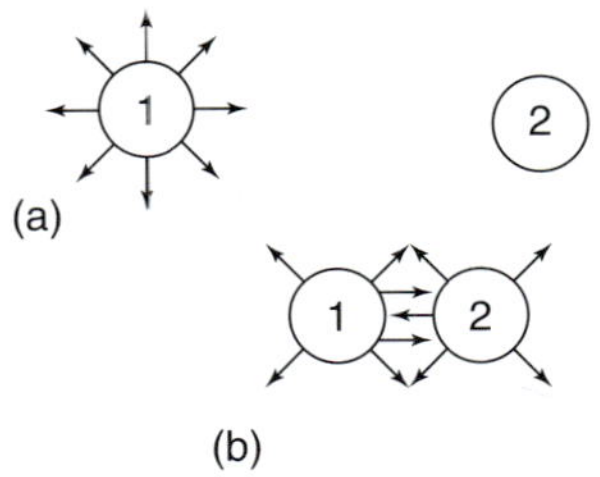
>
> **Figure 13.12** Radiation of two oscillators. (a) $d \gg d_0$, radiation by the primary excited oscillator 1. (b) $d \ll d_0$, both oscillators show the same radiation probability. This would result in fluorescence depolarization. Note that this example refers to the situation of energy transfer between identical fluorophores. At distance d_0, radiation and transfer are equal.

(Continued)

> **Box 13.3 (Continued)**
>
> In the case of perfect resonance, the value for d_0 is given by the calculation of Perrin:
>
> $$d_0 \approx \frac{c}{\omega} = \frac{\lambda}{2\pi} \quad (13.55)$$
>
> in which c is the speed of light, and ω the circular frequency, that is, the fundamental frequency of the oscillator divided by 2π.
>
> Radiation and transfer are determined by the pulsating electrical dipoles that the oscillators with a given energy content possess. For photons, this energy is equal to $\hbar\omega$, where $\hbar$ is the Planck constant divided by 2π.
>
> For radiation, the average duration of radiation τ (the radiative lifetime) is given by Fermi's golden rule (omitting a few constants):
>
> $$\tau \approx \frac{\hbar c^3}{M^2 \omega^3} \quad (13.56)$$
>
> where M is the dipole moment. For transfer, the interaction energy U of both oscillators at distance d should equal the electrostatic interaction energy between both dipoles:
>
> $$U \approx \frac{M^2}{d^3} \quad (13.57)$$
>
> This amount of energy is transferred between the oscillators for each oscillation period, or, more precisely, during the time period of $1/\omega$. In the case of resonance, these contributions accumulate such that the energy $U\omega$ is transferred per unit time.
>
> The time (t_0) needed to transfer energy with a magnitude of the energy content of a photon $\hbar\omega$ therefore becomes
>
> $$t_0 = \frac{\hbar}{U} \approx \frac{\hbar d^3}{M^2} \quad (13.58)$$
>
> At the critical distance d_0, the transfer time t_0 equals the duration of radiation τ, such that from Eqs (13.56) and (13.58) follows Eq. (13.55).
>
> However, this relationship returns critical distances that are very much overestimated. The insufficiency of the Perrin formula is caused by the fact that the assumed condition of perfect resonance is not fulfilled.
>
> This is already obvious from the fact that fluorophores do not posses perfect line spectra, but rather a broad band, caused by the coupling of the electron movements with slower atomic oscillations in the molecule itself and in the surrounding solvent. Furthermore, the fluorescence spectrum undergoes a Stokes' shift to higher wavelengths with respect to the absorption spectrum, leaving little overlap between them.
>
> During the long time of energy transfer, every frequency occurs as often as is given by the intensity in the fluorescence or absorption spectrum, respectively. The condition for the resonance case is fulfilled only in the short time in which the frequencies of both molecules are identical within the coupling width, that is, the frequency interval of size $U/\hbar$.

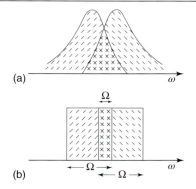

Figure 13.13 Absorption and fluorescence spectra. (a) Real distributions and (b) schematized distributions. The spectra are shown on the basis of the circular frequency ω.

To visualize his probabilistic argument, Förster schematically represents the overlapping absorption and emission spectra as two rectangles with width Ω, which overlap in a strip with width Ω' (Figure 13.13).

The conditions for resonance dictate that the frequency of one of the oscillators lying within the frequency interval $U/\hbar$ matches another whose frequency lies within Ω'. The probability W of these two conditions matching is therefore

$$W = \frac{U}{\hbar} \frac{\Omega'}{\Omega^2} \qquad (13.59)$$

As transfer is possible only within the corresponding fraction of time, the time required to transfer the energy, as given in Eq. (13.58), has to be divided by this probability W, returning

$$t_0 \approx \frac{\hbar \Omega^2}{U^2 \Omega'} \approx \frac{\hbar^2 d^6 \Omega^2}{M^6 \Omega'} \qquad (13.60)$$

We find d_0 by equating transfer and radiation, $t_0 = \tau$, and eliminating M using Eq. (13.56) (with $\lambda = 2\pi c/\omega$), which is Eq. (13.2):

$$d_0 \approx \frac{\lambda}{2\pi} \sqrt[6]{\frac{\Omega'}{\tau_0 \Omega^2}}$$

Förster then proceeds to show the error in Perrin's estimation of the critical distance. Perrin had neglected the difference between absorption and emission spectra, and had assigned the spectral broadening Ω solely to collisions of the excited fluorophore with the solvent molecules. Under Perrin's assumption of $\Omega' = \Omega \gg 1/\Delta t$, in which Δt is the average time between two collisions, Eq. (13.2) takes the form of the formula that Perrin had published before, which is Eq. (13.1):

$$d_0 \approx \frac{\lambda}{2\pi} \sqrt[6]{\frac{\Delta t}{\tau}}$$

Förster closes his early paper with a discussion on the reasons for the loss of fluorescence at high fluorophore concentrations and the relevance of energy transfer in photosynthesis.

Source: Förster 1946 [6]. Reproduced with permission of Springer.

13.6 Summary

The extreme distance dependence of FRET between fluorophores, correctly described for the first time by Theodor Förster, provides an important tool for the determination of molecular-scale interactions and structural rearrangements of biomolecules.

Especially, imaging of the occurrence of FRET inside living cells by various forms of FRET microscopy continues to generate valuable information on the workings of cellular mechanisms, as these rely on protein actions that can be visualized by FRET. The discovery of autofluorescent proteins that can be attached to virtually any protein of choice, and that can be genetically linked by simple molecular biological tools, has contributed to the astounding increase in popularity that this technique has seen in the last two decennia.

FRET is conceptually simple, but its practical implementation is far from trivial.

In intensity-based approaches, spectral contaminations between the fluorophores that are caused by their broad fluorescence spectra in combination with the requirement for a significant spectral overlap are possible causes for error. Quantification of FRET in these techniques is also rather complicated, as the signals tend to be dependent on the concentrations of the fluorophores. Approaches that investigate the changes in depopulation rates from the excited state of donor fluorophores do not suffer from these complications and can provide highly quantitative and robust measurements. Among these, the different forms of FLIM are the most useful.

New technologies and novel analytical tools have been – and are being – developed to help the researcher gain higher acquisition speed, sensitivity, and levels of information from FRET/FLIM experiments to obtain a clearer view of the molecular machines of life.

References

1 Clegg, R.M. (2009) in *Laboratory Techniques in Biochemistry and Molecular Biology*, FRET and FLIM Techniques (ed. T.W.J. Gadella), Elsevier, Amsterdam, Boston, MA, pp. 1–57.

2 Clegg, R.M. (2006) in *Reviews in Fluorescence 2006*, vol. **3** (eds C.D. Geddes and J.R. Lakowicz), Springer, New York, pp. 1–45.

3 Van der Meer, B.W. (2013) in *FRET – Förster Resonance Energy Transfer: From Theory to Applications* (eds I. Medintz and N. Hildebrandt), Wiley-VCH Verlag GmbH, Weinheim, pp. 23–62, http://doi.wiley.com/10.1002/9783527656028.ch03 (accessed 5 April 2016).

4 Förster, T. (1948) Zwischenmolekulare Energiewanderung und Fluoreszenz. *Ann. Phys.*, **437** (1-2), 55–75.

5 Perrin F.) Théorie quantique des transferts d'activation entre molécules de méme espèce. Cas des solutions fluorescentes. *Ann. Chim. Phys.* 1932;**17**(283).

6 Förster, T. (1946) Energiewanderung und Fluoreszenz. *Naturwissenschaften*, **33** (6), 166–175.

7 Förster, T. (1951) *Fluoreszenz organischer Verbindungen*, 1st edn, Vandenhoeck & Ruprecht, Göttingen.
8 Haugland, R.P., Yguerabide, J., and Stryer, L. (1969) Dependence of the kinetics of singlet–singlet energy transfer on spectral overlap. *Proc. Natl. Acad. Sci. U.S.A.*, **63**, 23–30.
9 Jares-Erijman, E.A. and Jovin, T.M. (2003) FRET imaging. *Nat. Biotechnol.*, **21** (11), 1387–1395.
10 Stryer, L. and Haugland, R.P. (1967) Energy transfer: a spectroscopic ruler. *Proc. Natl. Acad. Sci. U.S.A.*, **58**, 719–726.
11 Grecco, H.E. and Verveer, P.J. (2011) FRET in cell biology: still shining in the age of super-resolution? *ChemPhysChem*, **12** (3), 484–490.
12 Wouters, F.S. (2006) The physics and biology of fluorescence microscopy in the life sciences. *Contemp. Phys.*, **47** (5), 239–255.
13 Hell, S.W. (2007) Far-field optical nanoscopy. *Science*, **316** (5828), 1153–1185.
14 Huang, B., Bates, M., and Zhuang, X. (2009) Super-resolution fluorescence microscopy. *Annu. Rev. Biochem.*, **78**, 993–1016.
15 Patterson, G., Davidson, M., Manley, S., and Lippincott-Schwartz, J. (2010) Superresolution imaging using single-molecule localization. *Annu. Rev. Phys. Chem.*, **61**, 345–367.
16 Swaminathan, R., Hoang, C.P., and Verkman, A.S. (1997) Photochemical properties of green fluorescent protein GFP-S65T in solution and transfected CHO cells: analysis of cytoplasmic viscosity by GFP translational and rotational diffusion. *Biophys. J.*, **72**, 1900–1907.
17 Shimozono, S., Hosoi, H., Mizuno, H., Fukano, T., Tahara, T., and Miyawaki, A. (2006) Concatenation of cyan and yellow fluorescent proteins for efficient resonance energy transfer. *Biochemistry*, **45** (20), 6267–6271.
18 Esposito, A., Gralle, M., Dani, M.A.C., Lange, D., and Wouters, F.S. (2008) pHlameleons: a family of FRET-based protein sensors for quantitative pH imaging. *Biochemistry*, **47** (49), 13115–13126.
19 Bunt, G. and Wouters, F.S. (2004) Visualization of molecular activities inside living cells with fluorescent labels. *Int. Rev. Cytol.*, **237**, 205–277.
20 Lackowicz, J.R., Gryczynski, I., Cheung, H.C., Wang, C.-K., Johnson, M.L., and Joshi, N. (1988) Distance distributions in proteins recovered by using frequency-domain fluorometry: applications to troponin I and its complex with troponin C. *Biochemistry*, **27**, 9149–9160.
21 Van der Meer, B.W. (2002) Kappa-squared: from nuisance to new sense. *Rev. Mol. Biotechnol.*, **82**, 181–196.
22 Stryer, L., Thomas, D.D., and Meares, C.F. (1982) Diffusion-enhanced fluorescence energy transfer. *Annu. Rev. Biophys. Bioeng.*, **11**, 203–222.
23 Lakowicz, J.R. (2006) *Principles of Fluorescence Spectroscopy*, 3rd edn, Springer, New York, pp. 519–522.
24 Berney, C. and Danuser, G. (2003) FRET or no FRET: a quantitative comparison. *Biophys. J.*, **84**, 3992–4010.
25 Bastiaens, P.I.H., de Beus, A., Lacker, M., Somerharju, P., and Eisinger, J. (1990) Resonance energy transfer from a cylindrical distribution of donors to a plane of acceptors: location of apo-B100 protein on the human low-density lipoprotein particle. *Biophys. J.*, **58**, 665–675.

26 Kuhn, H. (1972) in *Physical Methods of Chemistry* (eds A. Weissberger and B. Rossiter), John Wiley & Sons, Inc., New York, pp. 579–650.
27 Bastiaens, P.I.H., Majoul, I.V., Verveer, P.J., Soling, H.D., and Jovin, T.M. (1996) Imaging the intracellular trafficking and state of the AB5 quarternary structure of cholera toxin. *EMBO J.*, **15**, 4246–4253.
28 Wouters, F.S., Bastiaens, P.I.H., Wirtz, K.W.A., and Jovin, T.M. (1998) FRET microscopy demonstrates molecular association of non-specific lipid transfer protein (nsL-TP) with fatty acid oxidation enzymes in peroxisomes. *EMBO J.*, **17**, 7179–7189.
29 Wouters, F.S. and Bastiaens, P.I.H. (2001) Imaging protein–protein interactions by fluorescence resonance energy transfer (FRET) microscopy. *Curr. Protoc. Cell Biol.*, Chapter 17 (Unit 17.1).
30 Hoppe, A., Christensen, K., and Swanson, J.A. (2002) Fluorescence resonance energy transfer-based stoichiometry in living cells. *Biophys. J.*, **83**, 3652–3664.
31 Jalink, K. and van Rheenen, J. (2009) in *FRET and FLIM Techniques* (ed. T.W.J. Gadella), Academic Press, Burlington, pp. 289–349.
32 Van Rheenen, J., Langeslag, M., and Jalink, K. (2004) Correcting confocal acquisition to optimize imaging of fluorescence resonance energy transfer by sensitized emission. *Biophys. J.*, **86**, 2517–2529.
33 Chen, H., Puhl, H.L., Kousnik, S.V., Vogel, S.S., and Ikeda, S.R. (2006) Measurement of FRET efficiencies and ratio of donor to acceptor concentration in living cells. *Biophys. J.*, **91** (5), L39–L41.
34 Zal, T. and Gascoigne, N.R. (2004) Photobleaching-corrected FRET efficiency imaging of live cells. *Biophys. J.*, **86**, 3923–3939.
35 Cheung, H.C. (1991) in *Topics in Fluorescence Spectroscopy* (ed. J.R. Lakowicz), Plenum Press, New York, pp. 127–176.
36 Hirschfeld, T. (1976) Quantum efficiency independence of the time integrated emission from a fluorescent molecule. *Appl. Opt.*, **15** (12), 3135–3139.
37 Szaba, G., Pine, P.S., EWeaver, J.L., Kasari, M., and Aszalos, A. (1992) Epitope mapping by photobleaching fluorescence resonance energy transfer measurements using a laser scanning microscope system. *Biophys. J.*, **61** (3), 661–670.
38 Harpur, A.G., Wouters, F.S., and Bastiaens, P.I.H. (2001) Imaging FRET between spectrally similar GFP molecules in single cells. *Nat. Biotechnol.*, **19** (2), 167–169.
39 Esposito, A. and Wouters, F.S. (2004) Fluorescence lifetime imaging microscopy. *Curr. Protoc. Cell Biol.*, Chapter 4 (Unit 4.14).
40 Gerritsen, H.C., Draaier, A., Van den heuvel, D., and Agronskaia, A.V. (2006) in *Handbook of Biological Confocal Microscopy* (ed. J.B. Pawley), Springer, New York, pp. 516–532.
41 Gaviola, E. (1927) Ein Fluorimeter. Apparat zur Messung von Fluoreszenzabklingzeiten. *Z. Phys.*, **42** (11-12), 853–861.
42 Lakowicz, J.R. and Berndt, K.W. (1991) Lifetime-selective fluorescence imaging using an rf phase-sensitive camera. *Rev. Sci. Instrum.*, **62**, 1727–1734.
43 Mitchell, A.C., Wall, J.E., Murray, J.G., and Morgan, C.G. (2002) Measurement of nanosecond time-resolved fluorescence with a directly gated interline CCD camera. *J. Microsc.*, **206**, 233–238.

44 Esposito, A., Oggier, T., Gerritsen, H., Lustenberger, F., and Wouters, F. (2005) All-solid-state lock-in imaging for wide-field fluorescence lifetime sensing. *Opt. Express*, **13** (24), 9812–9821.

45 Esposito, A., Gerritsen, H.C., Oggier, T., Lustenberger, F., and Wouters, F.S. (2006) Innovating lifetime microscopy: a compact and simple tool for life sciences, screening, and diagnostics. *J. Biomed. Opt.*, **11** (3), 34016.

46 Esposito, A., Gerritsen, H.C., and Wouters, F.S. (2007) Optimizing frequency-domain fluorescence lifetime sensing for high-throughput applications: photon economy and acquisition speed. *J. Opt. Soc. Am. A Opt. Image Sci. Vision*, **24** (10), 3261–3273.

47 O'Connor, D.V. and Phillips, D. (1984) *Time Correlated Single Photon Counting*, Academic Press, London.

48 Buurman, E.P., Sanders, R., Draaijer, A., Gerritsen, H.C., Vanveen, J.J.F., Houpt, P.M. *et al.* (1992) Fluorescence lifetime imaging using a confocal laser scanning microscope. *Scanning*, **14** (3), 155–159.

49 De Grauw, C.J. and Gerritsen, H.C. (2001) Multiple time-gate module for fluorescence lifetime imaging. *Appl. Spectrosc.*, **55**, 670–678.

50 Wang, X.F., Uchida, T., Coleman, D.M., and Minami, S. (1991) A two-dimensional fluorescence lifetime imaging system using a gated image intensifier. *Appl. Spectrosc.*, **45**, 360–366.

51 Ballew, R.M. and Demas, J.N. (1989) An error analysis of the rapid lifetime determination method for the evaluation of single exponential decays. *Anal. Chem.*, **61**, 30–33.

52 Esposito, A., Gerritsen, H.C., Wouters, F.S., and Wolfbeis, O.S. (2008) in *Standardization and Quality Assurance in Fluorescence Measurements II: Bioanalytical and Biomedical Applications* (ed. U. Resch-Genger), Springer, Heidelberg, pp. 117–142.

53 Verveer, P.J. and Bastiaens, P.I.H. (2003) Evaluation of global analysis algorithms for single frequency fluorescence lifetime imaging microscopy data. *J. Microsc.*, **209**, 1–7.

54 Digman, M.A., Caiolfa, V.R., Zamai, M., and Gratton, E. (2008) The phasor approach to fluorescence lifetime imaging analysis. *Biophys. J.*, **94** (2), L14–L16.

55 Wouters, F.S. and Esposito, A. (2008) Quantitative analysis of fluorescence lifetime imaging made easy. *HFSP J.*, **2**, 7–11, http://www.tandfonline.com/doi/abs/10.2976/1.2833600 (accessed 3 July 2013).

A

Appendix A: What Exactly is a Digital Image?

Ulrich Kubitscheck

Rheinische Friedrich-Wilhelms-Universität, Institut für Physikalische & Theoretische Chemie, Wegelerstr. 12, Bonn 53115, Germany

A.1 Introduction

This book deals with the acquisition of images using a great diversity of microscopic setups. From the very beginning, we considered explicitly the digital acquisition of images. Here, we want to provide the basic knowledge of what digital images are and give a first glimpse into digital image processing. For people needing and developing image analysis tools, we recommend computer programs like ImageJ or MATLAB, or the bioinformatics platform Icy, or the bioinformatics platform Icy. For these programs, a number of excellent introductions and internet resources exist [1–4].

A.2 Digital Images as Matrices

In microscopy, photodetectors measure the light intensity in a primary or secondary image plane. Usually no reference to the color or polarization of the detected light is given. The color is actually defined by the band-pass filter that is placed before the photodetector. Similarly, insertion of analyzers allows the measurement of light with a specific polarization direction. In this manner, the intensity values for certain wavelength regions and/or polarization directions can be determined. The information on the used color channel or polarization direction is then usually stored in the meta-data of the image file, or just the lab book. The image as such, however, contains only intensity or *gray values*. Therefore, we will concentrate here on the properties of gray values of images.

A.2.1 Gray Values as a Function of Space and Time

To start out, we will examine the process of image acquisition. Already this reveals a number of important features. Photodetectors either measure intensity primarily as a function of time – like in all types of scanning microscopes – or as a function of space in a given time interval like in light sheet microscopy. First we discuss the data acquisition process when it is a function of space. This is the so-called parallel data acquisition process.

Fluorescence Microscopy: From Principles to Biological Applications, Second Edition.
Edited by Ulrich Kubitscheck.
© 2017 Wiley-VCH Verlag GmbH & Co. KGaA. Published 2017 by Wiley-VCH Verlag GmbH & Co. KGaA.

A.2.1.1 Parallel Data Acquisition

Parallel data acquisition is performed with digital cameras. The technical features of digital cameras are far more complex than would appear at a first glance, therefore the choice of the correct camera for a specific application in optical imaging is far from trivial. Some details were given in Chapters 2 and 3, but there is much more. For in-depth and up-to-date information, we refer the reader to the often very informative websites of the manufacturers [5].

In optical microscopy, mostly digital charge-coupled device (CCD) and scientific complementary metal–oxide–semiconductor (sCMOS) cameras are used (see Chapters 2 and 3). They contain two-dimensional (2D) arrays of $N \times M$ photodetectors, where N and M denote the total number of columns and lines, respectively. Usually, N and M are multiples of 2, for example, 128, 256, up to 4096. The number of columns and lines is often, but not necessarily, identical. For simplicity, we assume here that $N = M$.

Each of the single detectors has a fixed total surface area, which is in the size range of 6×6 to $24 \times 24\,\mu m^2$. The overall size of the light-sensitive area of the camera chip is in the range of several millimeters to centimeters. During data acquisition, in most cases the complete array of photodetectors is simultaneously exposed to the incoming light for a specific time duration, the so-called image integration time. Photons hitting the light-sensitive area of a single detector element produce with a certain probability electrical charges, which are stored below its light-sensitive surface. After the image integration time, these charges are read out and *digitized*, which means they are translated into integer numbers.

It should be noted that there exist many different types of camera chips and many ways how exactly the charges on the chip are read out. For details, we refer the reader to the specific technical descriptions of cameras, for example, [5].

The conversion factor that specifies how many electrons make out one digital unit is often definable using the camera acquisition software. Usually it is greater than unity. In this way, we obtain an integer intensity value h_{ij} for each single element of the detector field, which are labeled by i and j for the column and line number, respectively. Columns and lines are usually counted from 0 to $N - 1$. The maximum number of electrons that can be stored in one detector element is limited, and therefore the maximum intensity value is also limited, again to a power of 2, for example, $2^8 = 256$ or $2^{16} = 65\,536$. The exponent of the 2 designates the *bit depth*. The array of the intensity values obtained from a 2D camera field represents mathematically a matrix, which we write as

$$H = H_{ij} \tag{A.1}$$

We refer to this array of integer intensity values as the *digital image*. Each array element is designated as a picture element or abbreviated as *pixel*. The intensity values of the individual pixels will be designated as h_{ij} (Figure A.1).

It is important to note that the pixels are not little square areas [6]. They are array elements that contain values that are approximately proportional to the average light intensity that was registered by the respective light-sensitive area of the photodetector during image acquisition. Indeed, digital images are only a rough representation of the light distribution on the camera detector. There are

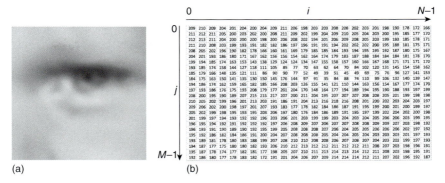

Figure A.1 Gray values as function of the position. Creation of digital images requires twofold discretization process.

two levels of digitization involved when translating the light distribution into the image: first, a spatial discretization by integrating the intensity over the area of the single detector, and, second, an intensity discretization by translating the electronic charge in the detector into an integer value using the electron-to-digital unit conversion factor. Of course, during both processes we lose information. We reduce the continuous spatial coordinates into discrete ones. The spatial discretization is a consequence of the data acquisition process itself. It should be adjusted such that the important information contained in the image produced in the microscopic image plane is retained. A single photodetector element can store usually several ten- or hundred thousand electrons, which are translated into intensity values ranging from zero to 255, or from zero up to 65 536. It is clear that some information is lost during this reduction, but indeed the loss is not as extensive as it may appear, because most of the information is "empty." Intensity discretization is necessary in order to reduce the required storage capacity for the images.

A.2.1.2 Sequential Data Acquisition

Detectors measuring intensity as a function of time are, for example, used in confocal laser scanning microscopes. Here, the data acquisition process is completely different from that in video microscopy. For acquisition of an x–y image, the confocal volume is moved by the scanning mirror across the sample. The scanning direction is designated as the x-direction. The light intensity is measured during this scanning process by a photomultiplier or avalanche photodiode, and the signal of a certain time period Δt – usually in the range of microseconds – is averaged and stored as one element of the image matrix. The integration of the next time period begins, and the next average intensity value is stored in the adjacent matrix element of the line. The scanning process takes place with a specific user-specified speed, called the scanning speed v. Hence, the intensity value stored in one matrix element corresponds to the average intensity measured in the confocal volume while moving the distance $\Delta x = v \times \Delta t$. After filling one line of the matrix with a total of N entries, the acquisition of the next line begins. To this end, the confocal volume is moved back by the scanning mirror to the beginning of the line, and the next line is scanned slightly vertically shifted in the y-direction, Δy. The instrument is programmed such that the distance of the vertical shift in the object plane is exactly a large as Δx such that $\Delta x = \Delta y$. Only then the final matrix is a geometrically correct representation of the object field. The image acquisition stops if the predefined number of M matrix lines has been filled. Assuming that the imaged object did not change during the data acquisition, this matrix is now designated as the image. As we have already noted, a pixel is *not* a little square [6].

A.2.2 Image Size, Bit Depth, and Storage Requirements

The size of each single detector element, d_d, should be well compared to the optical resolution of the microscope, d_{res}, such that no image detail is lost by the spatial digitization. This was already stressed in Chapter 2, where it was recommended that the image of the Airy disk radius in the image plane should exceed the size of the detector element by a factor of 2, at the minimum. This is stated by the famous Nyquist theorem: "A periodic structure with a given wavelength can only then correctly be reconstructed from discretely sampled values, if it is sampled at least twice per wavelength."

The Airy disk radius in the image plane has a size of $M_{tot} \times d_{res}$, where M_{tot} is the total magnification of the optical path between the sample and the detector. Thus, the magnification should be such that

$$M_{tot} \geq \frac{2d_d}{d_{res}} \tag{A.2}$$

Thus, depending on the extension of the field in the microscopic sample and the amount of detail that shall be retained in the image, we can determine the number of pixels of a given size that must be used to represent the information.

The size of the image matrix and its bit depth define the storage capacity that is finally required. This might become significant when not only a single plane of a sample but a three-dimensional stack of images is imaged at different axial positions like in confocal or light sheet microscopy. Also, often sample planes or complete volumes are observed as a function of time, adding a fourth dimension. Thus image size, stack size, bit depth, and number of frames in an image sequence have to be considered carefully.

The usual size of a confocal image is 512×512 or 1024×1024. With 8-bit depth corresponding to 1 byte, the small image size would be $512 \times 512 \times 1$ byte, yielding about 0.26 MB. The large image size with 16-bit depth accounts for $1024 \times 1024 \times 2$ byte ≈ 2.1 MB. Using the large image size at 16 bits for acquisition of an image stack with 128 planes requires already 268.4 MB. This can still be handled by current desktop computers. Things become involved when light sheet fluorescence microscopy is used. Here, 16-bit cameras with 2048×2048 pixels are employed. Just let us assume 64 planes are imaged over 2 days, with one stack per hour. This produces 25.7 GB of data in a single experiment. Similarly, super-resolution microscopy techniques can easily produce huge amounts of data.

Figure A.2 Look-up tables. (a) Typical gray scale LUT, (b) glow LUT, (c) spectrum LUT. (d) Image in gray scale, (e) glow, and (f) spectrum. The visual contrast in the images is enhanced when colors are used, but the image may lose its intuitive understandability.

A.3 Look-up Table

The digital image is just an array of integer values. In itself, it has no color information, but only integer values at the respective positions as defined by the column and row numbers. The so-called look-up tables or LUTs are used to define the rules for the actual representation of the image data. The simplest lookup table is a gray scale. Here, the intensity values from 0 to 255 – for an 8-bit image – are represented by gray shades from black to white (Figure A.2a–c). Another frequently used LUT is a modification of that scale. Here, 0 is represented by blue, and 255 by red. Intermediate intensities are shown in a respective gray shade. This has the effect of clearly marking image regions that are under- or overexposed. Obviously, in such image regions the quantitative intensity information is nonexistent. LUTs may be changed. This alters the visual appearance of the image completely, but not the underlying information, which is represented by the numerical intensity values of the image matrix (Figure A.2d–f). The human eye can discriminate about 70 different gray levels. Using color coding allows the discrimination of subtle intensity differences, because we can discriminate many more colors. This can be seen, for example, in the sky in Figure A.2d,f. In Figure A.2e, the sky appears practically with a constant intensity, whereas the color shades in Figure A.2f reveal subtle intensity variations.

A.4 Intensity Histograms

The intensity values of the image cover a certain range of numbers. During the digitization process they are usually adjusted to 8- or 16-bit integer values. Ideally, the complete range of values from just above 0 to nearly 2^8 or 2^{16}, respectively, should be used to fully exploit the dynamic range of the representation. The limiting values should be avoided in order to prevent under- or overexpression. A good way to evaluate the actual coverage of all intensity values is by examining the image intensity histogram (Figure A.3). It shows the frequency distribution of the intensity values. The histogram displays immediately which intensity values occur in the image. Also, it allows us to judge whether the choice of the LUT

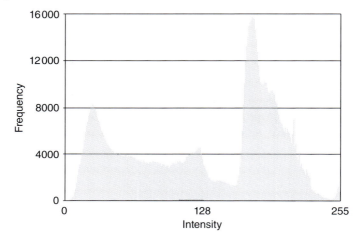

Figure A.3 Intensity histogram. Distribution of the intensity values of the image shown in Figure A.2. The first maximum corresponds to the dark areas of the dome, and the second to the extended bright areas of the sky.

is suitable for representing the important image details. In case the visual contrast is low, the LUT may be redefined to increase the contrast of the image. This can be done by changing the contrast options of the image processing software used.

A.5 Image Processing

Digital images are arrays of integers. Thus it is straightforward to use mathematical operations to process and analyze images. Such processing may serve many different purposes: for example, the correction of image defects, detection and quantification of structures, or reversing the effect of the optical imaging system by image deconvolution (Section 5.2.4). Also, the processing of localization (Chapter 8) or structured illumination microscopy images (Chapter 9) requires extensive processing of the acquired images in order to obtain the desired super-resolution information. Almost any application of modern microscopy uses some sort of image postprocessing.

All operations existing in the mathematical field of matrix algebra may be used. They can roughly be grouped into operations on single pixels and operations involving groups of pixels of the input image. In this short introduction, we can only give a few examples of image processing. Interested readers are referred to the image processing literature. As entry points, refer to [1–3, 7].

A.5.1 Operations on Single Pixels

The simplest operations are algebraic operations like $+$, $-$, $\times$, or $/$, which modify all pixel values by the same operation. A simple example is the subtraction of a constant intensity offset. In that case, all pixel values h_{ij} are reduced by a constant value c and yield new values g_{ij}:

$$g_{ij} = h_{ij} - c \tag{A.3}$$

Another useful example is bleaching correction. Often, the sample illumination during the image acquisition process reduces the average intensity due to photobleaching by a constant factor b each time an image is taken. That effect can be reversed by multiplication with the inverse of b:

$$g_{ij} = h_{ij} \times \frac{1}{b} \tag{A.4}$$

A common operation by which not all pixels are modified in the same way is a *flatfield* correction. It corrects for inhomogeneous illumination intensity or detection efficiency over the object field. In that case, one needs to acquire a fluorescence image of a sample with completely homogeneous label distribution, for example, a thin fluorescent glass slide or gel, yielding a reference image R_{ij}. The corrected image values are then obtained by

$$g_{ij} = \frac{h_{ij}}{r_{ij}} \tag{A.5}$$

Actually, the latter operation exemplified an operation using two separate images. Of course, we can also multiply, add, or subtract images – pixel by pixel. In all these examples, the intensity values at position (i, j) are modified without considering the neighboring values. This is different in the following, more complex, image processing operations.

A.5.2 Operations on Pixel Groups

In these operations, the original intensity value h_{ij} at position (i, j) is substituted by a new value g_{ij}, which is calculated considering the intensity of the neighboring pixels. Typical applications are image smoothing operations in which the intensity value of a central pixel is replaced by the mean value of the surrounding ones, such as

$$g_{ij} = \frac{1}{9}(h_{i-1,j-1} + h_{i-1,j} + h_{i-1,j+1} + h_{i,j-1} + h_{i,j} + h_{i,j+1} + h_{i+1,j-1} + h_{i+1,j} + h_{i+1,j+1}) \tag{A.6}$$

Here, the intensities of all neighboring pixels and that of the central pixel are summed up with equal weight. The division by 9 yields the mean value. Let us introduce the matrix

$$K_{ij} = \frac{1}{9}\begin{pmatrix} 1 & 1 & 1 \\ 1 & 1 & 1 \\ 1 & 1 & 1 \end{pmatrix} \tag{A.7}$$

With this definition, we can write

$$g_{ij} = \sum_{u=-1}^{1} \sum_{v=-1}^{1} k_{u,v} \cdot h_{i+u, j+v} \tag{A.8}$$

which can be abbreviated as

$$G_{ij} = K_{ij} \otimes H_{ij} \tag{A.9}$$

where the symbol $\otimes$ denotes a *convolution*. It is related to but not identical with a normal matrix multiplication. K_{ij} is designated as the *kernel* of the convolution. Thus, G_{ij} is the result of the convolution of K_{ij} and H_{ij}. This kernel is an example of a 3×3 digital filter.

Obviously, we can define completely arbitrary filters, for example, by considering more pixel values in the neighborhood of the central one and/or by using different weights. Especially important are the so-called Gaussian filters, which consider the neighboring pixel values weighted approximately according to a Gaussian function, for example:

$$K_{ij} = \frac{1}{16} \begin{pmatrix} 1 & 2 & 1 \\ 2 & 4 & 2 \\ 1 & 2 & 1 \end{pmatrix} \tag{A.10}$$

Gaussian filters are used to smooth images as with the mean filter shown above, but they pronounce the central value stronger. Therefore, the averaging effect is reduced.

In programs like ImageJ or Icy, it is easy to define such kernels and to test their effect. Further, kernels can be used to calculate derivatives or curvatures of images. They are often used to detect special features, for example, edges, or to identify objects.

A.5.3 Low-Pass Filters

Image filters like the mean filter or the Gaussian filter are also designated as *low-pass filters*. They average out sharp edges and suppress fine details in images. This is exactly what happens when we image an object with fine details using a microscope: the point spread function smoothes fine object features with sizes below or near the resolution limit of the imaging system. In that sense, the optics of a microscope is nothing but a low-pass filter. In case the details of the filter are precisely known – and the filter result, the image, is given – the effect of the low-pass filter can computationally be reversed. This operation is called image deconvolution and was outlined for confocal imaging in some detail in Section 5.2.4.

A.6 Pitfalls

The acquisition of digital images using light microscopy has revolutionized cell biology for many years. Digital images represent quantitative data. However, the acquisition of quantitatively valid images is by far not trivial. The titles of respective news features on digital imaging "The Good, the Bad and The Ugly" or "Seeing is Believing?" speak for themselves [8, 9]. The study of such articles is obligatory for serious microscopists.

A.7 Summary

The key points of this appendix are the following:

- Digital images are matrices in mathematical terms.
- Digital cameras are parallel data acquisition devices.
- In scanning microscopes, images are acquired sequentially.
- The recommended pixel size, the magnification factor, and the optical resolution are related to each other.
- The actual image representation is defined by lookup tables.
- Intensity histograms are a good tool to optimize the contrast of the presentation.
- Images are 2D arrays of integer numbers and can be modified or improved by digital filters.
- Microscopic imaging is nothing but low-pass filtering of the object function.
- There exist numerous practical pitfalls in digital image acquisition.

References

1 Broeke, J., Mateos-Perez, J.M., and Pascau, J. (2015) *Image Processing with ImageJ*, 2nd edn, Packt Publishing, Birmingham.
2 Vaingast, S. (2009) *Beginning Python Visualization: Crafting Visual Transformation Scripts*, Apress, New York.
3 Reyes-Aldasoro, C., Bhalerao, A., and Reyes-Aldasoro, C.C. (2015) *Biomedical Image Analysis Recipes in MATLAB: For Life Scientists and Engineers*, Blackwell Publishing, Oxford.
4 de Chaumont, F., Dallongeville, S., Chenouard, N., Hervé, N., Pop, S., Provoost, T., Meas-Yedid, V., Pankajakshan, P., Lecomte, T., Le Montagner, Y. et al. (2012) Icy: an open bioimage informatics platform for extended reproducible research. *Nat. Methods*, **9**, 690–696.
5 Web sites of camera manufacturers with detailed information on cameras and imaging: http://www.andor.com/learning-academy (accessed 19 October 2016), http://www.hamamatsu.com/eu/en/community/life_science_camera/index.html (accessed 19 October 2016).
6 Smith, A.R. (1995) A Pixel Is Not A Little Square, A Pixel Is Not A Little Square, A Pixel Is Not A Little Square! (And a Voxel is Not a Little Cube). Microsoft Technical Memo 6.
7 Wu, Q., Merchant, F.A., and Castleman, K.R. (2008) *Microscope Image Processing*, 2nd edn, Elsevier Science Publishing Co., Inc..
8 Parson, H. (2007) The good, the bad and the ugly. *Nature*, **477**, 138–140.
9 North, A.J. (2006) Seeing is believing? A beginners' guide to practical pitfalls in image acquisition. *J. Cell Biol.*, **172**, 9–18.

B

Appendix B: Practical Guide to Optical Alignment
Rainer Heintzmann[1,2]

[1] *Friedrich Schiller-Universität, Nanobiophotonik, Institut für Physikalische Chemie and Abbe Center of Photonics, Helmholtzweg 4, Jena 07743, Germany*
[2] *Leibniz Institute of Photonic Technology, Albert-Einstein Str. 9, 07745 Jena, Germany*

Lasers are often used for the alignment of optical systems, because they remain parallel over large distances. They can specifically be introduced into a setup for alignment needs or are sometimes already present for other purposes. Such a laser usually defines the optical axis, which should traverse the precise center of all lenses and other optical elements.

B.1 How to Obtain a Widened Parallel Laser Beam?

A widened parallel laser beam is often needed for alignment purposes and for other microscopy applications. The beam typically leaves the laser cavity with a width of 0.2 to ~2 mm. First, we need to enlarge its diameter and then make it exactly parallel. The enlargement is achieved by the use of a telescope, consisting of two lenses of different focal lengths adjusted such that the focal point of the first lens coincides with that the second. The laser beam will thus be focused to that focal point, where one can place a small pinhole of diameter of 20–100 μm. This pinhole serves to "clean up" the laser beam profile and remove unwanted distortions of the beam (Figure B.1). The expansion ratio of the laser is defined by the ratio of the two focal lengths. For microscopy applications, one typically uses 1 inch optics to which the width of the widened laser beam can then be matched. To achieve optimal performance, one should use plano-convex lenses and orient the planar surfaces toward the central focus. Minimizing the incidence angles at all optical surfaces is a general rule of thumb for optimizing performance.

It is difficult to make a widened laser beam exactly parallel to produce a planar wave front. Luckily there is a useful tool, the *shear plate,* which is commercially available.

The shear plate is a wedge of a defined slope and thickness (Figure B.2), which is inserted into the beam path typically at an angle of 45° with the slope of the wedge oriented perpendicular to the beam. The laser is reflected at both the front and the back side of the plate. These reflected beams will interfere with each other.

Fluorescence Microscopy: From Principles to Biological Applications, Second Edition.
Edited by Ulrich Kubitscheck.
© 2017 Wiley-VCH Verlag GmbH & Co. KGaA. Published 2017 by Wiley-VCH Verlag GmbH & Co. KGaA.

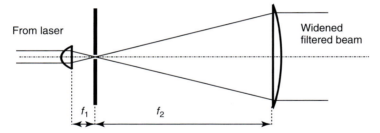

Figure B.1 A telescope with a pinhole in the central focus for cleaning up and widening a laser beam.

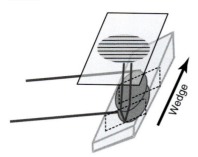

Figure B.2 A shear plate. Shown here is the principle of operation. A precision-manufactured glass wedge oriented sideways to the beam will lead to a series of parallel bright and dark interference stripes along the direction of propagation. Convergent of divergent light beams will lead to a tilted set of lines.

With an incident parallel beam, the reflection from the front surface will thus generate one parallel beam, whereas the reflection from the wedged back surface will generate another parallel beam of slightly different orientation. Thus, a fringe pattern with straight lines will be produced, as shown in Figure B.2. The fringes are ideally oriented along the direction in which the laser beam was originally traveling. If the beam is converging or diverging, this fringe pattern will tilt sideways and change its fringe spacing. Such a shear plate can be inserted in any parallel part of the beam as a useful tool for diagnosing errors in focusing or lens positioning.

If the beam passes through the lenses at oblique angles, or plano-convex lenses are inserted with the wrong surface facing the incoming beam, this can lead to higher order aberrations, which are visible as curved fringes generated by the shear plate.

For the shear plate to work, we need two different paths to interfere, for which a laser with sufficient longitudinal and lateral coherence lengths is required. The coherence length of a light source describes the maximum optical path difference over which interference can be observed. Sometimes, one also observes fringes with the contrast fluctuating with time. This indicates that the longitudinal coherence length of the laser is changing over time ("mode hopping").

Another useful tool is a set of precisely manufactured posts with two holes of diameter 1 mm, one at the nominal beam height and the other 3 mm above that. This tool can be used to check the beam height above the table at various places in the optical setup. The two holes serve to define two beamlets that should stay parallel over a large distance when illuminated by a single parallel beam. However, if the illuminating beam is converging or diverging, the distance between the two spots of light will change along the beam propagation (Figure B.3).

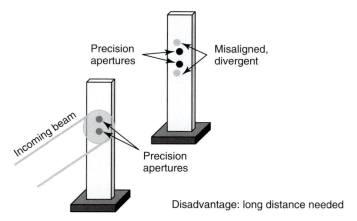

Figure B.3 An alternative way of determining whether the light is parallel using two sets of holes at predefined distances.

B.2 Mirror Alignment

Many optical setups contain mirrors to fold the beam path on the optical table. It is useful to have mirrors in a setup because they allow easy adjustment of the beam path. With two successive mirrors, one can do a *beam walk* and adjust the position as well as the angle of one part of the setup with respect to the next part. In a beam walk, the first mirror will have a larger influence on the position of any point downstream in the beam compared to the second mirror. The second mirror can be used to adjust predominantly the angle. Even though either mirror will affect both the angle and positions, it is nevertheless useful to keep this predominance in mind: the first mirror affects mainly the position, and the second mirror mainly the angle. For example, to adjust the beam to pass through two pinholes, the first mirror should be used first to get the position right on the first pinhole and then the second mirror should be used to change the angle, to hit the second pinhole. Obviously, this beam walking has to be repeated a number of times until the goal is achieved.

Mirrors are also quite useful when they can be flipped in and out of the beam path. This allows one to quickly and reproducibly switch between different setups sharing some common component, for example, the laser or the microscope. Often, magnetic-base precision repositioning elements are more reproducible than magnetic flip mounts.

Despite these advantages, each mirror will also lead to intensity losses. The losses are usually higher than the reflectivity specified in the catalog, as it is difficult to keep the mirror entirely free of dust. When dealing with setups that need to preserve the polarization of the light, the situation gets even more problematic. Mirrors will almost always alter the polarization state when it is neither p- nor s-polarized. This is because they usually introduce a phase shift between p- and s-polarizations. To minimize this effect, one should aim to use the mirror as close to the perpendicular beam incidence as possible and prefer a metallic front-surface mirror over a dielectrically coated mirror.

In fluorescence microscopy, one often uses a *dichromatic beam splitter*. If one has complete freedom of design here, one should also try to use this at as close to normal incidence as possible because this will not only avoid polarization issues but also increase the steepness of the spectral cutoff characteristics. However, one has to be aware that this also shifts the design cutoff wavelength of the dichromatic beam splitter.

B.3 Lens Alignment

Optical systems often consist of multiple lenses. A lens performs best for rays passing through its center close to the optical axis and at small angles. Thus the individual lenses have to be well aligned with respect to an optical axis. This means that all the lenses have to be well centered, with the optical axis being orthogonal to the lens surface at its center. As the centering of lenses is of utmost importance, the lenses are often placed in X–Y translational holders. To achieve the correct tilt of the lenses (usually adjusted without using special holders), the residual back reflection of the lens surface can often be used.

B.4 Autocollimation Telescope

A very useful tool is an *autocollimation telescope* comprising a grid, an illumination unit, and a camera. Autocollimators are used for the precise alignment of machine parts, precision-manufactured, and therefore quite expensive. With such a device rigidly mounted on the optical table to define the optical axis, one can quickly change between various focal planes and center each lens using a built-in cross-hair target. By focusing on the cross-hair that is reflected back from an optical surface, the tilt of that lens can be optimized. Ideally, the autocollimated image is displayed on a little video screen. In this way, the user can stand near the lens and align it while looking at the result of the autocollimation on the screen (Figure B.4). However, in many cases, one does not have access to an autocollimator, and the alignment is done with a laser beam.

B.5 Aligning a Single Lens Using a Laser Beam

Starting with a parallel beam (checked that it is exactly parallel using a shear plate, see above), shifting a lens sideways will move the focal point sideways by the same amount. If the illumination beam is significantly smaller than the lens size, the shift will also alter the beam direction, as the narrow laser beam then behaves almost like a ray as drawn in a ray diagram.

Steps to place a single lens into a widened parallel laser beam are as follows:

1. Make sure that the laser beam is parallel.
2. Place an iris aperture A_1 closed down to ~1 mm in the middle of the beam to define its center. Place a second iris A_2, closed down to ~1 mm, where the focus of the lens should be.

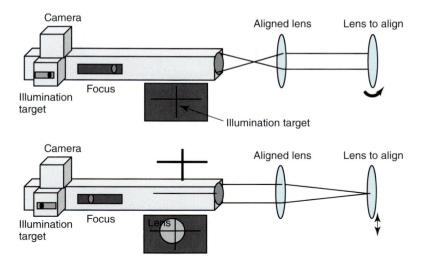

Figure B.4 The principle of operation of an autocollimation telescope. It is mounted in a fixed position to define the optical axis. Focusing on different optical surfaces allows centering of the optical elements. Focusing on the illumination cross-hair that is reflected by optical surfaces helps to precisely adjust their tilt to be perpendicular to the optical axis.

3. Adjust the iris A_2 such that the 1 mm beam traverses it; this defines the optical axis along which the lens should be aligned.
4. Insert the lens such that the 1 mm beam generated by A_1 aims approximately at the center of the lens.
5. Finely adjust the lens position by aiming the beamlet at the center of aperture A_2.
6. Adjust the lens tilt to make the lens perpendicular to the beam (see method below) and repeat steps 5 and 6 until satisfaction.

Tilting the lens from normal incidence should never be considered as a valid means of alignment of its focus. Always place the lens as perpendicular to the optical axis as possible and align its X–Y position. Then use the following trick to ensure that it is at normal incidence to the light beam.

A lens generates a small amount of reflected light from each of its surfaces. The light from the first surface is usually diverging (for biconvex lenses) and only useful for tilt alignment if the beam is small with respect to the lens size and the lens surface is not strongly curved. In this case, one should ensure that the reflected light returns through the first iris aperture A_1, defining the small center beamlet. One can also use the light returning from the second lens surface. This focuses at roughly half the focal distance before the lens (since it passes the lens twice). This focused spot is sometimes better visible and easier to use for the tilt adjustment. Also here, this spot should be in the middle of first aperture positioned at half the focal distance before the lens. It is useful to use a white business card with a ~1 mm hole punched on it, such that the back-reflected beam is visible when it misses this hole, indicating a reflection from a tilted surface (Figure B.5). An even better method is to use a pellicle beam splitter inserted before aperture A_1 such

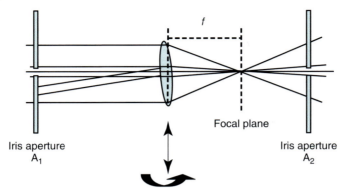

Figure B.5 Image demonstrating the use of iris apertures and back reflections from optical surfaces to adjust the position and tilt of optical elements.

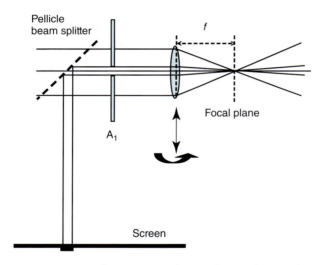

Figure B.6 Image demonstrating the use of a very thin membrane mirror (pellicle) and a pinhole to precisely adjust the tilt of an optical element.

that the back-reflected light returning through A_1 is visible without disturbance from the illumination light (Figure B.6). It is important to use only this beam splitter type because only the minuscule thickness of the pellicle beam splitter ensures that the optical axis defined by the laser will not be altered by this method.

When two lenses with a common focal plane are inserted into the beam path, shifting either of the lenses will alter the direction of the emerging beam. Thus, it is possible that the beam after the second lens points to the correct direction although the beam is not centered correctly on the first lens. This could lead to suboptimal performance of the lens system.

Therefore, optical systems need to be aligned by inserting one lens at a time and keeping the already aligned lenses fixed.

B.6 How to Find the Focal Plane of a Lens?

It is often required to determine the focus of a lens. When this problem arises, the first thing one has to know is whether one is dealing with a single lens or a lens system. With a single lens one can assume the reference plane, which is the plane in which the rays of ray diagrams can be assumed to be bent, to lie inside this lens. When the position of the focal plane has been determined (see below), the focal distance is then approximately given by the distance between the middle of the lens to the focal plane. Flipping the lens around should yield the same result. With lens systems such as tube lenses of some microscope manufacturers, the situation can be much more complicated. These systems can have two different reference planes, one for each input from either side, which can lie far outside the actual lens system itself. The focal distances need to be measured from these reference planes. Here, we consider only the situation of a single lens.

The easiest method for determining the focal plane of a lens is to look at the image of an object placed at a great distance. If there is a window in the room, the objects in the sky, such as clouds, can be assumed to be at infinite distance compared to the lens focal length. An image of the clouds will, therefore, be formed at the focal plane of the lens. For a first estimate, one can also use the lamp on the ceiling for this purpose but only if the focal length of the lens is relatively short, for example, smaller than 20 cm.

For a lens with very a short focal length, it can be difficult to determine the plane of focus in this manner, because the image may be too small. However, when using an extended plane wave for illumination, such as a laser expanded to the full lens diameter, there is a useful trick to determine the focal distance precisely (Figure B.7). If a piece of matte surface (such as the matte side of a tin foil)

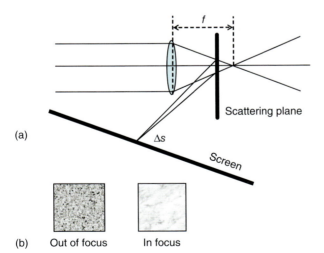

Figure B.7 Focus determination. (a) Reflection at a rough surface to find the precise focus of a lens. (b) The focal point can be determined by watching the size of the speckles generated by the reflection off the rough surface. The largest speckles occur when the scattering plane is located in the focal plane.

is mounted and moved along the optical axis to find the focus, a strong speckle pattern is observed in the back-reflected light. One has to be very careful when working with lasers because the reflection off a matte surface can be dangerous. Thus, the laser intensity should be very low. This speckle pattern stems from the interference of the waves scattered back by various heights of the surface with its micro-roughness. The size of these speckles is critically dependent on the optical path difference Δs between different scattering positions on the matte plane, as shown in Figure B.7. Therefore, the size of the speckles depends on the illuminated surface area. A small illuminated area will generate large speckles, whereas a larger illuminated area will generate smaller speckles. The position with the largest speckles in the diffuse reflected light therefore very accurately gives the correct focus position (method courtesy of Mats G. L. Gustafsson). This method can also be useful when focusing through the pinhole in a beam-expanding telescope.

B.7 How to Focus to the Back Focal Plane of an Objective Lens?

In many setups, for example, total internal reflection fluorescence or structured illumination microscopy, one needs to focus a laser beam to the back focal plane (BFP) of an objective lens. Strictly speaking, this should be designated as the front focal plane of the objective when the laser is used for illuminating the sample. However, with the term *back focal plane*, which is commonly used, one refers to the imaging path where the light is emitted by the sample.

There is a simple trick to achieve this focusing to the BFP: when a laser beam is focused onto the BFP, a parallel beam spanning the field of view will leave the objective lens (Figure B.8). At a large distance from the objective, for example, the ceiling of the room for an inverse microscope, the beam should still be small. Minimizing the size of the beam leaving the objective at a larger distance, such as

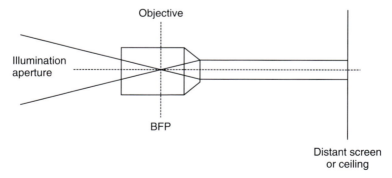

Figure B.8 Focusing a laser beam to the back focal plane of an objective lens generates a parallel beam of light exiting the objective. At a larger distance, the smallest spot generated when altering the focal point indicates the correct focus position in the back focal plane of the objective. For laser safety reasons, great care has to be taken with this method. An eye-safe laser should be used, and laser eye protection must be worn.

the ceiling of the room, allows determining whether the focus was adjusted correctly to the BFP. One has to be *extremely careful with this type of adjustment for laser safety reasons* because the laser beam exits the objective as a parallel beam with a small diameter and can easily point to any direction of space. Therefore, this procedure should be used only with lasers in the intensity range deemed safe for the human eye. For this adjustment, it is also necessary to focus the beam onto the BFP under a relatively large solid angle; that is, one should not use a small aperture defining the optical axis but use the full aperture of the optical illumination system for finding the focus position at the BFP. If one only focuses a ~1 mm laser beam in this way, the field of view in the sample will be very small and thus the spot on the ceiling (inverted microscope) will be very large because of diffraction.

To adjust the beam to the precise center of the BFP, one has to ensure that the optical axis is marked at a far distant point with a pen with the objective not in place. Then the objective is inserted and the position of the focusing lens adjusted until the smallest possible spot is visible at the same position.

Index

a

absorption   14, 16–18, 61, 67, 85–89, 106, 136, 140, 141, 149, 153, 157, 166, 188, 208, 212–214, 218, 223, 226, 230, 236, 255, 274, 276, 348, 389, 394, 397, 408, 410–412, 416, 419, 420, 448, 449
Abbe's diffraction limit   368
absorption dipole moment   416
acceptor density   426–428
acceptor emission filter   431
acceptor fluorophores   413
acousto-optical modulator (AOM)   345
Airy disk   38, 41–44, 48
Airy disk radius   458
Airy pattern   173, 174
angle of incidence   376
angle of refraction   377
animalcules   165
astigmatic imaging approach   281–284
astigmatism   21
ATTO647N   325, 337
autocollimation telescope   468
autocorrelation function (ACF)   188, 189, 191, 192, 199
autofluorescence   381
Avalanche photodiode (APD)   114
Avogadro's number   412
axial resolution   45, 47, 173–179

b

back focal plane (BFP)   249, 472–473
ballistic light   377
bandpass filter   324, 326
Bayesian statistics   404
beam aberrations   332
beamlets   466
beam propagation   466
beam splitter   311
O^2 benzylcytosine (BC)   145
Bessel beam   262
Bessel function   38
bioluminescence   95
binning   106
biplane imaging approach   284
birefringence   13
bleaching   120
blue-shifted dye   334
Bragg condition   33
Braun tubes   181
bright field   31–32, 44, 45, 61, 70, 71, 73, 74
brightness   28, 29, 57–62, 67, 74, 76, 80

c

camera acquisition software   456
carbocyanines   275, 281
cell membrane glycans   283
centrosomes   287
charge-coupled device (CCD)   26, 104, 107, 112, 273, 386–387
Chinese hamster ovary (CHO)   381
chromatic reflectors   18–19
chromophore   274
click chemistry   281–283
cluster size   397
coherent beams   295
colliding-pulse mode-locked (CPM)   218

complementary
metal–oxide–semiconductor
(CMOS)   111, 112, 388
confocal effect   346
confocality   302
confocal laser scanning microscope
(CLSM)   342
confocal scanning microscopy   165, 247, 261, 429
conjugation process, synthetic fluorophores   142
continuous fluorescence microphotolysis (CFM)   343
continuous fluorescence photobleaching   343
continuous illumination   358
continuous wave (CW)   167, 172, 329
conventional dual-color imaging scheme   334
conventional fluorescence microscopy techniques   264
convolution   41
convolution theorem   294, 298
copper-catalyzed azide–alkyne cycloaddition   282
Coulombic coupling   411, 414
Coulomb's law   414, 418
coupling FRET   425
critical angle   378
critical illumination   28–30
crystallography techniques   422
cut-off frequency   295, 302
cyan fluorescent protein (CFP)   152, 423
cylindrical lens   246, 248

d
DAOSTORM   287
Davson–Danielli–Robertson model   341
defocused imaging   395
Deinococcus radiodurans   150
depth of field (DOF)   49, 81, 392
depth of focus   49, 256, 313, 392
detection, PSFs   175
development, confocal microscopy   169

dichroic mirror   96, 97, 102, 375
dichromatic beam splitter   468
dielectric mirrors   16–17
differential interference contrast (DIC)   13, 23
  interpretation   81
  optical setup   77
differential interference microscopy   77–80
different lens configurations   259–260
diffraction   1, 7, 323, 324, 326, 327
  coefficients   341, 342, 346
  limit   269, 270
  pattern   33, 34, 36
diffusion-enhanced energy transfer   424
diffusion-reaction equation   351, 355
diffusion time   188
digital camera, fluorescence microscopy   113
digitally scanned light sheet microscopy (DSLM)   246, 248
4,4′-diisothiocyanatostilbene-2, 2′-disulfonic (DIDS)   352
dipole–dipole distance   414
dipole moments   414, 416
Dirac delta distribution   298
discovery, fluorescence of   91–92
discriminator   172
dispersion   9
donor fluorophores   425
donor emission spectrum   412
donor excitation filter   431
donor quantal spectrum   412
donor quantum yield   412, 417, 431, 438, 440
donor quenching (DQ)   428–429, 432
double helix PSF   284
double-sided illumination   253, 254, 258
doughnut-shaped depletion beam   335
dSTORM   275, 282
dual-color imaging   334
dual-view system   375
dynamic imaging   373

e

electrically tunable lenses (ETLs)   248, 256
electromagnetic radiation   391, 408
electronic delay   332
electronic filters   103
electron-multiplying charge-coupled device (EM-CCD)   109, 387
emission dipole   416, 417
emission spectrum   90, 99, 411
epifluorescence   98
epi-illumination   31–32
equilibrium properties, fluorescence emission   88–89
excitation, PSFs   175
excitation filter   101
excitation light distribution   311–313
excited state   325, 327
exciting light intensity   99, 122–123
exciting photon   89

f

Fermi's golden rule   414, 416
field of view (FOV)   247, 248
finite optics setup   25
FlAsH   148
fluid mosaic model   341
fluorescein   410
fluorescence confinement   326
fluorescence correlation spectroscopy (FCS)   172, 354
fluorescence cross-correlation spectroscopy (FCCS)   190
fluorescence depletion   326
fluorescence emission   272, 300
  depolarization of   428
fluorescence excitation   293
fluorescence filter   246
fluorescence intensity, measurements of   119–121
fluorescence labeling   368
fluorescence lifetime changes   437–441
fluorescence markers   157
fluorescence microscopy   86, 90–95, 299
fluorescence photobleaching recovery (FPR)   341

fluorescence photobleaching techniques
  CLSMs-assisted photobleaching methods
    implementation   358–359
    opportunities   359
  concepts and procedures
    complexity from bottom up   346–347
    principle and several modes   340–345
    set up   345–346
  continuous fluorescence microphotolysis (CFM)
    background and data evaluation   354–357
    combination of   357
    variants   357–358
  fluorescence recovery after photobleaching (FRAP)
    binding   350–351
    diffusion measurements   347–350
    membrane transport   351–354
fluorescence photon   88
fluorescence quantum yield   136, 147
fluorescence recovery after photobleaching (FRAP)   160, 341
fluorescent labels
  characteristics   134
  properties   133, 135, 142
  sensors in cell   160
fluorescent protein (FPs)   152, 156, 324, 329, 337
fluorophore   270, 407
  classes of   337
  emission spectrum   326, 336
  excitation   326
  kinetics   328
  photoactivatable/photoconvertible   272
  photobleaching   270, 355
  photon number   280
  photoswitchable   273
  property of   342
  reaction-induced photoswitching   273

fluorophore (*contd.*)
  rhodamine   275
  ROP   354
  for STED microscopy   337
  stimulated emission depletion   324
flux equation   356
focal lengths   251
focal plane   250, 260, 471–472
Förster distance   413, 420
Förster equation   414–418
Förster radius   399, 400
Förster resonance energy transfer
    (FRET)   155, 191, 369
Fourier frequency   36
Fourier theory   41
Fourier-transformed illumination
    pattern   299
4Pi microscopy   357
Fraunhofer diffraction pattern   38
frequency-domain FLIM   443–444
frequency spectra   415
fringe distance   315
fringe period   313
fringe projection, *see* two-beam
    interference
full quantitation   433–434
full width at half-maximum (FWHM)
    175, 372
functional magnetic resonance imaging
    (fMRI)   237

g

Gaussian beam   250–252
Gaussian filters   462
Gaussian function   272, 394
Gaussian intensity profile   348
Gaussian photon distribution   372
Gaussian-shaped signal distribution
    390
geometric-optical term   176
global fitting approaches   446
gratings   34, 37, 38, 57, 330
  constant   8, 33, 35
  distance   299, 312
  structure   34
gray values   189, 317, 455

green fluorescent protein (GFP)   150,
    281, 368, 410
grid projection   295

h

hetero-FRET   413
high-pressure mercury vapor
    arc-discharge lamps (HBO)   99
  burners   100
high-resolution information   300–301
homo-FRET   414
Huygens–Fresnel integral equation   35
Huygens wavelet   37

i

illumination intensity (I)   381
illumination microscopy   296
illumination pattern   298–301,
    303–307, 313
illumination photons   324
illumination time   381, 382
image brightness   57
image contrast   90–93
image deconvolution   184, 462
image integration time   109
image processing operations
  focal plane   37
  pixel groups   461–462
  single pixels   460–461
imaging artifacts   278
imaging contrast, measures improving
    285
immersion media   62–65
immunofluorescence   281
immunoglobulin G antibodies   147,
    282
induced acceptor emission   429–432
infinity space   24
instrument response function (IRF)
    172
integrator   172
intensified charge-coupled device
    (ICCD)   104, 107, 109
intensity-based measurements, FRET
    432–433
intensity distribution   173
intensity histograms   288, 459–460

intensity zero   332
interference effects
　light rays   11
　pattern   304
interference microscopy   77
interferometer   303, 304, 309
　plane   306
　setup   308
interferometric imaging   284–285
intermediate image   25
intrinsically photoswitchable probes   274
inverted microscopes   32, 473
in vivo/in vitro measurements, FRET   401
IrisFP   154
isoSTED configuration   336

j
Jablonski diagram   92, 325

k
Köhler illumination   28–30, 379, 383, 384
kinetic measurements, FRET   401
kinetics, fluorescence emission   88–89

l
Lambert radiator   58
Lampyris noctiluca   93
laser scanning microscopes (LSMs)   180, 362
label size *vs.* structural resolution   280–282
lateral resolution   41, 175
　coherent light sources   43
　incoherent light sources   41
　optical units   47
lens alignment   468
lens classes   61–62
lens design   53–57
light-emitting diodes (LEDs)   100, 101
light microscopy   372
　angular aperture   27
　apertures   50
　components   23
　contrast   67
　dark-field   68
　DIC   77
　interference contrast   74
　optical contrast methods   68
　phase contrast, *see* phase contrast
　phase objects   67
　entrance pupil   50
　exit pupil   50, 51
　field-of-view   28
　high resolution objectives   65
　illumination system   28
　　beam path   30
　image path   24
　magnifications   26
　　CCD   26
　　focal length   27
　　magnifying glass   27
　multiview imaging   257–259
　numerical aperture   27
　telecentricity   51
　wave optics and resolution   32
light propagation   377
light sheet fluorescence microscopy (LSFM)   246
light sheet illumination   254, 259, 263
light sheet microscopy
　construction and working   248–249
　photobleaching and toxicity   247
　polarization   253
　principle of   246–247
　3D imaging   255–357
　water-dipping illumination lens   250
light sources, STED   332–333
light waves   2
　circular polarized   4
　on interference   2
　plane wave   7
　transverse wave   3
　description   3
linear   299
　polarization   378
lipofuscin   316, 318
live-cell dynamics   160
live-cell labeling   282
LivePalm   286
localization errors   280
localization precision (LP)   280, 374

localization-based super-resolution microscopy
  experimental setup for 275–277
  fluorescence labeling 280–285
  imaging contrast, measures for improving 285
  intrinsically photoswitchable probes 274
  optical resolution and imaging artifacts 278–280
  organic fluorophores, photoswitching of 275
  photoactivatable and photoconvertible probes 274
  single-molecule localization microscopy (SMLM)
    quantification of 288–289
    reference structures for 287–288
    software 285–287
lock-in techniques 384
Lorentzian model function 327
low-pass filters 462

m

magnification, optical system 48–49
maleimides 143
mean-square displacement (MSD) 400
metal halide lamps 99
Michaelis–Menten equation 353
microchannel plate 107
Micrographia 165
microtubulin 277, 287, 288
mirror alignment 467–468
modulation contrast 314, 315
Moiré effect 300, 301
molecular brightness 136, 196
molecular complexes, stoichiometry of 371
Monte Carlo simulations 425
multicolor imaging 373
multidirectional SPIM (mSPIM) 255
multiparameter fluorescence detection (MFD) 193–195
multiple acceptors 424–425
multiple donor–acceptor pairs 424

n

nanocrystals 140
nanometers 377
near-field scanning optical microscopy (NSOM) 383
near-infrared (NIR) region 86
negative sign, image formation 35
N-hydroxysuccinimide (NHS) esters 142, 143, 146–149
Nipkow disk 181, 183
nitrilotriacetic acid 144
NMR 422
noise detector, EM-CCD 389
noninvasive measurements 402
nonlinear microscopies 203, 239
nonlinear optics 206
non-natural amino acids 146
nonscanning applications 186
normalization factor 178
number and brightness analysis (N&B analysis) 195
numerical aperture (NA) 165, 247, 372
Nyquist–Shannon sampling theorem 278
Nyquist theorem 384, 458

o

objective lens 326
objective-type TIR 378
object size estimation 313
occupancy errors 434
oblique angles 466
optical aberrations 20
optical components, light microscopes 26
optical delay 332
optical elements 13–19
optical filters 17–18
optical lenses 13–15
optical metallic mirrors 15–16
optical parametric oscillator (OPO) 217
optical path length difference 37
optical projection tomography (OPT) 262
optical resolution 126, 278
optical sub-diffraction techniques 422

optical transfer function (OTF)   40–41, 294, 298
organic dyes   138
organic fluorophores, photoswitching of   275
orthogonal-plane fluorescence optical sectioning (OPFOS)   245
oscillation probability   418
oscillator strength function   416
oversampling   50

p

PA-green fluorescent protein (PA-GFP)   234
Palm3d   286
paraffin   316
particle   1, 2, 6
parallel data acquisition   455
patterned techniques, application of   315
Perrin formula   448
Perrin's equation   414, 415
Perrin's estimation   411
phase contrast   69
  conjugate planes   70
  focal plane   73
  phasor diagrams   72
  phase ring   73
  properties   74
  Zernike's experiments   70
phase fluorimetry   443
phase modulation technique   443
phasor plot   447
phosphorescence   88
photoactivatable fluorescent proteins (PA-FPs)   277
photoactivatable (PA)   151, 274, 369
photobleaching   124, 138, 329, 337
  CLSMs   358
  curve   397
  effects   299
  localization microscopy   374
  quantum efficiency   342
  technique   435
photoconvertible probes   274
photodetectors   104, 346

photomultiplier tube (PMT)   113–114, 169, 342, 432
photon antibunching   397
photon noise/shot noise   117
photoswitchable fluorophores   273
phototoxicity   125, 137
phycobiliproteins   149
physical grating   306
picture elements, camera   105
pinhole   169
pivoting light sheet   254
pixel
  oversampling   384
  size   384
  undersampling   384
pixelation noise   390
planar distribution   425
Planck constant   414, 448
plane of incidence   11
plano-convex lenses   466
pointillism   271
point spread function (PSF)   38, 40–42, 48–50, 65, 67, 294, 297, 314, 323, 357
Poissonian signal fluctuations   390
polarization microscopy   395
polarization vector   305
precise measurements   399–400
primary image plane   25
principles, confocal microscopy of   166
prism-type TIR   378
protein–lipid ratio   341
pulsed interleaved excitation (PIE)   191
pulsed irradiation   358
pulsed lasers   328
  diodes   332
  laser systems   332
pulse synchronization   332

q

Quantenspektrum   416
quantification methods   425
quantitative measurements   401
quantum efficiency   89
quantum yields   399, 412, 413, 415, 417
QuickPALM   284

r

radial resolution   173–179
radiative lifetime   440
radical anion   275
read noise   118
rapid diffusion limit   424
rapid lifetime determination   445
rapidSTORM   286
raster image correlation spectroscopy (RICS)   198
rate of transfer   413, 414
ratiometric measurements (pH)   121
Rayleigh distance   294
Rayleigh length   251
reaction-induced photoswitching   273
reactive oxygen species (ROS)   137
ReAsH   148
redox chemistry   275
red-shifted dye   334
reducing and oxidizing system (ROXS)   337
reflected light image   97
reflection   10–11
refraction   9–10
refractive index   9
region of photolysis (ROP)   342, 343, 346
relaxation   325
resolution limit, microscope   35
reversible photoswitching   274
reversible saturable optical fluorescence transitions (RESOLFT)   161
Rhodococcus rhodochrous   145
rigorous approach, optical resolution   33

s

saturated structured illumination microscopy (SSIM)   161
saturation effects   299, 372
scanning confocal imaging   179, 195
scanning tunneling microscopy (STM)   269
scientific complementary metal-oxide-semiconductor (sCMOS)   247, 277
secondary image plane   25
sensitive detection   94
selective plane illumination microscopy (SPIM)   206, 246
sensitized emission   429
sensors
　CCD   107
　FRET   431
sequential data acquisition   457–458
Shannon/Nyquist sampling frequency   333
shear plate   465
signal fluctuation   389
signal-to-noise ratio
　detector noise   389
　signal fluctuation   389
silicon–rhodamine (SiR) dyes   145, 337
sine condition   59–60
single-molecule emitter   397
single-molecule fluorescence resonance energy transfer   400
single-molecule localization microscopy (SMLM)   324
　principle of   271, 272
　quantification of   288
　reference structures for   287
　software   285
single-molecule methods, mobility/anisotropy   371
single-molecule microscopy
　collection efficiency   377
　dual-view system   375
　dynamic imaging   373
　magnification/resolution   376
　multicolor imaging   373
　photoactivation/photoswitching   373
　photobleaching localization microscopy   374
　single-molecule/-particle tracking   374
　super-resolution imaging   372
single-molecule signals
　brightness   394
　color   395
　defocused imaging   395
　intensity pattern   393
　orientation   394

point spread function   391
  polarization microscopy   394
single-molecule tracking   260
single-photon avalanche diode (SPAD)
    401
single-sided illumination   254, 258
solvent relaxation   88
small objects, misrepresentation
    127–128
Snell's law   377
spatial frequency   36, 294
spatial resolution   270
  fluorescence microscope   280
spatially modulated illumination
  excitation light distribution   311
  object size estimation   313
  overview   309
  setup   311
specificity, fluorescence labeling
    93–94
spectral emission bands   431
spectrofluorometry   428
spinning disk confocal microscope
    (SDCM)   181, 184
stage scanning   179–180
standard deviation   372
stimulated emission depletion (STED)
    155, 324, 357
  applications
    fluorophore, choice of   336
    labeling strategies   337
  experimental setup
    axial resolution, improving   335
    multicolor imaging   334
    scanning and speed   333
  fundamental concepts   324
  key parameters in   328
stochastic optical reconstruction
    microscopy (STORM)   161,
    275, 422
stochastic switching processes   324
Stokes' shift   409–411, 420, 448
structured illumination microscopy
    (SIM)
  high-resolution information   300
  illumination pattern   303
  image generation in   297

interference pattern   304
  optical sectioning   301
super-resolution microscopy
  resolution limit   294
  spatially modulated illumination
    (SMI)
    excitation light distribution   311
    object size estimation   313
    overview   309
    setup   311
superman cape function   419
synchronization, STED   332
synthetic grating   306

t
tetracysteine   144
thiols   275
transfection   159
three-beam interference   295, 302
three-dimensional SMLM   283
three-dimensions, fluorescence
    microscopy   121–122
time-correlated single-photon counting
    444–445
time-domain FLIM   442, 444
time-of-flight camera   444
total internal reflection (TIR)   378
total internal reflection fluorescence
    (TIRF) microscopy   354
  collimation   384
  illumination time (t_{ill})   382
  intensity   380
  polarization   382
  uniformity   379
  wavelength   383
two-beam illumination   302
two-beam interference   295, 297, 298,
    301
two-dimensional (2D) grating   301
two-photon excitation microscope
    203, 329
  absorption   212
  advantages   220, 238
  caged compounds   233
  continuous wave   208
  depth of imaging   230
  description   205

two-photon excitation microscope (*contd*.)
  detection strategies   219
  focal volume   210
  history and theory   207
  in vivo image   229
  lasers   216
    CPM   218
    OPO   217
    pulse widths   218
  limitations   222
    diphenylhexatriene   224
    heating, laser powers   223
    hyperspectral image   229
    photobleaching   227
    spatial resolution   223
  PA-GFP   234
  photochemistry   231
  photobleaching   211
  quantum dots   239
  scattering   213
  UV excitation   231
  UV fluorophores   231
Tyndall effect   85

u
undersampling   50

v
Venus-YFP (yellow fluorescent protein)   423
von Bieren condition   33, 35, 60

w
water-dipping lenses   259
wave   1–11, 13–18, 20, 23, 32–50, 57, 67, 70–80
wavelength effects   330
wavelength fluorophores   419
widefield fluorescence microscope   271
widened parallel laser beam   465–467
  single lens   468–470
Wollaston prism   375

x
X-ray crystallography   372

z
zero-order diffraction   307